THE *Gresham Lectures* OF John Flamsteed

THE *Gresham Lectures* OF John Flamsteed

Edited and Introduced by
ERIC G. FORBES

MANSELL 1975

Mansell Information/Publishing Ltd
3 Bloomsbury Place
London WC1A 2QA
England

International Standard Book No: 0 7201 0518 8

text set in 11/11½ pt Linotron typeface
at The Universities Press, Belfast,
printed by photolithography and bound in Great Britain at
The Scolar Press, Ilkley, Yorks.

Contents

Preface

Apart from Francis Baily's mere mention of 'Flamsteed's Lectures on Astronomy read at Gresham College in 1681–1684. Small thick Quarto, bound in a marbled paper cover' on p. lxvii of *An Account of the Revd. John Flamsteed* . . . (London, 1835), I know of no other printed description of the manuscript now published for the first time in this work. Up to the present day its contents have remained entirely unknown; consequently much of the information it provides about the early phase of the first Astronomer Royal's scientific career has never been incorporated into any account (including Baily's) of his life and work.

The reasons for this unwarranted neglect are not difficult to find. The Preface and Supplement to Baily's book were composed with the prime intention of vindicating Flamsteed's character and reputation against attacks by eighteenth- and early nineteenth-century commentators on the scientific achievements of Isaac Newton who took sides with their hero in his controversy with Flamsteed ensuing from the unauthorized publication of the latter's observations. Baily's introduction contains autobiographical notes, memoranda, and a diary of events selected from the vast quantity of papers now preserved in the archives of the Royal Greenwich Observatory, Herstmonceux Castle, Hailsham, Sussex (viz. the R.G.O. MSS.) which constitute, in a sense, Flamsteed's posthumous defence of his position in that famous dispute. The 159-page Appendix contains about 230 items of correspondence with Newton, Abraham Sharp, and a score of other contemporaries plus a posthumous exchange of fifty letters between Sharp and Joseph Crosthwait relating to the preparation for the press of Flamsteed's *Historiae Coelestis Britannicae* (3 vols., Londini, 1725), on which his claim to fame with posterity rests. Only eleven letters, however, pertain to the period prior to the year 1684 when the Gresham College lectures end. Baily's main interest in perusing Flamsteed's manuscript writings was to glean from them archival information to assist in his revision and correction of the star-catalogue printed in the last volume of that work, which he duly incorporated into his own catalogue of 8,377 stars prepared under this auspices of the British Association for the Advancement of Science and published in 1845.

The remarkable edifice of celestial dynamics which Clairaut, Euler, Lagrange, Laplace, and other eminent continental scientists had built upon the axioms of Newton's mechanics and his inverse-square distance law of universal gravitation, had contributed to a switch in the focus of nineteenth-century astronomical research towards the dynamical explanation of perturbations in the solar, lunar, and planetary theories; while James Bradley's observational discoveries of the phenomena of stellar aberration and nutation had encouraged Friedrich Wilhelm Bessel to develop a rigorous science of positional astronomy.

Flamsteed's Gresham Lectures are concerned with the size of the solar parallax; the theory of cometary motion; the theory of telescope optics; the application of telescopic sights to astronomical instruments; the errors in ancient observations; the stereographic ecliptic projection of the celestial sphere; the physical constitution of heavenly bodies; the evaluation of the precessional constant; and the theory of astronomical refraction. A century of research in these fields had, however, made many of his opinions either redundant or out-of-date, while the accuracy of his observations was quickly transcended through the introduction of more sophisticated instrumentation and techniques.

The apparent irrelevance of this collection of thirty-nine Gresham Lectures to Flamsteed's labours at the Royal Observatory, Greenwich, may alone suffice to explain why Baily paid no heed to them. Even if he had taken the trouble to browse through this 646-page manuscript—an assumption which is rather doubtful in view of the author's difficult orthography—he would have found little in it to appeal either to his own astronomical interests or to those of the majority of his contemporaries.

Other factors responsible for the lack of attention hitherto shown to these Gresham Lectures are the first Astronomer Royal's failure to allude to them in any of the autobiographical memoirs in Baily's book or in hitherto published correspondence. Even in his unpublished letters (which still outnumber those already in print in a ratio of over three to one) they are seldom mentioned, and then only as an excuse for his delay in replying to a colleague's inquiry or request. It would also appear that they were delivered to but 'a slender auditory' so, with the exception of some brief reports in the *Philosophical Transactions* pertaining primarily to his observations of celestial phenomena, and 'The Doctrine of the Sphere' (1680), the subject matter which they contain was never widely disseminated. Certain principles and proofs from his theory of telescope optics were communicated to William Molyneux of Dublin in the course of an extensive though still unpublished correspondence during the decade prior to the appearance of the latter's *Dioptrica Nova* (Dublini, 1692), the nature of which can be inferred from the acknowledgements made by that author in his Preface. The prevalence of a positivistic attitude towards scientific progress during the nineteenth and early twentieth centuries, and the slowness of history of science to assert itself as a recognized academic discipline, may also be regarded as general influences opposed to the subsequent investigation of the archival materials which Baily had left unexplored.

However, the present generation of scientifically minded historians is much more eager to gain a deeper insight into, and clearer understanding of, the problems confronting practical astronomy in the very decade when Newton conceived his *Principia Mathematica Philosophiae Naturalis* (Londini, 1687). This interest alone justifies the publication of these manuscript lectures which are designated as R.G.O. MSS. 38 in the Public Record Office's system of classification. A further justification for the publication of these lectures is that recent philosophically orientated writings on the seventeenth-century astronomical revolution have completely ignored the contributions of the inventions of telescope, screw-micrometer, and pendulum clock; the gradual refinement of observational methods; and the application of new instrumental techniques. Another stimulus to the appearance of this work in the year 1975 is, of course, the Tercentenary of the Royal Observatory at Greenwich. What could be a more fitting way of marking this occasion than to pay a literary tribute to the first Astronomer Royal which gives new insights into his youthful intellectual development?

Flamsteed, in common with most of his English contemporaries, generally adopted the Old Style (O.S., or in the Latin abbreviation *st.vet.*) of dating his letters and documents. This style was based upon the Julian Calendar, which supposed the length of a solar year to be *exactly* $365\frac{1}{4}$ days whereas in fact it is a little less; moreover, since the Julian year began on 25 March, dates between 1 January and 24 March 1672, for instance, were denoted by a double year 1671/2. However, most Catholic countries on the continent of Europe were already adopting the calendar introduced by Pope Gregory XIII in 1582 which took account of the cumulative error in the length of the Julian Year and defined 1 January as the date on which the new calendar year should begin. Several of Flamsteed's correspondents, including Giovanni Domenico Cassini in Paris and Johann Hevelius in Danzig, customarily used the New Style (N.S., or in Latin *st.n.*) dating of this Gregorian Calendar which, during the latter half of the seventeenth century, was ten days ahead of the corresponding Julian date. Thus, for example, 1 January 1672 [N.S.] was identical with 22 December 1671 [O.S.], while conversely 1 January 1671/2 [O.S.] was identical with 11 January 1672 [N.S.]. Unless a statement has been made to the contrary, dates cited in this text are in the Old Style.

The Gregorian Calendar did not become officially adopted in England until the year 1752, long after Flamsteed's death.

The transcription of Flamsteed's Gresham Lectures was begun in 1970 from a Public Record Office (P.R.O.) microfilm, and completed during the following year from a xerographical volume of those same papers kindly lent to the author by the National Maritime Museum; this volume is normally kept at the Old Royal Observatory at Greenwich along with similar copies of the other seventy volumes of Flamsteed's papers. The P.R.O. references to items contained in several of these which it was necessary to cite among the annotations, were invariably used. Typescript copies of the lectures prepared from the transcriptions were checked against the original R.G.O. MSS. at Herstmonceux Castle, and the appropriate corrections made, shortly before Easter 1972. Great care has been taken to retain as far as possible the original spelling and punctuation. Unnecessary repetitions stemming from Flamsteed's practice of copying the last word on one page at the beginning of the next, and from the accidental rewriting of the same word twice elsewhere throughout the text, have been eliminated. The syllables, words, or phrases which were introduced by the editor for the purpose of clarifying the meaning are placed inside square brackets; the existence of those in the manuscript which were indecipherable, or obliterated by a mark, a cut, or a tear in the paper, is indicated by three ellipses. The same convention was also applied to indicate that material has been omitted from quotations. Regrettably, a considerable number of errors were found to be present in certain of Flamsteed's theoretical demonstrations (especially in Lectures 21, 35, and 37), and the correct designations of angles and angular distances have been substituted for those which appear in the original text. These changes, along with his own deletions, alterations, and additions, are recorded in the annotations. The diagrams have been redrawn and those that were missing (if indeed they ever existed) have been reconstructed: like the footnotes, they have been given a numerical suffix (e.g. Fig. 1.2 refers to the second *figure*, and note 1.2 to the second *footnote*, in Lecture 1). The diagrams have been grouped together at the end of each lecture, except for those in the Introduction which have been reproduced within the text. The list of contents gives the page numbers in Flamsteed's original manuscript, the corresponding official P.R.O. volume and folio numbers (*recto* and *verso*), for the lectures and a brief outline of their content as an aid to the rapid location of lectures dealing with specific themes.

Before beginning work on the Introduction, an entire year was spent collating and systematically arranging the large quantity of Flamsteed's unpublished correspondence which, being indispensible to a proper understanding of the early phase of his career, is of considerable relevance to the question of when and why he came to be interested in the particular aspects of astronomy which appear in his lecture syllabus. The Introduction was then composed during the winter of 1973–4 at the Deutsches Museum, Munich. It is intended to reveal the influences upon Flamsteed's early work at Derby and Greenwich and to account for the nature of the astronomical researches embodied in these lectures. At the same time, it provides the basis for a critical appraisal of his ideas and methods. The accompanying annotations are numbered consecutively.

The most frequent abbreviations to be found in this publication are:

o^{r} = our	y^{em} or y^{m} = them	☌ = conjunction
w^{ch} = which	y^{n} = than or then	☍ = opposition
w^{t} = what	y^{t} = that	♁ = Earth
w^{th} = with	y^{u} = you	
y^{e} = the	★ = star	

d, h, m, s = day, hour, minute, and second of time.

gr *or* o, ′, ″ = degree, minute, and second of arc (although ′, ″ are sometimes used ambiguously instead of m, s).

s = sign (i.e. 30° interval in celestial longitude) or sine (the trigonometrical function).

c, t = cosine and tangent, respectively.

The astrological symbols for the Sun, Moon, and planets—also used to denote the days of the week:

☉	☽ or ☾	♂	☿	♃	♀	♄
Sun	Moon	Mars	Mercury	Jupiter	Venus	Saturn
Sunday	Monday	Tuesday	Wednesday	Thursday	Friday	Saturday

The twelve signs of the Zodiac:

♈	♉	♊	♋	♌	♍	♎
Aries	Taurus	Gemini	Cancer	Leo	Virgo	Libra
(Ram)	(Bull)	(Twins)	(Crab)	(Lion)	(Virgin)	(Scales)

♏	♐	♑	♒	♓
Scorpio	Sagittarius	Capricornus	Aquarius	Pisces
(Serpent)	(Archer)	(Goat)	(Water-carrier)	(Fish)

The three principal astronomical works of Giovanni Battista Riccioli frequently consulted by Flamsteed and often cited in the annotations to his text as:

Riccioli 1 = *Almagestum novum*, i (Bononiae, 1651)
Riccioli 2 = *Almagestum novum*, ii (Bononiae, 1651)
Riccioli 3 = *Astronomiae reformatae* (Bononiae, 1665).

Acknowledgements

I would like to thank the Faculty of Arts Research Grants Committee of Edinburgh University for reimbursing the costs incurred in the transcribing and typing of Flamsteed's manuscript, and the travel and subsistence expenses of my two-week visit to the Royal Greenwich Observatory during March 1972 in connection with this project. I am also greatly indebted to the Alexander-von-Humboldt-Stiftung in Bonn, and to the British Academy in London, for their respective awards of a Lecturing Fellowship and a European Exchange Grant which enabled me to complete my research at the Deutsches Museum in Munich between August 1973 and March 1974. I should like to take this opportunity of expressing my gratitude to the University of Edinburgh for granting me sabbatical leave to complete this and begin another research project during that period.

Finally, I would like to thank the Controller of Her Majesty's Stationery Office for his kind permission to publish the text of the Lectures.

ERIC G. FORBES

History Department
University of Edinburgh

Introduction

The association of John Flamsteed, F.R.S. (1646–1719), First Astronomer Royal, with Gresham College in London, is an aspect of his life and work which has hitherto remained totally outwith the scope of his biographers and historians of astronomy. Even if they had been aware of it, they would probably still have chosen to ignore it on account of its tenuous nature and apparent irrelevance to his dedicated labours at the Royal Observatory in Greenwich during the last forty-four years of his life. The scanty evidence suggests that Flamsteed was initially asked by the Gresham Professor of Astronomy, Dr. Walter Pope, to deputize for him during an illness by which Pope nearly lost his sight.[1] Precisely when this occurred is not clear, but it is likely to have been several months prior to the date of his first lecture on 27 April 1681 although this arrangement was not formally ratified until more than three months after it.[2] A statement in John Ward, *The Lives of the Professors of Gresham College* (London, 1740) is misleading inasmuch as it implies that Flamsteed undertook this commitment some five years later, and that his obligation ceased with Pope's retirement in September 1687.[3]

What induced Flamsteed to agree to such an onerous and time-consuming task in the first instance is a matter for speculation. He may have done so merely out of friendship for a man whom he admired, stimulated perhaps by the promise from Pope of some pecuniary remuneration for his services; for the death of his influential friend and patron Sir Jonas Moore on 27 August 1679 had put Flamsteed in an uncertain financial position of which he bitterly complains in a letter to Dr. Seth Ward, Bishop of Salisbury, on 31 January 1680.[4] Another factor influencing his decision may have been that, when approached, he had almost completed his laborious literary task of preparing, for the posthumous edition of Jonas Moore's *A New Systeme of the Mathematicks* (London, 1681), his treatise on 'The Doctrine of the Sphere', new solar tables, and a revised version of the lunar tables which had originally appeared in John Wallis's edition of Jeremiah Horrox's *Opera posthuma*

Notes on pages 70–78.

mathematica (Londini, 1673). Moreover, the mural arc that had been made by Robert Hooke on Sir Jonas Moore's instruction and at his expense, had proved to be useless;[5] and with it being out of commission Flamsteed's observations were consequently more restricted until the new 140° mural arc, which he himself both designed and to a large extent constructed, was installed in 1683.[6]

Flamsteed makes no allusion to his Gresham College lectures in any of the three autobiographical extracts written in June 1685, June 1695, and November 1707 published in Francis Baily, *An Account of the Revd. John Flamsteed* (London, 1835) although each of these incorporates the relevant period, from April 1681 to November 1684, while he was engaged in this commitment.[7]

Nor does he allude to them in any of the correspondence published to date which, on the whole, is concentrated either on an earlier or a later phase of his life. Several brief references to this activity among unpublished correspondence with William Molyneux of Dublin[8] and Richard Towneley of Towneley in Lancashire, England,[9] are, on the whole, quite uninformative, his lectures being merely mentioned as an excuse for his delay in replying to a letter. On one occasion, however, his comment that he had little leisure since the Term was approaching and he had to provide for 'ye almost ruined Gresham lectures'[10] produced the following reaction from Molyneux:

> Truly tis Pitty So Noble a designe should fall to the Ground, And tis a shame for the Gentry of London to suffer the Great Professors of that College to Read sometimes to almost Bare Walls. Were a seditious balling Fanatick in the Pulpit he would have a Thick Audience to hear his Informall Doctrine, Whilst the Caelestiall Discourses of a Learned Astronomer or Other Mathematician are heard but by a Few, and perhaps by them Neglected.[11]

The 'Noble design' here referred to is that specified in the Will of Sir Thomas Gresham (1519–79) on 5 July 1575 by which his mansion house, garden, stables, etc., were entrusted jointly to the City of London and the Company of Mercers on condition that these bodies would elect seven unmarried lecturers to live in these premises and 'daylie to reade' lectures for the benefit of the citizens of London in divinity, astronomy, music, geometry, law, physic, and rhetoric for an annual salary of £50.[12] This wish was executed after the death of his widow Dame Anne Gresham in December 1596, and provisional rules were duly drawn up by a Grand Committee for Gresham Affairs on 16 January 1597.[13] This committee consisted of twelve attorneys representing the interests of the City of London (viz. the past and present Lord Mayors, two aldermen, and eight commoners), and twelve representatives of the Mercer's Company (viz. the Master, three Wardens, and eight life-members of the court of assistants). *Any* twelve of these gentlemen constituted a quorum over which the Lord Mayor presided. There was also a Sub-Committee comprising six members from either side, of whom three from the City and three Mercers were empowered to execute any decisions taken.

Two of the principal problems initially facing the Grand Committee were in what language the lectures should be delivered, and at which times of day and year, neither of these having been specified in Sir Thomas Gresham's Will. Reasons which were given for the lectures being in Latin were that the fame of the founder of Gresham College, and the honour of the City of London, would thereby be more widely spread; thus learned foreigners would not only benefit but perhaps also be encouraged to follow this example. Besides, it was still common practice to read lectures in Latin at Oxford and Cambridge, from which the professors themselves had necessarily to be drawn. On the other hand, Sir Thomas's concern appeared to have been more for the citizens of London, few of whom understood Latin; and more local people would consequently be attracted if the lectures were read in the vernacular, which was in keeping with the current trend to translate works out of Latin and Greek into modern European tongues. There would be less feeling of competition with the two ancient universities, and the common people might thereby acquire more respect for learning. The outcome of these early deliberations was the compromise solution that both Latin and English should be used.

It was proposed at the committee's first meeting on 16 January 1597 to hold the astronomy lectures on three afternoons per week during five terms, beginning on the Monday before the Trinity Term. Just under a year later, after consultation with the professors, their number was reduced to two per week (on Fridays from 8–9 a.m. in Latin, and 2–3 p.m. in English) during four terms beginning on the Monday before the St. Michaelmas Term, thus making the College Terms coincide with those in the Law Courts—a practice which was adopted right through the seventeenth and early eighteenth centuries. The programming of Flamsteed's lectures for Wednesdays rather than Fridays may have been arranged to suit his own convience. His desire to rid himself of the above-mentioned literary obligations or the ill-timing of Pope's request for assistance may suffice to explain why he began his course in the Trinity Term of 1681 instead of at Martinmas.

Regarding the content of the astronomy lectures, the prescription recommended by the Grand Committee on 16 January 1597 and approved without change when the revised 'Ordinances and Agreements' came into force almost a year later was as follows:

> The astronomy reader is to read in his solemn lectures, first the principles of the sphere, and the theoriques of the planets, and the use of

> the astrolabe and the [cross-] staff, and other common instruments for the capacity of mariners; which being read and opened, he shall apply them to use by reading geography and the art of navigation in some one terme every year.[14]

Despite the fact that the seventeenth century had witnessed a revolution in observational astronomy through the introduction of the telescope and micrometer-eyepiece, the same agreement was apparently still being honoured in Flamsteed's time, as is suggested by his statement in Lecture 25 that:

> The Astronomicall Lecturer is obliged to explaine the Theorys of ye Planets by a mutuall stipulation with ye Honorable trustees of ye colledge made 86 years agone: when it was first settled. before I understood this I had happily falne upon subjects not disagreable with this stipulation.[15]

His previous lectures had, in fact been concerned with the accurate determination of: the Earth–Sun distance (or solar parallax), the theory of cometary motion, the theory of telescope-optics including the application of this instrument to quadrants and sextants, the stereographic ecliptic projection of the sphere, the constitution of the heavenly bodies, and the various errors incurred by earlier observers. In the remaining fourteen lectures, Flamsteed does not fulfil his intention of rectifying his omission. Much of his time was, in fact, spent in discussing the phenomena of precession and refraction. Indeed, he appears to have finally abandoned the aim of lecturing on planetary theory in deference to the necessary but for him distasteful additional obligation of delivering the same lectures in Latin. Thus the penultimate lecture is a repetition in Latin of Lecture 33; and the last, a Latin demonstration of a geometrical analogue to the trigonometrical problem of solving a plane triangle when two of its sides and the contained angle are given, appears to have no integral connection to the rest of the syllabus.

Flamsteed's reference to having had 'but a slender auditory'[16] at his very first lecture, and his comment 'No body came not read'[17] on that prepared for 20 June 1683 which was twice postponed before being ultimately delivered on 31 October, reveal that this undertaking was far from being a publicity stunt! Apart from the Gresham Professor of Geometry Robert Hooke and three other gentlemen, Reeve Williams, Edward Pagitt, and Robert Wood who taught mathematics to the 'blue-coat boys' of Christ's Hospital, London, it is hard to name anyone who may have been attending his lectures regularly. It is even more difficult to conceive that even these teachers or any of their pupils might have conscientiously followed the entire course which was spread throughout thirteen terms over a period of three and a half years. A hint of sabotage also seems implied by the comment appended to Lecture 13 for 4 May

1682:

> present onely Mr. Williams Mr. Paget Dr. Wood with ye Blewcoates Mr. Hookes designe![18]

However, when Flamsteed spoke of the 'almost ruined Gresham lectures',[19] he was probably referring not only to his own, but also to the others by George Gifford (divinity), Sir Thomas Barnes (music), Roger Meredith (civil law), John King (rhetoric), Henry Paman (physic), and Hooke himself. The fact that the numbers attending these lectures were consistently low was accepted by the Sub-Committee for Gresham Affairs as the main cause of apathy shown by these professors in failing to read their lectures over 'many months, nay sometimes whole years together'.[20] Flamsteed reveals a greater sense of moral responsibility on the occasion of his last lecture, when he tells his audience:

> Let the others break faith, or excuse themselves, or evade their duty; as for myself, if I only fulfil no more noble task than obeying the laws of the place and your injunctions, I shall consider myself never to have done anything more distinguished or praiseworthy than to obtain your acknowledgement of my having acted in accordance with the wishes of the testator.[21]

His collective condemnation of the other professors followed by an abrupt end to his astronomy lectures, suggests that a tense situation had been developing between him and these resident professors during the summer of 1684. Unlike himself, the latter had literally to live with this depressing situation which apparently persevered in spite of the Grand Committee's decision on 24 March 1685 to advance each of the lecturing-times by one hour.

Thus one is unable to avoid the conclusion that Flamsteed's Gresham College lectures had virtually no influence upon his contemporaries. Their historical importance derives entirely from the converse possibility of establishing the influence of his contemporaries upon him. At the same time, they enable one to gauge the extent of his originality and to appreciate the rôle of his own astronomical observations in moulding his early intellectual development. Although it is unfortunate that there exist no records of Pope's lectures,[22] the strong links that will be shown to exist between the contents of Flamsteed's syllabus and his earlier work at Derby and Greenwich are entirely opposed to the assumption of an interdependency in their respective teaching programmes. One can, moreover, discount the idea that the topics were deliberately selected to appeal to a popular audience.

Early influences on Flamsteed's work

One of the most important of his literary sources in the preparation of his entire course of lectures was the work by Giovanni Battista

Riccioli, *Almagestum novum* (2 vols., Bononiae, 1651), which contains a wealth of information on astronomical theories and methods from ancient times up to the first half of the seventeenth century. Flamsteed acquired the first volume of this book in 1666, and read it through systematically during the course of the following winter transcribing some of the sections which particularly interested him.[23] He refers to it frequently, and at times critically, in his lectures. As for the author himself:

> Riccioli [is] a p[er]son who by his laborious methodicall collections had gained very ample commendations. if his wresting some his inadverting some and misapplying of other good observations had not detracted from his praises.[24]

He had 'frequently shewd himselfe' to be an ill-manager of celestial observations.[25] The Jesuit father's prime motivation, however, was (according to Flamsteed) not the love of truth but the desire to plead the cause of the theologians and to defend the literal meaning of the scriptures at a time of serious conflict between science and religion. Indeed, one finds statements of little apparent relevance to the history of astronomy such as John the Baptist was born at the summer solstice and Jesus Christ at the winter solstice in order to fulfil the former's prophesy: *Oportet illum crescere, me autem minui* ('It is right for him [Christ] to grow, and for me to shrink').[26] Thus one requires to take note of the spirit in which this book was written in order to understand why Riccioli adopted particular viewpoints.

An excellent example illustrating the dependency of Riccioli's astronomical ideas upon his religious preconceptions is his treatment of Aristarchus's method of finding the sun's distance by the moon's dichotomy; namely, by the time at which the moon's surface appears to a terrestrial observer to be exactly half-illuminated by the sun's rays. The twenty-year old Flamsteed, while writing an autobiographical account of his childhood years, had paused to reflect upon this as follows:

> he introduced an arc for correcting the apparent dichotomy, and reducing it to the true; which cannot be admitted. For he supposes the moon's parallax to cause her to appear hollow at the dichotomy next succeeding the new moon, and more than half full at [the] dichotomy preceding the change; which I shall prove not to be so in my astronomical works. In the meantime, whoever will but read what he hath written (page 733, the third problem) will find how groundlessly he introduceth it, if they but seriously consider that the difference of the parallax of the moon's centre and her superior horn is equal to the difference of the parallax of her centre and her inferior horn; with a very small difference, which will scarce ever arise to half a second, were her diameter double the breadth it is.[27]

Flamsteed's choice of the ancient problem of estimating the Sun's and Moon's distances as the logical starting point for his Gresham College lectures caused him, some fourteen years later, to reconsider this early critique of Riccioli's modification of Aristarchus's method. It is evident from his further remarks on this theme in Lecture 2, that he had in the interim changed his mind about the groundlessness of Riccioli's motive. The point with which he takes issue can best be explained with the aid of Riccioli's own diagrams and notation (cf. Figs. 1 and 2).[28]

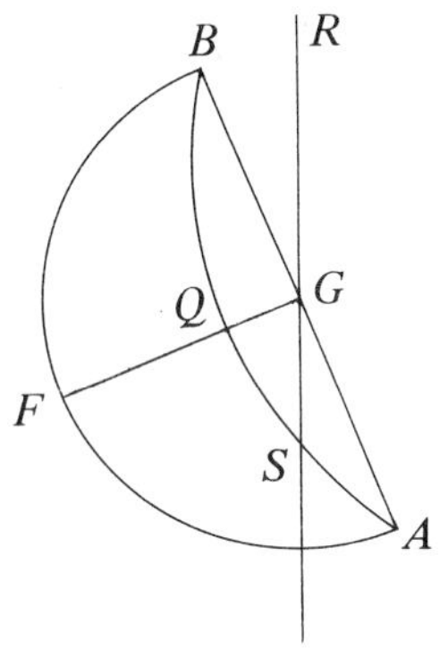

FIG. 1. Last quadrature. Sun on east side of ecliptic (i.e. to left of diagram).

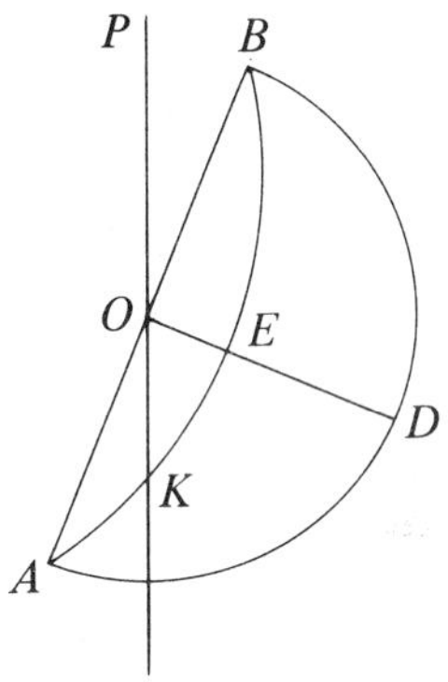

FIG. 2. First quadrature. Sun on west side of ecliptic (i.e. to right of diagram).

Riccioli had defined the time of dichotomy as that at which the illuminated parts near the centre of the disc at the point (*O* or *G*) where it crosses the vertical circle (*POK* or *RGS*), appear on the line of cusps (*AOB* or *AGB*); instead of that at which they appear on the thin ellipse, or meridian line (*AKEB* or *ASQB*). By so doing, he effectively advanced the time of the first quadrature and retarded that of the last quadrature by about an hour—corresponding to the distances (*OE* or *QG*) along the line of motion (*DO* or *FG*) between the limit of the shadow (*E* or *G*) and the vertical circle. Thus he over-compensated the contrary effect of the lunar parallax and ascribed the remaining time-difference to that of the solar parallax. By proceeding in this manner, argues Flamsteed, the Jesuit father avoided the alternative conclusion that the solar parallax is insensible—that is, that the sun is at an infinite distance from the Earth. He thereby forestalled conflict with the statement of the Conclave of seven cardinals based on the Holy Scriptures (published by Riccioli)[29] to the effect that the Sun moves daily from east to west. Only by postulating that this motion is apparent, and produced by the contrary rotation of the Earth about its axis, could one ignore the solar parallax and still save this appearance.

Flamsteed's own telescopic observations at Derby in the spring of 1671 had first made him conscious of the practical difficulty in estimating the apparent times of the dichotomy on account of the irregularities in the moon's surface;[30] thus a year or so later, he resolved to abandon that method for finding the solar parallax and to seek another. By April 1672 he had discovered what he was looking for in Johann Kepler, *Astronomia nova* (Pragae, 1609), namely, the method of Tycho Brahe involving observations of the apparent angular distances between Mars and the Sun, by day, and between the stars and Mars at night.[31] For reasons explained below, the former, when compared with the tabulated true distance, gave the sum of the parallaxes of Mars and the Sun. The latter gave the distance of Mars in terms of Earth radii, thereby permitting the Sun's distance to be expressed in similar units and thus in absolute measure. We shall now follow through the principles of this method, and exemplify them using the results cited by Flamsteed in Lecture 2.

Let us first consider the stellar observations. Flamsteed was rightly sceptical of the validity of any conclusions based upon measurements of star positions as inaccurate as many of Tycho's were, but well knew that he himself now had a double advantage over 'The Noble Dane' which would ensure the viability of this new method: namely, an astronomical telescope and a micrometer eyepiece, which enabled him to measure small angular distances in the heavens to the hitherto unobtainable accuracy of about one second of arc (1″). The screw micrometer had initially been devised by William Gascoigne, a native of Yorkshire, who had used it around 1640 for determining the diameters of the Sun, Moon, and planets.[32] The circumstances by which Flamsteed came to possess this instrument, and to have access to its inventor's scientific correspondence, are described as follows in his letter of 10 May 1690 to William Molyneux of Dublin with whom he had been corresponding extensively for several years:

> Before or late civil wars no country yt I know of in the World had so many ingenuous men in it at one time as there were in Lancashire & its confines. At Broughton neare Manchester there lived one William Crabtree a factor for Blackwell Hall who kept a Correspondence with Mr Jeremiah Horrox then resideing neare Liverpoole in ye same county. In ye Autumne of ye yeare 1640 he visited Mr William Gascoigne who then lived with his father at Middleton neare Leeds in Yorkshire. & saw his instrumts & Contrivances for Measureing small Angles in the heavens. and from this time till the wars Interrupted it kept a freindly Correspondence with him Mr Crabtree had desired from him an explication of ye Doctrine of Glasses in a letter dated Dec. 2. 1640. Mr Gascoigne shews him in 4 figures how ye raies of light parallel to ye Axis are carried through a plano convex or plano

concave glasse when the plaine side is towards ye object. as also how an inclined ray is carried through a like posited plano convex glasse into the distinct base, & through ye convex eye glasse to ye superficies of the eye. & this he does very briefly useing allways ye usuall simbolls in the roome of words. hee meddles not with ye double convexes or double concaves nor with ye convex or concave side of a glasse turned to ye object. he sayes nothing of ye composition of ye severall sorts of telescopes, nor compounding more yn 2 glasses in a tube. nothing of the Course of the raies through ye eye nor ye picture projected on ye Retina, nor of the Rectificatio[n] of Instruments for takeing heights or distances with telescopicall sights. these are all yu well know the Naturall Consequences of his doctrine & the superst[r]ucture of yr Freind. tho I am psuaded yt he was not wholly Ignorant of them For I have seene at Townly the Iron Carkasse of a sextant which Mr Towneley told me was Mr Gascoignes. . . .

This correspondence continued till September 1642 Mr Gascoigne in his letters gives him an account of all his observations made with his glasses of the Suns & Moones diameters the distances of neare fixed stars Eclipses and Appulses from Dec 1638 till this time. I have excerped them from his letters & intend to publish them with my owne as soone as I shall have a competent number of observations of ye Meridionall distances of ye fixed stars from the Vertex taken on My Murall arch to accompany their Intermutuall distances measured formerly with ye sextant which I hope a very few yeares will afford me if God spare me health Mr Crabtree died in ye yeare 1643. Mr Gascoigne survived to the next when haveing engaged on ye Kings side he was slaine in his service at Marston Moore fight July ye 2. 1644 sometime after Mr Christopher Towneley of ye Car in Lancashire & Sr Jonas Moore heareing that all Mr Crabtrees papers were still reserved by his wife went to visit her at Broughton & had all his Mathematicall papers bestowed on them. which they packt up & carried home betwixt them as each had conveniencys they sorted them after & Mr Towneley changed his for others more acceptable to him with his Newphew Mr Richard Towneley of Townely in Lancashire. amongst these were Mr Gascoignes letters to Mr Crabtre[e] which conteined the description of his Micrometer. his Dioptricall propositions & observations aforementioned. Mr Townely finding the Contrivance of ye Micrometer might be much betterd caused a new & lesse troublesome one to be made by an ingenious Watchmaker his tenant yu have ye description of it in the Philosophicall Transactions of November 11. 1667. Num. 29. When I was at London in ye yeare 1671. Sr Jonas Moore presented me with one of these. And on my returne some occasions of my fathers calleing me into Lancashire I paid a visit to Mr Towneley who shewed me how he applyed it to his tubes. & frankely lent me Mr Gascoignes letters from which I deduced those two first Dioptricall propositions I sent yu. this is the plane history of that Invention. & the begining of my observations at Derby for the first valewable Measures that I tooke in ye heavens were got with that Micrometer . . .[33]

The question of Flamsteed's debt to Gascoigne for his knowledge of telescope-optics will be discussed later on, when we come to consider the contents of Lectures 4–8.[34] The 'first valewable [micrometer] Measures' to which he is here referring are those of Mars's retrograde motion with respect to three bright stars in the constellation Aquarius taken at the end of September 1672 when that planet happened to be near its perigee position.[35] The angular distances of the stars from Mars were so small that all could be seen in the fields of his 7-foot and 12-foot telescopes which was, of course, the basic reason why the micrometer could be applied to this investigation of Mars's—and hence the Sun's—parallax. The opportunity was ideal, and Flamsteed's prayers that neither poor health nor bad weather would spoil his plans on this occasion were fortunately answered.

A new determination of the solar parallax

The following sequence of line diagrams (Fig. 3) drawn using the information supplied in Lecture 2, summarize the results of observations which Flamsteed made with the above-mentioned instruments at Derby on 26 September 1672 after returning from a visit to

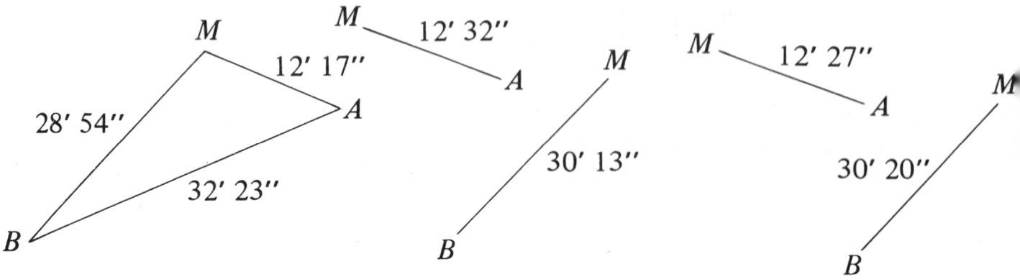

FIG. 3. Sequence of line diagrams illustrating the results of Flamsteed's micrometer measurements of Mars's changing position, *M*, with respect to two fixed stars, *A* and *B*, on 26 September 1672.

the Towneley residence. *A*, *B*, denote two of the bright stars[36] and *M* the apparent position of Mars with respect to them, the distances changing as indicated while the time progressed.

Note that the positions of the two stars do not vary, but only that of Mars for which the celestial longitude and latitude at 7h 03m p.m. were unambiguously fixed by the three (almost) simultaneous micrometric measurements. In practice, Flamsteed customarily based his micrometric readings on the distance between each star and the more remote limb of the planet, and then subtracted the latter's semi-diameter (in this case 15″) to obtain the distance to the centre of its observed disc.

Now consider the variation in the distance *AM*, as a function of the time-intervals between the observations:

Difference in *AM*	+15″	+10″	?
Cumulative time intervals	5h 37½m	5h 50½m	6h 10m

Flamsteed assumed that the decrease in this quantity is *uniform*, arguing that if *AM* decreases by 5″ between 5h 37½m and 5h 50½m (viz. in 13m), it will decrease by a further (19½m÷13m)×5″=7½″ between 5h 50½m and 6h 10m (viz. in 19½m). Thus the required change in *AM* is +10″−7½″=+2½″ over the whole 6h 10m time-interval, yielding *AM*=12′ 19½″ at 13h 13½m. At 13h 13½m p.m. (or 1h 13½m a.m. on 27 October 1672), Flamsteed knew by direct measurement the invariable distance *AB*, and the distance *BM*; also, by the combination of measurement and calculation just described, the distance *AM* (Fig. 4). Having established the celestial

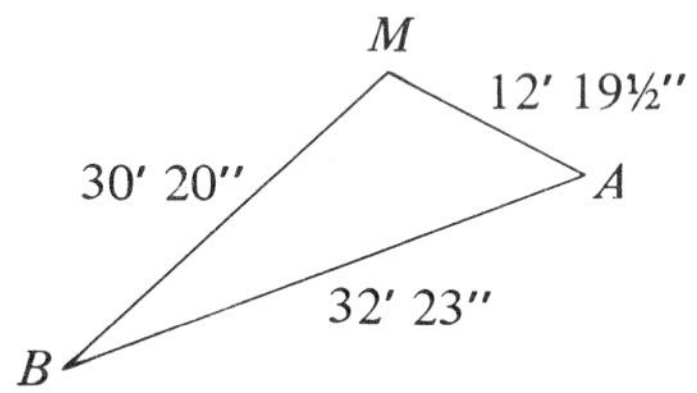

FIG. 4. Mars's position with respect to the stars *A* and *B* at 1h 13½m a.m. on 27 September 1672.

co-ordinates of *A* and *B* two nights later from independent observations with his 3-foot radius wooden quadrant which revealed the presence of substantial errors of several minutes of arc in Tycho's catalogued values, he was consequently able to deduce those of Mars 'by a strict calculation'.[37] He found by this means that its variation in longitude over the 6h 10m time-interval in question was 37½″ retrograde, and in latitude 1′ 19½″ decreasing.

The corresponding values of Mars's true motion had now to be interpolated from tables in order to control whether diurnal parallax may have affected this result. The effect anticipated is shown in Fig. 5, in which *E* represents the centre of the Earth, *O* the position of an observer on its surface, M_1, M_2 the positions of Mars at the initial and final times of observation respectively. The reason why they appear to be so widely separated is that for purposes of illustration *O* is regarded as fixed and Mars, together with the whole sphere of fixed stars, appears to be carried round it in the opposite sense to the Earth's diurnal rotation; that is, from east to west. X_1, X_2 denote the corresponding true (geocentric) positions of Mars projected upon the uniformly rotating celestial sphere; and Y_1, Y_2 the apparent

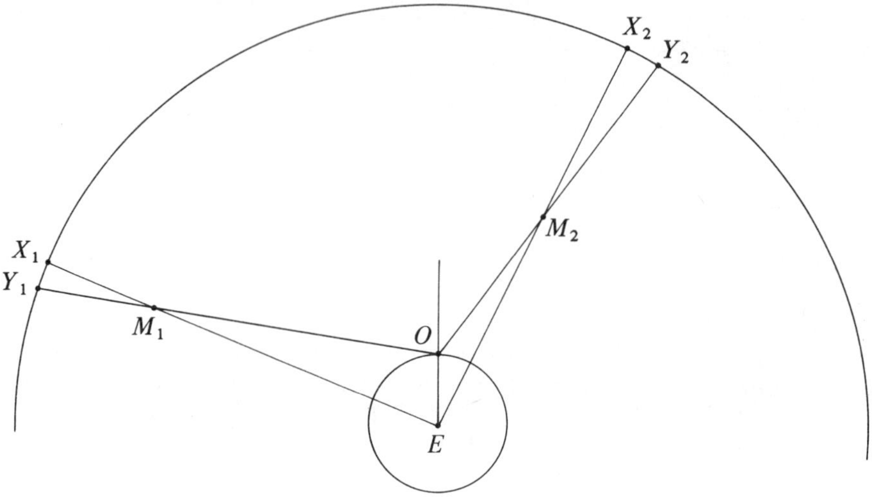

FIG. 5. Diagram illustrating the effect of parallax.

(observed) positions. It is evident that, owing to the diurnal parallax, the planet seems to move from Y_1 to Y_2, instead of from X_1 to X_2, and thus the actual motion will be over-estimated in the observations by an amount equal to the sum of the arcs X_1Y_1 and X_2Y_2. In other words, the true motion ought to be less than the apparent observed motion by the positive quantity $X_1Y_1+X_2Y_2$. By an analogous process of reasoning, but with M_2 now denoting the Sun's position at the *same instant* as Mars was at M_1, this quantity can be alternatively identified with the combined effects of the parallax in altitude on the two celestial bodies. Since he neglected astronomical refraction, Flamsteed identified it with the difference between the observed and calculated values of the Sun–Mars distance.

Flamsteed obtained the values of Mars's celestial longitudes and latitudes on 26 and 27 September from Joannes Hecker's *Ephemerides* for 1672,[38] then scaled down their differences by a factor of 6h 10m/24h = 37/144 to discover the changes that had occurred in these respective co-ordinates during the 6h 10m interval between the initial and final observations. Rules and tables are given in Hecker's Introduction for correcting the times of observation to compensate:

(*a*) The effects of the Sun's irregular motion in its quasi-elliptical orbit, the axis of the latter being inclined at approximately $23\frac{1}{2}^\circ$ $(=\varepsilon)$ to that of the diurnal rotation.

(*b*) The difference in longitude between the meridian of Uraniburg—the site of Tycho's observatory—on which the tables were based, and that of Derby where his own were obtained.

Although Derby itself does not feature among the towns listed in the 'Catalogus Civitatum . . .', we know from Flamsteed's appendix to John Wallis's edition of *Jeremiae Horroccii . . . Opera Posthuma* (Londini, 1673), composed about the same time, that he deduced its longitude by adding 5m to the Uraniburg-London difference of 48m that had been adopted by Horrox.[39] He no doubt applied the theoretical corrections for (*a*), calculated in his own tract 'De Temporis AEquatione Diatriba' which is also appended to that same volume,[40] in preference to the empirical values of Hecker.

The above procedure yielded the retrograde motion of 20″ in longitude, and the decrease of 1′ 13½″ in latitude, which are given in Lecture 2. Significantly, both figures are lower than those (cited above) derived from observation, and thereby justify the belief that the diurnal parallax has been detectable by this means. The values of the *sum* of the two parallaxes are $37\frac{1}{2}'' - 20'' = 17\frac{1}{2}''$ in longitude, which Flamsteed approximates to 18″, and $1'\,19\frac{1}{2}'' - 1'\,13\frac{1}{2}'' = 6''$ in latitude.

The next requirement was to determine which proportions of these values are inherent in the observation at 7h 03½m p.m. on 26 September, and at 1h 13½m a.m. on 27 September, respectively; which was sufficient to fix the amount of Mars's parallax. For this purpose, Flamsteed introduced (without proof) precepts taken from the traditional nonagesimal method of calculating parallaxes in celestial longitude and latitude, that he could have found in any one of a number of books on mathematical astronomy.[41] Since, however, an understanding of this part of Flamsteed's procedure is necessary for an appreciation of the remainder of this investigation, and of central importance in his following analysis of cometary motion (Lecture 3), we have decided to present a mathematical derivation of the formulae by which the position of the nonagesimal on the celestial sphere can be specified.

In Fig. 6, the great circle *AB* defines the celestial horizon, and *AN* the ecliptic intersecting it at A. *PZQ* is the observer's meridian through the poles *P*, *Z* of the celestial equator and horizon respectively, cutting the ecliptic at *Q* (the culminating point). The nonagesimal *N* is defined as that point in the ecliptic whose celestial longitude measured eastwards along the cliptic from the Vernal Equinox ♈ is 90° greater than that of A. Since *Z* is also 90° from *A*, the great circle *NZ* is perpendicular to the ecliptic. Producing it to *K*, where $NK = 90°$, therefore defines the pole of the ecliptic which, though actually situated to the east side of the meridian is drawn upwards above the celestial sphere for the purpose of obtaining a clearer illustration. Producing *NZ* in the opposite direction, it intersects the celestial horizon in *B*. Thus, since $BZ = NK = 90°$:

(1) The altitude of the nonagesimal

$$= BN = BZ - ZN = BZ - (NK - ZK) = ZK.$$

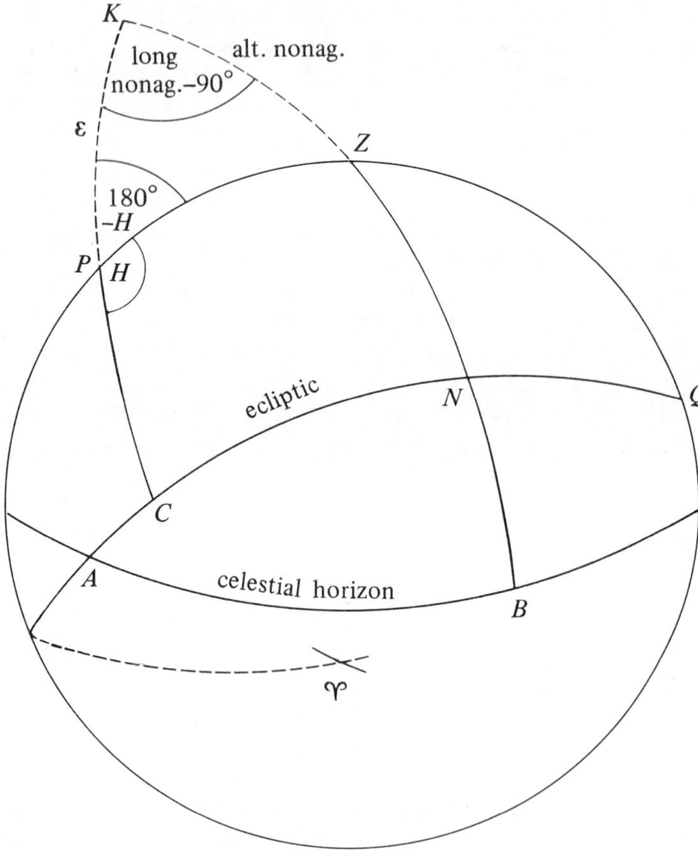

FIG. 6. Diagram to illustrate the method of calculating the position of the nonagesimal point (N).

Now draw the great circle arc KP and produce it to cut the ecliptic in C. Since it lies on the line joining the poles of the ecliptic and equator, C must lie halfway between their two points of intersection (viz. the equinoctial points). Its longitude, measured in the ecliptic from A towards N, is therefore 90°; or the arc CN = the longitude of the nonagesimal $-90°$. Thus we obtain:

(2) Long. nonag. $= CN+90° = C\hat{K}N+90° = P\hat{K}Z+90°$.

To find (1) and (2), one requires to solve the spherical triangle PKZ for KZ and $P\hat{K}Z$, given: $PZ=90°-\varphi$, where φ is the latitude of the observer; $PK=\varepsilon$, the obliquity of the ecliptic; and $K\hat{P}Z=180°-H$, where H is the hour angle measured westwards from the observer's meridian.

The solution is easily effected by dividing the spherical triangle PKZ into two right-angled triangles by means of the perpendicular

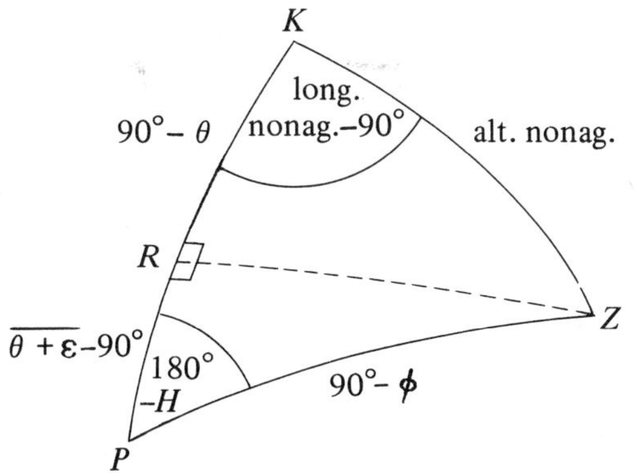

FIG. 7. Auxiliary diagram for calculating the nonagesimal point.

ZR from Z to PK (cf. Fig. 7). If we put $CR=\theta$, $KR=KC-CR=90°-\theta$ and $RP=KP-KR=\varepsilon-(90°-\theta)=(\theta+\varepsilon)-90°$. Applying Napier's Second Rule to the right-angled $\triangle PRZ$, one gets:

$$\sin(90°-\overline{180°-H})=\tan(\overline{\theta+\varepsilon}-90°)\tan\varphi$$

or

$$\tan(\theta+\varepsilon)=\tan\varphi\sec H.$$

This equation enables the values of the auxiliary angle θ to be calculated from the known angles ε, φ, and H, for a given place and time of observation.

Applying the same rule to $\triangle KRZ$, one next obtains:

$$\sin(90°-\theta)=\tan(90°-\overline{\text{long. nonag.}-90°})\tan RZ$$
$$\cos\theta=-\tan(\text{long. nonag.})\tan RZ.$$

But from $\triangle PRZ$:

$$\sin(\overline{\theta+\varepsilon}-90°)=\tan(90°-\overline{180°-H})\tan RZ$$

Eliminating $\tan RZ$:

$$\frac{\cos\theta}{-\cos(\theta+\varepsilon)}=\frac{-\tan(\text{long. nonag.})}{\tan(H-90°)}$$

$$\tan(\text{long. nonag.})=\frac{\tan H\cos\theta}{\cos(\theta+\varepsilon)}.$$

Similarly, applying Napier's First Rule to $\triangle KRZ$, one obtains:

$$\sin(90°-\text{alt. nonag.})=\cos(90°-\theta)\cos RZ$$

and from $\triangle PRZ$:

$$\sin\varphi=\cos(\overline{\theta+\varepsilon}-90°)\cos RZ.$$

Eliminating cos RZ:

$$\frac{\cos(\text{alt. nonag.})}{\sin\varphi} = \frac{\sin\theta}{\sin(\theta+\varepsilon)}$$

or

$$\cos(\text{alt. nonag.}) = \frac{\sin\varphi\sin\theta}{\sin(\theta+\varepsilon)}$$

These formulae for the longitude and altitude of the nonagesimal point uniquely specify its position. This is an essential prerequisite to the calculation of the parallactic angle—defined as the angle between the ecliptic and the vertical circle from Z through the position of the celestial body (in this case, Mars).

Flamsteed's statement that Mars was 15° above the horizon, short of the nonagesimal in the first observation (at 7h 03½m) and past it in the last one (at 13h 13½m), is a guide to the drawing of Fig. 8

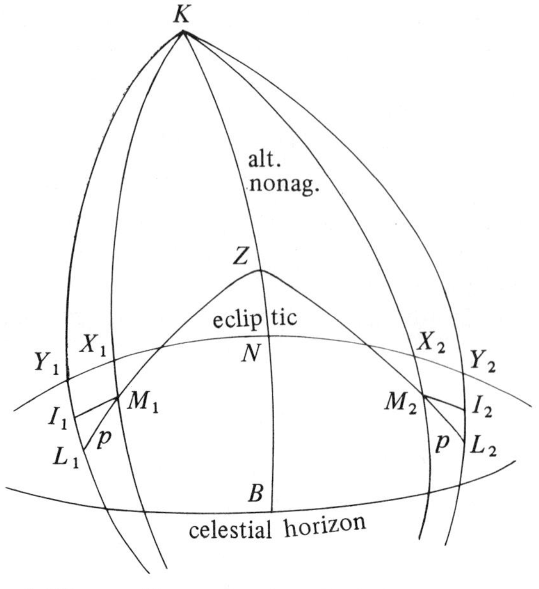

FIG. 8. Diagram illustrating the calculation of parallax.

showing the connection between the parallactic angle and the parallaxes in longitude and altitude. In addition to the notation already defined, we have represented on Fig. 8 the true positions of Mars at the times of the initial and final observations M_1, M_2; the vertical circles ZM_1, ZM_2; and the longitude circles KM_1, KM_2. The complements of the parallactic angles, $K\hat{M}_1Z$, $K\hat{M}_2Z$ can now be found from a solution of the spherical triangles KM_1Z and KM_2Z since the true celestial co-ordinates of Mars were known from

Hecker's *Ephemeris* (as outlined above). $Z\hat{K}M_1$, $Z\hat{K}M_2$ are the differences between the respective longitudes of the nonagesimal and those of Mars at the times in question. KM_1, KM_2 are each equal to 90°±the latitudes of Mars at those times (positive if that planet is south of the ecliptic); and $KZ=NB$, the altitude of the nonagesimal. The results of Flamsteed's calculations of the parallactic angles, cited in Lecture 2, are 86° 07′ and 40° 24′ respectively.

In order to deduce the connection between the parallactic angles and the parallaxes in longitude, it is necessary to produce the vertical circles to L_1, L_2 to take account of the effect of parallax in increasing the zenith distances of celestial bodies. For the same zenith distance, the parallaxes in altitude $M_1L_1=M_2L_2=p$, say. The parallaxes in longitude are, by definition (cf. Fig. 5), the arcs X_1Y_1, X_2Y_2 on the ecliptic where that great circle is intercepted by the meridians of celestial longitude through the true and apparent positions of the planet M_1, M_2 and L_1, L_2 respectively. In Fig. 8, let M_1I_1, M_2I_2 be small circle arcs parallel to the ecliptic, cutting the meridians KY_1L_1 at I_1 and KY_2L_2 at I_2. Since Mars's latitude is essentially constant, $M_1X_1=M_2X_2=\beta$, say. The parallactic angles are equal to $I_1\hat{M}_1L_1$ and $I_2\hat{M}_2L_2$, the complements of $K\hat{M}_1Z$ and $K\hat{M}_2Z$ respectively.

Hence Flamsteed's first precept:

> *As ye summe of the Cosines of the Parallactick Angles:*
> *to ye cosine of ye parallactick angle at 13h 13′:*
> *so ye summe of ye parallaxes in longitude 18″*
> *to ye parallax of longitude in ye observation then made*[42]

may be derived as follows:

$$\frac{\cos I_1\hat{M}_1L_1+\cos I_2\hat{M}_2L_2}{\cos I_2M_2L_2}=\frac{\dfrac{I_1M_1}{p}+\dfrac{I_2M_2}{p}}{\dfrac{I_2M_2}{p}}$$

$$=\frac{X_1Y_1\cos\beta/p+X_2Y_2\cos\beta/p}{X_2Y_2\cos\beta/p}=\frac{X_1Y_1+X_2Y_2}{X_2Y_2}$$

Substituting the known values for the first three terms of this proportion:

$$\frac{\cos 86^\circ\,07'+\cos 40^\circ\,24'}{\cos 40^\circ\,24'}=\frac{18''}{X_2Y_2}$$

he deduced $X_2Y_2=16''$.[43]

His second precept:

> *As ye cosine of the Parallactick Angle: to ye Radius :: so the parallax of Mars in Longitude: to his parallax in Altitude*[44]

corresponds to:

$$\frac{\cos I_2\hat{M}_2L_2}{1}=\frac{X_2Y_2}{p} \quad \text{(for the last observation)}$$

or

$$p=X_2Y_2 \sec I_2\hat{M}_2L_2=16'' \sec 40^\circ\, 24'=21''$$

Since, however, $p=I_2M_2 \sec I_2\hat{M}_2L_2=X_2Y_2 \cos\beta \sec 40^\circ\, 24'$, the precept is valid only when $\beta=0^\circ$ i.e. when Mars is situated in the plane of the ecliptic. This fact probably implies that it was originally derived for the Sun itself, but the error incurred through its application to the problem of estimating Mars's parallax would have been of little significance compared to that arising from Flamsteed's neglect in this instance of the phenomenon of astronomical refraction at the rather low altitude of 15°.

The relationship between Mars's horizontal parallax (π) and its parallax in altitude (p) is illustrated in Fig. 9, where E denotes the

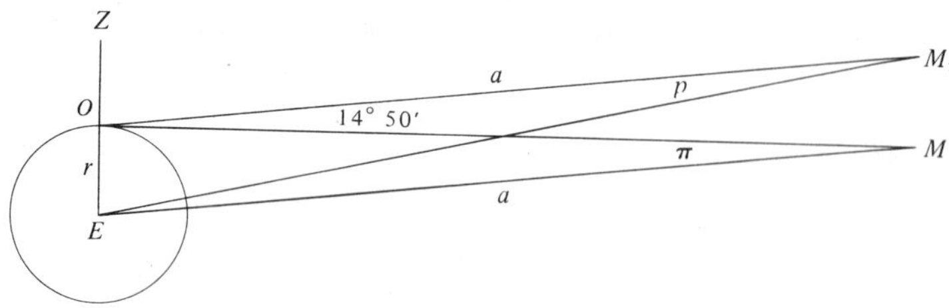

FIG. 9. Diagram illustrating the dependency of parallax upon altitude.

Earth's centre, and OE the radius from it to the position of the observer O. M_2 is the true position of Mars, and M a fictitious position on the horizon.

$$M_2\hat{O}E=90^\circ+M_2\hat{O}M=104^\circ\, 50' \quad \text{and} \quad O\hat{M}_2E=p.$$

Put

$$OE=r, \qquad O\hat{M}E=\pi.$$

Since p is so small, and the dimensions of the Earth and Mars also small in comparison with the Earth–Mars distance, one may suppose that to a sufficient approximation

$$OM_2=EM_2=OM=EM=a, \quad \text{say.}$$

Thus

$$\frac{\sin p}{r}=\frac{\sin 104^\circ\, 50'}{a} \quad \text{or} \quad \sin p=\sin\pi \cos 14^\circ\, 50'$$

Hence $p=\pi \cos 14^\circ\, 50'$ and $\pi=p \sec 14^\circ\, 50'=22''$ approximately.

The dependency of p upon the altitude indicates that neither of Flamsteed's precepts would be strictly valid if the planet's altitude should exhibit a marked variation during the time-span of the night observations. The results contained in the first volume of his posthumously published *Historiae coelestis Britannicae* (3 vols., Londini, 1725) reveal that this quantity did indeed change, from 15° 00′ at 7h 03m p.m. increasing to a maximum of 19° 10′ at 7h 44m p.m. then decreasing to its lowest value of 14° 40′ at 13h 13m p.m.;[45] but such small differences are of no significance in this context. It might, however, be noted that the value of 14° 50′ adopted in Lecture 2 is the mean between the first and last observations.

To find the Earth's distance from the Sun (= *d*, say) in terms of the same units, two further pieces of information are required; namely, the angle $M\hat{E}S$ between Mars and the Sun, and $E\hat{M}S = 180° - (M\hat{E}S + E\hat{S}M)$ subtended at the Sun by the Earth and Mars (cf. Fig. 10). The former was measured by Flamsteed with his 3-foot

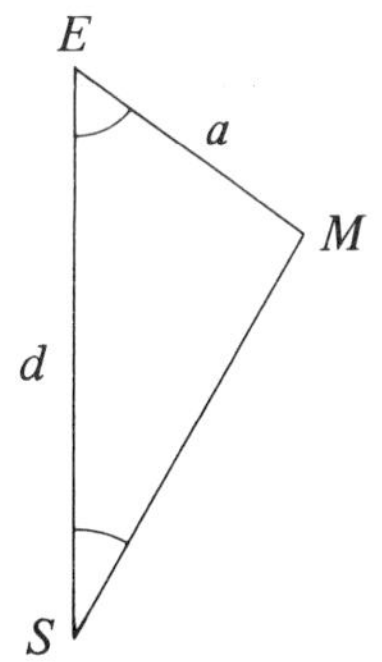

FIG. 10. Diagram illustrating how the solar parallax is derived from that of Mars.

radius wooden quadrant. $E\hat{S}M$ was probably approximated by interpolating the difference between the Earth's longitude (= 180° ± Sun's longitude) and Mars's for the instant of that daylight observation from Hecker's tables of motion—equivalent to supposing $\beta = 0°$, as in the second precept. Thus, applying the sine formula to *MSE*:

$$\frac{d}{\sin(M\hat{E}S + E\hat{S}M)} = \frac{a}{\sin E\hat{S}M},$$

from which Flamsteed deduced that $\frac{a}{d} = \frac{455}{1{,}000}$ (both angles being known, although their values are not stated in the text). Since

$$\sin \pi = \frac{r}{a}, \quad a = \frac{1}{\pi} \operatorname{cosec} 1''. \, r = \frac{206{,}265}{22} \text{ Earth radii}$$
$$= 4{,}688 \text{ Earth diameters.}$$

Thus $d = \frac{a}{0 \cdot 455} \gtrsim 10{,}000$ Earth diameters, corresponding to a value of 'almost 10″' for the Sun's horizontal parallax.

This completes our commentary on the earliest known successful application of Tycho's method for determining the parallaxes both of Mars and the Sun. Flamsteed repeated this investigation at Greenwich on three subsequent suitable occasions between 1676 and 1681, taking the Sun–Mars distance both in the morning and in the evening with his new 6-foot 9-inch radius sextant which was equipped with a telescopic sight. Following Henry Oldenburg's death in 1678, the *Philosophical Transactions* ceased temporarily to exist, before being resumed under the auspices of the Royal Society in January 1683. In consequence, Flamsteed was unable to avail himself of this organ of publication to report the golden opportunity that he and his young assistant Edmond Halley had on 14 January 1679 of repeating the whole procedure with an even higher degree of precision. However, from the outline of it given in Lecture 2 it is apparent that on this occasion he was paying particular attention to the effects of astronomical refraction, and carefully selected his comparison stars to assist in the elimination of this phenomenon. His result of about 10″ for the Sun's horizontal parallax was nevertheless just the same as that 'collected before from ye Derby observations of ye yeare 1672'.[46] Another set of measurements using a different comparison star yielded the slightly higher value of 12″. Such a level of agreement is a tribute both to the consistency of the method and to the accuracy of his observations. He was nevertheless cautious in his final conclusion that this parallax was *no more than* 15″, signifying that the Sun was over ten times more remote than most astronomers had hitherto believed.

A new theory of cometary motion

The topicality of the event, coupled with the importance of parallax determination in its interpretation, would appear to account for the transference of Flamsteed's interest to the comet of 1680 which is the main topic of Lecture 3. The views expressed here on the nature, formation, and motion of comets generally, were drawn mainly from the second volume of Riccioli's *Almagestum novum* and from some contemporary accounts of observations printed in the *Philosophical Transactions* for the years 1665, 1668, 1672, and 1677.

However, Flamsteed himself had been busy making observations of the comet of 1680 during the course of the previous winter and as we shall now see it was these which gave rise to the 'philosophical thoughts' on this subject which he communicated to Halley in a letter dated 17/27 February 1680/1[47] and his 'amicable controversy' with Newton.[48] As an aid to understanding, we have broken up his calculation into five stages, each of which is illustrated by the values given in Lecture 3 and diagrams constructed using these data.

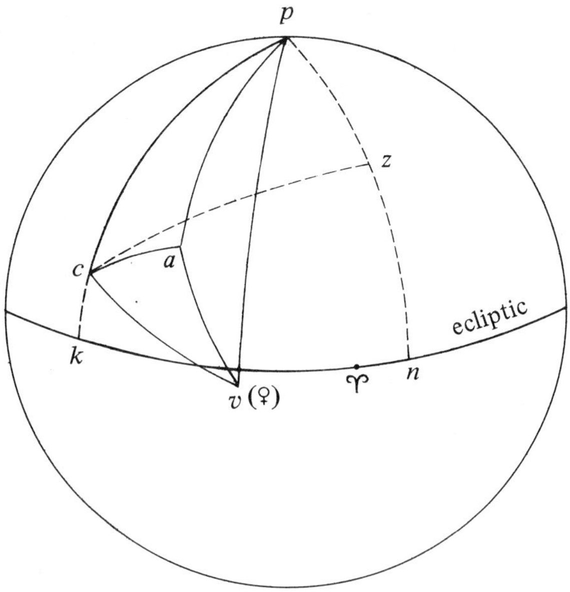

FIG. 11. Illustration for Flamsteed's theory of cometary motion.

In Fig. 11 (≡Fig. 3.1), p is the pole of the ecliptic,
$pv = 91°\ 35\frac{1}{3}'$, the distance of Venus (♀) from it.
$pa = 60°\ 39\frac{2}{3}'$, the distance of p from the star Aquila (a).
$a\hat{p}v = 1°\ 10'$, the longitude difference between Venus and Aquila.
$ac = 28°\ 15\frac{1}{2}'$, the distance between the comet (c) and Aquila.
$vc = 23°\ 58\frac{1}{6}'$, the distance between the comet and Venus.
ck is the latitude of the comet (measured northwards from k, the point of intersection of the meridian pc with the ecliptic).
long. ♀ = ♑ $28°\ 27\frac{2}{5}'$, the longitude of Venus.
n is the nonagesimal point, which lies in the ecliptic at a longitude 90° eastwards (or ahead) of the ascending node at which the ecliptic cuts the horizon.
z is the zenith point of the observer, positioned such that pz= altitude of n.
♈ is the first point of Aries or the spring equinox, where the ecliptic cuts the equator: its position on this diagram is determined by the condition that c is in the constellation of Capricorn (♑).

1. Calculation of the comet's celestial latitude and longitude

Applying the cosine formula to Δpav,

$$\cos av = \cos pa \cos pv + \sin pa \sin pv \cos a\hat{p}v$$

The quantities on the right-hand side all being given, one finds $av = 30°57\frac{1}{5}'$.

Applying the sine formula,

$$\frac{\sin p\hat{v}a}{\sin pa} = \frac{\sin a\hat{p}v}{\sin av}$$

$$\sin p\hat{v}a = \frac{\sin a\hat{p}v}{\sin av} \,.\, \sin pa, \quad \text{which is known.}$$

Hence

$$p\hat{v}a = 1°\ 58'\ 55''.$$

Applying the cosine formula to $\triangle acv$,

$$\cos ca = \cos av \,.\, \cos vc + \sin av \,.\, \sin vc \,.\, \cos a\hat{v}c$$

or

$$\cos a\hat{v}c = \frac{\cos ca - \cos av \,.\, \cos vc}{\sin av \,.\, \sin vc},$$

which is known; from which

$$a\hat{v}c = 62°\ 16'\ 53''$$

and

$$p\hat{v}c = p\hat{v}a + a\hat{v}c = 64°\ 15'\ 48''$$

Applying the cosine formula to $\triangle pcv$,

$$\cos pc = \cos pv \,.\, \cos cv + \sin pv \,.\, \sin cv \,.\, \cos p\hat{v}c,$$

which is known. Hence $pc = 81°\ 18\frac{1}{2}'$ and so

the latitude of the comet $ck = 90° - pc = 8°\ 41\frac{1}{2}'$ N.[49]

Similarly,

$$\cos cv = \cos pc \,.\, \cos pv + \sin pc \,.\, \sin pv \,.\, \cos c\hat{p}v$$

or

$$\cos c\hat{p}v = \frac{\cos cv - \cos pc \,.\, \cos pv}{\sin pc \,.\, \sin pv},$$

which is known. Hence

$$c\hat{p}v = 23°\ 43\frac{2}{3}'.$$

Whence it follows that *the longitude of the comet*, measured from west to east in the plane of the ecliptic is:

$$\begin{aligned} \text{long. } c &= \text{long. } v - cpv \\ &= ♑\ 28°\ 27\tfrac{2}{5}' - 23°\ 43\tfrac{2}{3}' \\ &= ♑\ 4°\ 43\tfrac{2}{5}' \end{aligned}$$[50]

Flamsteed made the arithmetical slip of writing his result as ♑ 6° 43⅗′.

2. Calculation of the co-ordinates of the nonagesimal point (*n*)
For this, refer back to the previous analysis (cf. pp. 13–16) where, given the latitude of the observer (φ), the hour-angle of the observation (H), and the obliquity of the ecliptic (ε), we found

$$\tan(\text{long. } n)=\frac{\tan H \cos\theta}{\cos(\theta+\varepsilon)} \quad \text{and} \quad \cos(\text{alt. } n)=\frac{\sin\varphi \sin\theta}{\sin(\theta+\varepsilon)}$$

where the auxiliary angle θ is defined by

$$\tan(\theta+\varepsilon)=\tan\varphi \sec H.$$

3. Reduction of the comet's apparent place (*c*) to its true place (*c′*)
Assuming the circumstances of the observation to be known, it follows that the longitude difference $c\hat{p}z=kn=\text{long}\,.\,n-\text{long}\,.\,c=97°\,31'$ (according to Flamsteed) and the altitude of $n=pz$ $[45°\,23'^{51}]$ are both known from the same considerations as were outlined in the previous analysis of Flamsteed's solar parallax investigation. In $\triangle pcz$: $\cos cz=\cos pc\,.\,\cos pz+\sin pc\,.\,\sin pz\,.\,\cos c\hat{p}z$. Since pz, $c\hat{p}z$ have just been found (either from such considerations or, more probably, from consulting nonagesimal tables), and pc was derived in **1**, the right-hand side of this equation is known and one obtains $cz=87°\,45\tfrac{1}{2}'$. Also,

$$\cos pz=\cos pc\,.\,\cos cz+\sin pc\,.\,\sin cz\,.\,\cos p\hat{c}z$$

or

$$\cos p\hat{c}z=\frac{\cos pz-\cos pc\,.\,\cos cz}{\sin pc\,.\,\sin cz},$$

which is known. Thus, by definition, the parallactic angle $=90°-p\hat{c}z=37°\,44'$. Since the effect of refraction is to increase the true altitude of a celestial body, let the extension of the vertical circle zc, namely cc', denote the refraction in altitude (cf. Fig. 12).

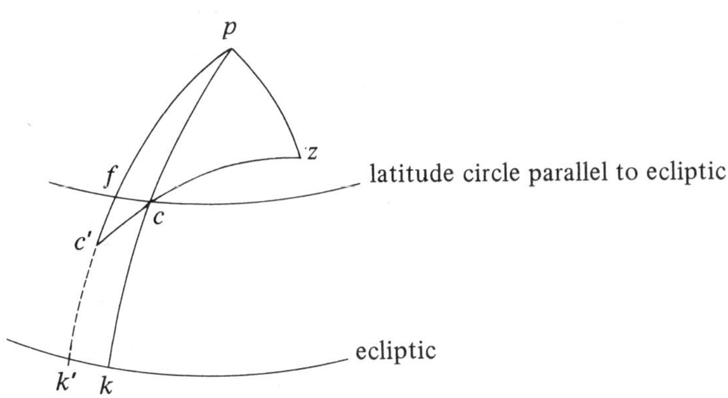

FIG. 12. Diagram illustrating the effect of refraction on celestial co-ordinates.

Let the longitude circle through c' (i.e. $pfc'k'$) cut the ecliptic in k' and the latitude circle fc in f. Then fcc' = parallactic angle $37° 44'$. From quite different observations, the nature of which will be discussed later (cf. pp. 49–50), Flamsteed derived $cc' = 17'$ at the comet's apparent altitude $90° - cz = 2° 14\frac{1}{2}'$.

$$\begin{aligned}\text{Thus the refraction in longitude} &= kk' = fc \,.\, \sec ck \\ &= cc' \,.\, \cos f\hat{c}c' \,.\, \sec ck \\ &= 17' \cos 37° 44' \sec 8° 41\tfrac{1}{2}' = 13\tfrac{3}{5}'\end{aligned}$$

$$\text{and the refraction in latitude} = fc' = cc' \,.\, \sin f\hat{c}c' = 17' \sin 37° 44' = 10\tfrac{2}{5}'.$$

The true place of the comet on 12 December 1680 is therefore:

$$\text{long} \,.\, c = \text{long} \,.\, k = ♑\ 4° 43\tfrac{2}{5}' - 13\tfrac{3}{5}' = ♑\ 4° 29\tfrac{4}{5}'$$
$$\text{lat} \,.\, c = ck = 8° 41\tfrac{1}{2}' - 10\tfrac{2}{5}' = 8° 31' \text{ N},$$

subtracting since the longitude increases in an easterly, and the latitude in a northerly direction.

Unfortunately, Flamsteed made the mistake of transposing the values of the longitude and latitude corrections, and slightly miscalculated the former as 15′. Using his erroneous value for the comet's longitude cited in **1**, he therefore deduced ♑ 6° 33′ and 8° 26′ N for the respective co-ordinates. A repetition of the foregoing procedure accepting (as we are obliged to do) his values for the parallactic angle and longitude of the nonagesimal point, has nevertheless revealed the presence of additional errors in his calculation of the comet's true co-ordinates, which should have yielded ♑ $5° 48\frac{2}{5}'$ and $6° 49\frac{1}{5}'$. Thus the net errors actually incurred by him in this case were $+\frac{3}{4}°$ in longitude and approximately $+1\frac{2}{3}°$ in latitude.

4. Calculation of the length of the comet's tail and its direction (with respect to the Sun)

From observations made about two hours after sunset Flamsteed, employing his erroneous value for the comet's longitude, found that the true value of that quantity had increased by 14′ to ♑ 6° 47′. The true longitude of the Sun (☉), interpolated from Hecker's tables, was ♑ 1° 58′. Hence, in Fig. 13,

$$☉k = 4° 49'.$$

The mean position of two nearby stars in Sagittarius, represented by o on the celestial sphere, is defined by the co-ordinates

$$\text{long.}\ o \equiv \text{long.}\ l = ♑\ 29° 20'$$
$$\text{lat.}\ o = ol = 39° 14' \text{ N}.$$

Produce the great circle arc ☉c to intersect the meridian of

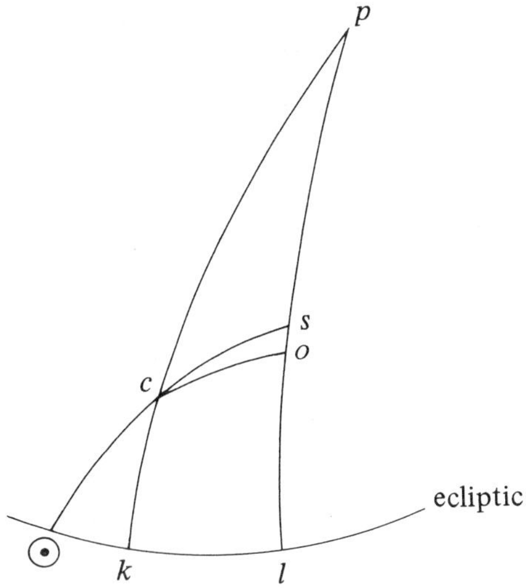

FIG. 13. Diagram illustrating Flamsteed's calculation of the length and direction of the comet's tail.

longitude through o in s, and let l be the point at which that meridian cuts the ecliptic. In the right-angled $\triangle\ \odot ck$,

$$\cos\ \odot c = \cos ck\ .\ \cos\ \odot k = \cos 8^\circ\, 26'\ .\ \cos 4^\circ\, 49'$$

Hence $\odot c$ can be obtained. It is $9^\circ\, 42'$ (according to Flamsteed). Then

$$\frac{\sin\ \odot k}{\sin\ \odot \hat{c}k} = \frac{\sin\ \odot c}{\sin 90^\circ} \quad \text{or} \quad \sin\ \odot \hat{c}k = \frac{\sin 4^\circ\, 49'}{\sin\ \odot c},$$

which is known; whence

$$p\hat{c}s = \ \odot \hat{c}k = 29\tfrac{1}{12}^\circ \quad \text{(according to Flamsteed).}$$

In $\triangle pco$, $\cos oc = \cos pc\ .\ \cos po + \sin pc\ .\ \sin po\ .\ \cos c\hat{p}o$ and since $pc = 90^\circ - ck = 81^\circ\, 34'$, $po = 90^\circ - ol = 50^\circ\, 46'$, and $c\hat{p}o =$ long. $l -$ long. $k =$ ♑$29^\circ\, 20' -$ ♑$6^\circ\, 33' = 22^\circ\, 47'$, one finds $oc = 37^\circ$ (to the nearest degree).

This distance is identified with the length of the comet's tail since Flamsteed observed that the tail ended at roughly the place where the two stars in question appeared. Also,

$$\frac{\sin po}{\sin p\hat{c}o} = \frac{\sin co}{\sin c\hat{p}o} \quad \text{or} \quad \sin p\hat{c}o = \frac{\sin 50^\circ\, 46'}{\sin 37^\circ}\ .\ \sin 22^\circ\, 47'$$

from which $p\hat{c}o = 29\tfrac{5}{6}^\circ$, and

$$s\hat{c}o = p\hat{c}o - p\hat{c}s = 29\tfrac{5}{6}^\circ - 29\tfrac{1}{12}^\circ \ \text{(according to Flamsteed)} = +\tfrac{3}{4}^\circ$$

Thus he concludes that the angle which the axis of the tail forms with the great circle passing through the centre of the Sun and the head of the comet is $\frac{3}{4}°$ south (Fig. 13 ≡ Fig. 3.2). A repetition of his calculations using the *same* figures as he himself adopted yields $p\hat{c}s = \hat{c}k = 29\frac{5}{6}°$; thus $s\hat{c}o = 0°$, and the tail of the comet is directed at 180° to the direction of the Sun. However, when the *correct* figures are substituted in their stead, one obtains $s\hat{c}o = -1°\ 13'$.[52] These results suggest that no significance ought to be attached to the apparent difference found by Flamsteed between the direction of the comet's tail and that of the great circle arc between its head and the Sun.

5. Calculation of the node and path of the comet

The further application of the method to the determination of the elements of the comet's path requires a repetition of the same observations on at least one other occasion. In Fig. 14 (≡ Fig. 3.3),

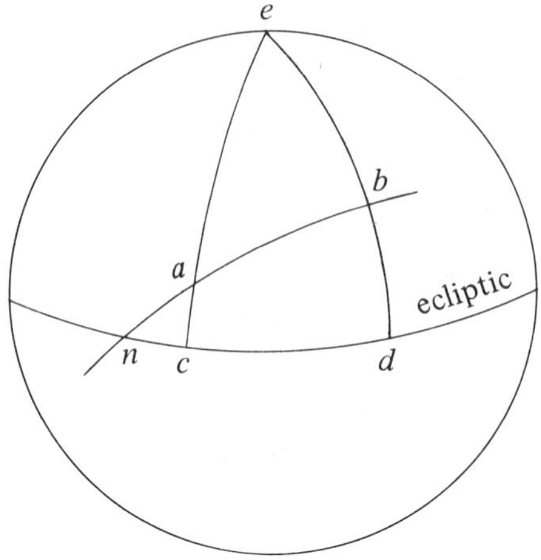

FIG. 14. Diagram illustrating the path of the comet.

ncd is an arc of the ecliptic; *e* its pole; *a*, *b* two positions of the comet on the celestial sphere; *c*, *d* their positions in longitude; and *ac*, *bd*, their latitudes at the respective times of observation on 12 and 21 December 1680. Flamsteed found (cf. **3**) that long. $a =$ long. $c =$ ♑ $6°\ 33'$, and lat. $a = ac = 8°\ 26'$ N ($\equiv ck$ in Fig. 11). He informs us in Lecture 3 that long. $b =$ long. $d =$ ♒ $5°\ 7\frac{2}{3}'$ and lat. $b = bd = 21°\ 45\frac{1}{2}'$. n is the point, or node, where the great circle through a and b intersects the ecliptic. In $\triangle aeb$:

$$\cos ab = \cos ae \,.\, \cos be + \sin ae \,.\, \sin be \,.\, \cos a\hat{e}b$$

Since, according to Flamsteed,

$$ae = 90° - ac = 81°\ 34',$$
$$be = 90° - bd = 68°\ 14\tfrac{1}{2}',$$

and

$$a\hat{e}b = \text{long. } b - \text{long. } a = 35^\circ\, 7\tfrac{2}{3}' - 6^\circ\, 33' = 28^\circ\, 34\tfrac{2}{3}',$$

$$ab = 30^\circ\, 33\tfrac{1}{5}'.$$

Also,

$$\frac{\sin ab}{\sin a\hat{e}b} = \frac{\sin be}{\sin e\hat{a}b}$$

or

$$\sin e\hat{a}b = \frac{\sin 68^\circ\, 14\frac{1}{2}'}{\sin 30^\circ\, 33\frac{1}{5}'} \,.\, \sin 28^\circ\, 34\tfrac{2}{3}'$$

from which it is found that:

$$e\hat{a}b = 60^\circ\, 55'\, 05''.$$

Applying Napier's rules of circular parts to the right-angled $\triangle anc$:

$$\sin(90^\circ - a\hat{n}c) = \cos(90^\circ - n\hat{a}c)\cos ac = \sin e\hat{a}b \,.\, \cos ac$$

Hence $\cos a\hat{n}c = \sin 60^\circ\, 55' \,.\, \cos 8^\circ\, 26'$, which is known; and so the inclination of the great circle ab to the ecliptic, viz. $a\hat{n}c = 30^\circ\, 10\frac{2}{3}'$. Similarly,

$$\sin ac = \tan(90^\circ - n\hat{a}c)\tan nc = \cot e\hat{a}b \,.\, \tan nc.$$

Thus

$$\tan nc = \sin 8^\circ\, 26' \,.\, \tan 60^\circ\, 55' \quad \text{and so} \quad nc = 14^\circ\, 46\tfrac{1}{3}'.$$

Consequently,

$$\begin{aligned} \text{long. } n &= \text{long. } c - nc \\ &= \text{♑}\, 6^\circ\, 33' - 14^\circ\, 46\tfrac{1}{3}' = \text{♐}\, 21^\circ\, 46\tfrac{2}{3}'. \end{aligned}$$

A repetition of this part of the calculations using the *correct* values of ae and $a\hat{e}b$, accepting Flamsteed's values for the co-ordinates of b, yields

$$\begin{aligned} \text{long. } n &= \text{♑}\, 5^\circ\, 48\tfrac{2}{5}' - 11^\circ\, 20\tfrac{2}{5}' \\ &= \text{♐}\, 24^\circ\, 28', \end{aligned}$$ [53]

or approximately $2\frac{2}{3}^\circ$ more than the values which he himself derived (cf. Tables 1 and 2).

An important question still left unanswered is: is one justified in identifying the direction of the comet's motion with that of the great circle ab? Flamsteed attempts to investigate this problem by repeating the whole process of calculation for each successive pair of observations obtained at Greenwich during the months of December 1680 and January 1681. If the angles of inclination of the various great circle arcs should prove to be equal to one another, then such an identification would be valid. The results cited in Lecture 3 are collated in tabular form below (Table 1).

TABLE 1

Flamsteed's data on the comet of 1680–1 (taken from Lecture 3)

	Year and time of observation (Old Style dating)	Angle between comet's tail and great circle through its head and the Sun	Position of node	Length of arc described	Angle of inclination with the ecliptic, of great circle arc joining successively observed positions
1680	12 Dec.				
	21 Dec.	>$\frac{3}{4}$°S	♐ 21°46$\frac{2}{3}$′	30°33$\frac{1}{5}$′	30°10$\frac{2}{3}$′
	24 Dec. (6h 12$\frac{1}{2}$m a.m.)	>1$\frac{3}{4}$	♐ 17°42$\frac{3}{5}$′	13°04$\frac{1}{3}$′	28°27$\frac{2}{3}$′
	26 Dec. (5h 14m a.m.)	>—	♐ 16°44$\frac{2}{3}$′	—	28°14$\frac{3}{5}$′
	29 Dec. (7h 55m)	>5$\frac{1}{2}$N	♐ 16°24′	55°	28°12′
	30 Dec. (8h 02m)	>6	♐ 15°38$\frac{1}{2}$′	50°	28°11$\frac{3}{4}$′
1681	5 Jan. (5h 51m)	>6$\frac{1}{2}$	♐ 14°28′	25°	28°26$\frac{1}{10}$′
	10 Jan. (5h 54m)	>—	♐ 15°47′	—	28°11$\frac{3}{4}$′
	9 Jan. (6h 49m)	>—	♐ 16°24′	—	28°01′
	13 Jan. (6h 56m)	>—	♐ 16°21$\frac{1}{2}$′	—	28°01′

In addition to these cometary observations, Flamsteed refers to others made at Greenwich on 1681 30 January (8h 07m), 2 February (6h 20m), and 5 February (6h 50m). His failure to quote similar results for these three times in Lecture 3 implies that the latter was actually composed between 7 February 1680/1, when he complained to Richard Towneley that he had not yet had time to calculate any of the angles listed in the last column of Table 1,[54] and 22 March 1680/1 when he sent that same correspondent the

TABLE 2

Flamsteed's revised data on the comet of 1680–1 (taken from his letter to Richard Towneley, of 22 March 1680/1).

	Tempori p m h ′	Cometae loca obs^r s o ′	Latitudo Bor[ealis]	Intersec [i.e. node] s o ′	Angulus Intersectionis
1680 Dec. 12	4 46	6 33	8 26		
21	6 32	5 07$\frac{1}{2}$	21 45$\frac{1}{2}$	21 47$\frac{1}{3}$	30 10$\frac{2}{3}$
24	6 12	18 49	25 23$\frac{1}{3}$	21 16	29 21
26	5 14	28 24	27 01	20 58$\frac{1}{2}$	28 54$\frac{1}{2}$
29	7 55	13 11$\frac{1}{4}$	28 10	20 37	28 22
30	8 02	17 37	28 11	20 30$\frac{1}{2}$	28 13
1681 Janu. 5	5 51	8 49	26 15$\frac{1}{2}$	20 01	27 31$\frac{1}{2}$
9	6 49	18 43$\frac{1}{3}$	24 12$\frac{3}{4}$	19 46$\frac{3}{4}$	27 11$\frac{3}{4}$
10	5 54	20 41	23 44	19 45$\frac{1}{3}$	27 08$\frac{1}{2}$
13	6 56	25 59$\frac{1}{2}$	22 17$\frac{1}{2}$	19 37	26 59
23	7 44	9 35$\frac{3}{4}$	17 57	19 29	26 48
30	8 07	13 19$\frac{1}{2}$	16 41	19 33	26 53
Febr. 2	6 20	15 13$\frac{3}{4}$	16 02	19 37	27 03
5	6 50	17 00	15 27$\frac{1}{3}$	19 47$\frac{1}{2}$	27 12$\frac{1}{2}$

improved table of his observations and results up to and including 5 February.[55]

Notice that the corrected times in Table 2 are the same as those copied from Lecture 3. There are, however, two important differences: corrections had now been applied for refraction and the revised calculations had been based upon a comparison of the first observation with each of the rest instead of upon that of each successive pair. Table 2 also contains a complete list of recalculated values for the comet's true ecliptic co-ordinates. Thus Flamsteed felt able to assure Towneley that 'these [data] you may confide in'.[56]

In the earlier letter of 7 February, Flamsteed had already sent him some 'very course' observations of the comet made during November 1680 by Jean Charles Gallet at Rome that he had recently received from Halley (who had actually been a witness to them) together with 'the more accurate ones of ye Gentlemen of the Observatory'[57] at Paris which were nevertheless of lesser significance because their times overlapped with those made at Greenwich. The drawing illustrating Flamsteed's own observations along with earlier and quite reliable ones made at Rome by Antonio Collio (received on 1 March 1681) and at Canterbury by Abraham Hill (Fig. 15) is reproduced from the later Towneley letter. Flamsteed likewise imparted the same information in a letter of 4 February 1680/1 to Edward Bernard, Savilian Professor of Astronomy at Oxford,[58] but with little commentary since he had previously sent him notes on how to determine the angles listed in the last column. Such was Flamsteed's faith in the accuracy of his method—and his calculations—that he regarded these angles as evidence for a gradual movement of the comet away from the pole of the ecliptic which became less as the comet receded further from the Sun; only at remote distances from that luminary was the motion along a great circle arc. He also believed that a *theory* of motion would have to explain why the angle made by the axis of the comet's tail was always found to be *ahead* of the great circle passing through its head and the Sun, as well as the less regular tendency of the node to regress and apparently to advance again with increasing distance from the Sun. It was certainly on account of these features of his revised (but, as we have seen, still uncertain) results that Flamsteed confided to Towneley that:

> It will be very difficult to find a Theory for these observations that shall answer them exactly I am certeine none in Hevelius his method will doe it.[59]

Johann Hevelius's views on the subject of cometary motion are contained in his *Prodromus Cometicus* (Gedani, 1666) where he favours the belief that the path is a conic section with the Sun at one focus, as Kepler had shown to be the case for planets, and in his

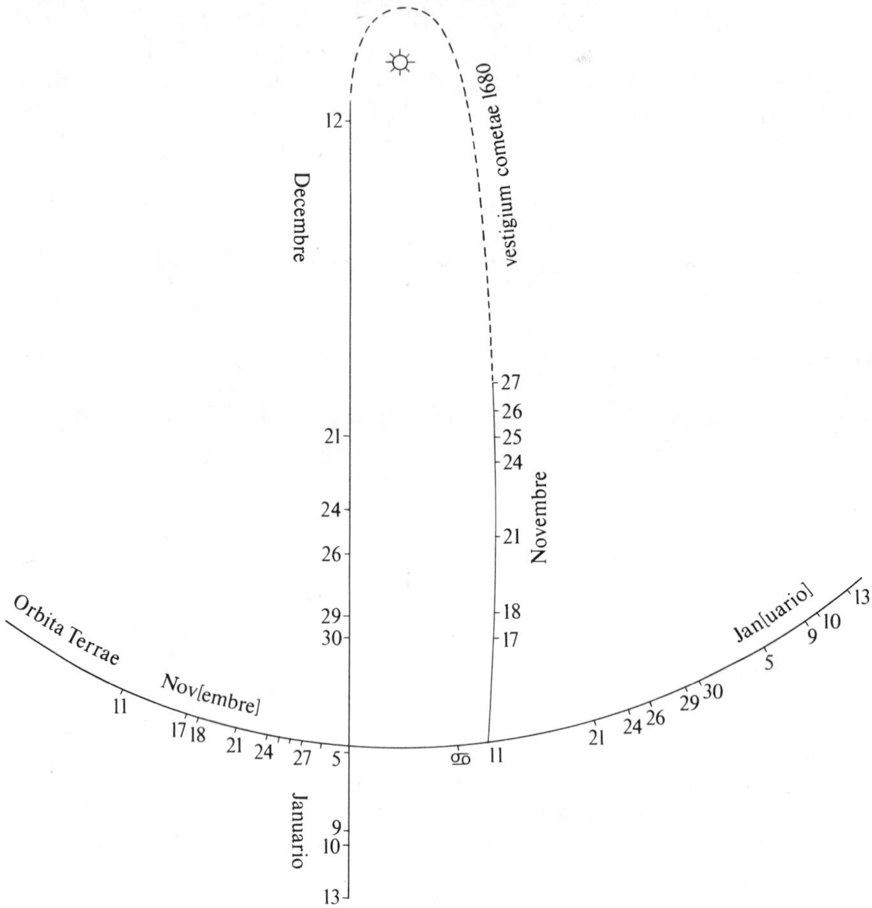

FIG. 15.

Cometographia (Gedani, 1668), where he preferred the belief that it was rectilinear. His physical justification of the latter view, which had also been held by Kepler, is described in the third volume of the *Philosophical Transactions* for 19 October 1668.[60] Comets were supposed to originate from matter ejected radially outwards (at constant speed) from an axially rotating celestial body which tends to condense after coming into contact with the latter's circulating atmosphere. This condensation acquires an ever-increasing velocity on account of the augmentation in the circular motion at greater distances from the centre of the parent body. The net effect of the two components is to produce a 'spin off' whereby the condensation—or comet, as it now becomes—flies off tangentially and continues along a straight line, always facing the Sun since it is largely composed of a similar substance. Despite this ingenious explanation, however, Flamsteed remained unimpressed since neither a straight line nor a Keplerian orbit was compatible with the above observational data.

What those data did enable him to assert was that the comet observed by Gallet and Collio in November and that observed by the Paris observers and himself in December, were identical; both being of the same appearance, with the same side always facing along a common direction of motion. Moreover, that comet appeared to be attracted by the Sun and to move towards it, while in the east; and to be repelled by the Sun when in the west, after having been carried around by what he calls the solar vortex. Whence he concluded that 'it will be evident that its appearances were affected with ye Earth's motion & not to be salved without it'[61]—interestingly echoing Hevelius's own belief in the *Prodromus* 'that without the annual Motion of the Earth, no rational Account can be given of any Comet, but that all is involved with perplexities, and deform'd by absurdities'.[62] A further inquiry, described briefly in Lecture 3, established that the comet of 1680 showed no appreciable parallax.

By claiming that a comet seen moving towards the Sun in November was the same as one moving in the opposite direction a month or so later, Flamsteed was rejecting the currently held view of Kepler, Hevelius, and others that comets moved in straight lines. However, from a letter of 15 December 1680 to Towneley[63] and another of that same date to James Crompton in Cambridge,[64] it is evident that this conviction did not stem from the observations described above, but from what he 'learnt from Others'. H. W. Turnbull, in annotating that Crompton letter for the second volume of *The Correspondence of Isaac Newton* (Cambridge, 1960), wrongly identifies these 'Others' as contemporary observers[65]—of whose results Flamsteed was at this time totally unaware. In fact, it appears from his introductory remarks in Lecture 3 that he was initially inclined to the belief that comets moved along a great circle arc from his reading Riccioli's account of what Regiomontamus and Apianus (in particular) had written on this subject.[66]

In a second letter to Crompton on 3 January 1680/1,[67] Flamsteed conjectures that an unexpected shortening in the length of the comet's tail might either be due to the weakening repulsion from the Sun failing to push the material out of the atmosphere surrounding the head, or to the exhaustion of a hypothetical energy source within the comet that produced its blaze. We know from an extract in Newton's hand accompanying his letter of 28 February 1680/1 'For Mr. Crompton to be sent to Mr. Flamstead',[68] that Flamsteed had written a third letter to Crompton on 12 February expressing the views that the atmosphere surrounding the head may not be gaseous but rather a translucent liquid, and that the tail appeared to exhibit a slight curvature backwards (or towards the west). The reply itself indicates that Crompton had handed on the correspondence to Newton for his perusal and opinion, and that the third (missing) letter must also have contained Flamsteed's interpretation of the comet's motion invoking similar principles of vortex motion and magnetic action to those described in his letter of 17 February to Halley.[69] Here, as at the end of Lecture 3, he was applying to *comets* the physical explanation of *planetary motion* propounded by Kepler in *Epitome astronomiae copernicanae* (Ulmae, 1618, 1620, 1621).

Newton's response to this idea in a letter dated 28 February 1680/1 is interesting:

> though I can easily allow an attractive power in ye ☉ whereby the Planets are kept in their courses about him from going away in tangent lines, yet I am the lesse inclined to believe this attraction to be of a magnetick nature because ye ☉ is a vehemently hot body & magnetick bodies when made red hot lose their vertue .... If then the Sun be a magnet, the axis of ye Comet ought to be strongly directed by him into such a position as the laws of magnetism require, and being so directed the Comet will be always attracted by ye Sun and never repelled.[70]

He questioned the generality of Flamsteed's inference that all comets verge from south to north, and expressed doubts that the atmosphere can be liquid. The irregular periodic fluctations in motion also made him adhere to his belief that there had indeed been two comets rather than one.

This letter from Newton, to which Flamsteed alludes in Lecture 3, initiated an exchange of correspondence which is, as Francis Baily has remarked, 'interesting, as showing us the state of the science at that period; and the good understanding that then existed between Newton and Flamsteed'[71] who were already, of course, no strangers to one another.[72] In his reply, dated 7 March 1680/1 and also addressed to Crompton,[73] the Astronomer Royal cites Gallet's, Collio's, and his own (uncorrected) observations as evidence in favour of his view that the comet seen during the previous year in November was the same as that seen in December. Newton's belief that the comet's motion was too irregular to permit such an identification was a result of his confusion of the new style dating used by Gallet with the old style still being adopted in England: the difference between these amounted at that time to ten days.[74] Flamsteed also states the position for 11 November 1680 communicated to him by Abraham Hill of Canterbury—important in being the earliest of all the observations—and quotes Halley's sighting of the comet's tail from near Bologna on 8 December during the journey from Rome to Paris. He encloses a diagram of the Earth's position on the days of Collio's and his own observations, thereby taking account of the effect of the annual motion (which Giovanni

Domenico Cassini had failed to do) together with the lines connecting these to the comet's path. This representation reveals that the comet moved towards the Sun practically in a straight line and away from it along a similar path—apparently confirming his previous proposal[75] that the comet was somehow attracted and repelled by the Sun. Newton's objection to the cause of this motion being magnetic in origin had been founded upon the assumption that the Sun was made of a material with properties similar to that of iron, which Flamsteed was not prepared to accept. Nor did he assent to Halley's opinion that the comet was attracted by the Sun's gravitation, because he felt that this would result in the dissipation of its material. He liked Newton's suggestion that the comet swings round the Sun and back again,[76] rather than approach it and be repelled by it before reaching it; but he could not see how to square this hypothesis with his observations, since it implied a considerable acceleration to the comet's motion.

On the basis of the new evidence thus provided, Newton became personally convinced that there had indeed been only one comet observed, and was mildly critical of Flamsteed's neglect of the effect of the comet's changes in latitude when calculating its orbital speeds.[77] He considered several difficulties facing the idea that the comet swings round the Sun, but came to the conclusion that

> all these difficulties may be avoyded by supposing ye comet to be directed by ye Sun's magnetism as well as attracted, & consequently to have been attracted all ye time of its motion, as well in its recess from ye Sun as in its' access towards him, and thereby to have been as much retarded in his recess as accelerated in his access. & by this continuall attraction to have been made to fetch a compass about the sun . . . , the *vis centrifuga* at [the point of closest approach, where the motion was momentarily at right angles to the direction of the Sun] overpow'ring the attraction & forcing the Comet there notwithstanding the attraction, to begin to recede from ye Sun.[78]

The reason why Newton now felt ready to accept the concept of the Sun's magnetic attraction was that he believed that he might 'have a way of determining ye line of a Comets motion (what ever that line be) almost to as great exactness as the orbits of ye Planets are determined, provided due observations made very exactly be had'.[79] Yet, when he wrote the last letter to Flamsteed on this subject, on 16 April 1681,[80] not only did he fail to account for this change of heart but also pretended to remain unconvinced that only one comet had been observed. The method which Newton had developed is surely that which he later expounded near the end of his *Philosophiae Naturalis Principia Mathematica* (Londini, 1687), which he goes on to exemplify using the data on the comet of 1680 that Flamsteed and others had provided.[81]

Our study of the above-mentioned correspondence and Newton's repressed remarks suggests that it might have been Newton's consideration of Flamsteed's explanation of that comet's motion which first made him appreciate that a central attractive force in the Sun is capable in principle of producing a Keplerian orbit. The inverse-square distance dependency of such a force is a consequence of this belief, as demonstrated directly in the first volume of the *Principia.*[82] The problem of deciding whether he ought to ascribe it to the directional as well as the attractive property of the Sun's magnetism, or to the effect of the Sun's gravity, may have been settled in Newton's mind by two factors: his belief that the Sun, being extremely hot, cannot also be magnetic;[83] and his acceptance of John Wallis's theory of the tides in the *Philosophical Transactions* for August 1666,[84] which contains the earliest rational demonstration that a gravitational force can be operative across apparently empty space and extend (at least) to the distance of the moon. Perhaps too little significance has hitherto been attached by historians of science to that particular influence on this important phase in Newton's intellectual development.

A new theory of telescope optics

The reason why Flamsteed now decided to continue during the next (Trinity) term with a series of six lectures on the theory of refraction through lenses, telescope optics, the application of a micrometer eye-piece to a telescope, and that of a telescope to an astronomical quadrant, is given by him in Lecture 6. Here he alludes to 'a famous p[er]son' who, after reading his recommendations of telescopic sights (published together with some micrometric measurements), had strongly urged against such a practice; with the result that 'some of o^{r} freinds were almost p[er]suaded to be of his opinion'.[85] Thus Flamsteed felt himself obliged to explain the principles of telescope optics in order to refute those imputations, 'on which account', says he, 'I resolved to make this the taske of the last terme'.[86] This resolution was strengthened by his awareness of the fact that very little on the subject had been published in English, and by his recognition that 'however skillfull any p[er]son be except hee be accommodated with convenient and fitting instruments, hee cannot put knowledge into practise'.[87]

The subject of this criticism was none other than Johann Hevelius of Danzig, a highly respected practical astronomer who was less concerned about waging a polemic over the use of telescopic sights than with defending himself against Flamsteed's attack on the reliability of his own observational data. Before citing the results of the above-mentioned micrometric measurements of Mars's distance from the two stars in Aquarius from which he deduced the solar parallax to be 'not more than 10 seconds' Flamsteed, in a letter to Henry Oldenburg dated 16 November 1672 printed in the next month's issue of the *Philosophical Transactions*, announced his intention of applying the same method to finding both the diameter and distances of the planets. He concluded:

> Tycho errs 5 minutes at least both in the Places and Latitudes of them [the two stars] compared one with another. And certainly he errs as much in many others; so that the labour of Monsieur Hevelius to rectify their places is very needful . . . But if M. Hevelius use not glasses . . . I fear, we shall but be where we were; and yet without this Restitution we cannot expect any thing certain, much less accurate, in Astronomy.[88]

Not surprisingly, Hevelius now felt the need to give a reasoned justification of his use of bare sights in his *Machina Coelestis pars prior* (Gedani, 1673).[89] After discovering that Flamsteed had again raised the same objection in an open letter of 7/17 July 1673 to Cassini on the subject of Jupiter's satellites,[90] Hevelius wrote to Oldenburg temperately defending his own instrumental methods and protesting against Flamsteed's opinion that nothing more could be achieved by his method of observing.[91] But now Robert Hooke, who was no friend of Flamsteed, took sides against Hevelius, vehemently attacking the latter's published opinions in his *Animadversions on the First part of the Machina Coelestis of . . . Hevelius* (London, 1674), a revised version of the second of his Cutlerian Lectures.[92] He maintained that Hevelius's instruments were no better than those of Tycho Brahe, being limited in accuracy by the resolving power of the human eye. Telescopic sights were capable of increasing this accuracy by a factor of ten. In his eyes, Hevelius's fears that these sights were less reliable (being more likely to suffer derangement) than open ones, and that they might produce unsuspected deviations of the rays refracted through them, were unfounded. Hooke also questioned Hevelius's claim—made in support of the high internal consistency of his measurements—that the successive intervals in right ascension or longitude of eight stars distributed round the ecliptic added up to exactly 360°, since the uncertain effects of refraction would introduce a discrepancy of many seconds.

The tone of this reaction hurt Hevelius very deeply, as is evident from his long letter to Flamsteed on 24 June 1676.[93] He quotes in this letter the following testimony by the councillor Bernhard Fullenius, son of a former Professor of Mathematics at the University of Frankfurt-on-Oder, dated 1 March 1675, to vindicate himself against Hooke's rash criticisms:

> As for your *Organa*,[94] there is no reason for anyone to doubt their truth, accuracy, and reliability, of which I have such considerable

> proof that had I not previously been an eye-witness of your observations, I should not have the least conception of the like. The most extensive possible range of observations to the fifth magnitude, made in my presence, constitute evidence for this, sufficiently proving and demonstrating the trustworthiness and precision which has never given rise to an error of as much as six seconds. I bear witness that this duly occurred whenever you had me as a companion while observing the fifth magnitude. I found the same thing while I had your very dear wife and Johann (my printer who observes for me) as companions. Thus, since you have innumerable observations of the greatest [degree of] correctness, you can deservedly boast of the great accuracy of your instruments . . .'[95]

Hevelius asks Flamsteed to defend him against Hooke until the fruits of his twenty-five years of observations are published, and voices his scepticism of that antagonist's unproven claims in the *Animadversions* . . . of his own ability to attain an accuracy of about $\pm 1''$ using a small sextant equipped with telescopic sights. By way of 'repayment' for this favour, Hevelius supplied Flamsteed with observations of Mars's position with respect to the same stars in Aquarius on which the latter had based his initial investigation of the solar parallax.[96] He himself had obtained these at Danzig on the same day (viz. 26 September 1672) by a different method not involving the use of either a telescope or a micrometer.

Writing to Richard Towneley on 6 July 1676, Flamsteed reports on the contents of this letter, remarking that Hevelius 'expresses his resentm[en]ts at Mr Hooke very frequently'.[97] Certain anomalous differences in those star positions which Hevelius had sent him, obtained during Mars's passage through the Pleiades, suggested to Flamsteed either that his correspondent's instruments were not exact or that 'distances cannot be so exactly measured with plain sights as with glasses'.[98] In his reply to Hevelius, written two weeks later[99] Flamsteed, as a result of his comparison of their respective observations, insists that his colleague had incurred an error of $\pm 2'$ in both celestial co-ordinates of the stars. He speculates that part of the discrepancies in latitude may have stemmed from Hevelius's over-correction of meridian altitude measurements for the effects of atmospheric refraction, and attributes the remaining differences to the errors inherent in the use of plain sights. It never seems to have occurred to him that appreciable errors could have been inherent in his own telescopic-cum-micrometric observations.

Not surprisingly, Hevelius reacted strongly to this renewed slight on his observing technique and data. In his indignant reply of 23 December [Old Style] 1676,[100] he stresses that the discrepancies between their respective results could well have arisen not in the act of observing but rather in Flamsteed's method of reduction.[101] The latter had used Tycho's erroneous positions for the comparison stars, and worked upon the incorrect presupposition that Hecker's solar ephemerides—in effect, an extrapolation from Kepler's Rudolphine Tables which were themselves dependent upon Tycho Brahe's observations[102]—contained no appreciable error. In fact, the solar theory stood in need of correction by as much as 7′, or 2h 48m.[103] There were many reasons—five are explicitly cited in this letter—why systematic errors could arise in taking the Sun's meridian altitudes with an astronomical sextant, and Hevelius was therefore prepared to concede an inaccuracy of about $\pm\frac{1}{2}'$; but no more. The internal consistency of his repeated measurements of this quantity was of the order of only ±1″. Yet Flamsteed was straining every nerve to slight all his observations, also to reveal to contemporaries and posterity that he could never derive angular distances in the heavens by his method to within ±2′.

Nor was Hevelius pacified by Flamsteed's promises of support and friendship. As he says:

> Anybody who reads your letter will undoubtedly say that you are doing with all your strength and in fact with better weapons the same thing as Hooke is trying to prove with lesser ones: namely, that I cannot measure distances with my sextants, that is with the naked eye, except to three or at most to two whole minutes . . .
> I well know now that I shall labour in vain with you, since I perceive abundantly clearly that you will never agree with me whatever other argument I produce and however many more observations I adduce to the contrary . . .
> However this may be, I will willingly allow you to persist in your opinion provided that you would also allow me mine. Equally, you will never persuade me to abandon my safer way and to embrace another which seems to me to be less safe, liable to so many obstacles, and must always be adopted with the greatest caution . . .
> It is clear that you attach so much weight to your single observation made with the micrometer, that immediately from this single observation of yours, before you have examined the matter a little more deeply and obtained observations of others on the same subject, you condemn and reject my observations. As if your observation alone was to be regarded as the norm for examining the data of all the others . . .
> My catalogue of stars such as it is, and my trifling observations, cannot serve you; since I already perceive clearly enough from your letter and from [your] very hostile modes of speech what you think of my catalogue and of my poor observations. Thus there is no reason why you should await it eagerly, nor why I ought to hasten my work such as it is for that reason; since you can very well do without all my contributions nor will any advantage to you result except that you may have something to mark with a pencil in order to eradicate the less certain things in which there may be some mistake.[104]

These extracts have been quoted not only to convey the mood of Hevelius's emotional response, but also to indicate the nature of his counter-objections to his critic's procedure.

Flamsteed did not take kindly to the moral advice and entirely justified reprimands of his older and more experienced contemporary but, in a letter to Oldenburg dated 5 April 1677,[105] asks the latter to make it quite clear to Hevelius that there was no question of any plot against him between Hooke and himself. He was already aware that there was a remarkable level of agreement between the Greenwich and Danzig observations. After perusing an advance copy of the second part of Hevelius's 1,500-page *Machina Coelestis*[106] presented to him in quires by the author just over two years later, he expressed his amazement to Towneley that their respective measures commonly differed by less than 20″, and seldom by more than $\frac{1}{2}'$.[107] Meanwhile, in May 1678, Edmond Halley had returned from his expedition to St. Helena and begun preparing his *Catalogus Stellarum Australium* (Londini, 1679) for publication. While on a visit to Greenwich in October of that year, he read the letter to Flamsteed of 23 August 1676 in which Hevelius expresses his desire to see those observations.[108] This induced him to write to Hevelius from Oxford on 11 November[109] within days of his catalogue being printed, sending him a copy and stating that he would willingly re-calculate it employing the fundamental star positions computed by Hevelius rather than those of Tycho Brahe. He would be greatly honoured if his work could be immortalized by inclusion in the still-unfinished part of the *Machina Coelestis*.[110] In conclusion, he announces his intention of shortly visiting Danzig to become personally acquainted with his correspondent, to study his instruments and his observing technique, and to consult him concerning the further advancement of astronomy. It would appear however, from Hevelius's subsequent correspondence with Flamsteed,[111] that neither that letter nor the catalogue reached its destination.

Hevelius's own account of Halley's visit, which lasted from mid-May to mid-July 1679, is contained in his *Annus Climactericus* (Gedani, 1685) along with some of the above-mentioned letters having a bearing on the controversy over the question of the accuracy of telescopic sights.[112] Halley apparently began on the very night of his arrival by testing the consistency with which Hevelius could measure the angular separation of the stars Regulus and Spica using his 6-foot radius bronze sextant; and subsequently observed in co-operation with Hevelius himself, his wife Elizabetha, with the printer, or with his host's relative Johann Olhoff. Although he himself was unskilled in using open sights, he found he could observe consistently to within $\pm 1'$. These experiences and conclusions were described by Halley in a letter of 7 June to Flamsteed,[113] where he signifies his surprise at the consistency of $\pm 5''$ obtained by Hevelius and says that he 'dare no more doubt of his Veracitye'.[114] In a testimonial to his host dated 8 July [O.S.],[115] the date of his departure from Danzig, Halley confesses to having formerly doubted whether observations made with open sights might not be several minutes of arc in error, and to having wondered what considerations had dissuaded Hevelius from employing telescopic ones. He writes:

> I voluntarily offer myself as a witness of the barely credible certainty of your instruments against all who in days to come may seek to call your observations in question.[116]

Thus when Flamsteed says that 'some of o^{r} freinds'[117] were almost persuaded to be of Hevelius's opinion, Halley was undoubtedly uppermost in his thoughts. If such an experienced astronomer should have been so greatly swayed by his experience of observing alongside Hevelius, the value of applying telescopic sights to levels, quadrants, and sextants required an explicit and immediate vindication. In Jean Picard, *La Mésure de la Terre* (Paris, 1671) where the procedure for rectifying astronomical instruments in this way is briefly described, the accuracy of the results for the length of a degree measured by triangulation on the Earth's surface had already provided experimental justification for Flamsteed's point of view, although few copies of this sumptuous treatise reached England[118] and French was a language which he himself had great difficulty in understanding. What he felt was still lacking was a rational geometrical proof of the principles of telescope optics, which would *demonstrate why* the conventional instrument embodying a pair of convex lenses is the most suitable one for astronomers to employ. The Gresham Lectures provided him with the occasion to advertise the hitherto little-known optical ideas of William Gascoigne and the conclusions which he had derived geometrically from them. Answers to the questions of how and when Gascoigne's optical papers fell into his hands have already been provided in the extract from the letter of 10 May 1690 to Molyneux cited above (cf. pp. 8–9). In another letter written to the same correspondent almost eight years earlier,[119] Flamsteed had openly acknowledged that he had learnt his optics

> from some loose papers of one Mr Gascoigne . . . in lesse y^{n} 2 hours time & all that I know of glasses is but a superstructure on y^{t} foundation Nor doe I doubt but if y^{u} will please to take some little paines in prosecuteing ye first principles y^{u} will find suitable successes. before I met with these I was, as y^{u}, tormented with readeing the discourses of Opticall writers, who I now find understood not the prime proposition & therefore onely gropeing the Darke themselves could never lead others aright. They were ignorant of that proportion Which the Angles of incidence beare to the Angles refracted or if they knew some little of it they perceived not how it might be applyed to glasses.[120]

He had already communicated Gascoigne's principles to Molyneux on 11 April 1682.[121] They were:

1. That any streight line drawne from ye center of a Circle to any point in the periphery is perpendicular to the tangent at that point.
2. that when ye Rayes of light passe from a rare medium to a dense they are refracted towards the perpendicular. but when from a dense into a rare from it.
3. that (not the Arches or Angles of Incidence themselves tho under 30 degrees but) the sines of the Angles of Incidence are Proportional to ye sines of ye Angles Refracted. Mon^{sr} De chart [Descartes] states them as 3 to 2. Mr Newton as 300 to 193 which is nearely as 14 to 9. & this proportion as most exact I most Commonly use.[122]

Newton's experiments on light and colour[123] had established 14 to 9 as a *mean* ratio for white light; each colour having a slightly different refrangibility and thereby producing the phenomenon known to us as colour aberration.

A fourth (unnumbered) principle, applicable to reflection, was that the angles of incidence and reflection of light from a plane surface are equal. From this is followed that rays parallel and close to the axis of a concave sphere would be focussed upon it at half the distance of the radius of curvature from the vertex; but it also followed that they would be even closer to the latter point, if further from the axis. The difference in focal length gives rise to the phenomenon which we call spherical aberration.

This, then, was Flamsteed's foundation—not Honorato Fabri, nor Père Cherubin, nor Thomas Hobbes, nor Jacques Rohault, nor Johannes Kepler, nor René Descartes, whose major optical writings Molyneux specifically begs him in a letter of 30 December 1682[124] to consult in order to prove that they were to be held responsible for his view (which Flamsteed had been criticizing) that light rays are propagated in a parallel beam from the eye-glass to the crystalline humour of the eye. Flamsteed, replying to this request some three weeks later,[125] explicitly rejects the work of Cherubin[126] and Hobbes[127] and confesses that he had not had the patience to peruse Kepler's[128] and Descartes'[129] treatises on this subject

> haveing benefitted my selfe more by a letter & figure or two of Mr Gascoignes with an houres serious thoughts then I could have done by ten spent in ye perusall of any of theire writeings y^{u} have the same principles to proceed on & if y^{u} please to employ y^{r} thoughts y^{u} will have as little need as I had of them onely if y^{u} desire to know how ye hyperbola & Ellipsis collect ye parallel rayes to a point De Chartes' demonstration is ye plainest & easiest.[130]

Unfortunately, the lecture which Flamsteed would indeed appear to have delivered at Gresham College on 15 June 1681 on this last-mentioned theme, is not bound together with the others (nor indeed does his system of numeration make provision for it).[131] The principal objection that he raises at the beginning of Lecture 5 and again later in a letter of 29 March 1684 to Molyneux[132] against the use of such lenses, which were very difficult in practice to make, is that the focus would not be unique because of the effect of colour aberration; hence they should not be expected to perform any better than spherical ones. Moreover, if the rays entering the telescope were not strictly parallel to the axis, there would still be a slight difference in the focal length.

Flamsteed begins his series of lectures on optical matters by adopting Gascoigne's choice of Descartes' ratio $\frac{3}{2}$ as a simple but close approximation to the refractive index of light through glass; then geometrically constructs, in turn, the path of refracted rays through a block of glass (Fig. 4.1), the segment of a sphere (Fig. 4.2), and a plano-convex lens (Fig. 4.3)—concluding that the focusing effect of the latter when the convex side faces towards the light source is greater than if it were placed the other way round. He next considers a double-convex spherical lens whose surfaces are of equal radii of curvature (Fig. 4.4), and calculates the position of the focus of parallel rays through it. His major conclusions from this investigation are that:

(1) No spherical lens will bring all parallel rays to a single focus, though others (with hyperbolically—and elliptically—curved surfaces) will.

(2) The plano-convex lens with the convex side facing the light-source is the best arrangement for providing a sharper focussing property, and thereby (in modern terminology) for obtaining a higher resolution.

His claim to have been the first to discover this latter property, and to make explicit the distinction between the conditions required for obtaining higher magnification and higher resolution, is supported by Molyneux in a letter dated 1 August 1682.[133] Flamsteed ends Lecture 4 by drawing attention to Honorato Fabri's definition of refraction as a function of the angle of incidence rather than its sine—the latter being fully confirmed by experiments of Gascoigne, Newton, and 'o^{r} Geometry Lecturer' (Robert Hooke)—and to the consequent uncertainty in the results obtained by Fabri in his *Synopsis Optica* (Lugduni, 1667).

In Lecture 5, Flamsteed stresses the fact that a telescope with two convex lenses, on account of its large magnification and the possibility of placing a micrometer or threads in the common focus, is the *only* kind for use in astronomical observations. He goes on to describe how the eye itself acts as an additional lens to invert the image of an object, referring to Descartes's *camera obscura* experiment with the eye of an ox[134] as evidence for this inversion. It would appear, however, that he had also learned the principles of catoptrics from his study of Gascoigne's letter of 8 June 1642 to William Crabtree. This contained a description of Gascoigne's own ox-eye experiments[135] and suggested that the latter himself was initially indebted to Claude Mydorge's *Prodromi catoptricorum et dioptricorum* (Paris, 1641) and Christopher Scheiner's *Oculus; hoc est, fundamentum opticum* (Oeniponti, 1619). Gascoigne had established that the refractive index of the vitreous humour was approximately the same as that of glass, but was unable to calculate this quantity for the other refracting surfaces because of their variable and complex curvature. However, these very same features ensured that the eye was capable of accommodating itself to compensate the effect of colour aberration produced by an eye-piece with a high degree of curvature, and the same phenomenon could be further reduced by using object-glasses with a greater focal length. Flamsteed repeats his claim to the discovery of the experimentally verifiable fact that magnification is not a measure of distinctness, the former depending upon the ratio of the focal lengths of the object- and eye-glasses, and the latter only upon the focal length (or radius of curvature) of the object-glass. He infers from this that no advantage would be gained through introducing more lenses into the telescope, and recommends the principle of Newton's reflecting telescope with a parabolic mirror bringing all rays to a single focus in a short tube.[136]

In applying a telescope to a quadrant, the object-glass should be positioned over the centre, or pivotal point, the eye-piece on the outer edge of the limb, and the focal plane with cross-wires on the inner edge of the limb. In effect, the width of the limb is therefore determined by the magnifying power of the telescope: it should be raised by a small amount (e.g. $\frac{1}{5}$ inch) above the plane of the instrument so that the plumb-line through the pivot is not obstructed. The zero point on the limb is obtained by inverting the quadrant and bisecting the difference between the two points where the plumb-line crosses the outer limb. A level with telescopic sights, shaped like a carpenter's square or like the letter 'T', could be rectified in the same way. To avoid the effect of astronomical refraction, which is capable of producing 3′ differences or more between morning and midday observations, this procedure ought to be carried out in day-time only in 'serene and settled' atmospheric conditions, or at night by sighting on stars near the zenith; the disadvantage of the latter alternative being that several nights' experiments would be required.

After a preamble in which he draws attention to the degree of skill and diligence of the astronomer, both in collecting and mathematically reducing his data, as being non-instrumental factors equally likely to affect the accuracy of his observations, Flamsteed comments in Lecture 6 upon Gascoigne's recommendation of a convex eye-glass as an aid to stellar observations. Its wider angle (or field) gave it a better light-gathering property and the common focus lay inside the tube which had not been the case for the telescopes with concave eye-glasses used by pioneering observers like Galileo and Father Christopher Scheiner. Gascoigne's discovery that the poles (or thickest sections) of lenses were seldom situated at the geometrical centre, for which he had no remedy, was a defect that had to be taken into account in a rigorous theory although—as Flamsteed later implies—it could to a large extent be circumvented by grinding thinner lenses.

In Lecture 7, Flamsteed deals with the nature and performance of concave glasses in the same manner as he had dealt with the convex. His first geometrical construction is that of finding the direction of a ray of light falling perpendicularly upon the plane surface of the lens after having emerged from the spherical concave surface. That direction, projected backwards, meets the polar axis of the lens at a certain 'point of divertion' (or direction). His second construction is to investigate whether other perpendicular rays nearer to or further from the polar axis will also have the same point of direction. He finds that this is not so; rays falling closer to the axis have a point of direction further from the pole, and vice-versa. The underlying reason for this result is that smaller differences between the sines of angles correspond to a constant difference in degrees. A corollary to the first construction is that a ray perpendicular to the plane side of the lens but coming from the opposite direction, would continue along the line towards the point of direction. Here Flamsteed makes no allowance for a second refraction at the plane side, but this would be small if the lens were thin and/or its curvature shallow. All the calculations are repeated for the case where the concave side of the lens is nearer to the light-source, from which it is established that the points of direction are further away than before. A consideration of the double-concave lens reveals that its performance was inferior to the second case. Since it had been shown that no one focal point can be common to all the parallel rays, Flamsteed concludes that Isaac Barrow's geometrical demonstration in one of his optical lectures, for finding both a focal point and a point of direction, is erroneous.

The properties of double-concave and double-convex lenses of equal dimensions and radii of curvature are examined in Lecture 8, and the distance between the points of convergency for the latter type of lens is found to be barely half as much as the separation between the points of direction for the same rays in the former type. The effect of having a 1:5 ratio in the radii of curvature of each type of lens is further investigated.

Flamsteed points out that it *is* possible to have a telescope with a concave eye-glass linked to a micrometer, provided that the object-glass is plano-convex with its convex side facing the object viewed. A pair of cross-threads may be placed on its plane side, or alternatively two intersecting lines may be cut on the glass surface with a diamond. An important distinguishing feature of this arrangement is that the eye-glass (on which the threads are drawn) must be firmly fixed; whereas with the convex eye-glass, the threads were firmly fastened and the eye-glass moveable. Flamsteed does not favour the adoption of a meniscus or limular object-glass even though the rays emerge from it with less inclination, since long focal-length lenses have enough depth of focus to make them suitable. At the beginning and end of this lecture there are two pages (which ought to have been placed together) containing the following six-point summary of Flamsteed's own original contributions to the theory of optics:

1. That both ye plano convex & plano concave have theire burneing points & points of direction equally distant from theire poles.
2. That in ye plano convex. when ye convex side is towards ye object the burneing point is nearer ye pole then ye point of direction in ye plano concave alike posited & theire differences nearely the same. this performeing best
3. that in ye double convex the burneing point is farther from ye pole then ye point of direction in ye double concave. & the spaces betwixt the convergeing points scarce in ye one halfe so wide as the points of direction in ye same Concave section for ye other.
4. of ye Telescope & ye proportions of Augmentation Distinctnesse.
5. why the convex of two glasses invertes. but two other a Convex & concave not.
6. the proportions for lengths how to find the length of a glasse wrought in two different tooles. for ye double Convex as ye summe of theire radij: to ye greater:
   so ye focall length of the lesser section: to ye focall length for both.
   For ye Meniscus
   As ye difference of ye two Radij to ye greater:
   so ye focall length of ye lesse. to ye focall length of ye glasse. of Colours in glasses.[138]

These were all matters which, he tells his audience, he had discussed in his lectures because he had not found them properly treated by any optical writers.

One might therefore conclude that most of the content of Lecture 4, with the possible exception of the calculation relating to the double convex lens, stem from Gascoigne's writings. The 'Naturall Consequences' of that gentleman's principles, or 'the superstructure' which Flamsteed had managed to erect upon these, included: the application of telescopic sights to astronomical instruments and the rectification of the latter for measuring altitudes (Lectures 5 and 6); the focusing properties of plano-concave and double-concave lenses and the effects of inverting them with regard to the direction of the light-source (Lecture 7); the comparison of the double-concave with a double-convex lens of both equal and unequal radii of curvature, and the properties of meniscus lenses (Lecture 8).

The following summary, in Flamsteed's own words, of the contents of all his lectures up to this point—that is, between 27 April and 22 June 1681—is to be found in an (as yet) unpublished letter to Richard Towneley written at the Greenwich Observatory on 30 July 1681:

> In my first lectures I shewed how a necessary account of ye true lenght of a yeare occasioned all the enquirys in Astronomy, first the Suns motion & its inaequalitys, next the lunar Periods. & her Anomalies then her distances. & how all were found by very simple & Naturall Methods, then how the Suns distance was attempted how it came to be mistated & what I found it by ye observed pa[ra]llax of Mars, Then I read on ye appearances of the Comet. of which haveing wrote fully to you formerly I shall now say nothing. Next Terme I set upon Dioptricks & haveing explained the principles & showne how it might be found by Calculation where any ray not falleing on the Pole of the Glass would be collected with another ray proceding from y^e same point of the same object throughe the pole. I found & shewd by a calculation sufficiently prolix that if the plane side of a plano convex glasse wrought in a sphere of 4 inch diameter were obverted to the Sun all y^e raies falleing upon it within y^e distance of $\frac{1}{2}$ inch from y^e middle would be collected on y^e Axis & intersect it all within y^e space of $\frac{1}{4}$ of an inch but if the convex side of the same glasse were turned to y^e Sun then they would meet the axis all of them within y^e space of $\frac{1}{16}$ of an inche. What great thinge thinke y^u I gather hence? Ile tell y^u. I find it now more convenient when I goe to light my pipe of tobacco at y^e Sun with a plano convex glasse, to turne the Convex side towards him then the plaine, for then all the parallel raies meeting the Axis in a lesse lenght must needs burne faster then when the plaine side is towards him & I assure you in my opinion, as oft as I trie it, the experiment confirmes what I had learnt by speculation & Theory: whether ye discovery were worth my paines I refer it to you.
>
> Haveing dispatcht this, my next enquiry was concerning Telescopes how ye usuall & most usefull of two convex glasses performe. & here I found it necessary to distinguish betwixt ye amplification of an object & its decreased distinctenesse which no Author to my knowledge has yet treated of all supposeing grosly that the objects are rendred so much more distinct as they are amplyfied whence it should follow that a telescope of 3 foot with a Charge of 3 inches. should p[er]forme no better then an object glasse of 18 inches with a Charge of $1\frac{1}{2}$ inche. for each of these should amplifie 12 times & the object in both should appeare alike distinct but experience shews the contrary for tho each does amplifie alike yet if the Moone be viewed with each of these severally it will be found that y^u see double y^e number of spots with the 3 foot glasse y^t y^u did with ye 18 inches & those doubly distinct. I affirmed therefore that tho the proportion whereby glasses augm[en]t

> is as theire focall lengths are to each other yet the proportion whereby they render objects distincter is found by divideing the focall length of the obj[ec]t glasse by the depth of ye humane eye, & this y^{u} will find to answer severall phainomena which our carelesse writers of opticks have slightly passed over, for want of experience & practise. I shewed then ye way of rectifieing a telescopicall Quadrant or level. & this was the businesse of ye 2^{d} Terme.[139]

Observational errors and the value of the obliquity

In Lecture 9, Flamsteed returns to his initial theme by giving a brief survey of earlier methods of determining the annual periodic variation in the Sun's distance from the Earth from measurements of its angular semidiameter, his main sources being Riccioli's *Almagestum novum* and Gabriel Mouton's *Observationes diametrorum solis et lunae apparentium*... (Lugduni, 1670). However, the methods used by these two authorities themselves; viz. of estimating this quantity by timing the transits of the Sun's image between two threads or across the meridian, yielded values which were less accurate than those obtained by applying a micrometer to this type of observation—as evidenced by the similarity of the results obtained independently by Flamsteed himself, Towneley, and the Paris observers (who were using Adrian Auzout's micrometer).[140]

The prime significance of these measurements was that they gave strong empirical support to Kepler's belief that the eccentricity of the Earth's orbit—defined as the distance between its centre and the Sun, in terms of its mean radius—required to be bisected. Otherwise, the difference between the calculated apparent diameters of the planets at aphelion and perihelion would be twice as large as it is observed to be; the effect being most pronounced in the case of Mars, where an error of 1° would thereby be incurred. Ismael Bullialdus (or Bouilleau), in his *Astronomia Philolaica* (Parisiis, 1645) had provided a rational—though incorrect—justification of this effect by drawing a distinction between the centre of the Earth's orbit and that of its uniform motion (analogous to the equant point in Ptolemaic astronomy) and assigning the latter to the empty focus of the Keplerian ellipse. Thus the eccentricity was exactly half the distance between the Sun at one focus and the centre of uniform motion at the other. Bouilleau's application of the equant hypothesis to an elliptical rather than a circular orbit had the dual advantage of 'saving the appearances' and giving a physical significance to both foci; but the concept of the Earth rotating uniformly about the empty focus could not be reconciled with that of the Earth–Sun direction sweeping out equal distances in equal intervals of time, as required by Kepler's second law. However, when Seth Ward[141] could produce a sophisticated equant theory capable of predicting a planet's position to within ±1′, this incompatibility seemed more of an argument against Kepler than against Bouilleau.

Thomas Streete's *Astronomia Carolina* (London, 1663) and Vincent Wing's *Astronomia Britannica* (Londini, 1668–9) were books on which Flamsteed and Newton are known to have been largely dependent for their knowledge of planetary theory, and in both of these Bouilleau's hypothesis is adopted and Kepler's second law ignored.[142] It was his preconditioning to this erroneous conception, coupled with his belief (initially) in Descartes' vortex theory of planetary motion, which Newton had to overcome before he could effect the great synthesis in his *Principia* (1687) of all celestial movements in terms of his three axioms of motion and the law of gravity.[143] There he demonstrates that Kepler's law of areas follows simply from the assumption of the Sun's central attractive force; and if, in addition, the orbit is elliptical, that this force is inversely proportional to the square of the planet's distance from the Sun.[144] As we have suggested earlier, however, it may have been Newton's deliberations on Flamsteed's speculations regarding the comet of 1680 which finally made him realize that the postulates of a central attractive force and a conic section are sufficient for the description of the celestial motions,[145] and thus appreciate the significance of Hooke's letters to him a year earlier.[146]

Having discussed the Sun's motions in the ecliptic and established the eccentricity of its elliptical orbit by means of micrometric measurements of the annual variation in its angular diameter, Flamsteed proceeds in Lecture 10 to discuss the methods for determining the obliquity of the ecliptic (with respect to the celestial equator) with a view to establishing whether or not it had always been constant. That very problem was, in fact, one to which Hevelius had explicitly drawn attention in his important letter of 23 December 1676.[146] Riccioli's *Almagestum novum* is Flamsteed's source of information on the ancient Greek and Roman attempts to measure this fundamental astronomical quantity; he obtained his knowledge of Arabic and Persian observations from information which had 'not yet been published. or at least mad[e] common',[148] communicated to him in a letter dated 14 August 1681 from Edward Bernard, Savilian Professor of Astronomy in Oxford. No precise references to specific printed works are given either in the lectures or in Flamsteed's correspondence with his Oxford 'brother';[149] but allusions in Lecture 10 to 'Mr Graves' and handwritten entries relating to Flamsteed in a 'Catalogue of Books' preserved among the Rawlinson Manuscripts in the Bodleian Library, Oxford,[150] positively identify John Greaves's edition of *Shaw Cholgii Persae Astronomica* . . . (Londini, 1652) as one of the books that he consulted at this time. Two references are also made in Lecture 10 to the manuscripts of Mr. [John] Selden, presumably those belonging to the portion of this gentleman's collection of Oriental manuscripts which had been acquired by the Bodleian Library late in 1654 under the terms of his will.[151]

Flamsteed considers the results which he cites from the above-mentioned sources to prove nothing more than that the obliquity has remained constant throughout all preceding ages. Any differences in the results could be ascribed either to instrumental errors, or to some astronomers allowing for (an exaggerated value of) the solar parallax and others not. The method used was probably the same in all cases: viz. that of observing the greatest and least altitudes of the solstitial Sun and taking the mean difference.[152] In more northerly latitudes, astronomical refraction appeared to have caused a systematic decrease in the value of the obliquity obtained in this manner.

The main feature of Lecture 11 is a historical survey of how the Arabs came to be acquainted with the ancient Greek manuscripts,[153] which Flamsteed would appear to have taken from Josephus Justus Scaliger, *Juli Caesaris F: Opus de emendatione temporum* (Genf, 1629) to which he alluded in Lecture 1. He now cites not only Riccioli's *Almagestum Novum* but, for the first time, also its sequel *Astronomiae reformatae* (Bononiae, 1665) which he had nevertheless acquired for his library about nine years previously.[154] He explains how unnecessary allowances for solar parallax and the omission of corrections for refraction may have vitiated many of the results during ancient and medieval times.

In Lecture 12, Flamsteed continues the same theme into the renaissance period, placing particular attention upon the reliability of the instruments used by Regiomontanus and his pupils Bernhard Walther and Johannes Werner in Nuremberg. The source from which he took his description of the alhidade (alias parallactic ruler, or triquetrum) used by these observers was Willebrod Snell, *Coeli et siderum in eo errantium observationes Hassiacae . . . Quibus accesserunt J. Regiomontani et B. Walteri observationes Noribergicae* (2 parts, Lugduni Batavorum, 1618). Errors were liable to have arisen owing to the lack of a compass for setting the instrument in the plane of the meridian and a clock for recording the time-interval before or after midday. There is always a tendency to underestimate the values of the greatest altitude obtained from repeated observations; and warping could have easily produced an error of $\pm 5'$ or $\pm 10'$ in the value for the obliquity. From the figures published in Walther's catalogue of solar observations contained in the same volume Flamsteed finds, after making due allowance for the effect of refraction from a table based on considerations discussed later in Lectures 33, 35–38, that the obliquity of the ecliptic was $23° 28' 32\frac{1}{2}''$. His conclusion in Lecture 2 regarding the smallness of the solar parallax justified his neglect of that quantity. To the nearest minute of arc, he finds that the Nuremberg observations of Walther, Werner, and the Bologna observations of Copernicus and Dominicus Maria Novara *all* yield the same value of 23° 29′ for the obliquity. Lecture 12 ends with a discussion of Copernicus's error in supposing the obliquity to have varied with time from one epoch to the next, the object of which was to show how bad observations could lead to unnecessary geometrical artifices.[155]

Flamsteed goes on in Lecture 13 to consider the accuracy of the observations by William IV, Landgrave of Hesse-Cassel, and by Tycho Brahe. The latter first visited William's observatory in 1575, and it was as a direct result of a letter on his behalf from the Landgrave to King Frederic of Denmark that he was given the small island of Hven near Copenhagen in the Danish Sound as a site for his own observatory, Uraniborg.[156] The revenues which he was entitled to collect from its inhabitants provided the capital with which to pay for the construction of the building and the equipment he required. Tycho had many excellent instruments made to his design, including vertical quadrants (some of which were rotatable about a vertical axis) for altitude measurements, a 6-foot 9-inch radius mural quadrant to determine the declinations of the heavenly bodies to within 10″, sextants for measuring the angular distances between two stars, and armillary spheres.[157] Flamsteed sees in this royal sponsorship and the consequent lavish spending on instruments and assistants, the dawn of a new era for astronomy which did much to encourage improvements in the quality of the data acquired. The nominal precision of Tycho's observations was indeed high, since by repeating them several times and taking the average value he could quote figures for altitudes and inter-stellar arcs to the nearest 1″. However, the limited resolving power of the human eye, and the systematic errors incurred in constructing and mounting the instruments, reduced the accuracy considerably. Equally serious was the lack of a theory of astronomical refraction, uncertainties in parallax corrections, and (if observations made at different times were compared) the fact that the phenomena of aberration and nutation were totally unknown.[158] Thus, in practice, the altitudes measured with different quadrants were often found to differ by as much as ±2′.

An interesting inference is made by Flamsteed from his examination of Tycho's results for the latitude of his observatory. This quantity had been found in two ways: by taking the complement of half the sum of the Sun's observed meridian altitudes at the summer and winter solstices, and by observing the altitude of the pole star. In the former method, an unnecessary 'correction' of 3′ for the Sun's horizontal parallax resulted in the latitude being less than it ought to be, whereas in the latter method no parallax correction was (of course) made; hence results for the latitude obtained in 1583 were found to differ by 4′. Similarly, the obliquity of the ecliptic derived by subtracting the co-latitude found from the pole star observation, from the greatest meridional altitude with parallax 'correction' at the summer solstice, was found to be over 8′ greater than that derived by subtracting the least meridional altitude with parallax 'correction' at the winter solstice from the same value of the co-latitude. Flamsteed believes that it was this discrepancy that first led Tycho to believe that refraction was much more important than he had hitherto supposed, because the only way to remove the former was to make an allowance for refraction at the winter solstice observation. This was, of course, larger than it ought to have been because it had also to counteract the effect of the unnecessary parallax correction.

This discovery may have been the reason why Tycho began in 1585 to quantify the effects of refraction. Ignoring the solar parallax and reducing Tycho's observational data with the aid of his own estimates of refraction, Flamsteed finds the obliquity to be 23° 29′ 05″, and the latitude of Uraniborg 55° 54′ 05″; the latter figure being the same as that found by Tycho himself from his pole star observations (with the same 35″ correction for refraction at that altitude), and only 10″ less than that found by Jean Picard.[159]

The fact that a constant obliquity of 23° 29′ is also consistent with the later observations of other astronomers is verified in Lecture 14 by a similar reduction of results obtained by Edward Wright, Philip Lansberg, Albert Lineman, Giovanni Riccioli, Giovanni Cassini, Gottefrid Wendelin, and Johann Hevelius, which had hitherto appeared to be appreciably higher. Flamsteed obtained his information on the methods and data of all but Wright and Hevelius from Riccioli's *Almagestum Novum* and *Astronomiae reformatae.* Wright's astronomical observations were published in Joseph Moxon's edition of his *Certaine Errors in Navigation* (London, 1657) and Hevelius's in *Machina Coelestis pars prior* (Gedani, 1673). The ultimate confirmation lay in a comparison made at Greenwich between the Sun's meridian zenith distances at the summer solstice and the latitude of that observatory found by averaging the greatest and least altitudes of the pole star (duly corrected for refraction). There was therefore no doubt about the conclusions that 'the Latitudes & declinations of ye fixed stars observed by the ancients were scarce true within ye Compasse of a degree & . . . that the obliquity of the Eclipticke was invariable.[160]

A new graphical representation of celestial appearances

In the next seven lectures (Lectures 15–21), Flamsteed elaborated his 'Doctrine of the Sphere', which was a graphical representation of appearances in the heavens founded upon this newly established observational fact of the invariable obliquity of the ecliptic. The method of projection was stereographic rather than orthographic, since this had the important geometrical property that great and small circles on the celestial sphere would be projected into circles (not ellipses) in the plane of projection when the point of projection is chosen to lie at any point on the surface of the celestial sphere. This meant that all the necessary lines could be drawn with the aid of ruler and compasses only, in the classical tradition. The previous uncertainty about the constancy of the angle between the Earth's axis of rotation and the ecliptic, which Flamsteed in his previous two lectures had taken such pains to confirm, had formerly been a deterrent to the choice of the ecliptic as the fundamental plane of projection. William Oughtred, *Horologiorum Sciotericorum* (Oxoniae, 1663) for example, had projected from the south pole of the horizon—the nadir—on to the plane of the horizon, while Johann Stoeffler, *Elucidatio fabricae ususque Astrolabii* (Parisiis, 1585) had projected from the south celestial pole on to the plane of the equator. The most general treatise at that time on the method of stereographic projection, Christopher Clavius's *Astrolabium* (Romae, 1597), was unfortunately also the most prolix. Flamsteed referred to it several times, mainly to re-assert this fact; and his comments would seem to imply that he had arrived at the much simpler method described fully in the Gresham Lectures, after having made a thorough study of Clavius's work. This may indeed have been so, but a comparison of that text with Flamsteed's description of his method provides no indication of his having been directly indebted to it for any of his ideas. A more likely source of inspiration is John Collins's *The Sector on a Quadrant* (London, 1658) where Clavius's method is discussed and the geometrical principles of projection presented in a manner analogous to that used by Flamsteed.

The earliest clue to the genesis of the 'Doctrine of the Sphere' is to be found in the account of the above-mentioned manuscripts of William Gascoigne and William Crabtree, preserved among Flamsteed's unpublished papers at the Royal Greenwich Observatory.[161] Here Flamsteed gives an extract from Crabtree's letter to Gascoigne dated 21 June 1642 outlining the lunar theory of his decreased friend the Rev. Jeremiah Horrox.[162] Concerning its demonstration, Crabtree writes:

> I could never be fully satisfied nor doe I find it a jot explained any where among his papers which I must inervaile at onely I find a draught or 2 like this following rudely drawne upon loose papers. I conceive it thus. S. y^{e} ☉, T ye earth. The great circle ye earths orbe ye little circles y^{e} ☽^{s} orbes. A the true apogeon P ye Perigeon, C y^{e} ☌ wth y^{e} ☉, O y^{e} ☍. M ye meane apogeum, E ye center of y^{e} ☽^{s} orbe. ET ye excentricity. In ye 1, 3, 5, 7, figures ye meane & true apogeon are both

> one but as ye earth goeth from ye 3 & 7th figures, ye true Apogeon A is by degrees inclined to ye meane M. ye Angle ATM being ye equation of ye Moones Apogeum this equation from 3 to 5 and from 7 to 1 is to be added to ye meane Apogeon to find the true, but from 5 to 7 and from 1 to 3 is to be subtracted. as appeares by ye figure. so y^{t} ye sun doth alwayes attract to it selfe ye Apogeum or Perigeum of the Moone whether it is nearer unto. but in ye 1st & 5th figure being alike neare to both alters neather. as a loadstone y^{t} in ye equinoctiall circle hath no virtue but onely at its 2 poles. Allso in ye 3^{d} & 7th figures ye sun alters ye line of ye Absides nothing because it then points to ye sun. in ye figure 1 & 3 ye excentricity seemes by calculation to be least and at 3 & 7 greatest:
> The rest of ye calculation depends on ye ellipticall course of ye moone all but ye 7th praecept concerneing ye variation which I know not how to demonstratively to apply to this Theory. I should much desire some helpe therein. But for Calculations I find in all his last calculations it is as I have before described.

Flamsteed then remarks:

> This same method of calculation I find described in a manuscript committed into my hands by Mr Townly wth ye papers. what were in Latin but some other hand y^{n} Mr Gascoignes or Crabtrees. but findeing this theory not agree wth ye heavens & observations as I may prove elsewhere. I omit it.[163]

He subsequently identified the handwriting as being that of Jeremiah Shakerley, a brilliant young North-Country mathematician who had resided with the Towneleys from 1648 to 1650 prior to publishing Horrox's astronomical tables in *Tabulae Britannicae . . . the Doctrine of the Sphere etc.* (London, 1653) from which he may have borrowed the title of his own published dissertation.[164]

Flamsteed also voiced his dissatisfaction with the above description by Horrox of the moon's libration in a letter to John Collins dated 13 May 1672[165] written at around the same time as he transcribed Crabtree's letter on the subject. In place of 'Kepler's groundless notions', which Horrox had accepted, he proposes the following alternative physical explanation:

> The moon's system, being disposed about our earth, seems to be most likely poised upon its centre, and seing that the part contained betwixt the apogaeon and a line drawn perpendicular to the transverse diameter of the orbit, through the earth's centre, is more large and capacious than the part contained betwixt the said line and the perigaeon, it follows that the apogaeon part of the system may contain more of matter and celestial substances than the perigaeon region, and consequently (as the end of a false balance, which is longest from the centre) may supragravitate and incline the axis towards the syzigiacal line, where, by reason of the sun's directest action on the earth's sphere, or rather the moon's, the action of the earth upon the lunar

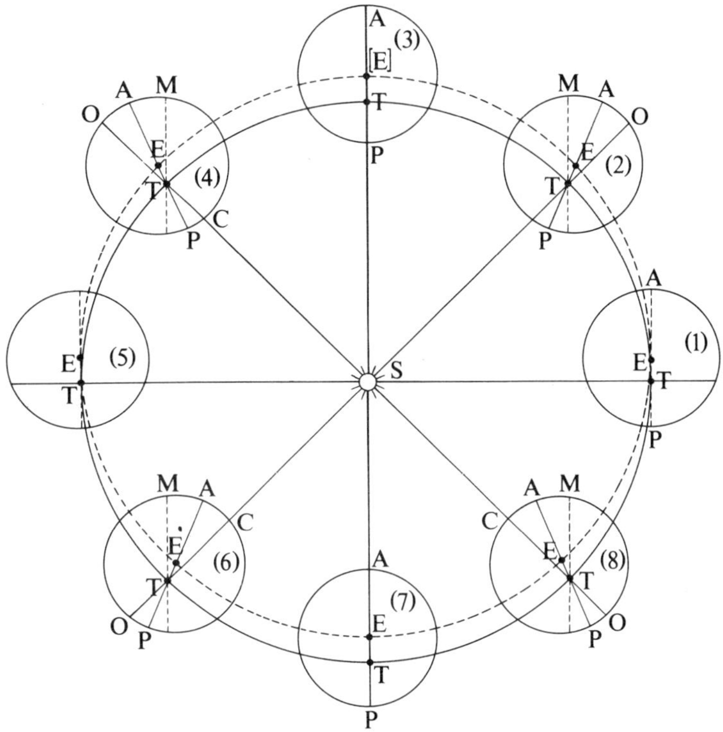

FIG. 16. Crabtree's drawing of Horrox's diagrammatic representation of the lunar theory.

> system, whereby the apogaeon and the ascending node are carried about, may be most broken, and the system may most easily receive and admit this libration.[166]

This somewhat obscure mechanical analogy is then exemplified with the aid of a modified drawing of Fig. 16.

Flamsteed completed his transcription of the Gascoigne-Crabtree papers a week later,[167] then passed them on to John Wallis who had already begun to prepare his Latin edition of Horrox's papers.[168] The precepts and accompanying diagram for the latter's lunar theory are faithfully reproduced in the text of 'Novae Theoriae Lunaris' (incorporating Flamsteed's new lunar numbers and calculations based on this theory), which is appended to that work.[169] However, Wallis seems to credit Crabtree with having *devised* the diagram (Fig. 16) as opposed to having merely copied Horrox's rough sketches. A comparison of this figure with Fig. 16.1 (cf. p. 264) strongly suggests that the former was the visual inspiration and concept upon which Part 1 of Flamsteed's 'Doctrine of the Sphere'—the only part presented in these lectures—is founded.

Both schemes are projections of appearances in the heavens into the plane of the ecliptic. The constant direction maintained by the line from the Earth to the Moon's mean apogee (TM) with respect to the background of stars is replaced in Fig. 16.1 by the constant angle made by the solstitial colure *pe* through the north celestial pole (*p*) and the pole of the ecliptic (*e*). The *monthly* west-to-east rotation of the Moon in its orbit around the point E (in Fig. 16) becomes transformed into the *daily* west-to-east rotation of a place (e.g. London) along a northern parallel of latitude around the pole *p*. Here, however, the analogies end.

The real novelty of Flamsteed's doctrine is that it was the first to reveal how the diurnal as well as the annual appearances in the heavens could be represented in terms of the Pythagorean postulates of the Earth's axial and orbital motions. (Like Horrox, Flamsteed preferred to attribute the discovery of the Earth's mobility to the Ancients rather than to Copernicus.) Moreover, this was one of the first known instances of the method of stereographic projection being explained in the English tongue, and had the added advantage of being derivable from only ten theorems hence of being very much simpler than the exposition of Clavius.

The motivation for Part 2 of this work—that concerning the application of the 'Doctrine . . .' to solar and lunar eclipses[170]—stems from a conversation which Flamsteed had with Halley before the latter's departure in November 1676 for St. Helena to observe the southern skies. Halley had then asked Flamsteed if he were able to represent Kepler's scheme of determining the magnitude and phases of a solar eclipse without recourse to calculation.[171] He himself did not appear to have had a proper method for doing this. In informing Towneley of that problem early in 1677,[172] Flamsteed says he had thought no more about it until he observed the Sun-spots,[173] which made him consider what the Earth might look like if seen from the Sun, and how the diurnal phenomena would appear if projected on it. He gives in this same letter a detailed explanation of the method which he evolved, considering the eye to be placed in the Sun and projecting the Earth's axis and the path of the zenith for a given place in the plane at right-angles to the line through the centres of these two bodies. He assumed that the polar axis remained constantly in the same direction. The projection yielded the Sun's zenith distance, the parallactic angle, the longitude difference between the nonagesimal degree and the Sun, and the altitude of the nonagesimal degree at any time of day, and thus enabled all the circumstances of the eclipse to be graphically determined.[174]

The occasion for elaborating upon this work and putting it into print arose about two years later when Flamsteed was asked by Sir Jonas Moore's sons-in-law to edit the navigational section of their late father's *A New Systeme of the Mathematicks*, and to supply another on astronomy. He told Towneley on 22 May 1680 that he regarded it as his duty to perfect his account of 'ye Diurnall Phainomena in ye Pythagorean or Copernican Hypothesis' (i.e. Part 1) and to demonstrate how it could be applied to the doctrine of eclipses and appulses with the aid of 'that Projection of ye sphere with which I acquainted y^{u} some two years since' (i.e. Part 2).[175] He added lunar tables and directions for calculating and projecting *lunar* eclipses,[176] and managed to finish it within the next two months. The death of his friend and co-editor Peter Perkins, a master of the mathematical school at Christ's Hospital, London, caused him to become twice as involved as he had intended in supervising the final stages of this literary obligation just at the time when he was being continually asked by his various acquaintances to give his views on the comet of 1680 (which we have already discussed). The book was finally published in February of the following year. In addition to Flamsteed's dissertation on 'The Doctrine of the Sphere' (1680) and his astronomical tables, it contains: rules and principles of arithmetic, practical geometry, plane and spherical trigonometry (all written by Sir Jonas himself); a concise geography, perused and completed by Halley; the section on navigation, which Perkins had managed to complete before his death; algebra and 'the most useful Books' (viz. 1–6, 11, 12) of Euclid's *Elements*.

The precise contents of Lectures 15–21 need only be summarized briefly, since the geometrical demonstrations and method are best understood from a study of the text itself. Flamsteed begins his analysis by showing first how all the circles of the sphere may be derived from the Pythagorean system (which supposes them fixed upon the Earth) using an *orthographic* projection, then explains the principles of the stereographic projection of the sphere on to the ecliptic for the case of London (lat. $51\frac{1}{2}$° N) with the south pole of the *ecliptic* as the central point. He demonstrates how to represent stereographically the equinoxes, solstices, diurnal and nocturnal arches, and uses theorems from Isaac Barrow's English translation of *Euclide's Elements* (London, 1660) which Sir Jonas had insisted should 'be every where follow'd'[177] to construct the equator or a latitude circle and hour-circles, and to divide these into equal degrees. Internal evidence enables one to identify each of the ten theorems which he ultimately employs for this purpose, and this has been done in the annotations. A distinction drawn in Lecture 17 is that in the Ptolemaic projection of the sphere the meridian is fixed and the parallels of declination moveable, whereas in Flamsteed's Pythagorean projection the meridian changes daily and all diurnal appearances are represented by one parallel only, viz. that of the path of the place for which the projection is made.

Up to this point, Flamsteed has shown what the appearances are at the four cardinal points: the summer and winter solstices, and the spring and autumnal equinoxes. Now he extends his method to include any intermediate time throughout the year and any moment during the day. In order to do this, three fundamental astronomical parameters are required: the right ascension (α) and declination (δ) of a heavenly body, and the obliquity of the ecliptic (ε). These will be the same at any given instant for all inhabitants of the Earth, so Flamsteed regards them as *general quantities* and the associated phenomena, *general phenomena.* He quotes, without proof, the various precepts by which the other dependent quantities may be calculated from them; but these are standard cases of the solution of right-angled spherical triangles[178] and require no commentary. Since the Sun's longitude ($\lambda_\odot$) is given in terms of $\alpha_\odot$, $\delta_\odot$ by the equation: $\cos \lambda_\odot = \cos \alpha_\odot \cos \delta_\odot$, and its latitude is zero by definition, its position is uniquely specified by means of only two co-ordinates, $\lambda_\odot$ and ε (cf. Fig. 17). In Lecture 18, it is shown how all diurnal

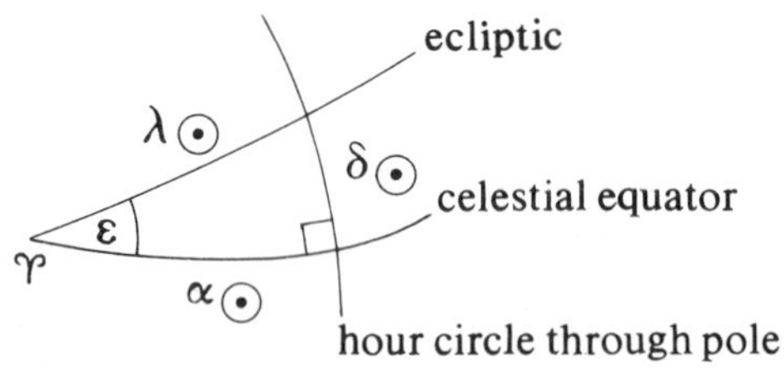

FIG. 17. Diagram illustrating the relationship between the celestial longitude and the equatorial co-ordinates of a heavenly body.

appearances at a place of known latitude (φ) can be represented for any given time throughout the year; the information supplied being sufficient for the graphical determination of the Sun's zenith distance, azimuthal distance from the 6 o'clock (i.e. East–West) meridian, and hour of day.

The analysis of the diurnal appearances suffers, however, from two limitations: it does not apply to cases where oblique-angled spherical triangles required to be solved, nor to celestial bodies lying outside the plane of the ecliptic. There was a need, too, to take account of the phenomenon of precession, but consideration of this problem is postponed until Lecture 32. In Lecture 19, Flamsteed shows how the ecliptic co-ordinates of a star ($\lambda_\star$, $\beta_\star$) may be plotted on his projection—this procedure had, in fact, already been published in his dissertation[179] for the case of London at 5 p.m. on 23 June, the Sun's azimuth and zenith distance being given. He goes on to explain how the longitude and altitude of the nonagesimal point, whose importance in the determination of parallaxes has already been demonstrated,[180] are graphically determined. Thus he could achieve everything which Clavius had achieved, using eight 'or 10 at ye Most' theorems. Lecture 20 is devoted entirely to the proof of his eighth theorem, stated in the text but alternatively formulated on a separate page as follows:

> If on ye globe a great Circle shall be strooke through the Poles of any other two great Circles described on it, & afterwards the Globe shall be cut by a plane Passeing through the said two poles & placed at right angles to ye plane of the great Circle so described, the plane so placed shall cut of equal arches in each of the great Circles through whose poles it passes betwixt it selfe & the great Circle passeing through the said their two poles.[181]

Flamsteed breaks up this rather complicated theorem into three parts and proves each part in turn; subsequently applying it, in Lecture 21, to an alternative construction of the equator and tropics of Cancer and Capricorn, something which 'the learned Clavius' had omitted to do. He explains how the Ptolemaic equatorial projections may be drawn, and how every circle (whether great or small) may be divided into regular intervals using ruler and compasses only. The same method could be used to divide 'the path of the vertex' (i.e. the locus of the place from which the celestial observations are made) in the case of his Pythagorean ecliptic projections.

His experience in this work appeared to him to confirm the belief shared with Horrox and many others of his own generation, that 'the Pythagorean susteme is the onely legitimate the rest but spurious', the one 'most consonant to reason',[182] and 'most agreable to experience'.[183] The greater facility with which he could determine parallax from the graphical representation of the nonagesimal point by employing his method rather than the Ptolemaic, convinced him that

> nature workes the shortest way . . . the most ordinary of her p[er]formances shew theire proper effects much more easily then the most ingenuous of humane contrivances can represent them. Hence I thinke too we may draw one probable Argument for the Verity of the Pythagorean systeme. for since as Nature workes the readiest way. & this representation of her answers all her appearances the most easily and Naturally tis the most likely to be the very truth it selfe & this its p[er]fect figure.[184]

A new interpretation of the physical constitution of celestial bodies

Judging by Flamsteed's remarks in Lecture 22 of 24 January 1683, he had recently become aware of his obligation, under the terms of the Ordinances and Agreements drawn up by the Grand Committee for Gresham Affairs in 1597, to incorporate lectures on the theory of planetary motion into his syllabus.[185] This knowledge was dependent upon observations of the angular distances between planets and stars, and upon the accuracy of the stellar co-ordinates themselves, so he maintains that it was incumbent upon him to examine, first of all, what had been written about stellar appearances, viz. the substance of stars, their different magnitudes and colours, their scintillation, their disposition on the celestial sphere, and their apparent motions.[186] He therefore devotes Lectures 22–28 to this vast range of topics, drawing analogies between the physical nature of the Sun, variable stars, novae, and comets. After pausing in Lectures 29 and 30 to consider the observational and instrumental errors vitiating many of the positions recorded in the star catalogues of Ptolemy, Ulugh Beigh, and Tycho Brahe, he goes on to investigate the effects of precession and particularly refraction in such detail that he ends his syllabus without ever fulfilling his obligation!

The philosophical reflections contained in Lecture 22 concerning the relationship between the knowledge of the natural world acquired through the senses and its association with a 'higher' realm of Truth acquired through Man's exercise of his reason, are to be found in the preface to Riccioli's *Almagestum novum*, although they stem from the writings of Plato, particularly from his *Republic* (Book 7) which that author cites. Ancient opinions regarding the nature of stars, the Renaissance speculations about the plurality of worlds, and the remarks regarding alternative cosmological systems, were also borrowed by Flamsteed from Riccioli's book.

His belief that stars are of the same nature as the Sun, on the other hand, came not from speculations but as a deduction from his own telescopic observations. He refers in Lecture 23 to his day-time measurements in 1672 of the diameters of Mercury and Sirius, using a 14-foot long telescope with a micrometer; from which he concluded both to be about 14″.[187] His failure to detect any effect of annual parallax in careful observations with his 3-foot quadrant of the angular distance between the bright star of Aries and Aldebaran, despite his precautions to reduce the effects of refraction, made him accept Copernicus's assertion that this quantity was less than 1′ and hence insensible.[188] His provisional assumption that the diameter of the Earth's orbit subtended exactly that angle at the distance of Sirius, coupled with his micrometric measurement of this star's diameter and his knowledge that the Sun's diameter is about $\frac{1}{220}$ of the Earth's orbital diameter, enabled him to deduce that Sirius would then be 94 times greater in volume than the Sun. If a star's diameter were only 1′, it would still be twice as bulky as the Sun. Taken together with his general observations that stars appear 'less hairy' when seen through a good telescope and must be very bright considering their vast distances from the Earth, this rough calculation was sufficient to convince him that the comparison with the Sun was a valid one.

Having thus established this important analogy, Flamsteed now described his observations of Sun-spots during the year 1676 with a view to sharing with his audience his insights into the constitution of stars. Sun-spots revealed themselves as being too stable to be clouds or emanations, thus he concluded that they must have been produced by volcanic eruptions from deep inside the Sun, the areas surrounding them (the faculae) appearing brighter only because they were close to a dark body and not because they were intrinsically brighter than the rest of the Sun. The existence of such spots on the Sun suggested that similar dark patches might exist on stars, obscuring part of their light; while the phenomenon of solar rotation implied that this obscuration would not be permanent, but that a star would appear periodically brighter or darker depending on the rate of its axial rotation (as well as, of course, on the size and distribution of its supposed spots).

Flamsteed cites, in support of this hypothesis, the description of a nova in the neck of Ophiucus (the Whale) given by Hevelius in his *Mercurius in sole visus* (Gedani, 1661–2) and subsequently reported in the first volume of Oldenburg's *Philosophical Transactions*[189] as part of an account of Ismael Bullialdus, *Ad Astronomos Monita duo* (Parisiis, 1667). Bouilleau's purpose in writing that tract was to incite astronomers to observe this star and another nova in the girdle of Andromeda. The former was thought to have a period of about 333 days, being visible to the naked eye for only 120 days: it appeared at its brighest, as a third magnitude star, for about 15 days. After investigating possible causes of this star's behaviour, that author concludes:

> there can be no other genuin, or at least, no other probable cause of its Emersion and Occultation, than this, That the bigger part of that round body is obscure and inconspicuous to us, and its lesser part lucid, the whole Body turning about its own Center, and one Axe; whereby for one determinate space of time it exhibits its lucid part to the Earth, for another, subducts it: it not being likely, that fires should be kindled in the Body of that Star, and that the matter thereof should at certain times take fire and shine, at other times be extinguisht upon the consumption of that matter.[190]

Regarding the other nova in the girdle of Andromeda, he notes that:

> whereas this Star hath been seen formerly, and that 150 years since, but yet neither observed by Hipparchus, nor any other of the Antients, that we can find; not also in the former Age of Tycho Brahe, nor in our Age, by Bayerus; and appear'd also in the Month of November last (wherein he wrote this Tract) much lessened and obscure, after it had, two years ago, shone very bright; that therefore it must needs appear and dis-appear by turns, like those in the Necks of the Whale and Swan.[191]

Accounts both of that nova in the neck (or bill) of the Swan, and yet another in its breast—two further examples referred to by

Flamsteed towards the end of Lecture 23—are contained in a later issue of the same periodical.[192] Hevelius's observations of the former are there extracted from a letter to Oldenburg written on 21 April 1671 [O.S.].[193] The other account, referring to both these stars, was translated from the French *Journal des Sçavans* for 22 June 1671.[194] The nova in the Swan's neck was thought to have a ten-month period of axial rotation, about a month shorter than that of the nova in the Whale's neck; the nova in the Swan's breast, on the other hand, appeared to require 'no less than Fourteen years'[195] to return to the same brightness. This behaviour indicated that a star's appearance or disappearance in the heavens did not necessarily mean that it had been newly created or destroyed. Flamsteed found, in these reports, evidence favouring his belief that stars are similar in nature to our Sun, and regarded the longer periods found for their supposed axial revolutions as an indication of their greater bulk.

Lecture 24 contains details of novae observed during the course of the previous 110 years, from the time when the famous one of 1572 that was largely responsible for Tycho Brahe's decision to dedicate the rest of his life to observing the heavens, first appeared in the constellation Cassiopeia.[196] Flamsteed's sources seem to have been Riccioli's *Almagestum novum*, Hevelius's 'Historiola Mirae Stellae',[197] Simon Marius's *Mundus Jovialus*, Galileo's *Sidereus Nuncius*, Huygens's *Systema Saturnium*, and the above-mentioned reports contained in Oldenburg's *Transactions*. This discourse serves merely to lend further observational support to his hypothesis. Lecture 25 is largely repetitive, the only fact worth noting from it being Flamsteed's rejection of the 'French' view (i.e. that of G. D. Cassini and Anthelme Voituret) that the nova in the Swan's neck had a ten-month period, since he had purposely looked for it during September and October 1680 during which time it ought to have appeared. Their conclusion that certain stars seen by the Ancients were now no longer visible, whereas other stars had changed in brightness, was not new.

The following lecture—Lecture 26—is much more interesting inasmuch as it contains Flamsteed's attempt to interpret the above-mentioned phenomena in terms of a physical model of stellar structure. Here he considers at some length, with the aid of five diagrams, the various possible explanations of novae, selecting that in the Whale's neck as a particularly good example in view of the reliability of the data collected on it. He argues the impossibility of reconciling the observations with the assumptions:

(1) That one hemisphere is dark and the other bright.

(2) That a lucid point or broad spot exists on the surface.

(3) That the inside is dark and the outside bright.

Only by supposing a star to be composed of a bright ball of lucid matter surrounded by a solid shell of dark matter, could a rational account of appearances (consistent with the initial premises) be provided. The rotation of a star having a gap in its dark outer shell could produce the brightness variation of novae, and there was no reason to reject *a priori* the possibility of periods as much as 332 times that of the Earth or 13 times that of the Sun. The Sun's own physical appearance could be accounted for by introducing the dynamical postulate that the outer matter is drawn through an orifice into the core and replaced by brighter matter from the inside; the reversibility of this process explaining why certain stars observed by the Ancients were no longer visible. Planets were considered to belong to this latter category of celestial object, and thus to be no different in nature from stars. Indeed, this is why Flamsteed is able to cite the gradual diminution in brightness of Jupiter's satellites in support of his hypothesis that stars may be enveloped by a dark atmosphere.

In Lecture 27, the applicability of the model is extended to comets, other novae, and (planetary) nebulae. Flamsteed regards the 'erratic' motion of comets as a consequence of their having experienced a transformation from the planet-type constitution to the star-type, through the flow of luminous matter from the outer shell into the inner core. In the case of the novae observed in Cassiopia and Serpentaria, he supposes that as a result of the active lucid matter pouring into the core through an orifice in the shell and coming into contact with sulphurous matter already present there, the crust surrounding the latter was broken and blocked that orifice, thereby rendering the body invisible although the fuel inside continued to burn and blaze. In the case of the nova in the Swan's bill, he thought that there was a gradual crumbling of the shell around the orifice; in its breast, there may have been more than one orifice otherwise it could not have remained both visible and weak for so long. That in the Whale's breast must have had a wide orifice initially, which remained open. The Andromeda Nebula may have resulted from the flow of outer luminous matter into a broken mass through a number of small holes out of which it still shone. An anonymous observer had seen this celestial object a hundred years before Tycho, although to the latter himself as well as to Johannes Bayer it had been invisible. The fact that Flamsteed's own observations always revealed it, implied (on this model) that some of the holes might have been obstructed at different times. The same assumption served to explain the irregularly shaped Orion Nebula.

A further consequence of the analogy between stars and the Sun was that stars may also be regarded as the central 'Suns' of a planetary system, and comets could be regarded as the planets of some vortex whose central 'Sun' is extinct. Although Flamsteed uses the term 'vortex', he does not accept the physical explanation of planetary motion in the third book of Descartes' *Principia Philosophiae*,[198] mainly because it failed to explain how the Sun can both attract and repel the planets while retaining them in their orbits. In his view, magnetism was the only physical cause capable of accounting for the elliptical form of the planetary orbits. Fundamentally, therefore, it was Flamsteed's faith in his analogies and adherence to this physical principle which governed his belief that the lucid matter in the interiors of the planets and in the exterior of the Sun must also possess such a property. Newton, not being bound by either of these presuppositions, felt himself under no obligation to admit the existence of magnetism in celestial matter. As soon as he had grasped the important fact that a central attractive force can produce an elliptical orbit, he felt free to accept gravity as the causal agency of planetary motion, as has been indicated earlier (cf. p. 34).

A revised estimation and explanation of the annual precession

Having summarized the views of different men on the cause of a star's twinkling, and affirmed his own belief that this phenomenon is produced by the *Earth's* atmosphere, Flamsteed goes on in Lecture 28 to recount ancient opinions of planetary motions and star positions. He gives a translation of a long extract from the opening chapter of Kepler's *Astronomia nova* (Pragae, 1609) where reference is made to the ancient distinction between the apparent east–west diurnal rotation of the fixed stars and the west–east motions of the Sun, Moon, and planets. He summarizes his initial discussion of the manner by which the Sun's and Moon's motions had been determined, and refers to the inaccuracy of the observing instruments and methods used by the Greeks. Differences in the declination of the bright star Spica ♍ observed by Tymocharis (*c.* 300 B.C.), Hipparchus (145 B.C.) and then by Ptolemy some three centuries later, seemed too large to be ascribed to such a cause. Hipparchus had noticed that its celestial latitude had remained the same, and therefore conjectured that the discrepancy was due to a gradual motion of the celestial sphere about the poles of its equator. The effect revealed itself as an increase in the star's longitude, or as a backward (precessional) movement in the points of intersection of the equator and ecliptic (viz. the equinoxes) from which that quantity is measured. This precession of the equinoxes had always to be reckoned with before comparisons could be made between star positions observed at different epochs; but the adoption of inaccurate values for the obliquity of the ecliptic and for the horizontal parallaxes of the Sun and Moon, the inadequacy of the lunar theory, and the use of erroneous precepts in calculations, all combined to vitiate the results obtained.

Tycho, though blessed with accurate instruments, adopted the much too high value of 3′ for the solar parallax and hence obtained wrong values for his meridional altitudes from which the celestial co-ordinates of both planets and stars were found. Moreover, he determined Venus's distance from the Earth with the aid of Copernicus's erroneous and inconsistent scheme for Venus's motion and, in his earlier observations, ignored refractions above an altitude of 20°. Flamsteed recommends pole star observations rather than meridional altitudes as a more reliable basis for determining the position of aphelion and the obliquity, pointing out that refraction and parallax can be minimized by observing the Sun and a planet (e.g. Venus or Jupiter) when both bodies are at approximately equal altitudes. A more suitable way of finding the equinoctial point than the use of a gnomon would be to observe the Sun twice near the equinox when it has equal (northerly and southerly) declinations.

In Lecture 30, Flamsteed ascribes the error of 1° found in Ptolemy's position of the equinox entirely to the effect of instrumental errors; but it might have arisen from the fact that he borrowed the value of the Sun's motion from Hipparchus who, by basing it on Callippus's 76-year luni-solar cycle, had obtained a mean value for the length of the solar year about 6 minutes too large. In the 300 years between Hipparchus and Ptolemy, this error cumulated to $1\frac{1}{4}$ days or over 1° in the longitude of the Sun, which would explain why Ptolemy's equinoxes are found to be too late by just about this much.[199] The implication for stellar observations is that the amount of the precession over the same period is underestimated. Ancient measurements of stellar latitudes were liable to err on occasion by nearly $\frac{1}{2}^{\circ}$, and meridian altitudes quite commonly by $\frac{1}{4}^{\circ}$; and these errors had given rise to the false impression that the obliquity had not remained constant. Contemporary observations, however, dispelled this idea (which Riccioli supported), as had been firmly established in Lecture 10. Flamsteed concludes that stars do not change their relative positions; hence their longitude differences and their respective latitudes remained invariable. The precessional effect, which was easily determined, applied equally to all.

These points being settled, Lecture 31 is devoted to a quantitative determination of the annual precession from observations of Spica ♍'s declinations. This star was especially suitable for such an investigation both because of its brightness and because it had been observed by Tymocharis, Hipparchus, and Ptolemy. Comparing each of their declination measurements with that of Tycho, Flamsteed obtains three values for the precession and selects the one deduced from Hipparchus's observation (viz. $51\frac{1}{3}''$) as the most reliable. Assuming this to be the true value, he finds that he requires to postulate errors of −15′ and −17′ in the respective observations of Tymocharis and Ptolemy, which he attributes to the manner in which they read their armillas. A very significant fact noted by Flamsteed, which seemed to owe its origin to the same considerations, was that the declinations of the fixed stars in the Alphonsine Tables[200] are *all* 17′ 08″ greater than those in Ptolemy's *Almagest*. Despite his own previous preference for an annual precession of 50″,[201] and Streete's adoption of only 48″,[202] he was now convinced that 51″ was a better approximation to the truth.[203]

Flamsteed still had to discuss the question of whether the secular increase in the longitudes of stars was uniform, and not periodically greater and less than the last-mentioned value. This he does in Lecture 32 (21 November 1683) and again, more clearly, in Lecture 34 (16 April 1684) which appears to be a redraft of that former lecture.[204] After discussing the nature of the uncertainties in the ancient and Arabic observations from which Copernicus had been led to believe in the existence of an irregular oscillation, or 'trepidation', in the movement of the equinoxes,[205] Flamsteed comments upon the weakness of the evidence cited by more recent supporters of that viewpoint, and the uncertainty of comparisons drawn between Tycho's stellar positions and the less accurate ones in the Alphonsine Tables, Ulugh Beigh's catalogue, and other individual works dating from a later period. There was, he concludes, no reason for supposing that the value of the annual precession was anything other than constant. Any theory had only to account for that fact, and could ignore the fictitious phenomenon of trepidation.

Flamsteed had already given an explanation of precession in 'The Doctrine of the Sphere',[206] which he now repeats. In Lecture 34, however, he explicitly states the philosophical premises listed in that printed tract which underlie his treatment of this subject. These are:

(1) A solar vortex, bearing the Earth and other planets round the Sun in their appropriate (observed) periods.

(2) A mobile Earth, revolving daily on its axis and annually about the Sun.

(3) A very subtle aetherial matter, pervading all space and transmitting light to us from the Sun and off the planets.[207]

Olaus Roemer had demonstrated, from a study of the eclipses of Jupiter's satellites, that the speed of light was finite though very great,[208] which lent observational support to Flamsteed's idea that the universal aether may be denser nearer the Sun and rarer farther away from it. Presumably by analogy with the speed of sound waves in air, he considered that the greater the density of the transmitting medium, the more rapid the speed of transmission.

Essentially, Flamsteed accounts for the phenomenon of precession by supposing that the Earth's axial rotation is reduced due to the matter in the solar vortex being both denser and swifter on the side nearer the Sun where the two motions are opposed, than on the farther side of the Earth where they are both in the same sense (viz. from west to east). The result of this differential retardation is that the 'line of direction' through the poles of the ecliptic and equator is borne backwards (i.e. east to west) contrary to the Earth's—and the vortex—motion.[209] Hence the parallelism of the Earth's axis with respect to the stellar background is not quite preserved, and the 51″ annual precession is thereby produced.

A new theory of astronomical refraction

The other two phenomena known at this time to be producing apparent changes in star positions were parallax and refraction, and some consideration had already been given to the *effects* of these in Lectures 13 and 14. Now, however, Flamsteed wishes to explain to his audience how a *theory* of refraction can be constructed. The main source on which he bases his discussion of this important astronomical problem in Lectures 33, 35–37 (38 being merely a Latin repetition of Lecture 33) is Kepler's *Ad Vitellionem paralipomena* (Francofurti, 1604). The first person to give a value for the horizontal parallax and the functional form of the decrease in refraction with increasing altitude was Tycho Brahe, whose method of finding the refraction in altitude from a knowledge of the latitude, hour angle, and declination is clearly described by Flamsteed in Lecture 33. Tycho's results for the horizontal refractions of the Sun, Moon, and fixed stars varied by several minutes of arc owing to the irregular fluctuations in the density and temperature of the Earth's atmosphere, thus his figures of 34′ for the Sun, 33′ for the Moon, and 30′ for the stars were uncertain average values. The progression of these estimates may be interpreted as an indication of the turbulence of the atmosphere at the times of observation rather than reflect, as Flamsteed here suggests, a compensatory allowance for a 3′ solar parallax. His own acceptance of these differences certainly stemmed from his recognition of the fact that the amount of refraction increases with the temperature of the atmosphere.[210]

The manner by which Tycho may have become conscious for the first time of the need to pay particular attention to the problem of refraction has already been referred to in the commentary on Lecture 13 (cf. pp. 49–50). The subsequent development of his thoughts on the subject can best be followed by reading the letters which he exchanged during the years 1585 and 1586 with William IV (Landgrave of Hesse-Cassel) and the latter's mathematician Christopher Rothman, contained in *Tychonis Brahe Dani, Epistolarum*

Astronomicarum libri . . . (Francofurti, 1610). The gist of that correspondence which Flamsteed probably borrowed from Kepler's summary[211] is presented in Lecture 35. Flamsteed recognized that what appeared to be a dispute between Tycho and Rothman was purely a semantic problem, Tycho distinguishing between an aetherial 'aether' and 'air' near the Earth, and Rothman between 'air' and a 'sphere of vapours' near the Earth. Neither suggested a proportion by which the refraction would decrease towards the zenith, and the refraction table which Tycho first published in his *Astronomia instaurata progymnasmata* (Pragae, 1610) was of a purely empirical nature.

Kepler, through his contact with Tycho in Prague during the year prior to the latter's death on 24 October 1601, was well aware of this problem long before that book was published; and, as Flamsteed remarks, 'his paralipomena ad Vitellionem seemes purposely wrote to cleare ye businesse of refractions'[212] by providing a theoretical basis for the phenomenon. Kepler's method for finding refractions in any given height, taken from that source, is explained in some detail in Lecture 35. Flamsteed provides a more rigorous geometrical proof of Kepler's result that the angles of incidence and refraction are not to be found at the foci of a hyperbola, and in Lecture 37 he does exactly the same for an ellipse. Kepler found that the curve representing the observations of refraction required to be 'more bent' than a hyperbola or 'less bent' than an ellipse, but Flamsteed does not fulfil his promise[213] that he would provide a proof of the fact that it is a circle.

By a method of trial and error, described in Lecture 37, Kepler finally managed to reconcile his belief that the resistance of the (atmospheric) medium to the angle of incidence (i) of a light ray entering it is proportional to sec i (proven in Lecture 35) with a choice of 6×10^{-5} Earth radii or $2\frac{1}{3}$ English miles for the height of the atmosphere. This was a crucial and rather sensitive parameter in his calculations, yet it seemed impossible to reconcile with the Ancients' determination of the height of the crepuscula—deduced as 50 miles from observations of the length of twilight—and also with the Torricellian experiments of Pierre Gassendi and others which pointed to an indeterminate value of over 5 miles.[214] Although Flamsteed makes no mention of the fact, Hooke had applied 'Mr Townly's Hypothesis'[215] (i.e. Boyle's Law) to the problem of estimating the height of the atmosphere and concluded that it 'may extend it self more than as high again, nay, perhaps indefinitely, or beyond the Moon'.[216] The same author further conjectured, primarily on philosophical grounds, that there may be 'a continual increase of rarity in the parts of the Air, the further they are removed from the surface of the Earth', thereby implying that a ray of light entering the Earth's atmosphere would be 'infinitely inflected, or bended, from a streight, or direct motion'.[217]

Flamsteed set himself the task, both in the remainder of Lecture 37 (11 June 1684) and in Lecture 36 (18 June 1684), of investigating this discrepancy. The length of twilight was largely a matter of definition, and Tycho, Rothman, Alhazen, and Vitello, Kepler, and Gassendi, all had different estimates for it. Riccioli even thought that evening twilight lasted longer than the interval between dawn and sunrise since, owing to the heating effect of the Sun's rays, the atmosphere expanded and its height increased during the day. Flamsteed adopts Kepler's definition that evening twilight ended when the Sun was 18° below the horizon, though stressing that the state of the atmosphere was as determinative a factor as the Sun's altitude. He demonstrates geometrically how to calculate the height of the atmosphere; his answer being over 49 miles or over twenty times larger than Kepler's. The reason why Alhazen and Vitello had obtained the much smaller value of just over 5 English miles was that they had wrongly supposed a ray of light entering the Earth's atmosphere to suffer a *single* refraction only before continuing in a *straight* line to the eye of the observer. Observations made during the month of June 1678 employing Hooke's 40-inch radius quadrant with telescopic sights[218] had convinced Flamsteed that, at very low altitudes of about 20′, morning refraction could exceed evening refraction—contrary to what Riccioli had conjectured. He interpreted this as evidence for vapours emitted from the Earth at night which were dispersed by the heat of the Sun's rays during the course of the day. The very existence of such vapours, fogs, mists, clouds, the Aurora Borealis, and the variability of the horizontal refractions from day to day, all constituted observational proof of the increase in density of the atmosphere closer to the Earth's surface; thereby implying multiple or continuous refraction, and that the ray would follow a *curved* path.

Thus when Newton later revealed to Flamsteed[219] that the 'Tabula Refractionum' communicated to him just over a month earlier[220] had been based upon a theory which presupposed a uniform density variation, the latter was quick in acknowledging it as 'the onely proper ground' for a table of refractions, which was 'altogeather rationall and agrees with ye nature of our Aire which is no equable medium'[221] as G. D. Cassini had wrongly assumed.[222] Meanwhile, Flamsteed's examination of other refraction tables had established that those constructed by Kepler,[223] Jean Picard,[224] and Charles Boucher 'an English gentleman now in Jamaica if living'[225] were also founded upon the false supposition of an 'equable' (or homogeneous) sphere of vapours about $2\frac{1}{2}$ miles high. Only those of Philippe de la Hire[226] did not conform to that principle. Vincent Wing's[227] and Thomas Streete's[228] tables were merely copies of Tycho Brahe's empirical table,[229] and likewise stopped at only 20° altitude instead of being extrapolated in the direction of the zenith.

Newton himself became dissatisfied with the theoretical basis of his table as soon as he became aware that it implied a refracting power at the top of the atmosphere which was as big as that at the base, and refused to concede to Flamsteed's view that the diurnal and annual changes in the amount of the horizontal refraction were due to vapours. The rarefaction and condensation of the lower region of the terrestrial atmosphere by the respective actions of heat and cold seemed to him to constitute a sufficient reason for this phenomenon.[230] During the early months of 1695, Newton constructed a new table of refraction using the observational data which Flamsteed had recently been supplying to him (including those obtained during June 1678, mentioned above). This time he postulated that the density of the air was proportional to the compression produced by the action of the inverse-square law of universal gravitation upon the particles from which it was composed,[231] an alternative theoretical basis to which Flamsteed was also willing to assent.[232] In communicating his second table to Flamsteed on 15 March 1694/5,[233] Newton claimed that the revised figures were correct to within 1″ for all altitudes above 10°, and to between 2″ and 3″ for those between 3° and 10°. As far as Flamsteed was concerned, the validity of Newton's approach to this problem, the practical importance of which he was well aware, was justified by the minuteness of the differences between the two sets of tabulated results.[234] He took pride in the fact that neither table would have existed without the observations that he had been able to provide, and based his future corrections to measurements of angular distances and celestial altitudes upon them. Regrettably, however, he did not follow Newton's suggestion[235] that thermometric and barometric readings ought to accompany every astronomical observation.

Flamsteed announced in Lecture 38 that it had been his intention to continue his Gresham Lectures with a discussion of refraction measurements, but had not in fact done so since he had been obliged to lecture in Latin instead. He further stated that he had delivered a Latin lecture at Gresham College a week ago and (subsequently) that he had dealt with two geometric problems in the last lecture of the past term. It may therefore be supposed that both statements refer to one and the same Latin lecture of 12 November 1684, which could be tentatively identified with the undated Lecture 39. Flamsteed goes on in Lecture 38 to say that he had touched again on one of the two geometrical problems in his last lecture delivered in English, and would mention the other in his next afternoon lecture (in English). Since it is unlikely that he would have troubled to rewrite his Latin lectures in English, it does not necessarily follow that English versions once existed but are no longer extant. On the other hand, Lecture 38 is itself an indication that the converse should not be taken for granted since—apart from containing the introductory remarks to which allusion has just been made—it merely constitutes a Latin translation of Lecture 33.

In order to end this survey of Flamsteed's syllabus on what happens to be an anti-climatic note, it only remains to summarize the content of Lecture 39. After praising numerous mathematicians and astronomers who had contributed to the development of trigonometry, and the invention of logarithms as an aid to calculation, he gives a simplified and improved proof of an analogue of the following problem: to find the remaining angles in a triangle which has two sides and a contained angle known, given those angles in terms of the sum and difference of the two known sides.

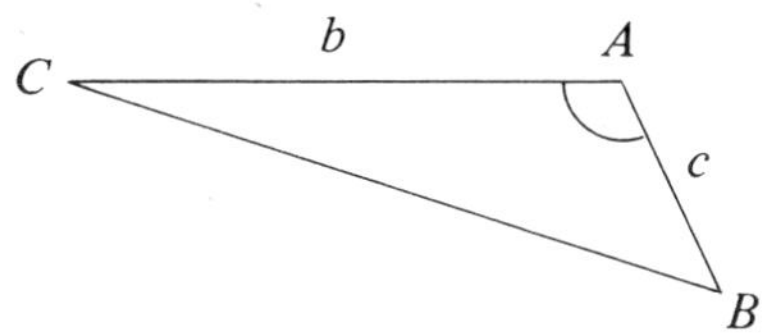

FIG. 18

Consider $\triangle ABC$ (Fig. 18), with $\hat{A}$ and the sides b and c known. Flamsteed had first to demonstrate the identities:

$$\tfrac{1}{2}(b+c)+\tfrac{1}{2}(b-c)\equiv b \quad \text{and} \quad \tfrac{1}{2}(b+c)-\tfrac{1}{2}(b-c)\equiv c.$$

He does this both for the case where b and c are straight lines, and where they are circular arcs. Then he proves with the aid of a geometrical construction:[236]

$$(1)\ \frac{c}{b}=\frac{\text{Radius}}{\tan 45^\circ}$$

$$(2)\ \frac{\text{Radius}}{\tan(90^\circ-45^\circ)}=\frac{\cot\hat{A}}{\tan\frac{1}{2}(\hat{B}-\hat{C})}.$$

By eliminating the radius from these two analogies, he obtained the trigonometrical formula

$$\tan\tfrac{1}{2}(\hat{B}-\hat{C})=\frac{c}{b}\tan\hat{A},$$

from which the difference $\frac{1}{2}(\hat{B}-\hat{C})$ is easily found by logarithmic calculation. Since the sum of the three angles in a plane triangle = 180°, the sum $\frac{1}{2}(\hat{B}+\hat{C})=\frac{1}{2}(180^\circ-\hat{A})$, which is known.

From these two trigonometrical relationships the respective angles $\hat{B}$, $\hat{C}$ were found by addition and subtraction respectively, as a special case of the above-mentioned identities.

NOTES

1. In John Ward's 'Extracts from the Repertory for Gresham affairs', B.M. Add. MS. 6196, f. 23 it is stated that Pope, though virtually blind, had arranged for his lectures in astronomy to be given by 'Mr. Flamstead a very able man in that art.'

2. At a joint committee meeting of the Court of Gresham College at Mercer's Hall, London, on 7 August 1681. Ibid., f. 22^{v}.

3. Op. cit., p. 113, where Ward remarks: 'In the year 1686 Dr. Pope was very ill at Salisbury of an inflammation in his eyes . . . During this illness Mr. Flamstead officiated for him, in reading the astronomical lectures at Gresham College; and upon the 21 of September 1687 he resigned that professorship.'

4. This letter is published in Francis Baily, *An Account of the Revd. John Flamsteed* . . . (London, 1835), pp. 118–23.

5. Ibid., p. 46.

6. Ibid., pp. 51–52, 54. This instrument is usually referred to as Flamsteed's *first* mural arc, to distinguish it from that constructed in 1688 by Abraham Sharp.

7. Ibid., pp. 44–66.

8. Preserved in the Southampton Civic Record Office, under Ref: D/M 1/1, ff. 1–145.

9. Preserved in the Royal Society Library, London, under RS MSS. LIX. c. 10. with the description 'MS Letters Flamsteed to Towneley 1672/3 to 1686/7' on the spine of the folder.

10. Flamsteed to Molyneux; 12 October 1681 (op. cit. [note 8], f. 5^{v}).

11. Molyneux to Flamsteed; 3 December 1681 (ibid., f. 6^{r}).

12. Sir Thomas Gresham's Will is published in John Ward, *The Lives of the Professors of Gresham College* (London, 1740), pp. 19–25.

13. These, together with the information on Gresham College in this and the following three paragraphs, are contained in John Ward's manuscipt 'History of Gresham College', B.M. Add. MS. 6204 (rebound in 1972). Cf. in particular, ff. 122–51.

14. Ibid., f. 145.

15. R.G.O. MSS., P.R.O. 38, f. 233^{r}; now published on p.339 of this volume.

16. Ibid., f. 93^{v}; or p. 189.

17. Ibid., f. 265^{r}.

18. Ibid., f. 131.

19. See note 10.

20. Op. cit. (note 1), f. 125^{v}.

21. Translated from op. cit. (note 15), f. 348^{r}; or p. 464.

22. Edward Sherburne, in the astronomical appendix to *The Sphere of M. Manilius made an English poem* ([London] 1675), p. 113 states that Pope had spent a great deal of time observing the motions and appearances of the heavens together with his predecessors Christopher Wren and Laurence Rooke at Gresham College 'the results of which he hath delivered in his astronomical lectures there read'. These were, however, never published.

23. Op. cit. (note 4), p. 22.

24. Op. cit. (note 15), f. 134^{v}; or p. 234.

25. Ibid., f. 125^{v}; or pp. 225–6.

26. This is one of several instances cited in J. J. Delambre, *Histoire de l'Astronomie Moderne*, ii (Paris, 1821). Cf. p. 276.

27. See note 23.

28. Cf. G. B. Riccioli, *Almagestum novum*, i (Bononiae, 1651), pp. 731–4.

29. Ibid., p. 733.

30. The first of this is found in the letter from Flamsteed to Oldenburg, dated 13 May 1671, published in A. Rupert Hall and Marie Boas Hall, *The Correspondence of Henry Oldenburg*, viii (Madison, Milwaukee, and London 1971), pp. 45–48. Cf. p. 47.

31. Kepler was concerned with the problem of finding the Mars-Sun distance from Tycho's observations of Mars's diurnal parallax and of the annual changes in the Sun's apparent angular diameter; whereas Flamsteed accepted the theories of these bodies' motions derived by Kepler from Tycho's data and sought the Earth–Sun distance.

32. This instrument, which came into Richard Towneley's hands after its inventor's untimely death, is described and illustrated by Robert Hooke in Henry Oldenburg (ed.), *Philosophical Transactions*, ii no. 29 (1667), 541–4.

33. Op. cit. (note 8), ff. 142, 143.

34. Cf. pp. 119–87.

35. Flamsteed's prognostication of this event was published in *Phil. Trans.* ii, no. 86 (1672), 5040–2; the observations and results being duly reported in no. 89 (1672), 5118–24.

36. These are identifiable with the stars numbered 26 and 27 in the earlier of the two papers cited in note 35 above, where the respective right ascensions and declinations derived from Tycho's observations are stated on p. 5042 as 11° 44′ 00″, 3° 58′ 0″ S, and 12° 12′ 0″, 4° 10′ 30″ S.

37. Flamsteed's use of the word 'strict' may imply that he worked with the formulae of spherical trigonometry; otherwise, if he wished merely to stress that he did not use a graphical method of estimating Mars's co-ordinates, the words 'by calculation' would have sufficed.

38. *Ioanni Heckeri ephemerides motuum coelestium ab 1666 ad 1680, ex observationibus correctis Tychonis Brahe et Io. Kepleri hypothesibus physicis, etc.* (Gedani, 1662).

39. 'Lunares Numeri ad novum lunae systema ab Astronomo Acutissimo Jeremia Horroccio excogitatum . . . a Johanne Flamstedio, Derbiensis', in Wallis, op. cit., p. 490.

40. Ibid., pp. 443–64.

41. E.g. Vincent Wing, *Astronomia Britannica* (London, 1669), or other contemporary ephemerides, contained tables for obtaining the longitude and altitude of the so-called nonagesimal point from the known circumstances of an astronomical observation. Flamsteed was later to develop a graphical construction for this purpose (cf. Lecture 19).

42. Op. cit. (note 15), f. 22^r; or p. 98.

43. The calculation, using the figure of 18″, actually yields a value of just over $16\frac{1}{2}''$; but since $X_1Y_1+X_2Y_2$ was in fact $17\frac{1}{2}''$ rather than 18″, the slightly lower result of 16″ for the parallax in longitude at the last observation still stands.

44. See note 42.

45. Op. cit., p. 16.

46. Op. cit. (note 15), f. 24^r; or p. 100.

47. An extract from this letter which omits Flamsteed's statement of his opinion of comets, their formation, motion, etc., is printed in op. cit. (note 4), pp. 123–4. The missing details are supplied in H. W. Turnbull (ed.), *The Correspondence of Isaac Newton*, ii (Cambridge, 1960), pp. 336–40.

48. This was conducted through the intermediary of James Crompton (1648–94), with whom Flamsteed had become acquainted during his student days at Jesus College, Cambridge. Four letters concerning the comet of 1680 from Flamsteed to Crompton, dated 15 December 1680, 3 January, 12 February, and 7 March 1680/1 found their way into Newton's hands, and provoked an exchange of ideas on this subject of which the main results are summarized below (cf. pp. 31–3).

49. My own calculations, using the values given in Flamsteed's text, yield: $av = 29° 24'$, $p\hat{v}a = 2° 04\frac{1}{3}'$, $p\hat{v}c = 68° 45\frac{2}{3}'$, lat. $c. = ck = 6° 59\frac{2}{3}'$ N.

50. Recalculation yields: $c\hat{p}v = 22° 25\frac{3}{5}'$, long. $c = 6° 02'$.

51. This bracketed value was not given by Flamsteed. It has been calculated from his data by applying the sine formula to the spherical triangle *pcz* with $cz = 87° 45\frac{1}{2}'$, $p\hat{c}z = 90° - 37° 44' = 52° 16'$ and $c\hat{p}z = 97° 31'$.

52. My own calculations, using the correct values derived from Flamsteed's data, yield $\odot c = 7°56\tfrac{1}{3}'$, $p\hat{c}s = 30°57'$, $oc = 38°33\tfrac{4}{5}'$ (or 39°, to the nearest degree), $p\hat{c}o = 29°\,44\tfrac{1}{5}'$; hence $s\hat{c}o = p\hat{c}o - p\hat{c}s = -1°\,13'$.

53. As in 52 (above), I found $ab = 34°\,51\tfrac{1}{2}'$, $e\hat{a}b = 59°\,22'$, $a\hat{n}c = 31°\,18\tfrac{3}{5}'$, $nc = 11°\,20\tfrac{2}{5}'$; hence long. $n =$ ♑ $5°\,48\tfrac{2}{5} - 11°\,20\tfrac{2}{5}' =$ ♐ $24°\,28'$.

54. Op. cit., (note 9), letter F.1.51.

55. Ibid., letter F.1.22.

56. Ibid.

57. See note 55. Halley had communicated this information to Flamsteed in a letter dated 22 January 1680/1, which would no longer appear to be extant.

58. A copy of this letter is preserved in MS. Smith 45, pp. 45–47 at the Bodleian Library, Oxford. See note 149.

59. See note 54.

60. Cf. op. cit., pp. 807–8.

61. See note 54.

62. Quoted from the 'account...' published in *Phil. Trans.* i, no. 6 (1665), 104–8. Cf. p. 105.

63. Op. cit. (note 9), letter F.1.50.

64. Flamsteed to Crompton, 15 December 1680, in op. cit. (note 47), pp. 315–17.

65. Op. cit. (note 4), p. 317.

66. Op. cit. (note 28), ii, pp. 8–10.

67. Op. cit. (note 47), pp. 319–20.

68. Ibid., pp. 340-7.

69. Ibid., pp. 336–40.

70. Ibid., pp. 341–2.

71. Op. cit. (note 4), p. 51.

72. Flamsteed first became personally acquainted with Newton in the year 1674 while he himself was still an undergraduate student at Jesus College, Cambridge. Cf. ibid., p. 29.

73. Op. cit. (note 47), pp. 348–56.

74. Cf. Preface, p. xiii.

75. Flamsteed to Halley, 17 February 1680/1, in op. cit. (note 47), pp. 337–8.

76. Newton to Crompton for Flamsteed, 28 February 1680/1, in ibid., p. 341.

77. Newton to Crompton, April 1681, in ibid., pp. 358–62. Cf. pp. 358–9.

78. Ibid., p. 361.

79. Ibid., p. 366. This statement occurs in an autograph draft dated 12 April 1681, preserved in the University Library, Cambridge; but it was *not* included in the letter sent to Flamsteed four days later (cf. note 80).

80. Newton to Flamsteed, 16 April 1681, in ibid., pp. 363–7.

81. Op. cit., pp. 480–90.

82. Ibid., Prop. XI, pp. 50–51.

83. This belief is expressed in the letter to Crompton. Cf. op. cit. (note 77), pp. 359–60.

84. 'An Essay of Dr. John Wallis, exhibiting his Hypothesis about the Flux and Reflux of the Sea', in op. cit., pp. 263–81.

85. Op. cit. (note 15), f. 61^v; or p. 149.

86. Ibid.

87. Ibid., f. 59^v; or p. 147.

88. *Phil. Trans.* vii, no. 89 (1672), 5119.

89. Op. cit., pp. 294 et seqq.

90. *Phil. Trans.* viii, no. 96 (1673), 6094–100.

91. Hevelius to Oldenburg, 7 April 1674 [N.S.], published in *Annus Climactericus* (Gedani, 1685), pp. 46–48 (cf. note 93).

92. These lectures had been delivered four years prior to this publication.

93. Op. cit. (note 91), pp. 67–74.
94. I.e. the results of observations published by Hevelius in ibid.
95. Translated from ibid, pp. 69–70.
96. Hevelius was under the impression that Flamsteed had measured the positions of the three stars numbered 26, 27, and 28 for which the co-ordinates are given in the prognostication (op. cit. 35, p. 5024). Perhaps he had, but it would seem that only the first two were necessary for establishing Mars's parallax (cf. note 36).
97. Op. cit. (note 9), letter F.1.18.
98. Ibid.
99. Flamsteed to Hevelius, 20 July 1676, in op. cit. (note 93), pp. 74–78.
100. Ibid., pp. 79–89. This letter is actually dated 2 January 1677 [N.S.].
101. This view was essentially correct although, not being supplied with the details of Flamsteed's calculations, Hevelius did not dare to challenge these (which our present study has shown to be erroneous), but only the unreliability of the tables that had been employed in the reduction.
102. See note 38.
103. Since the Sun's apparent sidereal motion is approximately 1° per day, an error of 7′ is equivalent to a time interval of about $\frac{7}{60}\times 24\text{h} = 2\text{h}\ 48\text{m}$.
104. Op. cit. (note 93), pp. 83–86. See note 100.
105. RS MSS, op. cit. (note 9), letter F.1.125.
106. *Johannis Hevelii Machinae Coelestis pars posterior*; etc. (Gedani, 1679).
107. Flamsteed to Towneley, 16 July 1679, in op. cit. (note 9), F. 1. 41.
108. Hevelius to Flamsteed, 23 August 1676, in R.G.O. MSS., P.R.O. 43, ff. $54^{r,v}$, $55^{r,v}$.
109. This letter is published in Johann Eric Olhoff, *Excerpta ex Literis . . . Dr. Johannem Hevelium . . . de Rebus Astronomicis . . . scriptis* (Gedani, 1683), pp. 182–4.
110. See note 106. Halley's catalogue was not, in fact, incorporated into this work.
111. Six letters were exchanged between these two astronomers during the year 1679. So far, only an extract of one of these, that from Flamsteed to Hevelius dated 9 January 1678/9, has been published in op. cit. (note 109), p. 182. In his reply of 14 April 1679, a copy of which is preserved among the R.G.O. MSS., P.R.O. 36, ff. $4^{r,v}$, 5^{r}, Hevelius states (transl.) 'I have not received any of the letters he has written to me; neither have I received his short work containing his observations of the fixed stars carried out on the island of St. Helena, which makes me particularly sad'. However, Halley was destined to arrive in Danzig just over a month later.
112. A brief account in English is to be found in E. F. MacPike, *Hevelius, Flamsteed, and Halley* (London, 1937), pp. 85–88.
113. E. F. MacPike, *Correspondence and Papers of Edmond Halley* (London, 1937), pp. 42–43.
114. Ibid., p. 43.
115. The Latin original is published in ibid., pp. 44–45, op. cit. (note 93), pp. 101–2, and op. cit. (note 109), pp. 187–8. An English summary of its contents is contained in op. cit. (note 112), pp. 87–88.
116. Translated from the Latin original of Halley's testimonial. See note 115.
117. See note 85.
118. See 'A Breviate of Monsieur Picarts Account of the Measure of the Earth', in *Phil. Trans.* x, no. 112 (1675), 261–72.
119. Flamsteed to Molyneux, 29 May 1682. See op. cit. (note 8), ff. $22^{r,v}$, $24^{r,v}$.
120. Ibid., f. 22^{r}.
121. Flamsteed to Molyneux, 11 April 1682, in ibid., ff. $14^{r,v}$, $15^{r,v}$, 19^{r}.
122. Ibid., f. 14^{v}.
123. *Phil. Trans.* vi, no. 80 (1672), 3075–87.

124. Op. cit. (note 8), ff. 47r–50v.
125. Flamsteed to Molyneux, 21 January 1682/3, in ibid., ff. 51r,v, 52r.
126. Le Père Cherubin, *La Dioptrique oculaire* ... (Paris, 1671).
127. Thomas Hobbes's 'Tractatus opticus' was first published in Père Marin Mersenne, *Cogitata Physico-Mathematica* (Parisiis, 1644). It is contained in William Molesworth's edition of Hobbes's *Opera Philosophica ... Omnia*, v (Londini, 1845), pp. 217–48. See F. Alessio's modern critique of it in *Riv. Storia Fil.* xviii (1963), 147–228.
128. Johannes Kepler, *Ad Vitellionem paralipomena* (Francofurti, 1604).
129. René Descartes, 'La Dioptrique', appended to *Discours de la Méthode* (Leyde, 1637).
130. See note 125, f. 51r.
131. At the end of Lecture 4, f. 45r (cf. p. 127), Flamsteed states his intention of showing how parallel rays are collected by a hyperbolic or elliptical lens; subsequently, in Lecture 5, f. 48v (cf. p. 135), he states that he had suggested and proved that (a parallel beam of) rays would be brought to a unique focus 'by segments of Hyperbolas or Ellipses whose transverse diameters to the distance of their Foci were in proportion as ye Sines of ye Angles of Incidence to ye Sines of ye Refracted angles'.
132. Op. cit. (note 8), f. 72r,v.
133. Ibid., f. 28r,v.
134. Op. cit. (note 129), p. 137.
135. Flamsteed's summary of these is preserved in R.G.O. MSS., P.R.O. 40, ff. 20v, 22r.
136. 'An Accompt of a New Catadioptrical Telescope invented by Mr. Newton ...', *Phil. Trans.* vii, no. 81 (1672), 4004–9.
137. Isaac Barrow, *Lectiones opticae et geometricae* (Londini, 1674).
138. Op. cit. (note 15), ff. 81v, 90r.
139. Op. cit. (note 9), letter F.1.53.
140. Auzout's original instrument was able 'to divide one foot into 24000 or 30000 parts', and therefore enabled him and Picard to measure the angular diameters of the Sun, Moon, and planets with an accuracy of 3″ or 4″. See *Phil. Trans.* i, no. 21 (1667), 373.
141. Seth Ward, *In Ismaelis Bullialdi Astronomiae Philolaicae Fundamenta, Inquisitio Brevis* (Oxford, 1654); and *Astronomia Geometrica* (London, 1656). Boulliau subsequently introduced an elegant refinement of Ward's treatment in *Ismaelis Bullialdi Astronomiae Philolaicae Fundamenta clarius explicata, & asserta. Adversa Clarissimi Viri Sethi Wardi Oxoniensis Professoris impugnationem* (Paris, 1657).
142. D. T. Whiteside, 'Newton's Early Thoughts on Planetary Motion: a Fresh Look', *British Journal for the History of Science*, ii (1964), 117–37; and 'Before the Principia: The Maturing of Newton's Thoughts on Dynamical Astronomy, 1664–1684', *Journal for the History of Astronomy*, i (1970), 5–19.
143. Eric G. Forbes, 'The Rational Basis of Kepler's Laws', *Journal of the British Astronomical Association*, lxxxii (1971), 33–37.
144. These are the famous Prop. I and Prop. XI of Book 3. A Pannekoek, *A History of Astronomy* (London, 1961), pp. 499–501, includes them as appendices under the respective titles: 'Newton's Demonstration of the Law of Areas' and 'Newton's Derivation of the Force of Attraction'.
145. Cf. p. 34.
146. This important correspondence has been discussed in Alexandre Koyré, 'An Unpublished Letter of Robert Hooke to Isaac Newton', *Isis*, xliii (1952), 312–37, and Johs. Lohne, 'Hooke versus Newton, An Analysis of the Documents in the Case of Free Fall and Planetary Motion', *Centaurus*, vii (1960), 6–52.
147. See note 100.

148. Op. cit. (note 15), f. 109[v]; or p. 205.

149. Eleven letters from Flamsteed to Bernard, and three from Bernard to Flamsteed, dating between 1678 and 1684, are preserved at the Bodleian Library, Oxford in MS. Smith 5, pp. 11–17; 12, p. 117; and 45, p. 31[a].

150. Ms. Rawlinson Q.b.3, f. 3.

151. William Dunn Macray, *Annals of the Bodleian Library Oxford* (2nd ed., Oxford, 1890), pp. 110–23; and *The Bodleian Library in the Seventeenth Century—guide to an exhibition held during the Festival of Britain 1951* (Oxford, 1951), pp. 18, 43–47 contain further details of Selden's manuscripts and indeed his library in general. The individual items comprising his collection are listed in Edward Bernard's *Catalogi Librorum* . . . (Oxoniae, 1697), pp. 157–67.

152. This procedure eliminates the latitude which is, however, easily calculable as the complement of the mean *sum* of the observed altitudes.

153. An authoritative modern book relevant to the same theme is Seyyed Hossein Nasr, *Science and Civilization in Islam* (Harvard University Press, Cambridge, Mass., 1968). Chapter 6, pp. 168–83 is concerned specifically with astronomy. Another classic work on this subject is Aldo Mieli, *La Science arabe* (Leiden, 1966).

154. Flamsteed acquired this book from Jonas Moore's library and took it with him when he left London for Cambridge at the end of May 1674.

155. Flamsteed's knowledge of Copernicus's procedure could have been obtained from his own copy of Rheticus's *Narratio prima* (Basileae, 1566).

156. J. L. E. Dreyer, *Tycho Brahe* (Edinburgh, 1890) is still regarded as one of the most readable and authoritative studies on this famous Renaissance astronomer. A very extensive and up-to-date bibliography is appended to the end of the article on Brahe by C. Doris Hellman in C. C. Gillispie (ed.), *Dictionary of Scientific Biography*, ii (New York, 1970), pp. 401–16.

157. Victor E. Thoren, 'New Light on Tycho's Instruments', *Journal for the History of Astronomy*, iv (1973), 25–45 contains a modern reappraisal of the constructional principles and accuracy of Tycho's sextants, mural and other quadrants, parallactic rulers, armillary spheres, etc.

158. These phenomena were first detected observationally in 1728 and 1748 respectively by one of Flamsteed's successors at Greenwich, the third Astronomer Royal, Dr. James Bradley, who duly publicized them in the *Phil. Trans.* xxxv, no. 406 (1729), 637–61, and ibid., xlv, no. 485 (1748), 1–43. See George Sarton's articles concerning these discoveries in *Isis*, xvi (1931), 233–65 and ibid. xvii (1932), 333–83.

159. Jean Picard, *Voyage d'Uranibourg, ou Observations astronomiques faites en Dannemarck* (Paris, 1680); reprinted among Picard's *Ouvrages de Mathématique* (La Haye, 1731). His result for the latitude of Tycho's observatory is stated on p. 94 of the latter work.

160. Op. cit. (note 15), f. 146[v]; or p. 246.

161. Op. cit. (note 135).

162. Ibid., ff. 18[r,v], 19[r].

163. Both this and the previous quotation are taken from ibid., f. 19[r].

164. Viz. 'The Doctrine of the Sphere, grounded on the Motion of the Earth, and the Antient Pythagorean or Copernican System of the World' (2 parts, London, 1680), 75 pp. published in Jonas Moore, *A New Systeme of The Mathematicks* (London, 1681).

165. Published in S. P. and S. E. Rigaud, *Correspondence of Scientific Men of the seventeenth century* . . . , ii (Oxford, 1841), pp. 139–44.

166. Ibid., pp. 139–40.

167. That is, on 18 May 1672. See op. cit. (note 135), f. 22[r].

168. *Jeremiae Horroccii Opera Posthuma* (Londini, 1673).

169. Ibid., pp. 469–71.

170. The reason why Flamsteed decided not to include this in his lectures was that 'a very ingenious friend' had dealt with that subject at Gresham College 'not long since'. This is probably a reference to Christopher Wren, who is credited by Delambre (op. cit. [note 174], p. 95), with having had the idea in 1660.

171. Johannes Kepler, *Tabulae Rudolphinae* (Ulmae, 1627) contains an explanation of an orthographic projection for calculating the Sun's and Moon's horizontal parallaxes and determining the path of the Moon's shadow during a solar eclipse.

172. Flamsteed to Towneley, op. cit. (note 9), letter F.1.23.

173. This statement (in ibid.) suggests that Halley had first raised the problem several months prior to his departure, since Flamsteed's Sun-spot observations had been made in June 1676 and published along with others in *Phil. Trans.* xi, no. 127 (1678), 662–4.

174. A critical analysis of this section of Flamsteed's 'Doctrine' is given by J. Delambre, *Histoire de l'Astronomie au Dix-Huitième Siècle* (Paris, 1827), pp. 97–100. However, he overestimates Flamsteed's indebtedness to Kepler (op. cit. 177) and gives him less credit for originality than he deserves.

175. Op. cit. (note 9), letter F.1.46.

176. 'Astronomical Tables for finding the true places of Luminaries, and their Eclipses, at all Times', in Moore, op. cit. (note 164), pp. 77–104.

177. Flamsteed himself informs us of this wish, in his Preface to Moore's book (op. cit. [note 164]).

178. A likely source from which Flamsteed may have derived a knowledge of the canons of spherical trigonometry, as well as the geometrical principles of stereographic projection, is John Collins, *The Sector on a Quadrant* . . . (London, 1658).

179. See note 164.

180. See the commentary on Lecture 2, pp. 10–20.

181. Op. cit. (note 15), f. 192[r].

182. Ibid., f. 233[r].

183. Ibid., f. 233[v].

184. Ibid., f. 185[v].

185. These 'Ordinances and Agreements' were revised and approved by that committee in 1598 before being set down on 16 January 1598 in the Register or Repertory of Gresham affairs. All twenty are listed in op. cit. (note 13), ff. 140–51. See ibid., f. 145 (viii).

186. The logic of this decision seems rather dubious, since accurate positional astronomy alone suffices for the purpose of improving planetary theory.

187. See the entry for Tuesday 22 October 1672 in Flamsteed's *Historiae Coelestis Britannicae*, i (Londini, 1725), p. 17.

188. Flamsteed later attributed annual variations of up to 40″ or more in the zenith distances of the pole star observed with his mural arc between 1689 and 1697 to the effect of parallax. See his letter to Wallis, 20 December 1698, in the latter's *Opera Mathematica*, iii (Oxoniae, 1699), pp. 701–8. His view was soon challenged by Jacques Cassini, 'Réflexions sur une lettre de M. Flamsteed, a M. Wallis touchant la parallaxe annuelle de l'étoile polaire', in Ulricus Junius, *Novae et accuratae motuum coelestium Ephemerides . . . 1701* [*–1703*] (Lipsiae, [1700]) who pointed out that the observed variation was out of phase with the expected one. Flamsteed had actually detected the still-unsuspected phenomenon of stellar aberration, which James Bradley was to have the honour of rediscovering and correctly interpreting in 1728. See note 158.

189. *Phil. Trans.* i, no. 21 (1667), 381–3.

190. Ibid. 382.

191. Ibid. 383.

192. Ibid. vi, no. 73 (1671), 2198–2202.

193. The original Latin letter, and a full translation, have now been published in op. cit. (note 30), pp. 3–8.

194. 'Observation d'une Étoile nouvellement découverte proche la Constellation du Cygne', *Journal des Sçavans*, 22 juin 1671 (Amsterdam, 1677), 631–8.
195. *Phil. Trans.* vi, no. 73 (1671), 2201.
196. The appearances and significance of this supernova are discussed at great length in *Tychonis Brahe Dani, Astronomiae instauratae progymnasmata* (Francofurti, 1610).
197. This tract was published in *Johannis Hevelii Mercurius in Sole visus Gedani... 1661... 3 Maji St. n...; cui annexa est, Venus in Sole pariter visa, Anno 1639, d. 24 Nov. St. V. Liverpoliae, a Jeremia Horroxio: etc.* (Gedani, 1662), pp. 146–71.
198. In the first manuscript inventory of his private library drawn up on 4 January 1676/7, Flamsteed lists a quarto Amsterdam 1656 Latin edition of this famous work, and prices it at 5 shillings. In a later inventory dated 6 March 1683/4, he quotes the date as 1658. See Eric G. Forbes, 'The Library of the Rev. John Flamsteed, F.R.S., First Astronomer Royal', *Notes and Records of the Royal Society of London*, xxviii (1973), 125 and 140 (note 31).
199. A fuller explanation of this viewpoint is to be found in a German letter from Tobias Mayer to Leonhard Euler, dated 22 August 1753, translated into English and published in my edition of *The Euler-Mayer Correspondence, 1751–1755* (Macmillan, 1971), pp. 73–76.
200. *Alfontij regis castelle illustrissimi celestiū motuū tabule: etc.* ([Venetii,] 1483).
201. By adopting this value, Flamsteed had found that he was able to improve the predictive ability of Horrox's theory of Jupiter to ±2′ or 3′. This fact is clearly stated in his letter of April 1674 to Sir Jonas Moore entitled 'A Preface to my Celestial Observations...'. See op. cit. (note 4), p. 109.
202. Thomas Streete, *Astronomia Carolina. A New Theorie of the Coelestial Motions* (London, 1661), p. 22.
203. The modern accepted value of the annual Precession is 50.″2564 + 0.″0222T, where T is measured in Julian centuries from the epoch 1900.0.
204. This raises doubts as to whether both were, in fact, delivered.
205. I. N. Veselovsky has recently pointed out that Copernicus's belief in this fictitious phenomenon may have stemmed from his reading of the Greek text of Proclus's *Commentary on the First Book of Euclid*, and not from the Arabic work by the Iranian mathematician Nasir al-Dīn al-Tūsī. See *Journal for the History of Astronomy*, iv (1973), 128–30.
206. Op. cit. (note 164), p. 3.
207. These concepts are implicit in his explanation of the Moon's libration, communicated to Collins on 13 May 1672 and quoted earlier in this Introduction (cf. pp. 52–3). See note 171.
208. An English translation of Roemer's own interpretation of this phenomenon was read by Flamsteed in *Phil. Trans.* xii, no. 136 (1677), 893–4.
209. Note that this explanation is analogous to that invoked by Horrox to account for the very much shorter periodic motion of the Moon's mean apogee. Cf. my earlier comments on pp. 52–4.
210. Flamsteed to Towneley, 23 November 1673, op. cit. (note 9), letter F.1.2.
211. Op. cit. (note 128), pp. 77–79.
212. Op. cit. (note 15), f. 320ʳ; or p. 428.
213. In ibid., f. 336ᵛ (or p. 444), Flamsteed states: 'the locus... should be a Circle that it is such I shall prove hereafter...'.
214. Gassendi's Torricellian experiments are described in a letter to Bernier dated Diniae, 6 August 1652, in *Opera Omnia*, vi (Lyon, 1658), pp. 317–19. On this theme, see Bernard Rochot, *Revue d'Histoire des Sciences*, xvi (1963), 53–76 and Pierre Humbert, *Sciences*, lxxiv (1947), 336–49.
215. Robert Hooke, *Micrographia* ... (London, 1665), p. 225.

216. Ibid., p. 228.

217. Ibid.

218. This quadrant, designed by Hooke, was the property of the Royal Society: it had been procured for him by Sir Jonas Moore early in 1677, since which time he had divided its limb and mounted it so that it could be rotated. It was one of the instruments recalled on 26 September 1679 shortly after Moore's death. See op. cit. (note 4), p. 45.

219. Newton to Flamsteed, 20 December 1694, in J. F. Scott (ed.), *The Correspondence of Isaac Newton*, iv (Cambridge, 1967), 61–63.

220. In a letter dated 17 November 1694, ibid., pp. 46–50.

221. Flamsteed to Newton, 31 December 1694, ibid., p. 63.

222. G. D. Cassini's table of refraction first appeared in Cornelius Malvasia, *Ephemerides novissimae motuum coelestium . . . ad longitudinem urbis Mutinae etc.* (Mutinae, 1662).

223. Op. cit. (note 128).

224. Presumably the table in op. cit. (note 159), pp. 111–12.

225. Flamsteed to Newton, 11 October 1694, op. cit. (note 219), p. 28. Boucher's observations of atmospheric refraction were possibly communicated to Flamsteed in response to the latter's request in unpublished letter of 24 January 1678, preserved among the R.G.O. MSS., P.R.O. 43, ff. 36^{r}–38^{r}.

226. Philippe de La Hire, *Tabularum astronomicarum pars prior* (Parisiis, 1687).

227. Op. cit. (note 41). Parallax and refraction are briefly discussed in 'Doctrina Spherica' (Book 3) and there is a postscript on refraction (in Book 5) on pp. 245–6.

228. Op. cit. (note 202).

229. A table of refractions for the Sun, Moon, and fixed stars, for altitudes from 0° to 45°, is contained in op. cit. (note 128), pp. 121–2. Another refraction table appears in ibid., p. 125.

230. Newton to Flamsteed, 15 January 1694/5, op. cit. (note 219), pp. 67–70.

231. Newton to Flamsteed, 16 February 1694/5, ibid., pp. 86–89. (Cf. p. 89).

232. Flamsteed to Newton, 2 March 1694/5, ibid., pp. 89–93. Flamsteed agrees that this proposition—Prop. XXII in Book 2 of Newton's *Principia* (Londini, 1687)—seemed 'a proper foundation for a Table of refractions' (cf. pp. 89-90).

233. Newton to Flamsteed, 15 March 1694/5, ibid., pp. 93–97; the revised table itself being on p. 95.

234. Flamsteed to Newton, 21 March 1694/5, ibid., pp. 98–100.

235. In a letter of 24 October 1694, ibid., p. 34.
See also op. cit. (note 4), p. 137.

236. Robert Anderson's independent demonstration of the following analogies was published for the first time in the third edition of Thomas Streete, *Astronomia Carolina* (London, 1716), pp. 58–59.

Lecture 1

Wednesday, 27 *April* 1681

The great distempers under which the learned Reader of Astronomy[1] now labors, have caused him to desueme to discharge his duty for him at this time. and because there is nothing more admirable tho perhaps lesse understood then those methods whereby the past and present ages have attained the knowledge of the motions distances and magnitudes of the Celestiall bodyes, I shall endeavor to explaine them and render them so easy to understand that the common people who measure all things by their own length & paces may no longer wonder at those vast proportions Astronomers have assigned them.

Hereby I shall open the way to a discourse concerning the admirable appearances of the late Comet of whose motions the Theory to solve them I intend God permitting to discourse in my following readings.[2]

The ingenuous & subtle Kepler has showed us in y[e] first Chapter of his Com. de Mot ★[e] Martis[3] how by the contemplation of the Diurnall revolutions of the heavens the first Astronomers might find, that the Moone was swifter than any of the other planets, that there were seven in number & that all the other stars were fixed; but how they came to y[e] knowledge of the periods of her revolutions, & how true distances from the earth, & how the proportions assigned by them amended by surrounding times & ages hee is silent; of those therefore I shall discourse in order, breifly first of the revolutions but of the distances more at large.

The antients accounted theire moneths by the Course of the moone whose meane returnes to y[e] sun being made in 29 dayes & something more y[n] an halfe[4] they usually allowed 29 dayes to one & 30 to the next & twelve of these moneths to y[e] yeare.

But finding that after three yeares theire spring was not so forward as when they began to account, by a moone, hence they conclude, that theyre was one revolution more made in three yeares, & therefore intercaled or added a moneth where they thought most convenient for theire affaires, this was performed amongst y[e] jewes by the sanhedrim, amongst the Romans by the Pontifex maximus, &

Notes on pages 89–90.

by ye seke Authority amongst the Greekes who called it their τριετηριζ.[5]

This put them upon enquiry into ye time of the same revolutions, who began it, appeares not by any monuments of Antiquity; for all the great pretences of the Egyptians & Chaldeans I am apt to believe the jews observed it before them. For the sacred story mentions the diall of Ahaz[6] the begining of whose Reigne Helvicus[7] places but two yeares nearer us than the commencement of Nabonassars era:

Now it cannot seene probable to me that the Jewes should have so much knowledge of the sphere as to make a diall which by the shadow should show ye hour of ye day without understanding both the course of ye ☉ & ye time of his returnes to ye same place in the heavens.

And this they were in a manner obliged to, for theire first fruites & other offerengs were to be paid on set moneths of the yeare particularly the feast of ye Ingathering was to be kept in ye Moneth Elul in ye latter end of ye yeare which contained part of our August & September. Now if they tooke not due care to intercale angles this moneth might happen before the fruites were ripe they considered therefore that ye shadow of anything perpendicular to the horizon were or ought to be of the same length at the same time of ye yeare. & therefore they might easily determine the due lenght of ye solar yeare by counting the Dayes interlapsed betwixt the two harvests, when the shadow of an upright gnomon came to be of the same length. and attending it carefully they found that it was after 12 of theire lunar moneths & 11 dayes nearely so that whereas the Lunar yeare consisted onely of 354 dayes the solar conteined 365.

The Jewes I meane the ten tribes were first carried Captive in ye 25th yeare of Nabonassar;[8] & from them we may presume theire Masters the Assirians, & the Chaldeans theire Neighbours, first had the knowledge of the solar & lunar revolutions: & it seemes some Confirmation of this opinion that the first lunar Eclipse wee find observed is two yeares after in ye first yeare of Mardokempadi or Merodack al Adan & ye 27th of Nabonassar. the founder of the Babilonian monarchy & Epocha whose yeares consist of 365 dayes precise & which were of use chiefly amongst the learned men of ye East. this epocha commenced in ye 6th yeare ab urbe Roma condita & 745 ante Christum.

About 130 yeares after this Thales Milesius first and after him Pythagoras travelling into the East learnt of the Chaldeans the true lenght of ye Solar yeares & lunar revolutions, Thales is said to have beene the first amongst ye Greekes that predicted an Eclipse of the Moone[9] & Pythagoras that taught the Motion of ye Earth. & the proportions observed in ye construction & constitution of the heavenly bodyes both which now may most justly supposed them to have got by theire acquaintance with the Chaldeans. these Greekes for any thing appeares in story before wholly strangers to the mathematicall sciences.

But it being noted that the length of the Midday shade might be the same twice every yeare, except at either solstice, Meton & Euctemon chose to observe the solstice by the shadow & thence to determine the length of the yeare rather then otherwayes. Ptolemy[10] tells us that they determined a solstice on ye 21th day of ye Moneth Phamenoth in ye morning when Absedus [i.e. Apseudes] was Prefect of Athens which Scaliger[11] & Petavius[12] make the 27th of June in ye 316 of Nabonassar or 432 yeares before the Incarnation. After which 152 yeares Aristarchus observed another solstice from which observations of theires, however difficult, it is likely the periods of the suns revolutions more stated by the Greekes.

Meton was hee who not onely ordered there calendar placeing the begining of the yeare on ye summer solst. as Festus Avienus Notes

> Sed primaeva Meton exordia sumpsit ab anno
> torreret rutilo cum Phoebus sidere Cancrum[13]

But hee contrived their Kalendar & Numerus Aureus whereby they found how to order the Phases of the New & full moone in it all the yeares of his Period or *ευυεαδεχατύρις*:[14] tho this was not so absolute but that Eudoxus found cause to alter it about 70 Yeares after whose Fasti were so well esteemed of the Greekes that the Poet makes Caesar when hed corrected the Calendar to rejoyce or applalude himselfe in this distick

> Nec meus Eudoxi vincetur Fastibus Annus[15]

The discovery of Metons Cycle[16] as it was easy so I conceave it to have occasioned the first correction of the lunar Theory. or rather the first formeing of it For the Phases of the Moone might often be observed to happen when ye shadow of the Gnomon was of the same proportion to the height it had in the same season 19 yeares before. there are no examples of this kind transmitted to us by ye ancients but wee need not run far back to find Eclipses of the Moone that have hapned at 19 yeares intervall on ye same day of the moneth: The Noble Tycho [Brahe] observed an eclipse of ye Moone at Uraniburgo october ye 18 1594. Longomontanus Octob: ye 18 1613 after 19 yeares at Hafnia [Copenhagen]: and in ye yeare 1632 another Eclipse was observed 19 yeares nearely after this:[17] againe Monsr. Hevelius noted an Eclipse of ye Moone October ye 19: 1659 & 19 yeares after a like eclipse was observed at Greenwich on ye 19th of ye same Moneth:[18] so that it is plaine Meton might easily state his period & from very course observations such as those times were capable of makeing.

In the meane time it was found that the solstices were the most difficult to be observed & therefore such ingenuous men as succeeded Aristarchus neare 150 yeares chose rather to determine the true length of the yeare from observations made by the Gnomon about the time of ye Equinoxes when the shade altered its length the fastest. whereby they Came to find that the Motion which they attributed either to the sun or more justly to the earth, was anomalous; that ye true lenght of ye yeare was not as they supposed first 365 dayes praecise, but more by almost 6 houres or more praecisely as Hipparchus has stated it. the dayes interlapsed betwixt the suns entrance into ♈ & his passage through y^e 6 northerne signes till hee cut the aequinoctiall againe 187 but thence till his returne to y^e vernall aequinox $\overset{d}{178} \cdot \overset{h}{05} \cdot \overset{\prime}{55}\tfrac{1}{2}$, & therefore the true length of the yeare $\overset{d}{365}\ \overset{h}{05}\ \overset{\prime}{55}\tfrac{1}{2}$.

Hence they collect that the suns motion was in one part of his orbite accelerated 2 degr. 23′. & retarded as much in the opposite pointe. & that the Middle wherein the inaequality ceased was $24\tfrac{1}{2}$ degr. short of ye summer solstice or in $5\tfrac{1}{2}$ of ♊ immoveable which they called y^e place of y^e $\odot^r$ Apog.

And now againe examineing theire Catalogue of lunar Eclipses they find that these had not alwayes hapned at 19 yeares praecise intervall. but sometimes a day earlyer or later, howe they raise arguments for the unequall motion of the Moone as well as the sun, & upon frequent experiments & trialls concluded that the greatest inequality in ye Moones motion was about 5 degrees in eclipses.

But Considering further that the Eclipses which hapned any yeare were repeated the next some ten dayes earlyer that if one of them was totall the other was partiall and alwayes made in antecedence or to the West in the heavens from those stars where they hapned the yeare before, & that the Moone was soone onely twice in ye Moneth to cut the Ecliptick but deviated at other times neare 5 degrees on each hand it. they hence collected that the pointes or places of intersection receded annually something more y^n 19 degrees. & Contemplateing the places of ye Moone not eclipsed that the inaequality of her motion was then sometimes more then ye degr. & the meane pointe where it ceased moved almost 40 degrees in a yeare & made its revolution in about 9 yeares.

Nor was it long ere this knowledge was Conveyed by ye Greekes to ye Romans. for wee find when Lucius Aemilius Paulus & M Primus Crassus were Consulls 585 yeares after the building of Rome & 166 before ye incarnation Sulpitius Gallus a Roman Tribune praedicted an Eclipse of ye Moone to Aemilius Generall then wageing war against Perseus ye Last of ye Macedonian Kings. & informeing the Army of the reasons of such eclipses praevented that panick feare which when it hapned it imprest in theire enemies thereby contributeing no little to the raine of ye Grecian Monarchy.

after which this arte become knowne to the Romans tho little esteemed by reason it contributed not much to the enlargem^{t} of theire Empire or encrease of theire military Glory the maine designe of that state. however some ingenuous persons amongst them embraced the Mathematicks & Caesar by the helpe of Sosigines Agrippa with what himselfe had learnt of the Aegiptians undertooke the Correction of the Kalendar brought it neare that account wee now use[19] & by the authority of his preisthood caused it to be enterteined in ye Conquered State.

I might here shew how Calippus before this has observed the faults of Metons & Eudoxus Cycles in which there were 19 of Nabonassars yeares with 4 dayes & 3/4 accordeing to his better working. to avoid the inconvenience of fractions he quadrupled the period that so in 76 yeares of Nebonassar & 19 dayes over hee might have precise 76 revolutions of the earthe, but this would be lookt upon as rather done for ostentation then otherwayes I shall therefore forbeare & haveing now showne how the periods of ye Revolutions of ye luminarys & inaequalitys of theire motions were first discovered I shall proceed to show how by a like easy methods the antients attained the knowledge of their distances from our Earth and parallaxes.

In ye serene aire & more upright spheres of Assiria Mesopotamia and Aegypt where the first inventors or improvers of Astronomy inhabited, they might frequently observe ye moone riseing neare bright fixed stars such as ye Bulls eye, ye Lions heart, Spica ♍^{s}, or Antares, when shee was neare her full, & visible above ye earth neare 12 houres sometimes more. considering therefore how motion towards the star or from it betwixt ye time of her rise & setting, they found it to appeare considerably lesse then what it ought to be compared with her remove from it when shee arose next Night, Considering further that her motion was swifter then any of ye other planets they conceaved shee might be nearest the earth of any this caused them to thinke of Contrivances for measuring the arches she dayly moved. Now the Armillaes & other instruments[20] were invented with which observeing the distance of the moone in antecedence from some of the aforementioned fixed stars they found that if it were about 4 degrees or betwixt 4 and 5 in her riseing she scarce ever passed beyond the star but onely reached or covered it by that time shee came to set. but attending her riseing next night they might find that her distances from ye said star that her diurnall motion was treble to that they had observed in 12 houres of ye night before whereas respecting the proportion betwixt the times of theire observations it ought to have beene onely double, or some little more. this difference therefore they attribute to some diversity of Aspect & finding it alwayes to happen, so often as they thus considered her, that the arches of her Motion about the earth were larger then the spaces she ran over under it they begin to argue further.

That since as they had learned by compareing the times of lunar Eclipses, observed different in distant places, & allwayes earlyer to the Westerne inhabitant then to ye Easterne, that ye earth had not a plane but sphericall superficies. this might proceed from the position of the observer on its superficies & that the proportionall motions were onely observable at the center for say they. let c as in ye first figure [Fig. 1.1] represent ye center of ye earth nbo its sphericall superficies, lem ye orbite of ye Moone nearer to it then afhAkgd the sphere of the fixed stars which they truely conceaved at a vast distance. the line anocd drawne through ye center of the Earth the true horizon. b perpendicular over ye center to it the place of an inhabitant or observer & flbmg the sensible horizon. with him let l be ye moones place in her orbite and m̄ the sensible horizon wherein to y^e^ observer at b she appeares under & in consequence of y^e^ fixed star h, with which viewd from c the center of y^e^ earth shee is in a straight line and really conjoyned. and above the true horizon the quantity of y^e^ angle hca = blc which they therefore termed the horizontall parallax or difference of aspect.

Hence they collect that in each place of the orientall quadrant of the Ecliptick the Moones visible longitude was forwarder than her true. & her distance from the vertex really lesse then it appeared to y^e^ observer except when it passed exactly over his head.

And considering further & in like manner the occidentall part of ye figure they find that contrary to what they had noted in y^e^ orientall, if y^e^ setting moone at m were really in conjunction with the star k or that y^e^ line cm produced from y^e^ Center of the earth by y^e^ Moones fell exactly on y^e^ star k yet to y^e^ observer at b shee would appeare beneath & in antecedence of it y^e^ quantity of y^e^ angle bmc = to y^e^ arch kg which is also insensibly different from kd by reason of y^e^ great distance of the fixed stars & smallnesse of y^e^ earths diameter bc compared with that of theire orbe cd.

And hence they collect that the horizontall parallax or y^e^ angle blc = bmc was nearely one degree for if y^e^ true semidiurnall motion of the moone were 6 degrees & y^e^ apparent as they observed it but 4 it must follow that the moones place was put forward in her rising one degree & as much backward in her setting which was the cause that y^e^ visible motion above y^e^ earth in 12 houres was observed nearely two degrees lesse then ye halfe of twelve degrees her apparent motion in 24 houres.

And now haveing found the horizontall parallax or ye angle blc about one degree, it followed that bc the Moones distance from ye earths center was about 60 times so much or so many of her semidiameters.[21]

Haveing thus determined the Moones distance from ye Earth the next they enquired and found out was ye suns but this wed may justly suppose not till some time after. For the suns greater light observeing that of ye planets and stars it was not possible for them to observe whether his motions above the Earth were lesse then these arches hed ran over under it, as they were in ye Moone, or not: they could therefore onely guesse at this, & that hee was more remote than ye moone, because she sometimes covered and eclipsed him at the time of ye New Moones: the first that wee find to have affirmed any thing concerneing his distance was Pythagoras who is said to have asserted Distantiam Solis esse vel Triplam vel ad summum sestuplam ad lunarem distantiam.[22] that the suns distance was but thrice or at most 6 times as much as ye Moones which being stated 60 of ye Earths semidiameters, it follows, that ye suns should be onely 180 or at most but 360 of ye like measures.

The ἀυτο[ς] ἔφη[23] of Pythagoras was sufficient argument for this limitation with his scholars, and the more ingenuous vulgar who knows no method whereby to examine it seeme to have acquiesced herein for about 2 ages and an halfe afterwards when Aristarchus ye Samian, considering the increase of the Moones light according to her removes from ye sun after her first emersion from him, to ye full, & then its decrease and waine till shee was covered under his raies, concluded his distance must be much farther. for argues hee let T be ye center of ye Earth BQA ye orbite of ye Moone whose Radius or semidiameter QT is 60 semidiameters of the Earth. & let TS be the suns distance from the Earth proposed by Pythagoras, [cf. Fig. 1.2] From S draw ye lines Sd which may touch the orbits ST joyneing the centers of ye sun & earth. Td perpendicular to ye touch line, & qT to the line connecting theire centers. its is evident that the Moones body being a Globe the sun can illuminate but one halfe of it at once, that at the Conjunction in B wee see it not, because the darke or unilluminated side is towards the earth, that shee shall appeare halfe illuminated when shee comes to ye point of Contingence in d. 9½ degrees short of the true quadrature; for as ST. 6 part to DT 1: so is ye sine of 90 degrees to the sine of ye angle DST 9° 35′ which is equall to ye angle QTD. the distance of the moone from ye true Quadrature for each are the complements of ye Angle DTS to 90 degrees.

Hence it will follow that the moone shall appeare halfe illuminated or bisected in D 18 houres or 3/4 of a day before shee is a full Quadrant Removed from ye sun whereas sayes hee by observation I find that then shee wants neare a whole digit of being halfe illuminated & that the light part of her diameter to ye darke is but as 5 to 7 at that distance. Aristarchus haveing found that the distance of the sun must thus be limited sets himselfe to observe the times of ye Dichotomy or halfe illumination but findes it difficult to determine the praecise time of it onely hee is sure that it was not till ye moone came within 3 degrees of the true Quadrature. Hee therefore states ye quantity of the angle DST 3 degrees. & thence collects Distantia quae sol a terra distat, major quidem est distantia lunari quam duo de vigintupla minor vero quam vigentupla as Riccioli translates him whose comment addes atque adeo solis est ad lunae distantiam undevigintupla.[24] So yt if ye Moones meane distance be proposed 60 of ye earths semidiameters the suns shall be 1140. & hence his horizontall parallax or the angle subtended by the earths semidiameter at yt Distance above 3 minutes.

But this method tho of good use at present was then more difficult on two accounts first for that the lunar numbers were not so correct at that time nor the instruments so exact as that it was possible for them to find the true time of ye Quadrature or when the Moone was onely 90 degrees removed from ye sun within halfe an houre next. by reason that they wanted the use of Optick glasses, whereby they might more exactly determine the time of the apparent Dichotomy or halfe illumination since the invention of which tis found that the difference betwixt the apparent Dichotomy & Quadrature is scarce more then halfe an hour and yet at the time when the Moone appeares halfe illuminated in ye Telescope shee seemes wanting and hollow to ye bare eye. even after the time of the true Quadrature by reason of the Naturall weakenesse of our unassisted sight & that fewer raies of light are reflected from the middle of the Moone then from her illustrated limbe. which observeing the times of ye Quadratures I have frequently noted my selfe & can give yu ye experience of others for it particularly Riccioli who tells us in the first part of his Almagest pag 732 Et si nudo oculo utaris, videbitur tibi ex hac ratione non raro cava etiam post primam et ante secundam Quadraturam, quod et Keplerus sibi accidisse fatetur[25] whence it is evident that if Aristarchus could have determined the time of ye true Quadrature exactly hee might have stated the suns distance much larger yet & his parallax lesse but I am apt to thinke that hee esteemed himselfe to have removed him sufficiently by this determination since it made the suns parallax almost insensible at least in the Instruments used by ye ancient in which 3 minutes was almost an imperceptible quantity & little esteemed whilest they could not determine the true longitudes either of the stars or planets much nearer then to a simple degree. especially before Hipparchus who was ye first of ye ancient Astronomers that attempted to gave us a Catalogue of the fixed stars.

This great lover of truth & labor as Ptolemey calls him further proved by the Phainomana of Eclipses that the distance to which Aristarchus had removed the Earth from ye sun was warrantable. & his demonstration has beene so well receaved by all that succeeded him that his diagramme has borne his name[26] & helpt to render his Memory gratefull even in our age. the some have either in favor of theire owne conceites or otherwayes ignorantly mistaken & misapplyed it. hee affirmes then that the semidiameter of the Earths shadow in ye place of the moones transit through it is equall to the summe of the horizontall parallaxes of ye sun & Moone lesse ye suns semidiameter which hee thus demonstrates Let S in ye 3^{d} figure [Fig. 1.3] represent the suns center, T ye earths, & TS ye line connecting them, gh an arch of the Moones orbite. SA ye suns semidiameter To ye Earths by ye Extremitys of ye earths and suns semidiameters let ye line aoe be drawne intersecting the line of Connexion TS in e. as also op a tangent to the y^{e} earths superficies parallel to TS & from y^{e} point of connexion o. the line oS then shall the Angle aoS be the suns apparent semidiameter at our earth from which takeing pos his horizontall parallax the remainder aop = aes shall be halfe the angle of y^{e} cone of y^{e} earths shadow.

And if y^{e} line op be continued backward till it intersect y^{e} moones orbite in f & og drawne to y^{e} place where it intersects y^{e} axis of y^{e} said cone or line connecting the centers of y^{e} sun & earth. then shall y^{e} Angle fog be y^{e} moones horizontall parallax from which if the halfe angle of y^{e} cone of y^{e} shadow oeT = foh be subtracted, the remain[d]er hog shall be y^{e} semidiameter of y^{e} shadow in y^{t} place where the Moone passes through it.

Which Hipparchus found if with Pythagoras he supposed the sun to be but about 6 times the distance of the Moone removed from y^{e} Earth would be too large by about 7 minutes for ye horizontall parallax of the sun would be 10 or ye 6th pte of a degree & consequently the times of ye duration of ye lunar Eclipses especially totall would be bound by calculation about halfe an houre longer then they really were observed & the partes eclipsed one third of the Moones diameter above what they had beene noted in partiall Eclipses. this caused them therefore to embrace Aristarchus his proportion nearely, as most agreeable to the limitations given by his probleme.

The Diameters were easily found & measured by ye antients more accurately then any of the formentioned requisites for when ye sun shone throug a narrow hold or cleft in a door or wall into an obscure Roome they saw his species projected on ye floore as an Ellipse whose transverse diameter exceeded ye other more or lesse according as his raies fell lesse or more obliquely on ye flore but if they were receaved on a tablet placed at right angles to ye species it appeared a perfect Circle & measureing the distance betwix the foramen or hole & the tablet that it was in proportion to ye semidiameter of ye species as ye Rad. to ye tangent of ye suns semidiameter.

I have now showed y^{u} how ye necessity of a good account of ye yeare produced both ye knowledge of ye periods of the Revolutions of the sun & Moone. how that was attended by ye inquiry into theire distances. how they were first Coursely got without any great apparatus of Instruments. how corrected by plaine & simple observations. I have been breife the subject deserves to be handled more at larger but may be I shall resume it at better leasure at present I end with Hipparchus. in my next lecture I shall let y[u] know what improvementes have beene made in 1800 yeares past since how Aristarchus his probleme has beene since the discovery of Optick glasses what distance of the sun wee have found by it. & wherein it is deficient and has beene misapplyed by some eminent persons. How it has beene better determine & defined by observations of the Parallax of Mars made lately & with larger & better instrument then have hitherto beene formed any where else that wee know of which will open my way & cleare our passage to y[e] Theory of the late Comet; I beg y[e] pardon. &t.

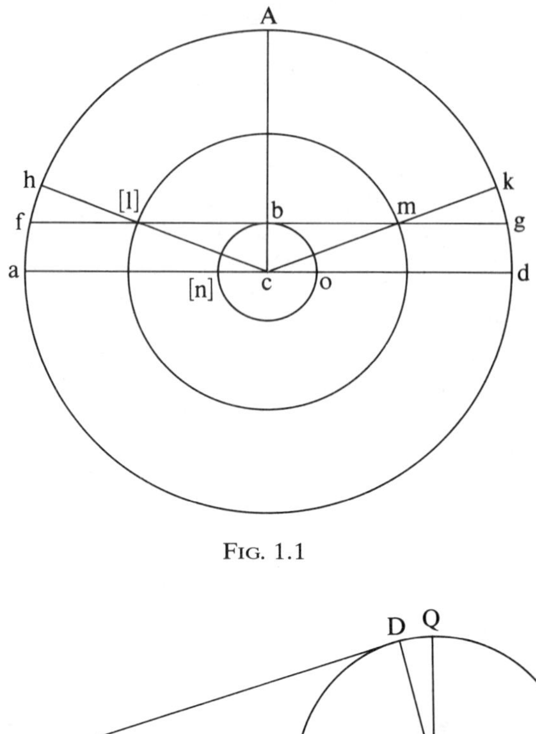

FIG. 1.1

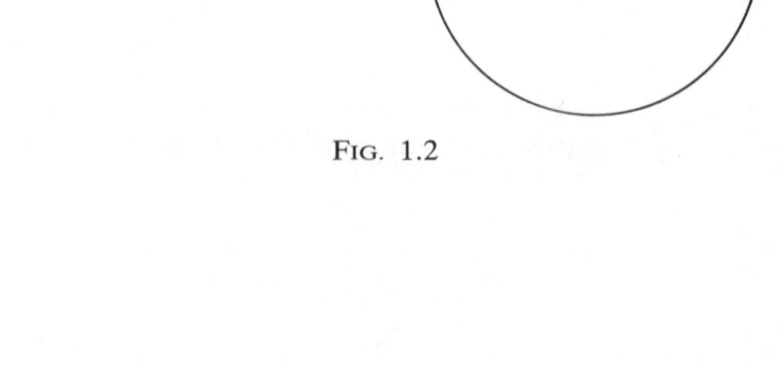

FIG. 1.2

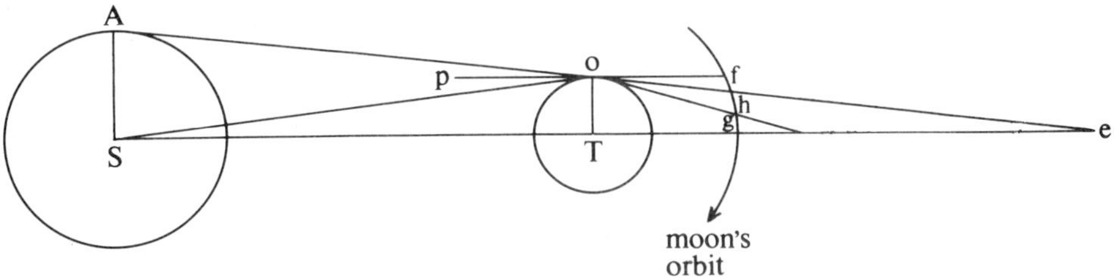

FIG. 1.3 [reconstructed].

NOTES

1.1. Walter Pope (d. 1714). It is stated in John Ward's 'Extracts from the Repository for Gresham affairs', B.M. Add. MSS. 6196, p. 23, that Pope (who was virtually blind) had arranged that his lectures should be given by Flamsteed who was 'a very able man in that art'.

1.2. In Lecture 3, Flamsteed discusses his own and others' observations on the appearance and motion of this comet of 1680. Cf. Introduction, pp. 20–34.

1.3. Johannes Kepler, *Astronomia nova ΑΙΤΙΟΛΟΓΗΤΟΣ, seu Physica Coelestis, tradita commentariis de motibus stellae Martis, ex observationibus G.V. Tychonis Brahe*... (Pragae, 1609), pp. 1–5.

1.4. The figure for the mean length of the synodic month cited in Riccioli 1, p. 241, is 29d 12h 44m 3s. The modern value is only 0s · 2 less.

1.5. 'Triennis', or three-yearly cycle.

1.6. 2 Kings, 16. Ahaz, the eleventh king of Judah, reigned from 742 to 726 B.C.

1.7. Christopher Helwich (Helvicus), *Theatrum historicum, sive chronologiae systema novum*. Flamsteed possessed a copy of the sixth edition (Oxford, 1662).

1.8. *Circa* 720 B.C. since, on the authority of Helvicus, Flamsteed regarded the reign of Nabonassar as having begun in 745 B.C. The presently accepted chronology places this event two years earlier.

1.9. Perhaps Flamsteed meant to write 'Sun', since Thales is reputed to have predicted a solar eclipse in 585 B.C. This conjecture is supported by the fact that, when again referring to Thales in Lecture 9 (cf. f. 94^{v}, pp.190–1), he first wrote 'Moone' then scored that word out and substituted 'Sunne' in its place.

1.10. Claudius Ptolemaeus, *Almagest*, lib. 3, cap. 2. Cf. Riccioli 1, p. 242.

1.11. Josephus Justus Scaliger, *Juli Caesaris F.: Opus de emendatione temporum*... (Genf, 1629).

1.12. Dionysius Petavius, *Uranologion sive systema variorum authorum* (Lutetiae Parisiorum, 1630), p. 247.

1.13. Literally: 'But Meton chose as his initial reference point the year when Phoebus [the Moon] was keeping Cancer [the Crab] warm with his shiny orb.' Cited in Riccioli 1, p. 243.

1.14. *ευνεαδεχατύρις* is the Greek term for the 'Cyclus Decennovalis' or 19-year cycle of Meton. Cf. note 1.16 and Riccioli 1, p. 242.

1.15. Literally: 'Nor shall my year be conquered by the Fasti of Eudoxus.'

1.16. This is the name given to the approximate relationship, reputedly discovered by Meton (fifth century B.C.), that 235 synodic months are equivalent to 19 tropical years.

1.17. By Martin Hortensius in Leyden on 27 October 1632. This information is contained in the same table from which Flamsteed took the dates (which he converted from New Style to Old Style) of the preceding observations by Tycho and

Longomontanus. Cf. Riccioli 1, p. 242. Since the particular cycle chosen begins with Tycho's and ends with Hortensius's observations, Flamsteed added those of Hevelius and his own of a different Metonic Cycle to complement the data cited by Riccioli (ibid.)

1.18. Details of this phenomenon are contained in John Flamsteed's *Historiae Coelestis Britannicae*, i (Londini, 1725), pp. 234–8.

1.19. In Flamsteed's time, the Julian Calendar was still being adopted in Britain although the Gregorian Calendar had long since replaced it in Catholic Europe. Accounts of the history of the calendar are found in many encyclopedias; see also, Joseph Delambre, *Histoire de l'astronomie moderne* (Paris, 1821), pp. 1–84, etc.

1.20. The use of gnomons, armillas, and parallactic rulers for observing the equinoxes and solstices is described in Riccioli 1, pp. 130–5.

1.21. In Fig. 1.1, $\frac{\mathrm{bc}}{\mathrm{lc}} = \sin \mathrm{b\hat{l}c} = \sin 1° = 60 \sin 1' = \frac{60}{3438}$; hence $\mathrm{lc} = \frac{3438}{60}$ bc or just over 57 bc.

1.22. 'Solis distantiam ad distantiam Lunae modo triplam, modo sescuplam ponens.' Cf. Riccioli 1, p. 108.

1.23. ἁυτὸ ἔφη means 'the authentic statement'.

1.24. The precise statement of Aristarchus's seventh Proposition and following comment as given in Riccioli 1, p. 110 is: 'Distantia, qua Sol a Terra distat, maior quidem est distantia Lunari a Terra, quam duodevigintupla; minor vero, quam vigentupla: Atq. adeo est ad Lunae distantiam fere ut 19. ad 1.'

1.25. The reference here is to Kepler's *Ad Vitellionem paralipomena, quibus Astronomiae pars optica traditur* (Francofurti, 1604), p. 237. Riccioli 1, p. 732 places 'tibi' and a comma after 'cava' in this quotation, and another comma after 'primam'.

1.26. This diagram is the subject of the fifth and longest of seven disputations published in John Wallis's edition of *Jeremiae Horroccii . . . Opera Posthuma . . .* (Londini, 1673). Cf. pp. 102–74. Its object was to illustrate how the semi-diameters and parallaxes of the Sun and Moon could be determined geometrically from the dimensions of the shadow cones cast by the Moon or Earth respectively during a solar or lunar eclipse. Horrox himself ascribed the concept to Aristarchus of Samos (cf. ibid., p. 104).

Lecture 2

4 *May* 1681

In ye conclusion of my last lecture I told you I intend to proceed in ye inquiry of those methods whereby the ancient & moderne Astronomers have arrived to ye knowledge of the true distances of ye Luminarys & theire parallaxes depending on them. I began as low and far of, as wee have any records of antiquity to guide us in this discovery: it will be necessary for the understanding of what I have further to say on this subject, to repeate ye summe of what I then told y^{u} which I shall doe with all requisite brevity.

I informed y^{u} then that it is most probable the necessity of a true account of ye yeare produced the discovery of these revolutions & distances that first men numbred theire Moneths & yeares by the Moone after 12 of whose returnes they found the Spring and fruites of the earth nearely in the same condition they were at the commencem[en]t of theire account, but that after 3 of these lunar yeares they noted that the Spring was backwarder then they conceaved it ought to be if the lunar yeare were equall to y^{t} of ye Sun whose influences they saw gave Maturity to ye fruites of the earth. & that another returne of the moone was made before they were so mature as 3 yeares before reckining 12 moones to a yeare they had found them that on this account the[y] ordered the intercalation of a Moone each three yeares at first and that this was usually performed amongst ye Jewes by ye Sanedrim, with ye Romans by the Pontifex Maximus, & by the Most Learned among ye Greekes.

I showed y^{u} that it was the most likely that the Jewes were the first who attained the knowledge of ye Periods of the Moones returnes to ye Sun that this experience . . . to ye Chaldeans & Assirians & I gave this reason for it that the first lunar Eclipse wee find recorded is in ye 27th yeare of Nabonassar & 1st of Marokempadi or Meredack Ball Adan which by the joynt consent of the ablest Cronology falls 2 yeares after ye Captivity of the ten tribes & that wee finde the diall of Ahaz mentioned in ye holy writ whose Reigne commenced scarce 2 yeares after Nabonassar's Era.

That it was probable the Greeks receaved this knowledge from ye Assirians for Thales Milesius & Pythagoras of whom the former is

Notes on pp. 101–103.

said to be the first that could foreshow the lunar Eclipse & the latter hee that asserted the motion of the Earth are both affirmed to have travelled & visited the Chaldeans & other people famed for theire knowledge of the heavens in theire times.

I told y^{u} that for determineing the true lenght of the solar yeare apart from the lunar. they observed that the shadow of an upright object or Gnomon measured at any time of the yeare they found to be of the same lenght againe after 12 of theire lunar moneths & 11 dayes & that hence they stated the true length of ye Solar yeare 365 dayes. that onely the longest & shortest shades were made once a yeare in theire hottest & coldest weathers, all the other intermediate lenghts twice. that therefore Meton & Euctemon chose to observe the times of the *solstices* for the determineing the lenghth of the solar yeares that Aristarchus made use of the same method 152 yeares after them & that tis probable these were the observations from which the Greekes limited theire solar yeares. that Meton might easily [have] stated their *ευυεαδηχατὲριυσ* from course notes upon which Euctemon & Challippus made the first Athenian Kalendars for 19 Yeares which Calippus afterwards corrected & produced to 76 from ye bare observations of lunar eclipses that had hapned at ye intervall of 19 solar yeares of which . . . late examples.

I showed y^{u} further that by observeing the lenght of the meridionall shade they found out the inequality of the Motion of ye Sun & Moone. & soone after by compareing the motion of the moone observed when full neare any bright fixed star whilest shee appeared above the earth in one night, with her apparent distance from it next night at her riseing they found her different of aspect about one degre[e] & consequently her distance from the Earth. to be about 60 of its semidiameters.

which being determined Pythagoras assumed the Suns about 3 or at most 6 times as much as the moones that is about 180 or at most but 360 of the Earths semidiameters. At which distance for ought we find related by any of the ancients it was esteemed till Aristarchus Samius considering that the moone receaved her light from ye Sun & was more or lesse illuminated according as shee was farther or nearer removed from him & observeing that shee was not fully halfe illuminated till neare the time of ye true Quadrature or when shee was 90 degrees distant from him concluded. that his distance from ye earth was 19 times more than the Moones which being stated 60 semidiameters of the Earth it will follow that the Suns was at least 1400:

Something more then which Hipparchus the next succedeing of ye eminent ancient Astronomers assumed it & demonstrated from ye observed durations of lunar Eclipses that it could be no lesse nor did any of them state it more except Possidonus who if my memory faile mee not removed him to as great a distance as most of ye Moderne

Astronomers suppose him if Ricciolus [derived] his conclusions from Possidonius his data may be credited.[1]

I concluded with Hipparchus in whose dayes a large new fixed star haveing appeared[2] he tooke occasion thence to correct ye whole Catalogue of them which worke by his testament he bequeathed to posterity. the last of his observations recorded by Ptolemy fall no nearer us then about 120 yeares before ye Incarnation.

Haveing thus shewed to what knowledge of the Coelestiall motions and distances the ancients had attained before our Saviours time I come now to enquire what progresse they have made since & by what steps & expedients wee have arrived to the limitations of ye present Age.

The necessity of which enquiry cannot but be evident to those who are acquainted with Astronomy since without the knowledge of ye Suns distances from ye Earth it will be impossible to determine what proportion either his [Globe] or the Globes of any of ye planets beares to o^r Earths what wonderfull wisdome and power the Almighty has showne in ordering the partes of ye Universe or how small a portion of it wee employ.

I told y^u in ye conclusion of my last lecture that Hipparchus consenting to the distance at whit Aristarchus had removed the Sun assumed it about 1140 of ye Earths Semidiameters. I shall now proceed with his successors and enquire what they have resolved on & first with those who pretend theire distances setled by his probleme. next with such as have made use of Aristarchus his Method since the discovery of the Telescope & lastly I shall let y^u know what has been found by late observations designed for finding the parallax of Mars which by consequence will give the Suns & his distance.

From Hipparchus his data Ptolemy the next of the Eminent ancient Astronomers collects the Suns Meane distance from our earth 1210 of its semidiameters & therefore his Horizontall parallax $2'.51''$.

Ptolemy lived about 140 yeares after y^e incarnation and for well meare 700 yeares after him wee meet with few or none recorded that were studious of Astronomy till Mahomet of Aracta or Albategnius examineing his obliquity of the Ecliptick falsly stated $23^{\circ}{-}51'\tfrac{1}{2}$ finds it much lesse & consequently the Solar Theory very different from Ptolemys whose Almagest being translated into y^e Arabian language. that people informed by him how the obliquity of the Ecliptick was different from what Ptolemy had stated it therein set themselves to observe it & the suns motions which seemes to have beene the cheife care of the following ages till Regiomontanus appeared in the world who bent his mind to ye improvem[en]t allso of ye other parts of Astronomy. in the meane time little alteration was made in ye Suns distances. Albategnius reteineing it a little lesse then Ptolemy who is followed by Alphonsus Alphraganus Fernelius Copernicus Ma[e]stlin Tycho & Longomontanus none of which supposed the Suns distance more ye 170 semidiameters of ye Earth different from another.[3] & all of them conceaveing their owne assertions confirmed by the Phainomena of Eclipses. of which tho the Noble Dane have left us no examples[4]. . . yet it appeares by his Progymnasmata pag. 472. of his first part of the Prag[u]e Edition.[5] that hee had severall by him & that hee relied not onely as some would seeme to intimate on ye concepts of Offusius. Lansbergius that hee might not seemes to build on other foundations has adventured to make this distance yet somethin[g] more. y^{t} is 1551 Semidiameters on ye Apoge. Bullialdus onely 1459[6] to whom o^{r} Countriman Mr Wing assents & contends that it may be proved by the Phainomana of lunar Eclipses especially of that observed by Bulliadus Dec 10. 1638.[7]

It will be too long a worke to goe about to examine what observation every Author has produced for his opinion. I shall not therefore attempt it at present tho it may be very well proved that theire observations duely considered would give either none or an insensible parallax & consequently an infinite distance to the Sun but I shall give y^{u} severall reasons why the deductions of o^{r} predecessors from lunar Eclipses were found so erroneous & faulty.

The first is that they were uncerteine of the Suns semidiameter which most of them stated lesse then the truth. The ½[8] Diameter of ye Apoge Sun by the most certeine observations made with ye Telescope & Measureing screws is found 15–50[9] this Albategnius better then most that have succeded him states 1540 Tycho Longomontanus & Kepler onely 15: Bullialdus & o^{r} Mr Wing 16′ : 09″[10]. above one intire minute more. which will cause some thousands of semidiameter error in the Suns distance defined by this method.

How difficult it was for ye Noble Tycho to determine ye true quantity of either ye Suns or Moones diameter Kepler best informe[d] us who tells us that on ye 5th of March the Suns diameter was found by severall observations betwixt 29′ : 50″ and 30′ : 50″ but the Moones observed with different instruments & by divers persons on February the 22 1591 twice 31 min, six times 32, seven times 33. six times 34 and once onely 36 minutes.[11] which differences might cause no lesse errors in taking the Suns distance then the faults committed in his observed diameters.

But these were not the onely impediments in the determinations of o^{r} praedecessors. what is more considerable they supposed the excentricity of the Moones orbite entire whereas the Telescope observations of this age plainely shew that not onely in ye rest of the planets but in her also it is bisected. they Conceaved the Perige moone in the Quadratures much nearer then ye Perige full moone whereas the telescope observations of her Diameters plainely evince the Contrary.

And if they erred so much in theire groundworke what hopes can theire be that the superstructure the true horary motions & horizontall parallaxes deduced from them should be any wayes more certeine or deserve more to be confided in then theire supposed Diameters of the luminary.

And yet considering the small helpes the Ancients had for the Makeing accurate celestiall observations theire want of optick glasses & good measures of th^{m} I thinke it deserves more of o^{r} woonder that they should ever attaine so neare ye truth rather then that they should fall something wide of it. the Demonstration of Hypparchus however holds firme & when wee shall have examined some more sure methods of proceeding I shall likewise shew that his probleme will when resolved as it ought, give ye same limitations for ye suns distances.

The first that[12] determined the suns distance from o^{r} Earth any wider was the subtle and ingenuous[13] Kepler who in his Ephemerides[14] removes him to 1800 Semidiameters & afterwards in his Epitome Astronomiae Copernicanae[15] & other workes[16] to 3470 that so hee might have his horizontall parallax but one minute[17] for which wee shall see his reasons by and by.[18]

Soone after the publication of his Comentary de Motibus Stellae Martis[19] the admirable invention of the telescope hapned whereby not onely the eye is carryed nearer the object but it is rendred much more distinct then could before be imagined. what this has contributed to the restitution of Astronomy I shall not need to intimate it is sufficient to tell y^{u} that now the learned Kepler excites all persons concerned in celestiall observations to repeate ye Experiment of Aristarchus & trie at what space of time the apparent Dichotomy of the Moone happens before the true Quadrature. Hee had found by observations with his naked eye. that shee was scarce halfe illuminate till after the first but before the latter quadrature which makes him suspect that the Suns distance was far larger then hee had yet adventured to state it.[20]

But the first telescopes were meanely contrived so as the[y] would scarce receave a whole moone & therefore hee found little returne to his admonitions or desires but some yeares after Ricciolus set upon it and Hevelius the former of which[21] finding that the apparent Dichotomy seldom was observable but after ye first & before ye last true quadrature & not understanding that this was caused by her parallax retardeing the appearance in ye occidentall but accelerateing it in ye orientall quadrant of ye Ecliptick, & well knoweing that these if taken as observed would make ye suns parallax insensible & his distance infinite which could never [have] stood with ye Earths stability & the Definition of the Conclave in ye Case of Galileus[22] to prevent any such true deduction give the following caution to ye observer. Praeterea Ita utendum Telescopio[,]ut tamen non fiat comparatio cornuum integrorum[,]cum intermediis partibus etc.[23]

that hee needs not attend till ye illuminate parts in ye middle be in a streight line with ye cuspes but onely when they are in a streight line one with another. whereby instead of mending the matter hee perplexes it more, for by reason that ye lines of illumination neare ye time of the Dichotomy are very narrow ellipses the partes in ye middle of the moone will appeare nearly in a streight line with each other an houre after the first & before ye second Quadrature which should the observer Chuse as hee may for any thing given in Ricciolus his Caution[24] would run[25] him upon a Ridiculous Absurdity.

But if an allowance be made for y^{e} ☽s parallax in his observations and if instead of the true hee use the apparent distance of the Moone from ye Quadrature at ye time of the observed Dichotomy it will be found that the time of the apparent halfe illumination will fall much lesse short of ye true Quadrature then hee has stated as is easy to prove from his owne observations & others since made with better glasses.

But I may not dissemble it that the observation of the Dichotomy is very difficult by reason of the inequality of ye Moones Surface which causes the line of illumination to appeare broken & renders it very hard to determine when the illustrated points are in a streight line with the Cuspes as I more then once experienced in my first observations particularly on ye 21th of June 1972[26] which caused mee to relinquish this method at that time & to thinke of some other more convenient.

Kepler in ye 2^{d} Part of his Commentary De Motibus Stellae Martis pag 64 has this notable Paragraph "Cum Autem Braheus etc. I shall give it y^{u} in English: Tycho haveing affirmed frequently to learned persons that by the observations of this yeare 1582 hee had found the Parallax of ♂ notably greater then the Suns.

I sayes hee that I might see & th[o]roughly understand the processe of his calculation turned the booke over carefully & found a title indeed which pretended an inquiry of the Parallax of ♂ from y^{t} yeares observations but the worke far different from what I expected. What think y^{u}? Why his Calculators had accommodated the place of ♂ found by observation to ye Copernican Theory. In theire figure they had bestowed a great deale of paines in resolveing all the triangles created by ye Double Epicycle in the Concentrick with large numbers and for this purpose that they might pronounce the Parallax of ♂ greater y^{n} ye suns. it appeares therefore that Brher [i.e. Brahe] proposed one thinge & his servants performed another hee required that compareing the morneing & evening observations of ♂ with each other they should thence find the Parallax. They on ye contrary enquired onely what Parallax the Copernican scheme would give him. And whether hee pronounced the parallax of ♂ so great on the sole Credit of his Calculators. Sayes Kepler modestly. I know not.[27] It appeares by this Paragraph that the Noble Dane was the first that proposed the Discovery of the Suns parallax by observations of ♂ & upon good consideration.[28] for at ye time of that opposition tho the sun was neare his Perige in 17 Deg of ♑[46] & ♂ within an octant of his Apoge in ye opposite pointe of ye Ecliptick yet his distance from ye earth was but 65 such parts as ye suns was 98 whence it would follow that if ye Suns horizontall parallax were 3 minutes as Tycho supposed, the horizontall parallax of ♂ would be one & an halfe of it of 4½ minutes & therefor his apparent place in his riseing would be put forward by about 3 minutes and as much backward in his setting, so that whereas his true ½ Diurnall Motion at that time was about 12 minutes in longitude retrograde, his apparent would be found neare 18 if the Parallax were so large as hee imagined it ought to appeare, Now hee knew well that 6 minutes was a very sensible arch on his instruments and therefore doubted not but the question of the suns parallax might be determined by his observations. how hee came to imposed upon[,]it is not difficult to guesse. but I had rather be silent of the slips of that great & deserveing person I had rather correct then expose them.

His Successor in ye Emporours[29] Service ye acute Keplers examining them & some of his owne over againe concludes the parallax scarce sensible but considering with himselfe that it might be something & that a minute was difficultly observable with Tychoes instruments concludes on that quantity & in his later workes uses that Parallax & the Suns distances collected from it.[30]

When I first applyed myselfe to celestiall observations I found frequently errors of 3 4 or 5 minute[s] in the places & distences of neare fixed stars in Tychoes Catalogues. this experience caused mee to doubt whether ye parallax of ♂ were observable wth his instruments or not & to resolve to make ye experiment when any convenient opportunity should present it selfe which was not long wanteing for turneing over the Ephemerides of the yeare 1672[31] I found that Mars in the September would passe Retrograde betwixt 3 bright fixed stars in ye water of ♒ at which time his distance from ye earth would be scarce more then ½ of the Suns distance. The Stars distances from each other not exceding the Capacity of my Telescopes one of which resolved above a degree of the other about 40 minutes.

No opportunity could have presented it selfe more conveniently. for in this all the measures I should need to take might be got by myselfe alone by ye helpe of my micrometer & tubes.[32] but least my other occasions might prevent mee from attending the appearance & that others might if accommodated joyntly make ye experiment I gave notice of this remarkeable appulse in ye Philosophicall Transactions[33] wth an Ephemeris of the Planets motions for the time of that appearance which I reck[o]ned might be easily observable from ye 15th of September to ye 3^d^ of October y^t^ yeare.

I shall be breife in my account of this because I intend a more full one of some later observations designed & made on the same account & for y^{e} same purpose. Some accidents hapned which hindred mee from attending [to] these observations till y^{e} 26 of September with at 7^{h}.03$'\frac{1}{2}$ after noone at Derby the distance of y^{e} planet was taken from y^{e} praeceding star 12$'$.17$''$ from y^{e} following 28$'$.54$''$[34] the distance betwixt y^{e} two stars 32$'$.23$''$. and at 12^{h}41$'$ from y^{e} praecedeing star againe 12$'$.32$''$, at 12^{h}.47$'$ from y^{e} following star 30$'$.13$''$. and at 12^{h}.54$'$ againe from y^{e} precedeing 12$'$,27$''$ with a telescope 12 foot longe & from y^{e} following 30$'$.20$''$ at 13^{h}13$'\frac{1}{2}$.

The height of the Planet at the time of these observations was 15 degrees so that I needed not feare the distances being corrupted with refractions or to make any allowance on y^{t} account.[35]

The places of the fixed stars I corrected by the observations of the 28th following whereby I found the place of the following star about 5$\frac{1}{2}$ minutes false in latitude & neare one in longitude in Tychoes Catalogue.[36]

And by a strict Calculation of the planets place from the observed distances at 7^{h}03$'$. and 13^{h}.13$'$ its apparent motion in longitude 37$''\frac{1}{2}$ Retrograde in latitude decreaseing 1$'$.19$\frac{1}{2}''$ in 6^{h}10$'$ interlapsed betwixt y^{e} two observations. But by tables corrected so as to answer these and other observations of the same yeare[37] his true motion in longitude 20$''$ 0 the true decrease of latitude 1$'$.13$''\frac{1}{2}$

The apparent motion exceeds the true onely 18$''$ & the apparent change of latitude [exceeds] the true but 6$''$.

Now because the Summe of the parallaxes of longitude in both observations extends but to 18$''$ to find what part of them is due to each observation the parallactick Angles at each observation must be sought which in the first will be found 86°.07$'$ the planet being short of the Nonagesime degree in the later 40°.24$'$ the planet being past it it will hold therefore.

As y^{e} Summe of the Cosines of the Parallactick Angles: to y^{e} cosine of y^{e} parallactick angle at 13^{h}.13$'$:: so y^{e} summe of y^{e} parallaxes in longitude 18$''$: to y^{e} parallax in longitude in y^{e} observation then made 16$''$. And againe

As y^{e} cosine of the Parallactick Angle : to y^{e} Radius:: so the parallax of ♂ in Longitude : to his parallax in Altitude, 22$''$[38]

The true height of the planet was then 14°.50$'$ which with y^{e} parallax in altitude gives his horizontall Parallax neare 22 seconds

This distance of ♂ from o^{r} Earth to its distance from the Sun was neareIy as 455 to 1000 & therefore the Suns horizontall parallax or the visible diameter of the Earth viewed from y^{e} Sun almost 19$''$: which will give his distance from us above 10 thousand Diameters of o^{r} globe.

The Diameter of y^{e} Perige Sun I have found by frequent observation 32$'$47$''$. that is an hundred times more y^{n} these observations allow o^r earth whence it will follow that the Globe of y^e Sun containes the Globe of o^r Earth a Million of times. For globes are in proportion to each others as the Cubes of theire Diameters and y^e Cube of one to y^e Cube of 100 is as 1 to 100.00.00

Since it has pleased his Ma[jes]^{tie} to encourage the study of Astronomy by raiseing an observatory and accommodateing it with better instrum^{ts} then have hitherto beene made elsewhere[39] thus wee know of no opportunity has beene omitted of repeateing this experiment. three oppositions of ye Sun and Mars have beene observed in 5 yeares & thrice has his parallax beene sought by observeing his evening & morneing distances from ye s^d stars with a sextant furnished with telescopicall sight & a very convenient movement. I shall not trouble you with an account of all Nor perplex y^u with figures, either would require more time then is allowed to this exercise but onely tell y^u what the observed distances were & what y^e result of y^e Calculus

On y^e 14th of January 167$\frac{8}{9}$ ♂ wanteing but some 6 degrees of y^e ☍ of y^e sun the aire being very serene & calme. was the golden opportunity very convenient for this purpose: hee rose before y^e stars of y^e Lion when hee was got above y^e reach of unlimited & doubtfull refractions his distances were taken and repeated often from ye Southerne in ye square of ye greater beare. from ye Lions heart the middlemost of his neck & the brightest. Mr Hall[e]y was present[40] & alwayes concluded y^e measures accurate whence we determined at 7.50 after Noone his distance from the brightest of y^e lions neck 11°.52′.25″ at 8^h.04′ from y^e Southerne of y^e precedeing stars in y^e square of y^e greater beare 40°.36′.40″

8.11 from y^e lions heart	12 .06 .20
8.46$\frac{1}{2}$ from y^e middlemost of his neck	9 .27 .30

I chose these narrow distance[s] before wider because they would be lesse liable to be corrupted by refraction & each with a respect to some particular conveniency. the bright star in y^e lions neck had y^e same declination with y^e planet & therefore in equall remove from ye Meridian the contraction by refraction would be the same & therefore needed not be regarded. the middlemost [star] in the lions neck had equall latitude w^{th} ♂ so y^t a small alteration of ♂ his latitude made no considerable change in the distance on that account. The lions heart was neare ye verticall passing by ♂ in ye evening observations & therefore in the morneing would be nearely in equall altitude with him, so that all the effect of refraction & parallax would be found in ye evening observation the Morneings free from them. & to determine ye latitude more exactly hee wanted not a degree of ye longitude of the star of ye great beare hee was then neare his limit & this will give us with y^e rest the inclination of his orbite. but not to stay on these. wee attended the convenient times of y^{e} next morneing & with like care repeated the distances which wee found counteing the houres still from the praecedeing Noone. and takeing y^{e} Middle of y^{e} Repetitions

$16^{h}.37'\frac{1}{2}$ from the lions heart	$12°.14'.45''$
16.50 from the middlemost of y^{e} neck	9 .35 .15
$18.17\frac{1}{2}$ from the brightest of his neck	12 .01 .30

Now seeing the Motion of the planet was retrograde & the parallax depresses or promotes his true longitude in y^{e} orientall quarter of y^{e} Ecliptick but retardes it in y^{e} occidentall it will follow that y^{e} computed difference of his distances from ye stars at ye times of ye evening and morneing observations shall be sensibly lesse than the observations if his parallax be sensible. Now haveing got the observed distances that I might not be deceaved by an amanuensis as Tycho had I calculated carefull first the place of ♂ to the severall times of ye evening observations from the measured distances. then by the Tables[41] which y^{u} may safely confide in for a dayes motion. how much the planet altered his longitude & latitude betwixt each of y^{em} & y^{e} morneings [observations] & thence how much that change altered his distance from each of y^{e} stars not omitting carefully to allow for refraction what was required. the effect in breife was this

Stateing y^{e} distance of ♂ from The bright star of y^{e} Lions neck 11.52 observed at $7^{h}50'$ in y^{e} evening it will follow that at $18^{h}\ 17'\frac{1}{2}$ or 17′ past 6 next morneing his true distance from it ought to be $12°.01'.39''$: without any consideration of any refraction I tould y^{u} before was excluded. So y^{n} if hee had any sensible parallax y^{e} distance ought to appeare wider. whereas it was observed almost $\frac{1}{6}$ of a minute lesse Againe. suppose y^{e} distance taken from Regulus or the lions heart as observed at 8.37 in y^{e} evening to be $12°.06'.20''$ at $16^{h}.37'$ next morneing it ought to be allowing for refraction $12°.14'.31''$ but it was observed then $12°.14'.45''$. So y^{t} y^{e} summes of parallax seemes to be on each side [of] y^{e} Meridian some 15 seconds whence his horizontall parallax will be found about 18″ & consequently the suns 12″.

And if wee state the distance of ♂ from y^{e} middlemost of the neck 9.27.30 as observed at $8^{h}.46'\frac{1}{2}$ in y^{e} evening, allowing what alteration will be made by y^{e} change of his place latitude at 50′ past 4 next morneing wee shall find that y^{e} said distance then ought to be $9°.35'.17''$: but correct by refractions onely 9.35.02 which by observations was then found $9°.35'.19''$ greater as it ought then y^{e} Calculated but hence the parallax of ♂ will be found no more y^{n} 15 seconds & the suns above 10 as I collected before from y^{e} Derby observations of y^{e} yeare 1672.[42] Indeed the observations made by The Lions heart yeeld not any but the observed differences of the Morneing and eve[n]ing distances are $\frac{1}{4}$ of a minute lesse whereas they ought to be more if y^{e} planet had any sensible parallax but so small an arch is no great part in y^e limbe of a 7 foot instrument[43]. and I conceave o^r observations confirme themselves sufficiently by so neare a consent certeinly hence wee may conclude that the suns Horizontall parallax is not more y^n $\frac{1}{4}$ of a minute lesse it may be justly be supposed nay I dare affirme with Wendelin[44] who with o^r Mr Horrox[45] & the Author of y^e Caroline Tables[46] supposes it no more then 15″ seconds *solis Hypothesin απχραλλακτικηυ non posse sensibilis erroris convinci*[47]

Y^u will see hereby what small or no ground there is for that opinion that ye sun in o^r later ages is approached nearer to ye earth then hee was formerly, the improvements of o^r instruments and senses convinceing us that hee both is & ever was above 10 times more remote then hee has beene commonly esteemed.

I had thought to have shewne how observations applyed to Hypparchus his theorem[48] would agree with this large distance of the sun but that I may nor trespasse on y^r patience I shall forbeare this till some better opportunity. & haveing found y^e Suns distance in my next lecture I shall make y^e same enquiry concerneing y^e distances motions & appearance of y^e late Comet.

NOTES

2.1. Riccioli 1, p. 111, infers that Posidonius's estimates of the mean Earth–Moon and Sun–Earth distances were $52\frac{15}{38}$ and 13,141 Earth-radii respectively.

2.2. Modern research has revealed that this object, observed by Hipparchus in 134 B.C., was a comet—not a nova.

2.3. These names, and the corresponding estimates of the Sun–Earth distance, are tabulated in Riccioli 1, p. 110.

2.4. Flamsteed wrote, but subsequently deleted, the following words at this point of the text:

> 'If wee may not except the Noble Dane [Tycho Brahe] who finding how difficult it was to determine the times of ye Principall phases of lunar Eclipses by observations made with his bare Eye or Naked Sights on his various instruments stands not upon Demonstration to prove it but upon the Harmonicks of Al Hasius [Alhazen] & because y^e Number 376 . . .'

2.5. Tycho Brahe, *Astronomiae instauratae progymnasmata* (Pragae, 1610).

2.6. The mean Sun–Earth distance tabulated in Riccioli 1, p. 110, is actually 1,460 Earth radii.

2.7. Vincent Wing, *Astronomia Britannica* . . . (Londini, 1669), pp. 123–4.

2.8. The fraction $\frac{1}{2}$ was inserted as an afterthought, since Flamsteed originally wrote down values of the Sun's angular diameter then subsequently deleted them.

2.9. Flamsteed is here referring to the result (in minutes and seconds of arc) of his own observations with Towneley's micrometer. The application of this instrument to the measurement of small celestial arcs had been explained by Richard Towneley himself in *Phil. Trans.* i, no. 25 (6 May 1667), 457–8, and a description of it given by Robert Hooke in ibid. ii, no. 29 (11 November 1667), 541–4.

2.10. The figures quoted for Albategnius, Tycho, and others are tabulated in Riccioli 1, p. 124. Wing, op. cit. (note 2.7), p. 193, gives 16′ 2″ as the value at apogee.

2.11. Kepler, op. cit. (note 1.25), p. 218. The equivalent Latin statement is quoted and underlined in a manuscript entitled 'Lectori Astro[no]miae Studioso

J.H. bene agere:' written in Flamsteed's own hand sometime after October 1671. Cf. R.G.O. MSS., P.R.O. 43, f. 4[r].

2.12. The words 'was bold enough to rem[ark]' followed, but were deleted.

2.13. Flamsteed habitually spelt 'ingenious' in this way.

2.14. Johannes Kepler, *Ephemerides Novae motuum coelestium 1617.1618.1619* (Lincii, 1616), p. 2.

2.15. Johannes Kepler, *Epitome astronomiae copernicanae* (Lentijs ad Danubium, 1618, 1620, 1621).

2.16. E.g. Kepler's *Tabulae Rudolphinae* (Ulmae, 1627). Cf. praeceptum 131, p. 92.

2.17. Riccioli 1, p. 111.

2.18. Lecture 17, f. 168[r].

2.19. Op. cit. (note 1.3).

2.20. Riccioli 1, p. 108, reporting a comment in Kepler's Reader's Notes' to the ephemerides for the year 1619 (cf. op. cit [note 2.14]).

2.21. An additional page (f. 20[r]) is interposed at this point to replace the following passage which Flamsteed deleted from f. 18[v]:

> 'finding that the observed Dichotomy generally fell neare the very time of the Quadrature to salve y[e] appearance conceaves that the observer ought not to attend but the illuminate parts in the middle were in a streight line with y[e] Cusps but onely when they were in a streight line with one another which by observation hee [Riccioli] determined was about an houre before y[e] first true Quarter & after the latter. whence the Angle QJD = DSJ subtended by y[e] radius of y[e] Moones orbite against y[e] sun will be found about 30 minutes & the suns horizontall parallax almost so many seconds. his words in pag 732 of his appendix [op. cit.] are Præterea ita utendum telescopio[,]ut tamen non fiat comparatio cornuum integorum-[,]cum intermediis partibus &c. whereby it is evident that he considered not what as hee ought how the Moones place was changed by the parallax casting her backward in the occidentall quadrant of ye Ecliptick and retardeing the time of this appearance. but formed in the orientall & anticipateing of it.'

2.22. This statement, signed by seven Cardinals, is published in Riccioli 2, pp. 498–9.

2.23. Riccioli 1, p. 732. This reference was cited in the passage deleted from the original draft (cf. note 2.21).

2.24. The phrase 'instead of mending y[e] matter at' has here been deleted.

2.25. The phrase 'the observer on an unavoydable' has here been deleted.

2.26. The first hint of Flamsteed's awareness of this observational difficulty is contained in his letter of 13 May 1671 to the Secretary of the Royal Society of London Henry Oldenburg. See A. R. and M. B. Hall (eds.), *The Correspondence of Henry Oldenburg*, viii (University of Wisconsin Press, 1971), pp. 45–48. Cf. p. 47.

2.27. Kepler, op. cit. (note 1.3), p. 64.

2.28. The last three words of this sentence replace the phrase 'not without reason', which Flamsteed deleted from the draft of his text.

2.29. Emperor Rudolph II (1552–1612) was the son of Maximilian II and his wife Maria, the daughter of the Emperor Charles V. He inherited the title of Roman Emperor on his father's death in October 1576. The political importance of his reign was unremarkable, but he was the patron of both Tycho Brahe and Kepler.

2.30. Riccioli 1, p. 111, states: 'in praefatione ad Ephemeridem anni 1619 [cf. note 2.14] ait se ex Martis parallaxium exilitate, & Harmonicis intervallis confirmari parallaxim Solis non esse maiorem minuto 1'.

2.31. This is almost certainly a reference to Joannes Heckerus, *Ephemerides motuum coelestium ab 1666 ad 1680, ex observationibus correctis Tychonis Brahe et*

Io. Kepleri hypothesibus physicis, etc. (Gedani, 1662) which Flamsteed frequently consulted.

2.32. Cf. note 2.9. A description of Flamsteed's modifications to this instrument and an accompanying diagram are contained in his letter of 28 March 1682 to Hevelius, a copy of which is preserved among his unpublished manuscripts at Greenwich (P.R.O. 42, ff. 17^{v}, $18^{r,v}$, 19^{r}).

2.33. 'Apographum Epistolae a Dn. Johanne Flamstedio, Astronomo Derbiensi, ad Editorem scriptae, de Transitu Planetae Martis ad tres Fixas in ♓, celebrando mens. Sept. & Octob. 1672', *Phil. Trans.* vii, no. 86 (19 August 1672), 5040–2.

2.34. We have corrected Flamsteed's value of 28° 54′ to 28′ 54″.

2.35. Very soon afterwards, Flamsteed recognized the need to take account of refraction at this and higher altitudes. He subsequently devotes a considerable amount of attention to this subject in Lectures 13, 33, 35, and 36.

2.36. Tycho Brahe, *Historia Coelestis* (Augustae Vindelicorum, 1666).

2.37. Presumably a reference to the ephemerides for that year, contained in Hecker (cf. note 2.31).

2.38. Cf. Introduction, pp. 17–8, for a demonstration of how these two precepts can be derived, and the extent to which they are valid.

2.39. The Royal Observatory at Greenwich was established by a Warrant from King Charles II dated 22 June 1675; its foundations were laid on 10 August 1675, and Flamsteed entered it in July 1676. The main instruments provided for him were a 6-foot 9-inch radius iron sextant with a brass limb and telescopic sights, and two 13-foot long pendulum clocks made by Thomas Tompion. He brought with him from Derby a 3-foot radius quadrant, and furnished himself with telescopes of 16-foot and 8-foot focus fitted with micrometer eyepieces. For further details, see Francis Baily, *An Account of the Life and Work of the Revd. John Flamsteed* (London, 1838).

2.40. Edmond Halley (1656–1742), who was later destined to become Flamsteed's successor as Astronomer Royal at Greenwich, had occasionally assisted with astronomical observations while still an undergraduate at Oxford. By now, however, he had gained considerable experience from his expedition to St. Helena to observe the southern skies (1676–8). Only a few months later, he was to leave England for Danzig in order to investigate personally Hevelius's claim that one could observe as accurately without telescopic sights as with them. Cf. note 4.2.

2.41. This is presumably an allusion to Hecker's *Ephemerides* (cf. note 2.31).

2.42. This result was published in *Phil. Trans.* vii, no. 89 (16 December 1672), 5118.

2.43. The arch corresponds, in fact, to a linear distance of only six thousandths of an inch, which is scarcely discernable to the naked eye.

2.44. *Gotifredi Wendelini Luminarcani Eclipses Lunares ab Anno 1573 ad 1643 Observatae:* etc. (Antverpiae, 1644). Cf. p. 33, where the author states: 'Parallaxis Solis tam Apogaei quam Perigaei, est secundorum 14″ semper insensibiliter eadem.'

2.45. Jeremiah Horrox, *Venus in sole visa* (Gedani, 1662). Horrox states his preference in this work for a solar parallax of 14″.

2.46. This is a reference to Thomas Streete, *Astronomia Carolina* (London, 1663), in which the author accepts a value for the solar parallax of 15″ 'as nearest the truth, and sufficiently exact' (cf. ibid. p. 34).

2.47. Translation: 'I am convinced that the hypothesis of no solar parallax cannot give rise to an appreciable error.'

2.48. Presumably a loose reference to Hipparchus's diagram (cf. note 1.26), or to its underlying geometrical principles. Horrox cites thirteen theorems in his 'Disputatio V. De Diagrammate Hipparchicho', op. cit. (note 1.26), pp. 102–74.

Lecture 3

11 *May* 1681

I promised y^u in my former lectures an account of the late appearances of the Comet seene in November, Dec, Jan, & Feb. last. I shall now discharge my selfe of that obligation, in this method; first I shall let y^u know what were ye opinions of o^r praedecessors concerneing Comets, and upon what reason they assumed them:[1] Next I shall give y^u an account of such as were made of this last in Novemb[er] at Rome and in Dec. at Greenwich, I shall enquire how they agree with the opinions commonly hitherto receaved, & because they will be found different I shall propose what I thinke most reasonable for salveing the appearances.

Seneca in the Natural questions in ye 7th book of his Nat qu[2] [tells us] that Epigenes & Appollonius [were] good naturall Philosophers, and who had both of them studied amongst the Chaldeans, differed in relateing of theire wise opinions concerneing Comets: Apollonius affirmeing that they numbred them with ye planets, & understood theire courses. Epigenes that the Chaldeans had nothing certeine concerneing them, but that they seemed to him to be raised kindled by a violent motion of ye aire which was the more probably theire opinion by how much it agrees nearer with Aristotles. the contrary of Appollonius seemeing to proceed from his affect[ion] for his Masters or his desire to heighten his owne repute by encreaseing theirs.

From this time downe to the last century wee have no more certeine [information] concerneing Comets delivered to us. Those ages were wanteing in observation, onely Regiomontanus[3] and Appian[4] have described the appearances of two or three that were seene in theire dayes, but with so few Circumstances, that the very yeare in which they appeared is Controverted betwixt Crugerus[5] & Kepler[6]: for observations wee find none, tho it is probable some they made, whence they collected what wee find affirmed concencerneing them as that

1 they moved nearely in the Arch of a great Circle

Notes on pp. 116–117.

2 that they had theire tayles allwayes opposite to the Sun, or in a streight line produced from him through the head of the Comet.
3 That they were nearer the Earth then the Moone

all which they might gather from very easy observation. For first haveing noted with what stars the head of the Comet lay in a streight line; severall nights, & thereby prickd downe its place in theire planispheres. for Globes were a later invention,[7] they might find hence that the severall places lay nearely in the Arch of a great Circle, or such an one as would cut the Periphery of the planisphere, or the superficies of a Globe into 2 equall parts.

And then strikeing a great Circle through the Sun and head of the Comet, they observed it to passe by those stars in the projection, by which the tayle of the Comet passed in the heavens.

And that they were nearer to the Earth then the Moone, they concluded, because the observed dayly motions of some were swifter y^n the Moones; whose dayly motions being swifter than any other of the planets, & her distance nearer to us, or horizontall parallax greater, they hence concluded that the distances of ye Comets from ye Earth were lesse then ye Moones, & this perhaps gave them occasion to suppose them composed of vapors exhaled from ye Earth fired by theire violent motion & carried forward as by a strong wind.

In ye Yeare 1577 the Noble Tycho observed a Comet from ye 13 of Nov: to ye 12th of January passeing through some of the same Constellations wherein the late Comet passed after its emersion from ye sun; and now all the forementioned opinions came to be questioned: the Angle which its way made with ye Ecliptick was found variable by the different dayes observations, as also ye place where the great circle passes by any two cut the Ecliptick; hence it followed that either it moved not in ye Arch of a great circle, or else the earth by its motion caused this effect. the Parallax hee found scarce sensible therefore the distance of ye Comet from ye earth was more then ye Moones. and the tayle Not allwayes in a streight line from ye Sun.[8]

Since then the like has beene observed in the Comets of ye yeares 1585. 1590. 1607 1618[9] 1652 1664 1665 1672. & 1677. which I can onely mention my intent being to limit my discourse upon that which appeared last & moveing over a greater arch then any of the aforementioned will afford us more matter of enquiry & give us better intimations of ye Nature of Comets then any of the praeceding

Tis probable y^t the Comet which appeared before ye Sun rise in November, was the same y^t was seene so large after his setting in ye Moneths of Dec. Jan. & Feb. following; I shall therefore give y^u some of those course observations of it I receaved from abroad first. then my owne, & next such arguments as induce me to thinke that both appearances were of ye same body

November y[e] 11: about 5 a clock in the Morneing at Canterbury it was seene & observed[10] in ♍ 12° with 2 degrees North latitude this is the earlyest observation I have met with of it.

It was seene afterwards by severall persons, but no where observed that I heare of so well as at Rome. whence Mons[r.] Gallet a Frenchman sent o[r] Friend Mons[r] Cassini these following [observations] Reduced to o[r] stile.[11]

Nov.: 17.18[h] pm:	Com. ♎ 8.00	fixte latitudine.	
18 . 17	♎13 . 00	1 . 00	
21 . 17½	♍ 0 . 00	4 . 00	
23 . 17	♍ 9 . 00	3 . 00	all South
25 . 18	♍16 . 15	2 . 00	
26 . 18	♍19 . 30	1 . 00	

other observations of the same comet made at Rome were by M Ant. Collio communicated to M[r] Cassini & from him by Mr Halley sent hither[12] agreeing well with Gallets in the first of y[e] 12th day & the following in longitude, but makeing the latitude never more y[n] one degree, & shewing a mistake in his account; for instead of y[e] 18th day 18[h], Collio has y[e] 18th in the morneing the same longitude, which shews the houres if counted from Noone ought to be reckned from y[e] 17th Day. & one more observation hee has on y[e] 27th in y[e] Morneing when hee gives the comets observed place ♍ 23°.35′ the praecise hour not mentioned.

Mr Halley in his jorney to Paris neare Bologne saw the tayle of the Comet riseing very small from y[e] horizon on Dec y[e] 8th in y[e] morneing,[13] whence it will follow that it had not then reached the sun, & had considerable north latitude. these Roman observations wee may reckon onely true within a degree, that they erre no more I am satisfied by a print I receaved lately from Lovaine[14] where in the places & latitudes of the comet are cut the same y[t] Collio gives them on the 18.20. & 23 of November. & confirme his account of the times correcting Gallets.

Clouds & o[r] frequent Fogges would not permit us to see the . . . [a whole line is cut off here] sixth of December after y[e] end of day light I first saw a very thin streame of light ascending up from y[e] horizon into y[e] Constellation of Aquila. from which I concluded the Tayle of the Comet was then past y[e] Sun. ♄ night the 11 it was much larger and so strong that it was visible before y[e] daylight was ended. the star appeared not yet but onely an illuminate beame, y[e] breadth of a moone, ascending almost streight up from y[e] horizon, like that which had beene seene before y[e] Sun rise & the appearance of y[e] Comet in y[e] yeare 1618 according to Gassendus his relation.[15] Next night ⊙ y[e] 12 of December the tayle appeared yet stronger, & brighter. placeing one of the Telescopes on y[e] tayle I followed it downe to the horizon whereby I met both y[e] head an obscure dull star much lesse then ♃ not round but like a square with y^{e} sharpe corners rudely rubd of. it was then neare 3 degrees high. wee therefore hasted to take such distances as might give its place & measured at

41 past 4 a clock its distance from ♀	23°-58′-45″
46$\frac{1}{2}$ [past 4 a clock] from y^{e} bright ★ of y^{e} Vultur[16]	28-15-30
49$\frac{1}{2}$ from ♀ againe	23-57-35

The Place of Aquila by the Greenwich observations is ♑ 27°-15′$\frac{1}{2}$ its latitude 29°-19$\frac{1}{2}$′, North, the true place of ♀ correct by the same ♑ 28°-24′, her latitude 1°.40′ South; but her visible place by reason of refraction ♑ 28° 27′$\frac{1}{3}$, & latitude 1.35 South, the visible place of Aquila ♑ 27°-17′$\frac{1}{3}$, its latitude 29.20$\frac{1}{3}$ North,

In y^{e} first figure [Fig. 3.1 ≡ Fig. 11, p. 21] let p̄ represent y^{e} pole of y^{e} Ecliptick pv the apparent distance of ♀ from it 91°-35′$\frac{1}{3}$, pa its distance from Aquila 60°.39′$\frac{2}{3}$, & the angle apv. the visible difference of longitudes of ♀ and Aquila 1°-10′, hence by the knowne resolutions of [an] oblique angled sphaericall triangle, theire visible distance the side av will be found 30°.57′$\frac{1}{5}$, & the angle pva 1°.58′.55″

Let further ac represent the Comets distance from Aquila 28°-15′$\frac{1}{2}$ vc its distance from ♀ 23°-58′$\frac{1}{6}$ in y^{e} triangle acv all the sides are knowne, whence y^{e} angle avc will be found 62°.16′.53″, which added to pva makes y^{e} entire angle pvc 64°.15′.48″

with which and y^{e} two sides pv and vc. before limited in y^{e} triangle PCV will be found PC y^{e} apparent distance of y^{e} comet from y^{e} Pole of y^{e} Ecliptick 81°.18$\frac{1}{2}$′, whose complem^{t} 8°.41′$\frac{1}{2}$, is its visible latitude North, and y^{e} angle CPV the difference of y^{e} longitudes of the planet ♀ and it 23°-43′$\frac{2}{3}$, which subtracted from y^{e} visible longitude of ♀ ♑ 28°-27′$\frac{2}{3}$, gives the apparent place of y^{e} Comet ♑ 6°-43′$\frac{2}{5}$ whose distance from y^{e} Nonagesime was then 97°.31′, from y^{e} vertex, 87°-45′$\frac{1}{2}$; therefore the parallactick Angle 37°-44′

The Refraction in altitude was 17′ therefore in Longitude 10′$\frac{2}{5}$; in latitude 15′, & the true place of ♀th ♑ 6°-33′. her latitude North 8°.26′

Nor need we thinke this place to be dubious by reason of the incerteinty of Refractions, it being found by frequent experience, that when the planet or star is elevated more y^{n} 2 degrees, the refractions are neareIy of the same quantity whether in ye Summer or Winter, morneing or evening observations.

About 2 hours after sunset the two neare stars in the middle of ye Sagitta shone through ye middle of ye tayle: the middle place betwixt these two stars is now ♑ 29°.20′ its latitude 39°.14′, The place of the comet about 14′ minutes forwarder then it was when observed, the Suns place: ♑ 1°.58′ in ye 2^{d} figure [Fig. 3.2 ≡ Fig. 13, p. 25] let ☉ represent y^{e} Suns place c the place of y^{e} comet, ck its latitude & ☉

k the difference of its longitude & y[e] Suns, in the triangle c ☉ k. the angle at c (which y[e] great circle passing by y[e] center of y[e] ☉ & Comet formes with y[e] circle of longitude Pk) will be found $29\frac{1}{12}$ deg, let o be y[e] ★★ of y[e] ♐. Pco the angle which y[e] axis of the tayle formes with y[e] Same Circle will be $29°\,\frac{5gr}{6}$, & therefore y[e] ∠sco which the Axis of the Tayle formes with y[e] great circle passeing by y[e] center of y[e] Sun & Comet not more y[n] $\frac{3}{4}$ of a degree to y[e] Southward & oc y[e] length of y[e] tayle 37d[egrees]

But I must confesse the tayle was very thin so that this observation I cannot affirme to be very exact, I onely mention it to shew by what method this deflection of the tayle is got which to avoyd prolixity I must forbeare to repeate hereafter

From y[e] 12 to y[e] 21th of December wee had such continually cloudy weather that I could never get an opportunity to observe y[e] head of y[e] comet tho in y[e] meane time the traine presented it selfe to view twice or thrice. but now, its head to y[e] bare eye appeared as a star of y[e] 3[d] light in the Telescope onely a bright haze like y[e] Neb[ula] and y[e] Girdle of And[romeda] from observations made this evening I state its distance from Aquila $10°.24'\frac{1}{2}$, from y[e] mouth of pegasus $20°\text{-}37\frac{1}{4}'$: whence I collect its longitude ♒ $5°.07'\frac{2}{3}$, its latitude north $21°.45'\frac{1}{2}$ past 6 a clock in y[e] evening

And now haveing gotten the observed places of the Comet twice, wee may inquire where the great circle strook through them cutts y[e] Ecliptick & what angle it makes with them thus

In y[e] 3[d] figure [Fig. 3.3 ≡ Fig. 14, p. 26] let ncdif represent a semicircle of the Ecliptick, c y[e] place of y[e] comet observed on y[e] 12 of December, ca its latitude whose complement to a quadrant ae is its distance from y[e] pole of y[e] Ecliptick 81°.34′; let also d be the place of the Comet observed Dec y[e] 21. db its latitude, & its complem[t]. be $68°\text{-}14'\frac{1}{2}$, its distance from y[e] said pole; the angle aeb the difference of its two observed longitudes $28°.34\frac{2}{3}'$;

Through a and b let y[e] arch of a great circle be strooke intersecting The Ecliptick in n. this shall be the place of the Node or intersection of the said Circle & Ecliptick which to find togeather with the arch of the great Circle the Comet has described betwixt the two observations in the oblique angled Sphericall triangle aeb, are knowne the sides ae, eb. and y[e] comprehended angle aeb. whence the angle at a or eab will be found 60°.55′.05″ & the arch of its motion ab $30°.33'\frac{1}{5}$

To find the longitude of y[e] point of intersection n, with the angle there formed anc, in that triangle are given the angle nac = eab 60°.55′.05″ ac the comets latitude at y[e] first observation 8°.26′, and the angle at c right. whence the inclination of the great circle strooke through the two observations to the Ecliptick anc will be found $30°.10'\frac{2}{3}$ & nc the longitude of the point of intersection in antecedence of y[e] comets place $14°.46'\frac{1}{3}$ which subtracted from ♑ 6.33 leave ♐ $21.46\frac{2}{3}$ the place of the said intersection.

After the same manner I have collected all the angles and places of y^e^ intersections of y^e^ great circles strooke through y^e^ observations hereafter mentioned I shall not therefore need to repeate the method but onely what I concluded from it.

I shall onely mind y^u^ that if the following observations shall give the same place & angle of intersection which I have found by this, it will prove that the Comet moved alwayes on the same plane of a great circle. or some continued streight line, if otherwayes not. I shall therefore enquire further.

Dec ye 24. wee had a cleare night againe & from the observations then taken I collect that at 12½ after 6 a clock at night The Comet was distant from ye Mouth of Pegasus 8°.31′.40″ and from ye pinion of the swans south wing 24°-18′-00″ the distance of the said two stars being 27°.31′ whence I conclude its place then ♒ 18°.49′ its latitude north 25°.23′⅖

The difference of the Comets longitudes observed this night & ye 21th past is 13°.41′½ with which & the complements of the two observed latitudes, proceedeing as before, I find the Arch of ye great Circle intercepted betwixt them 13°-04′⅓, the Intersection in ♐ 17°.42′⅗ at least 4 degrees different and in antecedence from ye former, the angle or inclination of the said great circle to ye ecliptick 28°.27′⅔ one degree & ¾ lesse. whence it is evident that either the Comet moved not in ye plane of a great circle passeing by the center of the Earth, or if it did the earth receded from ye line of its motion thereby causeing this alteration; whether of these is more probable wee may have occasion to examine hereafter,[16] at present I proceed with ye observations.

From ye observations of Decemb[e]r 26. at 14′ past 5 I state ye distance of ye head of ye comet from ye pinion of ye swans south wing 22°-46′⅓, from Markab in Pegasus 20°.23⅓′, the distance of the two fixed stars 36°.24′⅓, & hence its place in ♒ 28° 24′ its latitude North 27°-01′,

And compareing this observed place wth ye foregoeing ye arch of a great circle inter 8°.44′ between the place of ye intersection ♐ 16°-44′⅔ a degree short of that collected from ye observations of ye 21^t^ & 24th day compared togeather & the angle 28°-14′⅗ neare ¼ of a degree lesse.

The tayle passed over ye Star in ye Knee of Pegasus marked η in the Mapps of Bayer[17] but this I esteeme too neare the head to give ye deflection exactly.

Dec 29 was a cleare evening againe by observations then made at 7^h^-55′ its distances were from ye said Knee of Pegasus 9°.46′ from Markab 10°.14′.55″, the distance of y^e^ two stars 15°-49′-15″ hence the longitude of y^e^ Comet ♓ 13°-11′¾ its latitude 28°-10′ North & compareing this with y^e^ observations of y^e^ 26 the arch of a great circle moved betwixt them. 13°.12′⅕. the place of y^e^ intersection ♐ 16-24 y^{t} is $\frac{1}{3}$ of a degree short of y^{e} former & y^{e} angle 28°-12′ nearely y^{e} same. The tayle passed betwixt the two stars in the Northerne foot of Andromeda & seemed to be exactly comprehended betwixt them. it ended in the middle betwixt them & y^{e} Shoulder of Perseus The middle betwixt which wee may account in 89°.05′ with 36°.05′$\frac{1}{2}$ North latitude the ☉s true place was 19°.21′ of ♑ therefore the Angle which the Axis of the tayle formed with y^{e} line produced from the Centre of y^{e} ☉ & head of the Comet was 5$\frac{1}{2}$ to y^{e} North or in antecedence the breadth in y^{t} place 1°.50′ & its length 55 deg:

Next night being December y^{e} 30 the Comet was observed againe & from the measures then taken I find its distances at 8^{h}.02′ from Markab 8°.52′; from y^{e} head of Andromeda 19°.57′ the distance of y^{e} two stars 20°-12′ whence its place will be got in ♓ 17°-37′$\frac{1}{5}$ its latitude North 28°-11′$\frac{1}{5}$ its motion in y^{e} arch of a great circle since last nights observation 3°.54′ the place of y^{e} Intersection or point n̲ ♐ 15.38$\frac{1}{2}$ and y^{e} angle there formed 28-11$\frac{3}{4}$

The Southerne of y^{e} two Stars in y^{e} North foot of Andromeda was now in y^{e} Middle of the tayle whose place is 87°.58′ latitude 36°.05′ North. the suns true place ♑ 20°.22′ hence the deflection of y^{e} tayle neare six degrees its length something more then 50. On y^{e} 5th of January following at 51′ past 5 the distances of the Comet were from y^{e} bright star of ♈ 28°-14′: from y^{e} last of Pegasus Wing 14°-12′ betwixt y^{e} two stars 28°-03′$\frac{1}{3}$ hence its longitude ♈ 8°.49$\frac{1}{6}$′ its latitude North 26°-15′$\frac{1}{2}$ the Arch of a great circle intercepted betwixt y^{e} place observed on y^{e} 30th of December & now 18°.58′ the place of y^{e} intersection n̲ ♐ 14°.28′ & the angle 28°-26′$\frac{1}{10}$

The South edge of the Tayle touched y^{e} small star in y^{e} breast of Andromeda the North the middlemost of her girdle & it ended neare 2 small stars some 3 degrees short of her bright foot hence the angle of y^{e} deflection of y^{e} tayle will be found 6$\frac{1}{2}$ deg. & its length 25 degrees.

The distances of y^{e} Comet deduced from observation made Jan y^{e} 9th at 6^{h}.49′ were from y^{e} bright ★ of ♈ 19°.49′ from y^{e} last of Pegasus wing 17°.38′$\frac{1}{3}$ hence its place ♈ 18°.43′$\frac{1}{3}$ its latitude 24°.12′$\frac{3}{4}$

Next night being the 10 at 5^{h}.54′ the distance of y^{e} Comet was found after y^{e} like manner from y^{e} bright ★ of ♈ 18°.13$\frac{2}{3}$′ from y^{e} last ★ of Pegasus Wing 18-48$\frac{1}{6}$ whence its place by calculation is given ♈ 20°.41′ its latitude 23°.44′: compareing this observation with that of y^{e} 5th wee shall find the Inters[ection] in ♐ 15.57 One degree forwarder then the observations of y^{e} 30 of Dec & 5 of Jan compared togeather gave it so y^{t} whereas till then wee found it Retrograde continually it now becomes direct the same inclination remaineing by this observation, which gives it onely 28-11$\frac{3}{4}$. And if wee compare this observation with y^{t} of y^{e} evening precedeing it will give y^{e} Intersection in ♐ 16.24. the angle 28°.01′. agreeing well with y^{e} other.

I have beene more particular in this because wee Now have use of it. hitherto it commonly proved cloudy soone after wee had measured the distances, now wee had an evening that promised a setled serenity, & therefore I resolved to measure some distances & repeate them late to find its parallax. the bright star in ye Shoulder of Andromeda was neare the head of the comet this I pitched upon as most conveient & with all the care & diligence wee could wee measured the distances of ye Comet from it which at 6°.14′ from many others carefully repeated wee determined it 3°.07′ exact. afterwards wee waited till it should be removed at least 6 hours from ye Meridian but vapors begin[n]ing to appeare neare it (least wee should lose o^{r} opportunity) wee hasted to observe it againe & at 8^{h}.50′ wee found y^{t} distance encreased to 3°.18′.05″: from which wee shall enquire the parallax.

From the last nights observation to this the motion of the Comet in 23^{h}.35′ in y^{e} arch of a great circle is 1°.51′$\frac{1}{4}$ The comet was nearely in the same altitude both nights at the two different observations therefore its parallaxes in altitude y^{e} same. & because the motion in longitude dayly decreased the true horary motion 4′-30″:–so then in two houres 36 min. elapsed betwixt both observations it ought to have moved 11′ 42″ in the said great circle which passed 10$\frac{1}{2}$′ onely to the South of y^{e} Star from which these distances were measured. The Comet was in consequence of y^{e} Star. and 58° 13′ high in y^{e} first observation, but at the second onely 36° 59′ so that now the refraction contracts y^{e} distance & elevates y^{e} Star towards it about $\frac{1}{2}$ a minute wherefore if the Comet have no parallax wee ought at the latter observation to find its distance but 11′ 12″ encreased.- Accordingly the distance which at 6° 14′ was 3° 07′ at 8^{h} 50′ wee find to be 3° 18′ 05″ just 11′ 05″ more so that it is evident the Comet has no sensible parallax.

Had the horizontal parallax of the Comet beene as the Peripateticks suppose & some printed papers have affirmed[18] 60 minutes or the same with the Moones the Comet in the later observation ought to have appeared nearer to the star then in ye first by 3 or 4 minutes as if I had time I could easily demonstrate. but on the Contrary it appeared wider as it ought of the parallax were insensible which I therefore assume it to have been on ye certeinty of my observations. by which haveing determined this question I shall now proceed to recount y^{e} rest, & that breifly, afterwards I shall propose a theory to salve them Jan: 13 at 6.56 the observed distances of the Comet were from y^{e} bright ★ of ♈ 14°-08′-10″ from y^{e} last of Pegasus wing 22°-28′-20″ then its longitude 25° 59$\frac{1}{2}$′ latitude 22° 17′$\frac{1}{2}$ North:

The motion in y^{e} arch of a great circle intercected betwixt this place & that observed in y^{e} 10th in y^{e} evening is 5° 05′ 40″ the place where it cuts the Ecliptick ♐ 16° 21′$\frac{1}{2}$ the angle 28° 01′

Hence to y^{e} 25. I had no observation partly because of clouds &

some of y^{e} Moone but by the measures taken this evening I find at 7^{h} 44′ its distance from y^{e} North horne of the Bull 39° 40′ and from y^{e} Bright Star of ♈ 10°-07′-10″ the distance betwixt the two stars 44° 43′ 30″ whence the place of the Comet will by Calculation be found in 9°-35′ 48″ with 17°-56′-54″ North latitude. Jan y^{e} 30 at 7′ after 8 a clock its distance from y^{e} North Horne of y^{e} Bull 35° 54′ and from y^{e} bright side of Perseus 18° 43$\frac{1}{3}$′ and the distances of y^{e} two stars 31° 22′$\frac{1}{2}$ whence its place will be found in ♉ 13° 19′$\frac{6}{10}$ its latitude 16° 41′ North. Feb y^{e} 2° 20′ after 6 its distance again from y^{e} North horne of y^{e} Bull 33 58$\frac{1}{2}$ from y^{e} bright side of Perseus 18° 04′ hence its place ♉ 15 13$\frac{4}{5}$ its latitude 16° 02′. Feb y^{e} 5th at 50′ after 6 the distance of y^{e} comet from y^{e} north horne of ♉ was 32° 11′ 00″ from y^{e} bright side of Perseus 14° 00′$\frac{1}{2}$ hence its place ♉ 17° 00′ latitude 15° 27′ I could not now perceave any tayle y^{e} Comet had but onely a dusky light with my bare eye. after this I never observed it for it was growne so small that it was difficult to determine when it passed y^{e} measureing thrids in y^{e} Telescope.

Haveing thus acquainted. y^{u} with my observations I shall adde a remarke or two more & then shew y^{u} a Theory which will answer them.

Mr Newton ye learned professor of Astronomy at Cambridge, tells me in a letter of his,[19] that a Scholar who saw the Comet when orientall, informed him, its head appeared to ye Naked eye no bigger y^{n} a Star of ye 3^{d} light, but viewd with a perspective onely an haze, or like a nebulosa. the same hapned to me at first when it emerged from ye Sun on Dec ye 12th as I tould y^{u} before it appeared of a dull reddish colour so maid as any of the planets but on ye 21th of Dec. I was surprised to find no such star in ye head but onely an hazy light tho to ye Naked eye it appeared something bigger then a star of the 3^{d} magnitude. hence it seemes to be intimated that both were the same Comet for which I have this farther argument. Dec ye 21th at night by its observed longitude I find its Elongation from ye Sun 24 deg. the Roman observations of Gallet & Collio[20] . . . to ye same elongation the 26 of November in ye Evening the Middle betwixt these is Dec 9 at which time it ought to be in ☌ with y^{e} ☉. Now Mr Halley saw the tayle ye 8th day in ye Morneing erect from ye Easterne horizon. & on ye 10th it was seene in ye evening riseing streight up from ye West whence it follows that on ye 8th it was not arrived at ye Sun on the 10th it was past him haveing North latitude in both therefore y^{e} ☌. ye 9th day as ye observed elongations make it the like time ☌ may also be found by compareing the Roman of the 21 of Nov: with ye Greenwich observations of December ye 26 in which ye Elongations of ye Comet from ye Sun were equall; the 9 of December y^{e} day of y^{e} ☌ falls betwixt them.

It is further very remarkable that whereas in y^{e} Canterbury observations of y^{e} Comet made November y^{e} 11th its latitude was 2 degrees north the Roman [ones] of Gallet make it to have no latitude on y^{e} 17th of November Collios $\frac{1}{2}$ deg South. Gallets observations 4 dayes after make y^{e} latitude 4 deg South Collios but one. in which latitude hee makes it continue all the remaineing time of its appearance with whom y^{e} Lovaine print agrees, whereas Gallet make the latitudes againe decrease & but one degree on y^{e} 26th day with Collio. I suppose Collio might be something too little, Gallet too much, tho some observations from Straßburge[21] seeme to Counter him. It was o[n]ly Mishap that this Comet Could neither be observed at Greenwich nor Paris, else wee might have beene more certeine of those which give us a strong intimation of the Nature of its motion. For haveing described the orbit of ye earth as in the 4th figure [Fig. 3.4 ≡ Fig. 15, p. 30] & therein designed its places on ye dayes of ye Canterbury & Collios observations and layd of the prickt lines in which the Comet appeared I found that if I allowed the latitude to have beene North as observed at Canterbury on ye 11th, but one degree South as Gallets & Collios observations both make it on ye 18 of No: it ought afterwards to proceed into greater Southerne latitude. & not to keepe continually in ye Same, as Collio, nor to returne to it & peirce its plaine as Gallets & the Greenwich observations of it after y^{e} ☌ with ye Sun prove it did for I have given y^{u} a probable argument that both were the same appearance. This accident in its motion put one upon thinkeing what might be the cause of it, & the most probable reason seemed to be that as the Earth is a vast magnet attracteing all things within its little vortex.[22] So is the Sun to all that move in his greater, attracteing the planets to him & repelling them, accordeing as they have theire proper poles (by theire fixed inclinations & revolutions) turned toward or removed from him thereby causeing theire Aphelion & Perihelion distances for which otherwaies I can not find any likely reason. The Comet I conceave to have something of an answerable quality. whereby when it came within ye distance of the Diameter of the Earths orbit from ye Sun it became sensible of his attraction & inclined towards him its proper Course being from ye North into ye South side of ye plane of ye Ecliptick. but as it approached him nearer it was drawne more sensibly and strongly by his northerne pole, & would have run streight up to ye Sun as from ye pointe e, where it was observed on ye 24th of November. in ye line eZS but that by the Motion of ye Vortex from y by Z it was borne wide of it into h where it appeared the 27th day & tho the attraction of the Sun became more strong as it approached nearer him, yet the violent motion of the vortex also encreasing as it drew nearer y^{e} Sun it was still borne of till Comeing to γ where its acquired motion was not swifter then y^{e} motion of y^{e} vortex according to the succession of y^{e} letters $\alpha\beta\gamma\delta$ it was throwne round him & Car[r]ied on the other side

[of] him & tho its proper direction as I said before was from the north side of the Ecliptick into y[e] South yet by this attraction when it came within $\frac{1}{4}$ of y[e] Radius of the orbit neare it was changed & the Comet tendeing towards him returned and peirced y[e] Ecliptick againe changeing his Southerne into Northerne

I suppose that it kept allwayes the same side forward in the line of its motion. for it is found by experience that the side of y[e] bullet which lies next the muzzle of a true bored peeice when discharged falls first to the ground when it will follow that it is carried foremost in the line of its motion. betwixt the swiftnesse of which & that of a Comet ther[e's] no comparison,

Tis a well knowne quality of the Magnet that as it attracts one end of the Needle touched with it so it repells the Contrary, the like wee imagine to have hapned with y[e] Comet when it came round the Sun neare α (see y[e] 4th figure [Fig. 3.4≡Fig. 15]) had the opposite side to that whereby it had beene attracted turned towards his northerne pole whereby it was repelled most directly then it had beene attracted for as it approached the Sun near it imbibed more of his magnetick particles so had its owne faculty strengthened tho I conceave that it receded not the swifter from him on this account for it acquired that degree of velocity before, y[t] nature could not admit a greater.

And now being got to α where its contrary side was obverted to the Sun it proceeds thence in the streight line $\alpha\beta$ for as y[e] violence of y[e] laterall action of y[e] matter of the vortex grew lesse so allso the strenght of y[e] Repulsion diminished proportionably as it receded from y[e] ☉ whereby it was maintained in its streight lined motion so long as it continued in view haveing moved swiftly neare y[e] Sun proportionably slower at greater distances on this account.

To find the line it described in y[e] 4th figure [≡Fig. 15] haveing put S the Suns place in y[e] focus of the orbite of y[e] earth O♋T, & then layd downe the true places of the earth at the times of the severall observations I draw lines thence to the Sun. also others formeing angles with them equall to the difference of y[e] longitudes of y[e] Sun & Comet which last pointe into observed places. then if wee find in what place another line may be drawne [a]crosse these so as to the spaces cut it may be proportionall to the intervalls of time betwixt severall observations this shall be the line which perpendiculars let fall from y[e] place of the comet upon the plaine of y[e] Ecliptick shall forme there. after which manner the prickt lines of Elongation whereon y[e] places of y[e] comet observed in November are wrote in y[e] 4[th] figure [≡Fig. 15] were found & the line of motion abcdefgh but for December the points ik e[t]c. to x wherein y[e] spaces moved Neare y[e] Sun are shorter then the spaces moved in y[e] same times at farther distances & that this is the very line wherein it moved. the observed latitudes seeme to prove. for haveing divided the line ix in y[e] 5[th] figure[23] of the same length into the same spaces the line of the Comets motion is divided in the 4th figure [≡Fig. 15]. If makeing the distance betwixt each place of the earth in its orbite & answerable places of ye comets in its path ye Radij of tangent lines, erect upon it perpendiculars equall to ye tangent of ye Comets observed latitude at these severall places. y^u shall find there extremitys to fall on a streight [line] whence it is probable that the Motion of the Comet was wholly in a streight line after it emerged from ye Sun.

I may not omit to give an account of ye tayle because tho least admirable it was most taken notice of & is readily answered. if we consider first that those spots in the Moone which wee esteeme ye humid or Aqueous parts of it reflect little or no light but onely as ye Comet did What if wee say then that humid part of it overspread the rest which was the Reason that it never shone bright as other comets did but like a cloud in ∞ to ye bare eye. what if we say that whilest it was remote from ye Sun hee raised but few vapors from it but as it approached him & when it was in conjunction which being carried to a great distance by ye action of ye Sun upon them his raies reflected from them caused that appearance which was therefore longest at first shorter after as it receaded farther from him by reason his action upon it was lesse violent.[24]

How other Theorys may answer this appearance I may inquire hereafter haveing beene already too tedious tho I have spared & contracted my thoughts for want of time hereafter wee may have occasion to make a farther inquiry.

NOTES

3.1. This information was obtained from Riccioli 2, pp. 1–130.

3.2. This reference to Seneca's *Quaestiones Naturales*, citing the opinions of Epigenes and Apollonius, was taken from Riccioli 2, p. 3.

3.3. Ioannes Regiomontanus, *De Cometae magnitudine, longitudineque problemata 16* (Norimbergae, 1531).

3.4. Petrus Apianus, *Astronomicum Caesareum* (Ingolstat, 1540).

3.5. Peter Cruger, *Uranodromus cometicus* (Dantisci, 1619).

3.6. Johannes Kepler, *De cometis libelli tres* (Augustae Vindelicorum, 1619).

3.7. This statement is not strictly accurate. The first known globe is, of course, that which rested on Atlas's shoulders; and we are now aware of a celestial globe preserved-in Dresden showing 48 constellations, constructed in 1279 by the Persian mathematician and astronomer Mohammed Ibn Mu' aijad al Ordhi. Nicolas of Cusa possessed one about the year 1400. The first known papier-mâché globe was made by Gerhard Mercator, *alias* Kremer (1512–94) in 1551. Cf. O. Muris and G. Saarmann, *Der Globus im Wandel der Zeiten* (Berlin, 1961).

3.8. Riccioli 2, pp. 11, 12.

3.9. Riccioli's discussion (ibid.) ends with the comet of 1618–19; his *Almagestum novum* having been published one year before the next comet of 1652 mentioned by Flamsteed.

3.10. The name of the observer was Abraham Hill. Flamsteed's diagram of the comet's path illustrating this and other later observations made in Rome, Paris, and Greenwich is preserved in a letter to Richard Towneley dated 22 March 1680/1 (Royal Society MS, 243, F.1.52). Cf. Fig. 15, p. 30.

3.11. A copy of Jean Charles Gallet's observations was sent to Flamsteed by Halley, who had actually been present in Rome when they were made, in a letter from Paris dated 22 January 1681—together with those of Antonio Collio and G. D. Cassini and his associates at the Paris Observatory. f. 27ᵛ cf. p. 107.

3.12. Halley to Flamsteed; Paris, 22 January 1681. The original of this letter has unfortunately not been preserved, but its contents can be inferred from Flamsteed's reply of 17 February (Baily, op. cit. [note 2.39], pp. 123–4) and from his letter to Towneley five days later (cf. note 3.10).

3.13. Ibid.

3.14. This item does not appear to have been preserved among Flamsteed's papers.

3.15. Petrus Gassendi, *Opera Omnia*, iv (Lugduni, 1658), pp. 77–79.

3.16. Cf. f. 32ᵛ, p. 114 for Flamsteed's decision on this matter in favour of the former alternative. The manner in which he reached it is explained in the Introduction, pp. 27–9.

3.17. Joannes Bayer, *Uranometria omnium asterismorum* [51 plates] (Augustae Vindelicorum, 1603).

3.18. No fewer than fifty publications, mainly by German authors, on this Comet of 1680 are indexed in *The Catalogue of the Crawford Library of the Royal Observatory Edinburgh* (Edinburgh, 1890), pp. 127–30. It is, however, unlikely that Flamsteed was aware of most of these.

3.19. Newton to Crompton for Flamsteed; Cambridge, 28 February 1680/1 in H. W. Turnbull (ed.), *The Correspondence of Isaac Newton*, ii (Cambridge, 1960), pp. 343–4.

3.20. Cf. notes 3.11 and 3.12.

3.21. There is, unfortunately, no clue as to how Flamsteed came to possess these observations, which may have been made by Johann Caspar Eisenschmidt (1656–1712) who was active in Strassbourg at this time.

3.22. The magnetic hypothesis was first expounded in William Gilbert, *De Magnete* (London, 1600), and adopted by Kepler (op. cit. [note 15.5]) as a basis for his physical explanation of planetary motion. This concept was much in vogue during the middle decades of the seventeenth century, both in England and on the continent of Europe.

3.23. Probably the sketch enclosed with a letter from Flamsteed to Crompton for Newton, 7 March 1680/1, op. cit., ii, p. 355; (cf. Introduction, pp. 32–3).

3.24. At this point, Flamsteed has written but then crossed out the following passage:

> Yᵘ may inquire last what was yᵉ body of yᵉ Comet I shall suggest something to yᵘ may suggest & leave yʳ selves to determine. I defined its parallax insensible but by The theory on yᵉ 12ᵗʰ of December it was as far distant from yᵉ Earth as yᵉ Sun is & its diameter then about halfe a minute let us suppose it no more yⁿ yᵉ latter which in my last I told yᵘ was 20 seconds it will follow therefore that it was considerably bigger yⁿ oʳ earth it had yet it wanted part on yᵉ out side but with some prominences of more substantiall matter, as appeared by some vivid bright spot but exceeding small which were seene in yᵉ first nights observations.

Lecture 4

Wednesday, 8 *June* 1681

I doubt not but the better part of this Auditory is well acquainted not onely with the ages but the originall of optick glasses, to which wee owe the discovery of ye inequalitys of ye Moones Surface, the aemulous phases of ♀, the satellits of ♃, & his revolutions about his axis, the ring about ♄ and his Satellites; tho it has beene found by them that the Nebulous places in ye heavens, & the Milky Way were onely made up of greater Numbers of small fixed stars placed nearer to each other then elsewhere in the heavens,[1] & tho it has beene concluded y^t observations of coelestiall heights & distances might be much more accurately observed by them then by the common sights,[2] hitherto used on instruments, Yet such has beene the fault of o^r time that hitherto very little materiall on this subject has been published in o^r language.[3] the Raies of light passeing from a thin Medium, as Aire, into a thicker as Water, or Glasse, were refracted or bent inwards towards the perpendicular on the plaine of theire Incidence; but emerging from a dense or thicker Medium into a rare or thinner [one] they were turned of from the said perpendicular. & proceeded wider from it, As in the first figure [Fig. 4.1] if wee suppose egfh a solid peece of glasse whose superficies eg, fh are exactly polisht & plaine let ci be a perpendicular on the middle of this plaine and oc a ray of light falling upon ye said plaine in ye point c. Now if the medium of aire were continued it would proceed in ye said streight line oc to k, but falling on ye denser Medium of ye glasse at c its course is changed & bent inwards towards ye perpendicular ic continued. so y^t in ye glasse it moves along the line cl to n Againe let n be the point upon which it will emerge forth of the glasse from which conceave pn to be raised perpendicular to ye plaine fh, on which it emerges. had the medium of glasse beene continued it would have proceeded to l but here meeteing with the aire a thinner medium then glasse, it is againe refracted & bent not inwards & towards ye perpendicular pn as when it passed from aire to glasse, but outwards, & from it, so as that from n it runs of in y^e line nm wider from y^e perpendicular y^n y^e line nl.

Notes on pages 132–133.

The angle which y^e ray of light moveing in the free aire formes at y^e point c with y^e perpendicular ci = to acd is commonly called the angle of incidence. or inclination. but acn which y^e ray formes with y^e said perpendicular after it is immersed in y^e glasse the refracted angle & the angle ned intercepted betwixt the two courses of y^e refracted & unrefracted ray the angle of refraction.

The famous de Cartes.[4] & about ye same time with him our countriman Mr Gascoigne[5] were the first that determined that ye sines of ye angles of incidences were proportionable to the sines of ye angles refracted.

Concerneing the proportions betwixt ye sines of the sayd angles few Authors agree with each other by reason that they are difficult to be observed. The usuall way of observeing is by the prisme or triangular glasse through which if the Suns raies be transmitted it is found that his species receaved on a white paper held behind it will not be round as according to common reason it ought to appeare, if cast through a small hole placed betwixt it & ye Sun, but ovall, & longer then broad in all proportions betwixt 14 & 5 to 1. & spread over with different colours which as Mr Newton[6] has well showne prove that the raies of light are of different formes & that ye severall sorts are proper for representing severall colours; now before this discovery of his those that made experiments by the prisme for ye measures of refractions were doubtfull on which part of the Coloured oblong image of the Sun they ought to pitch some chose ye one end some the other. some the middle, hence proceeded those severall proportions authors have assigned betwixt the sines of ye Angles of incidences & the angles refracted Mr Newton makes them as 300 to 193.[7] others much lesse different.[8] I state them in round numbers as 3 to 2 which proportion will serve for use as well as any other.

And now the angle of incidence of any ray of light being given the refracted angle will easily be found let oc in ye first figure [Fig. 4.1] be produced through the glasse from c to k with any radius tc strike an arch intersepteing the line ck in d. & from d let fall da perpendicular to ct which divide into 3 equall partes take 2 of them betwixt y^r compasses & carrying one pointe along the arch td find the pointe r on which turneing the compasses about the other shall onely touch the line tc. if from c the pointe where the ray of light falls into the glasse y^u produce the line cr through this point it shews y^u the way of the refracted ray through the glasse.

In like manner is they way of the ray of light after its emersion from y^e glasse to aire againe determined. Produce y^e line of its way in y^e glasse cn to l. from n the point of emersion raise y^e perpendicular np with any radius nτ Strike an arch $\tau\rho\delta$ intersecting the line cl in ρ. from whence let fall $\rho\beta$ perpendicular to np. divide it into 2 equall parts & addeing one of them to y^e whole lenght take y^e Space betwixt y[r] compasses & finde a point in the said arch as at δ whereon setting one foot of y[e] Compasses the other turned about shall onely touch the line np. draw the line nδ this shall shew the course of [the refracted] ray after its emersion from the glasse nearely parallel to ocdk the true line of its course if it had not beene broken by the interposition of the denser but transparent medium.

Hence it is evident that the appearance of any object viewed through a peece of polisht glasse whose two superficies are parallel to each other shall not be sensibly different to its appearance if no glasse were interposed. The thicknesse of o[r] Common glasse being so inconsiderable that the Rayes of light after theire emergeing from it passe nearely in the same line they moved in before theire incidence. at least theire is no difference to o[r] senses.

Experience is the ground of all o[r] knowledge & from this experiment of refractions all the phainomena of Dioptricks and telescopes will be demonstrated. upon which account I have beene the larger on it the next & simplest figure after the plaine is the sphere which we may conceave to consist of infinite small plaines to each of which lines drawne from y[e] Center shall be perpendicular I shall therefore next shew y[u] how the raies of light passe from transparent solid, or Segm[t] of a Sphere.

In y[e] 2[d] figure [Fig. 4.2] let V represent y[e] glasse one of whose sides, du is a plaine the other egw a sphericall superficies the Radius of whose convexity <u>cw</u> is two inches <u>c</u> its center <u>w</u> its pole & therefore the line <u>cwi</u> produced through the center & pole of y[e] glasse the polar Ray, let du y[e] plaine side of y[e] glasse be observed in y[e] object, & let us conceave two raies of light procedeing from y[e] same point of a very remote object as the Sun, parallel to y[e] polar ray, the one of them at $\frac{1}{10}$ of an inch distant from it as <u>bf</u>, the other at $\frac{5}{10}$ as ad & let us suppose the Radius of the convexity cw to be precise 2 inches or 20 such partes as fu is one, and du five.

Conceave also that y[t] y[e] plaine superficie of the glasse du Stands at right angles to the polar ray cu which therefore can not be refracted in y[e] pointe <u>u</u> where it passes from y[e] rare medium of aire into y[e] denser of glasse because it is equally inclined to y[e] plaine on all sides but passes directly forward through it to <u>w</u> & thence into y[e] aire againe in a streight line towards i

For if wee draw a tangent line to y[e] pointe <u>w</u> where it emerges, that shall represent the plaine on which it passes forth of the glasse, which being perpendicular to <u>cw</u> & parallel to <u>du</u>, the ray can be no more refracted there, then it was at <u>u</u> where it entered y[e] glasse, & was not refracted at all: so y[t] y[e] polar Ray passes directly through the glasse as the line <u>cwi</u> without any deviation. to find where the ray <u>bf</u> which runs parallel to it shall meet with and intersect it. consider that because cu was perpendicular to y[e] plaine <u>du</u> & bf is parallel to it the said ray bf falls also perpendicular to y[e] plaine du, & can not be refracted in the point f, at its entrance into the glasse, but proceedes in the said streight line till it arrives at g where it emerges from y^e denser medium upon its convex superficies.

Produce the line bg or conceave it produced indefinitely to s. and from c. y^e center of y^e curved superficies through y^e point g draw y^e streight line $\overline{\text{cgt}}$ Then because cg is perpendicular to y^e point g or the plaine touching the curve on it, the angle bgc = gcw = tgs shall be the angle of incidence of y^e ray bf on y^e sphericall superficies at g the quantity of which angle may be thus found.

From g let fall gy perpendicular to y^e Axis of the glasse or polar ray ch. in y^e Right angled triangle cgy are knowne the hypothenusa cg 20 partes such as gy = fu is one whence y^e angle gcy = bgc, will be found 2°. 52′ and cy $19\frac{98}{100}$: so y^t y^e point ȳ falls onely $\frac{2}{1000}$ of an inch within y^e glasse from y^e pole w.

Now because when y^e rays emerge from a dense medium into a rare [one] they are refracted from the perpendicular the sines of y^e angles of incidence or emersion being to y^e sines of y^e Ref. ∠∠s as 2 to 3 it will hold as 2 to 3 so is y^e sine of y^e angle bgc = tgs 2°. 52′ to y^e sine of y^e angle tgq 4°. 18′$\frac{1}{15}$ whose difference from y^e angle tgs will be 1°. 26′. 06″ y^e angle of refraction = sgq:

Now because y^t cgt is a streight line if wee subtract this angle tgq 4°. 18′ from 180° y^e remain[d]er 175°. 42′ shall be the angle cgi. which togeather with y^e angle of incidence $\overline{\text{gci}}$ 2°. 52′ and the side gc 20 parts, being given, in y^e triangle $\overline{\text{cgi}}$ the side ci will be found 59 such parts & $\frac{9}{10}$ therefore wi $39\frac{9}{10}$ so that it is evident the raies of light emergeing betwixt g and w shall all be collected at y^e distance of twice the Radius of that sphere in which the glasse was formed or at the space of a diameter from y^e pole w. for y^e sagitta $\overline{\text{wy}}$ is insensibly small & if the glasse be supposed turned about on its axis ci the point g will describe a Circle & what is said of y^e ray bfg will hold of all the raies falling in that circle parallel to y^e Axis or Polar Ray ci.

Or if line gi into which y^e ray is bent after that it emerges from ye glasse may be found Geometrically thus. haveing produced the ray bg & perpendicular cg as far as shall be conceaved requisite with any convenient distance of ye Compasses setting one foot on g with ye other describe the arch tsq. from ye point s where it intersects the line of ye ray produced let fall sr at right angles to the perpendicular produced ct. bisect this line & adding the one halfe of it to ye whole take that space betwixt ye compasses & find a point in ye arch ts as atq: that one foot of ye compasses being set thereon the other turned about may just touch ye perpendicular produced as at p. a line drawne from g through ye point q. shall be that which the raie bf proceeds in after it has emerged from y^e glasse, & the point i where it intersects y^e Axis the nearest place to y^e pole w in which any of the parallel raies emergeing betwixt w̄ & ḡ can meet or intersect the polar as will appear by the following enquiry where y^e ray ad parallel to y^e polar cw but $\frac{5}{10}$ of an inch distant from it after its emersion from y^e glasse shall intersect y^e axis ci.

Haveing produced the ray ade to n̄. & the perpendicular ce to o. & let fall ex perpendicular to y^e Axis ci. I find in the triangle cex the angle of incidence ecx = cea = oen 14° 28′$\frac{2}{3}$, and y^e side cx 19$\frac{39}{100}$ therefore y^e Sagitta wx onely $\frac{61}{100}$ of an inch.

Then as 2 to 3, so mn y^e sine of 14°28′$\frac{2}{3}$ to kl y^e sine of 22°01′$\frac{1}{2}$ or y^e angle oel. which subtracted from 2 right angles or 180 degrees leaves the angle ceh 157° 58′$\frac{1}{2}$ with which, y^e side ec 20 partes & y^e angle of incidence ech 22-01$\frac{1}{2}$ in y^e triangle ech I find the side eh 57$\frac{14}{100}$ & therefore wh the distance of the point where the sayd ray intersects y^e Axis from y^e pole w 37$\frac{1}{4}$ partes nearer to it by almost $\frac{3}{10}$ of an inch then [the] place where y^e Ray bf did intersect it hence it followes that all the raies emergeing betwixt g and ē shall be collected or intersecte y^e Axis betwixt i and h and that in every plano convex glasse. The rayes that are remotest from y^e polar ci intersect it or are collected nearest the pole w but the nearer parallell Raies bg further of: yet none save those which are equidistant from ye Polar & parallel to it in the same point.

Thus y^u see y^t if the plane side of a single convex glasse be turned to ye Sun or any other object all the raies which fall on it parallel to ye Axis suffer but one refraction & are collected at less y^n one diameter from its pole or one & an halfe from its center c but if the convex side be turned to ye object wee shall find them to be refracted twice & ye point of concourse much different the Calculation more difficult or rather prolixer for if as in y^e 3^d figure [Fig. 4.3] wee suppose c y^e center of a single convex glasse whose Radius cw is 2 inches w its pole & wc its polar Ray and Axis, owP its convex Superficies converted to any object & subtending a quadrant, ag a ray of light moveing parallel to y^e axis at $\frac{1}{10}$ of an inche widenesse from it, be another ray $\frac{5}{10}$ of an inche wide from y^e same Axis, from c y^e center of y^e section draw y^e streight line cg. & from g let fall gy perpendicular to y^e Axis in y^e triangle gcy are knowne cg y^e Radius of y^e convexity & gy the distance of y^e point of incidence g from y^e Axis whence y^e angle of incidence gcy will be found 2°52′.

From g ye point of incidence draw gf through ye glasse parallel to ye Axis wc The angle fgc will be equall to gcy, the line gc is perpendicular to ye plane of incidence, & becaue ye raies of light passeing from a rare to a dense medium, as from glasse to aire, are refracted towards the perpendicular the sines of y^e angles of incidence being to y^e sines of y^e angles refracted as 3 to 2. it will hold as 3, to 2: so y^e sine of y^e angle of incidence fgc, 2°. 52′, to y^e sine of y^e angle refracted qgc 1°. 54′$\frac{2}{3}$, which subtracted from y^e said angle of incidence 2°52′ leaves y^e angle of refraction fgq 0°57′$\frac{1}{3}$

If y^e medium of glasse were continued from g to t the Ray of light would continue its progresse in the streight line gqs, but the medium being discontinued at q and y^e ray there emergeing from y^e dense medium into y^e aire it is a second time refracted. to find now the line of its progresse and where it shall meet with and intersect the polar or unrefracted ray wi.

First qu the distance of the point on which it emerges from y^e Axis of y^e glasse will be required. y^u may remember that at first I supposed the arch of y^e convexity op, 90 degrees wherefore ow = wP shall be 45 deg. uc 14|142 and uw 5|833[9] such parts as cw is 20.000

In the triangle qcy I found y^e angle at c 2°52′ therefore the side cy 19|975 from which takeing away cu 14142 the residue or difference yu 5|833 shall be = gf the thicknesse of y^e glasse in y^t place.

Which being given with y^e Angle fgq 0°.57′$\frac{1}{3}$ fq will be found 97 such parts as cw is 20.000 & therefore qu 903

From q draw y^e streight line qt parallel to ui, this shall be the perpendicular on the plaine to y^e point of emersion, & produce the line gq to s because at q the ray emerges from y^e dense medium of glasse to aire it will be Refracted from y^e said perpendicular & it will hold As 2: to 3: so y^e sine of y^e angle of emersion tqs 0°57′$\frac{1}{3}$ to the sine of y^e refracted Angle tqi 1°.26′.01″ = qui: & now in y^e ∟∟ [i.e. right-angled] triangle qiu wee have got the side qu: & the [angle]uiq whence iu the distance of y^e point of intersection i. from y^e plane of y^e glasse will be found 36.070 but from its pole w 41|900[10] or if wc be 2 inches wi shall be 4 & $\frac{19}{100}$ of an inch.

Haveing thus found how a ray of light parallel to y^e Polar or Axis of the glasse is carried through it & where it intersects or is collected with it when the distance of y^e point of incidence $\frac{1}{10}$ of an inche wide from y^e pole. I shall next enquire where a Ray falleing upon it, at a wider distance shall be collected or intersect it

Let be repre[se]nt another ray of light whose course in the free aire is parallel to y^e Axis wc, let ex[11] y^e distance of y^e point of incidence from y^e Axis be $\frac{1}{2}$ or $\frac{5}{10}$ of an inche Suppose allso be in d: the angle of incidence ced[12] will be found 14°28′$\frac{2}{3}$ & cx 19|365 Such as cw is 20.000 from which takeing cu 14.142 the remain[d]er ux shall be equall to ed. the thicknesse of the glasse at y^e place of incidence 5223 of y^e like parts

Because the ray falleing upon e passes from a rare into a dense medium the refraction shall be towards y^e perpendicular, and as 3: to 2:: so y^e sine of y^e angle of incidence ced 14°28′$\frac{2}{3}$ is y^e sine of y^e angle [of] refraction εec 9°35′$\frac{2}{3}$ whose difference from ced shall be the angle of refraction deε 4°53′:

Then in y^e triangle dεe right angled at d are knowne the side de 5223. and y^e said angle deε 4°53′ whence y^e side dε will be found 446 which subtracted from du = ex 5.000 leaves εu 4554:

From ε draw εn parallel to ui it shall be the perpendicular to y^e plaine of on y^e point of emersion at ε.

Conceave also eε. the line of the Ray[e]s waye in the glasse produced beyond it to m. the angle of emersion nεm be = to deε the refracted angle in the glasse & Now because the ray emerges from a dense medium into a rare [one] it will be refracted againe from the perpendicular. and it will hold as 2:3:: so y^e sine of y^e angle of emersion nεm = deε 4°.53′ to y^e sine of y^e refracted angle nεh = εhu 7°20′$\frac{1}{6}$ for uh & nε are parallel

Now then in y^e triangle εhu are knowne εu 4554 and y^e angle at h 7°20′$\frac{1}{6}$ whence uh the distance of the point of intersection h from y^e plaine side of the glasse will be found 35|370 but wh or its distance from y^e pole 41.230 such partes as an inch conteines 10 000

Wee found that the parallel ray of light which was but one tenth of an inch distant from y^e Axis was collected at the distance of 41290 from y^e pole w

So y^t all the raies of light which fall on y^e glasse betwixt y^e points g and e are collected within the space hi on y^e Axis which when y^e convex side of the glasse is turned to y^e object exceeds not $\frac{7}{100}$ or y^e 14 part of an inche whereas when y^e plaine side was obverted to the object the said space was found $\frac{25}{100}$ or $\frac{1}{4}$ of an inch at least:

whence it is evident that if the Convex side of the glasse be turned to the Sun any Combustible matter shall be sooner & more violently kindled then if the plaine were for y^t y^e parallel rayes falleing betwixt w and e are collected in one fourth of y^e space in y^t Case that they employ in this when ye plaine side is toward him & consequently the motion of action of the raies is much more strong & violent. but this effect is at a greater distance for wi, in y^e 2^d figure & y^t case is found 39|650 or lesse then twice the Radius of y^e Glasses convexity whereas in this case when y^e convex side is toward the Sun tis 41|930, that is $\frac{28}{100}$ of more y^n $\frac{1}{4}$ of an inch greater and above y^e diameter of y^e circle from y^e pole w.

Haveing thus shewed y^n how the rayes of light passe through a single convex glasse and where they intersect the Axis when either the plaine or convex side is toward y^e object. I shall next shew y^u how it may be found where they will intersect the said Axis when transmitted through a double convex of the like radius

In y^e 4th figure [Fig. 4.4] let W represent such a glasse cw = fu the Radius of its convexity 2 inches or 20.000 parts and uw the Thicknesse of the glasse 8000 of the like parts the line fc connecting the centers of the two convex superficies shall be the Axis of the glasse its length 32.000 = 2 cw − uw

let ad be a ray of light parallel to y^e axis and its distance from it $\frac{1}{10}$ of an inch or 1000, from the two centers c and f draw two lines thereto cd and df, as also from d let fall dx perpendicular to y^e Axis let also the line ad be continued till it intersects y^e other superficies of y^e glasse in g. then shall fd be y^e perpendicular to y^e point of incidence and y^e angle gdf = dfx the Angle of incidence which will be easily found for in the right-angled plaine triangle dxf are given dx 1000 and df 20.000 when dfx will be 2°.52′ and fx 19.975 which taken from cf 32.000 leaves xc 12 025

With which & dx in y^e right angled triangle dxc the angle at c will be found 4°46′ and y^e side cd 12|069

And because y^e Ray at d passes from a rare into a dense Medium the angle of incidence gdf being 2°.52′ the Angle of refraction gde will be found 57′20″ to which addeing the angle adc 4°46′$\frac{1}{6}$ the Complem^t of theire summe to 180° will be y^e angle cde 174°16′$\frac{1}{2}$; for $\overline{ag}$ is a streight line Which angle being given with y^e sides dc and ce found before, the side de will be collected thence 7|957 the length of the raies way thorow y^e glasse. & because ce is the perpendicular at its emersion, the angle of emersion dec = tes 3°.27′

From e let fall eg at right angles to dg or ch in the Narrow triangle dge are now given the side dc 7|957 and the angle gde 57′.20″ whence ge will be found 133 partes which taken from gy 1000 leaves ey 867. which with ce 2000 in y^e triangle ecy being knowne will give the angles ech 2°.29′$\frac{1}{10}$

The ray (at light) of light emergeing at e will be refracted from y^e perpendicular therefore the angle of emersion tes being as before found 3°27′$\frac{1}{10}$ the refracted angle teh shall be 5°10′$\frac{5}{6}$ & therefore ceh 174°49′$\frac{1}{6}$ which added to ecy the complem^t of y^e Summe to 180 degrees shall be y^e Angle ehc 2°.41′.$\frac{2}{3}$

And now in y^e triangle ceh all the angles with the side ec is knowne. whence the side ch shall be found 3.8410 or 3 inches and something more then $\frac{84}{100}$ of an inch. from which subtracting cw, 2 inches the residue wh shall be 1 inch & $\frac{84}{100}$ at which distance from y^e nearer pole w the ray ab $\frac{1}{10}$ of an inch distant from y^e axis & incedeing parallel to it in y^e free aire is, after it has passed the glasse collected with it as also will be all the raies incedeing parallel to it at that distance.

It will bee long & troublesome to repeate the worke of another raye which may passe $\frac{1}{2}$ inch or $\frac{5}{10}$ of one distant from y^e axis & Calculateing after the afore revised manner I find that if x be supposed 5.000 ch will onely [be] 36|853 or 3 inches and about $\frac{68}{100}$ of one So y^t all the raies which passe parallel to y^e axis at $\frac{5}{10}$ [inch] distance shall be collected in the point h nearer the next pole w then those which were onely $\frac{1}{10}$ of an inch distant from y^e axis by 15 Centesmes or y^e 7th of an inch. it is evident now that no sphericall glasse will Collect all the rayes incedeing parallell to the Axis in any one point. A single convex faileing near $\frac{1}{4}$ when y^e plaine side is toward y^e object the double convex of the same sphere almost halfe that Space, and y^e plano convex with y^e sphericall side towards y^e object not more y^n $\frac{1}{14}$th.

So y^t the plano convex is to be prefer[r]ed before the rest when y^e convex side is placed to y^e object which property has not hitherto beene minded by any writer of D[i]optricks I have met with: tis further observable hence that y^e plano convex glasse with y^e plaine side towards y^e object collects all the parallel raies into y^e Axis at lesse distance then y^e diameter of its sphere from y^e pointe <u>w</u> but when y^e Convex side is towards it at a greater for whereas the $\frac{1}{2}$ diameter of y^e sphere was supposed 2 inches in each & therefore the diameter 4. the remotest raye in y^e first Case met y^e Axis of 3 inches & 96 cents but in y^e latter at the nearest at 4 inches & $\frac{12}{100}$. but in the double convex when both sides are sections of the same sphere the place of collection will be lesse y^n y^e Radius distant from y^e pole more from the middle of y^e glasse for y^e remotest ray Converge at 1 inch & $\frac{84}{100}$ from the nearest pole w. to which adde halfe y^e convexity or $\frac{4}{10}$ of an inch y^e summe will be 2 inches & $\frac{1}{4}$: y^t is more y^n y^e radius by $\frac{1}{4}$ of an inch:

Honorato Fabri in the 43th proposition of his Dioptricks[13] tells us that when the convex of a plano convex glasse is turned to y^e object the rayes will be collected at something lesse then 2 y^e Radius distance from y^e pole whereas by a due (due) examination wee find y^e contrary let mee advise those that peruse this acute author to be very carefull how they admit any thinge on his authority for hee supposes the refraction in glasse to be $\frac{1}{3}$ of the angle of incidence whereas Me Gascoignes experiments[14] Me Newtons[15] & those of o^r Geometry Lecturer[16] all prove that Not y^e angles but their sines are proportionall

Wee see no sphericall glasses will Collect all the parallel rayes into one point but theire are other figures that will In my next lecture[17] I intend God Willing to shew how they are collected by the Hyperbola and Ellipsis & so proceed to explicate & shew y^e reason & proportys of Telescopes composed of two or more glasses and how they may be applyed to instruments & those rectified as I shall have opportunity:

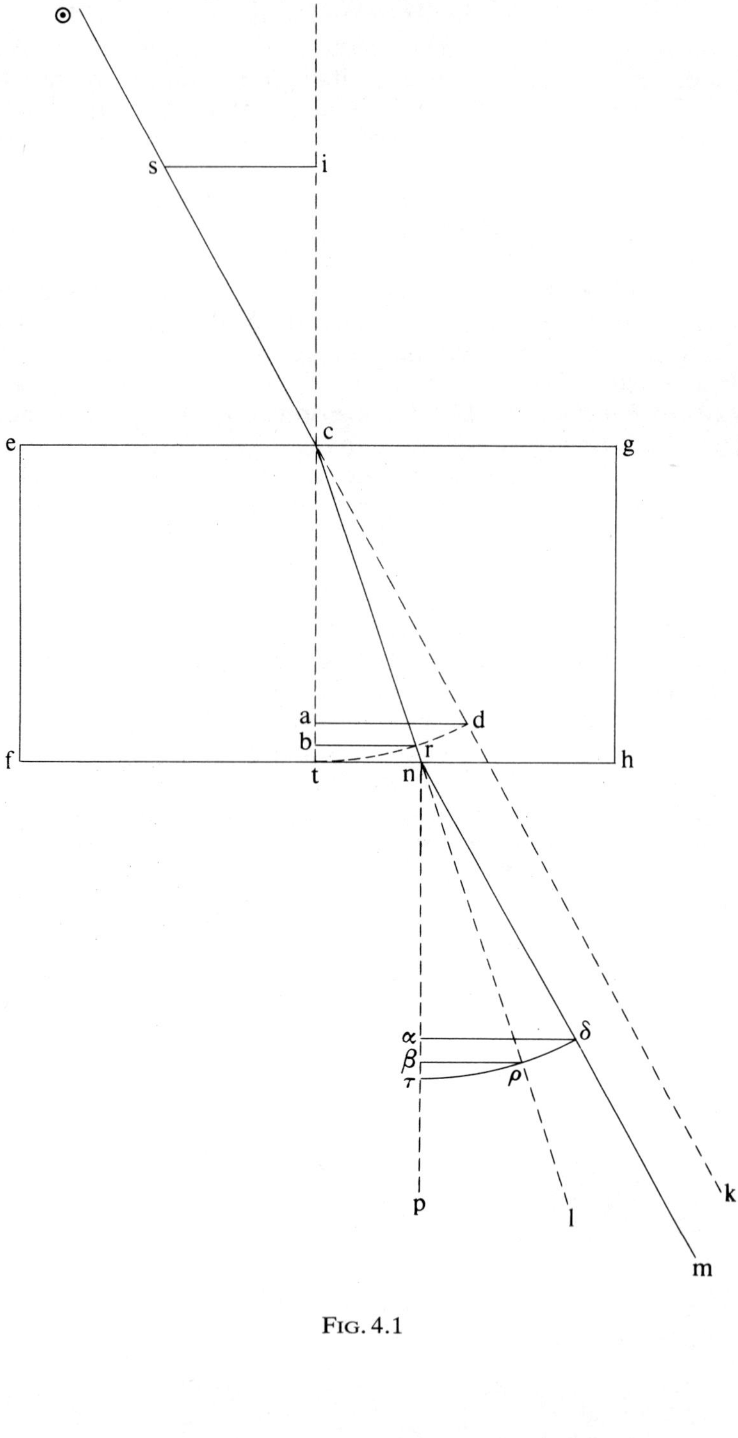

Fig. 4.1

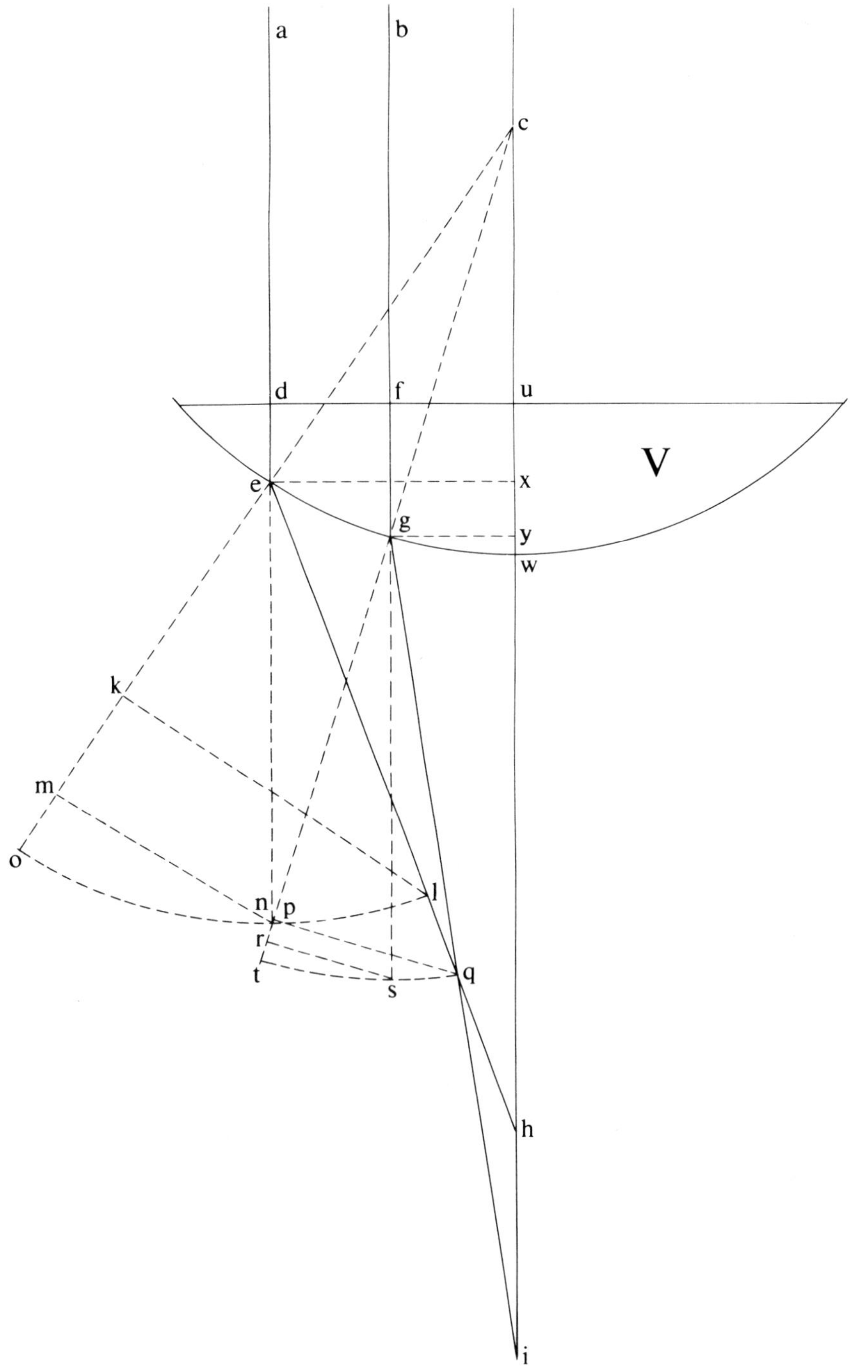

FIG. 4.2

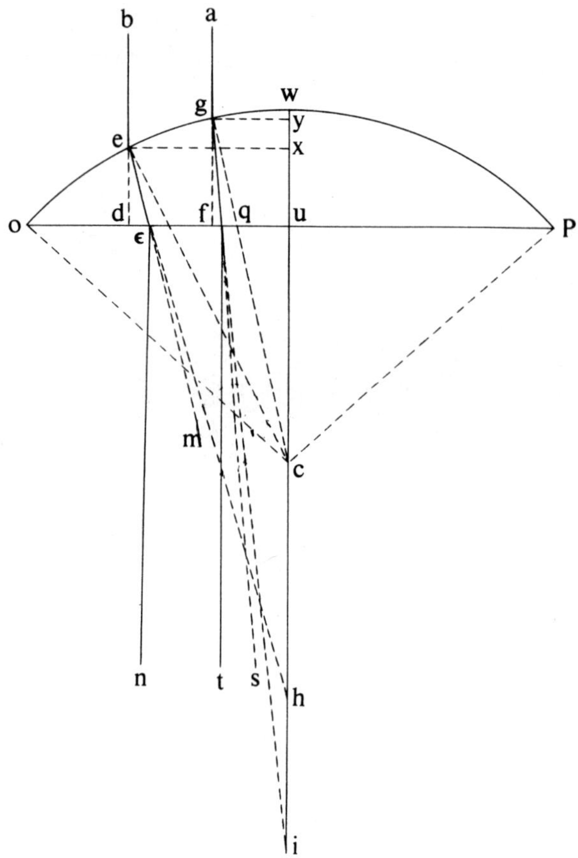

FIG. 4.3 [Reconstructed]

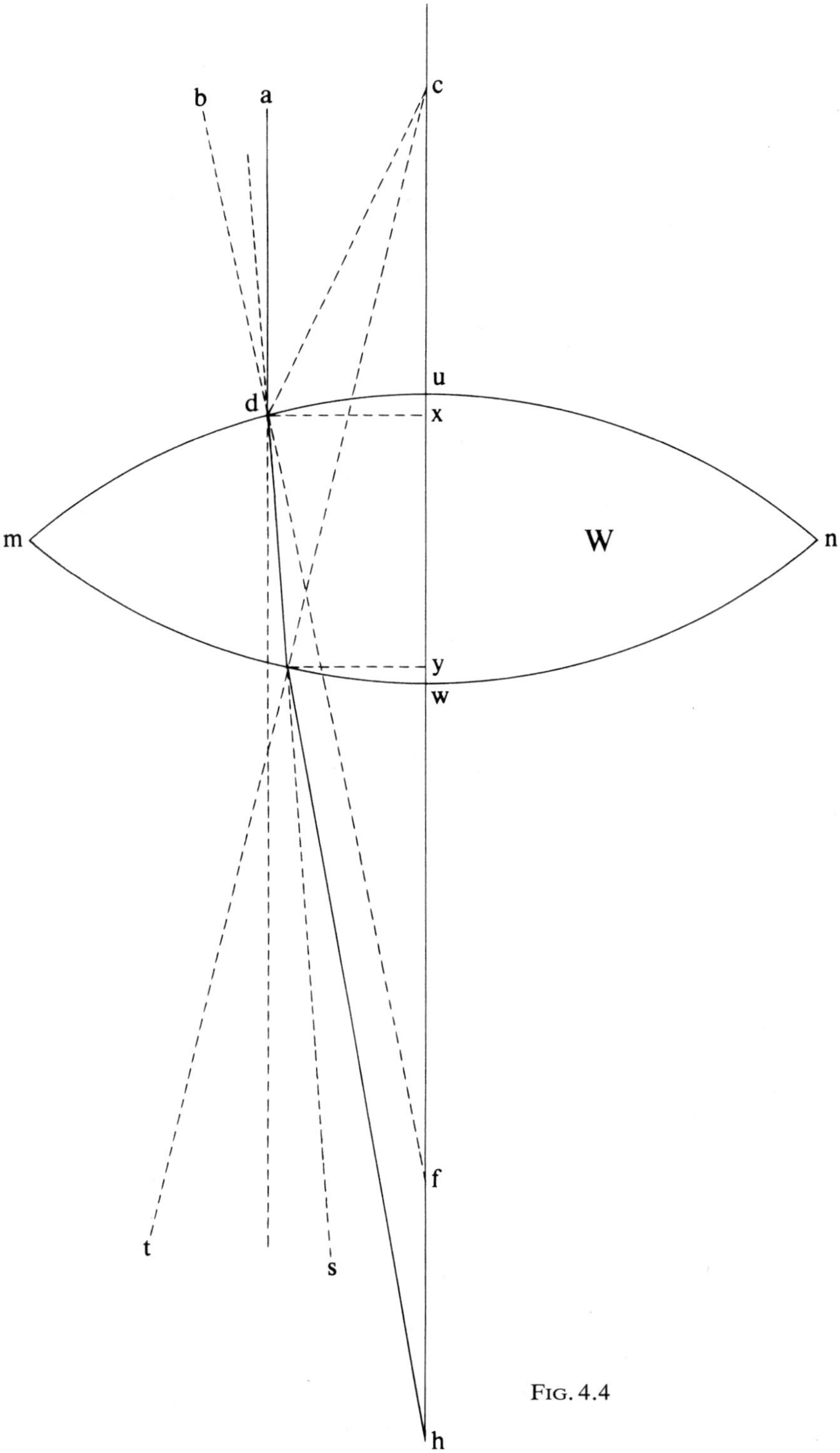

FIG. 4.4

NOTES

4.1. Up to this point, Flamsteed is summarizing the original discoveries made by Galileo and first published in the latter's *Sidereus Nuncius* (Romae, 1610).

4.2. This was accepted by most seventeenth-century astronomers, with the notable exception of Johann Hevelius of Danzig whose strong opposition to the use of telescopes in conjunction with quadrants and sextants provoked mild criticism by Flamsteed and a strongly worded attack by Robert Hooke in his second Cutlerian lecture entitled *Animadversions on Hevelius's Machina Coelestis* (London, 1674). The results of the comparative tests made by Halley during the early part of 1679 (cf. note 2.40) are contained in Hevelius's *Annus Climactericus* (Gedani, 1685).

4.3. The following passage formed part of Flamsteed's original text, but was subsequently crossed out:

> Tho severall learned persons have done well concerneing opticks in y[e] latine Tongue. Yet how glasses may be applyed to instruments & how the faults commonly committed in theire applycation might be amended or rather shund & how all the difficultys suggested by ingenious persons who had not the good to understand them aright might be avoyded the best authors of Dioptricks have been hitherto silent. the French indeed in theire late sumptuous treatise theire Measure de la Terre [by Jean Picard, Paris 1671] have touched on the subject of rectifications but transiently & that booke is rarely to be met with. I shall therefore make it my businesse in this & my following lectures of this terme fully to explaine the Nature of telescopes the reason of their performances, how they may be applyed to Levells, Quadrants, & Sextants. & how the instruments furnished with them may be so rectified & adjusted that they may be free from all suspicion of errors To cleare them from which I am the More obliged by reason that not onely all my owne observations of the heavenly bodyes but the most estimable of others have of late yeares beene made with ye telescope whose originall tho wee seeme indepted to a Mere accident or little triall for it yet is grounded in reason and experience. [Viz. that consisting of a pair of convex lenses] by which it was found long since that . . .

4.4. René Descartes' claim to this discovery has been a matter of debate among historians of science; the Dutch physicist Willebrod Snell being the person who is usually credited with having established the correct law of refraction experimentally in 1621. Descartes gave a logically unsatisfactory demonstration of it in his essay 'La Dioptrique' which is appended to his *Discours de la Méthode* (Leyde, 1637). Flamsteed's own belief was that Descartes found his inspiration in Kepler, op. cit. (note 1.25).

4.5. William Gascoigne (1612–44) made use of the sine law of refraction, which he probably took from Descartes' 'La Dioptrique' (cf. ibid.), in the unpublished optical papers whose contents were summarised by Flamsteed early in 1672 (cf. note 4.14).

4.6. Isaac Newton's 'New Theory about Light and Colors' was published in *Phil. Trans.* vi, no. 80 (19 February 1671/2), 3075–87. Correspondence and papers on this subject are contained in I. B. Cohen (ed.), *Isaac Newton's Papers & Letters on Natural Philosophy* (Harvard University Press, 1958), pp. 27–238.

4.7. Newton, ibid., p. 3077 gives $\frac{31}{20} = 1.55$; but in his *Opticks* (London, 1704) he adopts $\frac{300}{193} = 1{\cdot}51$.

4.8. Descartes' own estimate in op. cit. (note 4.4) was $\frac{3}{2}$.

4.9. Flamsteed, by identifying uw and uy, is effectively ignoring the difference $wy = 5{\cdot}858 - 5{\cdot}833 = 0.025$ corresponding to $\frac{1}{8}$ per cent of the radius of curvature of the lens's convex surface.

4.10. As a result of using the approximate value $uw = 5{\cdot}83$ (instead of $5{\cdot}833$) in his calculation, and neglecting $wy = 0.025$, Flamsteed finds $wi = 41{\cdot}900$ instead of $41{\cdot}928$. The error thereby incurred is only -0.7 per cent.

4.11. Flamsteed wrote bx instead of ex.
4.12. Flamsteed wrote bed instead of ced.
4.13. Honorato Fabri, *Synopsis optica* (Lugduni, 1667), pp. 110–16.
4.14. Summarized by Flamsteed in R.G.O. MSS., P.R.O. 40, ff. 9^{v}–22^{v}.
4.15. Op. cit. (note 4.6).
4.26. Robert Hooke, *Micrographia* (London, 1665).
4.17. This would appear to have been delivered on 15 June 1681, but it has unfortunately not been preserved. The subject has been treated by Claude Mydorge in op. cit. (note 15.15); an English translation of his treatise being appended to the fourth edition of William Oughtred's *Clavis Mathematicae* (Oxonii, 1667). Flamsteed possessed a copy of each of these books in his private library.

Lecture 5

22 *June* 1681

In my first lecture of this terme I shewed y^{u} the Nature of refractions, how ye raies of light were transmitted through a polisht peece of glasse whose two sides were parallel to each other, how the parallel rayes to ye Axis were brought to Converge upon it, not all in ye same point, but onely [in] those that were equally distant from ye Axis or pole in the places of theire incidences, I tooke notice of a remarkeable difference betwixt ye performance of the same plano convex glasse when ye plaine & when ye convex side was turned towards ye object, I shewed how the pointe wherein any ray would converge might be found by Calculation, whereby I made it evident. that if one side of a glasse were plaine, the other sphericall, the raies falleing upon it parallel to y^{e} Axis, would converge or make y^{e} burneing point, at lesse distance from y^{e} convex part of the glasse then the diameter of y^{e} sphere of which it should be a segment & two like parallel raies the one $\frac{1}{20}$ the other one fourth of y^{e} semidiameter of that sphere distant from y^{e} axis would not be brought to converge nearer then within $\frac{1}{8}$ of y^{e} said Radius or $\frac{1}{16}$ of the focall distance. but if y^{e} convex side of y^{e} same glasse were turned to the same object the distance of the place where they would converge would be something more then the diameter of y^{e} sphere. & y^{e} space betwixt the places where each would cut y^{e} axis not more y^{n} y^{e} 28th part of the Radius distant from y^{e} pole or Thickest part of y^{e} glasse. proceeding in the same Method by Calculation I found that if two like Rayes fell on a double convex glasse the Radius of whose convexity was y^{e} same: they would be collected on y^{e} Axis with y^{e} Space of about $\frac{1}{16}$ of y^{e} Radius or focall distance. something more then y^{e} length of y^{e} Radius from y^{e} middle of y^{e} glasse.

Hence it was evident that y^{e} segment of a sphere would not collect all the raies (that proceedeing from one point of a remote object fell on y^{e} glasse at different places) into one point on the Axis. I suggested then & proved in my next lecture[1] that this would be performed by segments of Hyperbolas or Ellipses whose transverse diameters to the distance of their Foci were in proportion as ye sines of ye Angles of Incidence to ye Sines of ye Refracted angles.

Notes on page 146.

I tould y^{u} then y^{t} wee had not yet found out any way of formeing Hyperbolar & Ellipticall sections so perfectly as sphericall, and further that if wee had yet by reason of different refraction of the rayes of light it could not be expected that if wee had such they should performe much better than sphericall.

Which being the onely figures that can be expected from o^{r} Artificers I proceeded then to shew y^{u} how ye enclined raies, or such as fall on the glasse oblique to ye Axis, would be transmitted; & because some very ingenuous persons have conceaved that the distinct image or representation of any object formed by ye transmission of the raies proceedeing from ye object, & transmitted through the glasse would be broader when ye one side of ye planoconvex was turned towards it then when ye other [was].[2] I shewed how the said image or picture was formed, & then proved by a Carefull & curious calculation that ye difference of the angles subtended by the said image, in either case, was insensible;

I promised y^{u} then to shew y^{u} ye construction and nature of the telescope, togeather with the uses of it; it is found by experience that a telescope or perspective made up of two glasses a convex one next ye object, & a concave at ye eye such as are commonly used and carried about tho they shew the object very cleare & distinct, yet take in but a small angle & is altogeather unfit for measureing. I shall therefore passe this by at present if I have time I may say something to it in the Conclusion of my lecture.

The Telescope consisteing of a payre of Convex glasses is the onely proper one for use because it takes in a large angle. and admitts an instrument, or thrids to be placed in ye Common focus of the two glasses. I shall treat. therefore of this & shew y^{u} how when applyed to a level Quadrant or other instrumt the said instrumt may be rectified.

In ye 10th figure [Fig. 5.1] let OB represent an object at a great distance from ye double convex glasse CD the Radius of whose convexity let be Ep. if wee suppose 3 rayes of light OC. OP. OD proceedeing from ye same point of ye object o & falleing upon ye glasse in ye pointes (in ye pointes) C, P, D, these after they have beene twice refracted once in theire entrance & againe at theire emersion from ye said object glasse shall be collected in E. let us suppose OPEL to be the axis of the said glasse and Bp, Bd, two other rayes proceeding from B another point in ye extremity of the said object. one of them Bp. passeing through ye pole of the glasse shall be the polar ray & shall meet with ye other BD after it has passed the glasse in ye point F wherefore the distinct base shall be in ye line FE where the image of ye object OB shall be painted inverted as I demonstrated in my last lecture on ye 9th figure for all the raies proceeding from ye point 0 wherever they fall on ye glasse shall be collected neare E & such as came from B neare F perpendicular to it in ye polar Ray BP produced through the glasse, for on it, in the distinct base, will all the laterall rayes BD, BC, be collected.

Let GH represent a double convex eye glasse formed in a lesser sphere, so ordered in a tube that its focus may joyne with ye focus of ye object glasse at E: the Rayes oc. od. being inclined after theire emersion from ye object glasse into ye lines CE : DE after they have passed this glasse GH are carried parallel to each other in ye aire in ye lines ot. qr. till falleing on ye utmost coat or tunica corneas of the eye. they are transmitted into ye aqueous humor AA & passeing through the pupill swimming in it the Christalline humor K, & ye vitreous uw they are againe Collected on ye Tunica Retina at the bottom of ye eye in ye extremity of the Axis at l. But the Rayes Bd. Bc. meeteing at F after the like transmission at m, are collected beneath l. wherefore the object is painted in ye bottom of the eye erect. as it stood in the open aire. which is the reason that it then appeares inverted to us: for the eye is but as the object glasse of a telescope which collects all the raies proceedeing from any one point of an object into some one point of ye tunica retina by meanes of its three humors, the watry at A, ye Christalline at K, & the vitreous or glasslike under it at u whereby ye picture of the object is represented theire exceeding distinct but inverted. as may be found if takeing a fresh eye of an ox or any other animall you Carefully take of a small part of ye utmoast coat or Tunica Sclerotis on ye back part of it. and afterwards place ye said eye in some hole made in ye windows of an obscure roome so as the forepart of ye eye may be outwards, the place y^u^ have opened on ye backpart inwards y^u^ will see the pictures of such objects as are in view of ye said eye projected on ye bared back part very livelike and distinct but inverted.[4] this experiment proves therefore that those objects which appeare erect to ye eye are really painted inverted in ye bottom of it, so then those which are painted inverted appeare erect & by Consequence those pictures which are projected on ye bottom of ye eye erect, appeare inverted. whence it is evident that object oB in ye 10th figure [Fig. 5.1] shall appeare inverted when viewed through a telescope of onely 2 Convex glasses becau[s]e the image or picture of it lm is painted in the same erect position in ye bottom of ye eye.

Hence the reason is evident why the objects appeare bigger but not more distinct, when wee apply a convex wrong in a small sphere then when wee apply one formed in a larger. For if ye Eye glasse GH had its side the sections of a lesser sphere the Raies that fall neare the extremitys of ye glasse would be more refracted & carried at a wider angle into ye eye & consequently the image lm would be projected broader in ye bottom of ye eye. but more indistinct or rather compassed with colours for tho the eye be such (such) an excellent compound both on the account of its figure & matter composeing the severall humors that it does carrie all the raies of light whether they be difforme, as Mr Newton[5] has excellently well endeavored to prove, or have different pulses and motions as o^r ingenuous professor of Geometry[6] seemes rather to assert, upon whethersoever of these accounts it is I say that they have different refractions yet the composition of the eye is such that it carryes all they raies of light however difforme or differently moved & proceedeing from the same point of the object into one & the same point of the image or representation painted on the bottom or tunica retina of it: so that the object to the Naked view appeares distinct and not at all coloured. but the simple & uniforme matter of glasse carrieing onely one of the sort of raies or such as are moved in the same manner to the same point the rest by reason of theire greater or lesse refractions fall wide of that point where the greater number are carried by the glasse & so represent a crowne of colours on the object, to avoyde & remedy which I have found it the best way to use glasses moderately convex, or wrought and formed in greater spheres, whereby the angles of incidences being lesse the refracted angles & the inclination of the raies entring the eye is also lesse. & consequently the difforme raies proceedeing from the same point of any object spread lesse from the point on which it is painted on ye Tunica retina of ye eye. therefore the Colours are few or small & narrower so that mixing togeather they compose the white colour or light that is they disappeare or are at most faint & little discerneable.

To find how much the object is amplyfied by the Telescope the writers of opticks have made use of this method. they Measure ye length of PE the distance of the object glasse from ye distinct base at E and the focall lenght of the eye glasse eq. by which divideing the lenght of PE the quotient shews how much the object appeares greater through the glasse then to ye bare eye. for example, if PE, the distance of the distinct base from ye glasse were 147 inches as I determined it would be in a plano convex glasse the Radius of whose convexity is 82 inches in my last lecture. & if Eq the Radius of y^e double convex charge were 3 inches. Divideing 147 by 3. the quotient will be 49 therefore say they the Telescope magnifies the object 49 times & wee see y^e object as distinctly as if wee were but $\frac{1}{49}$ parth of its true distance from us.

This proportion for finding how much any object appeares greater in the Telescope then to the bare Eye is Rationall & true enough but it shews not truely how much more distinct the object appeares through the glasse then without one. May wee not argue thus more reasonably. the Eye is as ye object glasse of a telescope which has its true focus on the tunica retina where every object is represented under its due & proper angle. that is the depth of ye Eye tl is in proportion to the breadth of the image projected in it lm when viewd without a glasse as the distance of the object from ye eye is to its true breadth, & because theire is the same proportion betwixt pe.

ye radius of the object glasse & ET ye breadth of ye picture of the object op may wee not more reasonably argue that the proportion betwixt the distinctnesse of ye object viewd with ye naked eye, & with ye glasse. is as the Focall distance PE to the depth of the eye. now if the Focall distance be as was before 147 inches. the depth of the Eye but one it followes that the object appeares 147 times more distinct then it would doe to ye naked eye. or as well as if wee were approached to ye 147th part of its true distance towards it tho at the same time it is magnified or seems bigger then to ye bare eye but 49 times as I tould you before this way of accounteing how much distincter objects appeare through glasses then to the bare eye may seeme strange no other before me haveing reckned after this manner but it will be confirmed by easy experiments. for if in ye Roome of a three inche Charge which is commonly used with a good 12 foot tube wee employ one of 6 inches. it will be found the object tho it appeares but halfe so broade as it did with ye 83 inch Charge is never the lesse distinct but rather more, better defined & free from Colours, which plainely proves that the proportion whereby objects appeare distincter through the telescope then to ye Naked eye depends not upon the difference betwixt the depth of the Charge or focall lenght of ye Eye glasse Eg. & the like distance of ye object glasse from its focus Pe which is mutable as y^{u} change or use different eye glasses. but as the said focall distance pe is to the depth of ye eye which changes not but is the same with all eye glasses of what spheres soever.

This doctrine of the distinctnesse of objects viewd wth telescopes will be further confirmed if y^{u} consider the Moone with a telescope of convex glasses about 18 inches long and againe with one of twice that length. but whose eye glasse shall burne at double ye distance of that you used with ye telescope of 18 inches. for by reason there is the same proportion betwixt the focall lengths of the object and eye glasses in both ye tubes. either of them ought to amplyfie the object alike & so they will be found really to doe but yet with this difference that the same object shall appeare much more distinct through ye 3 foot glasse & such small spots shall clearely appeare in the moones face as with ye 18 inche glasse were altogeather invisible or indistinctly. by reason that ye breadth of the image projected in ye distinct base of ye 18 inche glasse was not sufficient Nor could admit roome for ye figures of the small spots to be so distinctly painted as they would in ye distinct base of ye 3 foot glasse. where the breadth of the image projected was double to that of the other. this experiment is easily to be made & I doubt not but whoever shall trie it will find the Experiment confirme this assertion. that the distinctnesse of any object viewd with a telescope depends not on the greater or lesse depth or Radius of the Charge or eye glasse but on the length of the tube or ye Radius of ye object glasse.

So y^{t} wee are not to expect any improvem^{t} of Telescopes from ye multiplying of glasses in a tube. that contributeing nothing to ye amplyfieing the distinct representation of ye image. But if it were possible by any instrument to give ye exact figure of a parabola to a reflecting mirror or Speculum much might be expected & hoped for from Mr Newtons telescope[7] whereby all the raies proceeding from any one point of ye object would be collected into some one point of ye image which would consequently be represented as distinct as wee could desire it and a short glasse would therefore performe more and better then one of ye ordinary make of a much greater length.

But it is not my businesse at this time to informe y^{u} concerneing the reflecteing telescope. I shall confine my selfe to ye ordinary ones of two convex glasses. & accordeing to my promise shew y^{u} how they may be rectified or when they are applyed to quadrants or levells how the begining of the division may be found for the Quadrant, & the point on which the perpendicular must play in ye level:

It will be convenient when a quadrant is intended to be made that shall have telescope sights. first to provide ye glasses & to trie them by Celestiall objects as ye Moone ♃ or ♀. & findeing at what length they appeare best so to order the instrum^{t} that they may stand at that distance on ye index for y^{u} will find by experimente that that length with which y^{u} can see them best will not p[er]forme amisse for land objects & will serve also very well for all other in ye heavens

Next a payre of cells of brasse or other mettall must be provided in which the glasses must be so firmely fixed that they may not be moved or alter theire position by any accident the object glasse especially for the eye glasse it is not so necessary tho it were requisite that should be so firmed as not to be subject to any inclination. for if it be the objects on the one side will appeare darke & coloured on the other light & cleare besides these two cells another of the same make must be provided & so ordered that it may carry a paire of very fine thrids either of the smallest silver wier or the clew of a silke worme. So that one of them may be perpendicular (when put in its due place) to ye plane of the Instrum^{t} on which the telescope is to be fixed, which I call the *measurer* the other *parallel* to ye said plaine which I there call the *guide*

The Cell with the crosse thrids thus ordered must be so placed that it may lie in ye distinct Base of the Telescope. or over EF in the 10th figure [Fig. 5.1]: whose glasses if they were to be ordered upon a Ruler RR of the 11th figure [Fig. 5.2] which is to supply the place of the fixed index of ye quadrant the Cell which carries the object glasse being fixed at p̄ firme to ye said Ruler & that carrying the eye glasse at ḡ, the place of the Cell with the crosse thrids will be at e in the distinct base betwixt ye two glasses, which cell must be allways firmely screwed or otherwayes fixed on ye Ruler R : R that it may not if possible be removed or stray from its first position.

These glasses and cells thus disposed must be covered with a tube of tinne or any other substance to exclude the sky light which will otherwayes render the object viewed through them difficult to be found & lesse distinct

The radius of ye instrument may be contrived something lesse then the length of ye glasses. So a cell carrieing the eye glasse may stand on y^e outer & that with the thrids on y^e inner edge of y^e limbe which for this purpose may be contrived something broader then else where.& the object glasse may be fixed over the center.

It will be also convenient to place these cells & glasses not on the foreside of the $instrum^t$ but on the backside: that so if y^u desire to have the instrument also fitted for takeing distances y^u may apply another like telescope upon an index which may be moveable on y^e limbe of y^e quadrant being turned about on the center c without any impediment from the other or hindring of the sight by it when objects being neare a right angle distant from each other are to be observed.

For which reason as also that the perpendicular may play free when altitudes are to be taken it will be also convenient that the limbe be layd $\frac{1}{5}$ of an inch higher then y^e ribs of y^e instrument & y^e center pin as much which contrivances for instruments I therefore mention because I find that those made by o^r workemen hitherto for want of them are found very inconvenient I should have added some more but that It is besides my principall designe which was to shew y^u how instrument[s] allready made furnished with telescope sights & crosse hairs may be rectified.

In ye 12th figure [Fig. 5.3] ef represents the telescope fitted as before prescribed & placed on the backside of ye Quadrant whose plane is conceaved placed exactly perpendicular to the horizon the plumbe line ca. It neither resteing on the limbe nor playeing wide but just toucheing it.

Move the instrument till y^u see some object in the horizon just upon ye crosse thrids in the telescope note then where the perpendicular playes. suppose the object viewd to have been really above the level or horizon & let the point so noted in the limbe of the Instrument be at a in ye 12th figure [Fig. 5.3].

Invert the instrument so as ye telescope ef which before was ye hig[h]est part of it may become the lowest as in ye 13th figure [Fig. 5.4]. bring the same object y^u used before upon ye crosse thrids. then takeing ye plumbe line in ye hand move it slowly & gently on ye limbe of the Quadrant till y^u have found the point d on which being hung & let losse it will play exactly over the center from which it hung before. marke the point in the limbe. divide the space ad betwixt it & ye point first found into 2 equall parts in b to which from the center c draw the streight line cb. In this shall be the begin[n]ing of ye divisions. the Reason of which is plainely evident. conceave a line ck drawne from the center of ye instrument parallel to the Axis of the glasse or ye line which connects its pole & ye crosse of ye thrids. if the object be above ye horizon when the thrids are set upon it the angle kca which ye perpendicular formes with the said parallel line shall be so much lesse y^n a Quadrant as ye object viewd is above the horizon. for I proved before that ye line drawne from ye meeteing of the thrids through the pole of the glasse to the object it selfe was a streight line.

But when the Quadrant is inverted for the same reason the angle dck shall be so much more then a Quadrant as ye said object is above ye horizon. and therefore ye point b betwixt these two shall be a precise Quadrant distant from k or rather from the line drawne through the center of ye Quadrant parallel to another connecteing ye pole of ye glasse & ye crosse formed by the thrids in the Telescope

And therefore if y^u would find by ye Quadrant what distant point is level with ye eye: y^u need onely so order ye instrument that the perpendicular may play on ye point b & then applyeing y^r eye to ye telescope what ever objects y^u see on the crosse or measureing thrid shall be level with your eye. & by this meanes objects may be found level at some miles distance whereas by the levels with plaine sights tis impossible to see any object or to find a true level point at halfe a mile distance.

But care must be had when y^u thus rectifie your instrument first to chuse a remote object. 8 or 10 miles distance at least that so the angle subtended by ye Radius of ye instrument may be insensible otherwayes if y^u chuse one not more then a mile remote. y^u must so order ye station. that the telescope on y^r instrument when inverted may lie the same height it did when ye quadrant was erect whereby ye very Suspicion of error will be avoyded.

Next you ought to chuse a serene & settled aire for this Experiment & to repeate it as soone as y^u can. for it has been found by experience that sometimes those places which have appeared in [the] precise horizon in the Morneing, have at Noone appeared 3 minutes or more above or beneath. but this examination may be performed & repeated in one houres time in which space excepting alterations of weather the refractions change little. & if they should alter it would be easily perceived,

After the same Manner may a level with telescope sights be rectified whether it be made after the Manner of the Roman Capitall T or of Carpenters Square of both which formes I could equall approve if ye Ruler on which the Perpendicular playes be made as long as the glasse: whereby the perpendicular will be brought easily to its point & exactly adjusted.

I might here adde the Method of rectifyeing a Quadrant by ye fixed stars passeing neare our vertex whereby ye danger of refractions would be avoyded but this requireing severall nights experiments & not being much unlike the other at present I shall passe it by. & onely tell you that when 2 telescopes the one fixed the other moveable are applyed to an instrument for measureing angles. if y^u get any object upon the Crosse thrids of the fixed telescope, & there retaineing the instrument steady, alter onely the Moveable telescope till y^u have the same object in it upon its crosse t[h]rids likewise. a line drawne by the edge of ye Index shall shew ye begin[n]ing of ye divisions.

There is commonly a fault in glasse which ye Grinders seldom observe that one edge is thicker then another, whence it happens that the poles of very good glasses are seldome in y^e Middle & this may sometimes be the occasions of some inconveniency which I may perhaps shew how to remedy when I come to treat of some observations made with Telescopes if God permit next Terme.

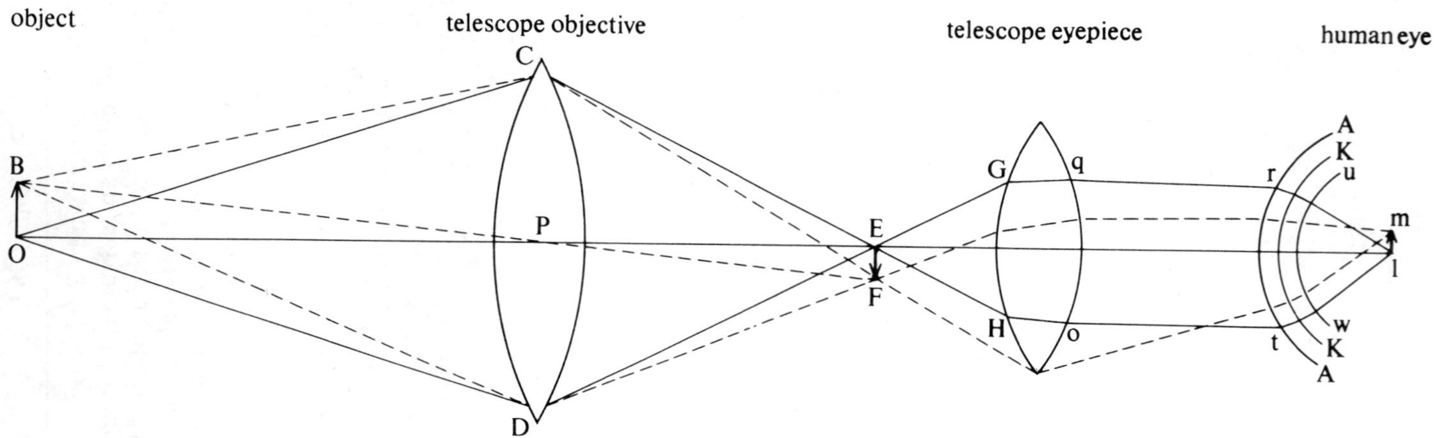

FIG. 5.1 [Reconstructed]

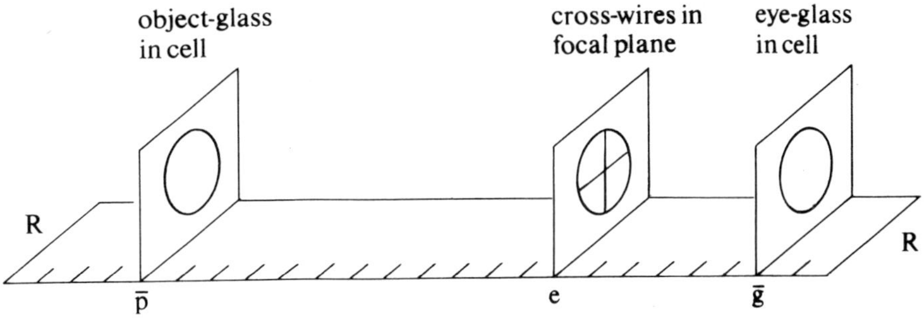

FIG. 5.2 [Reconstructed]

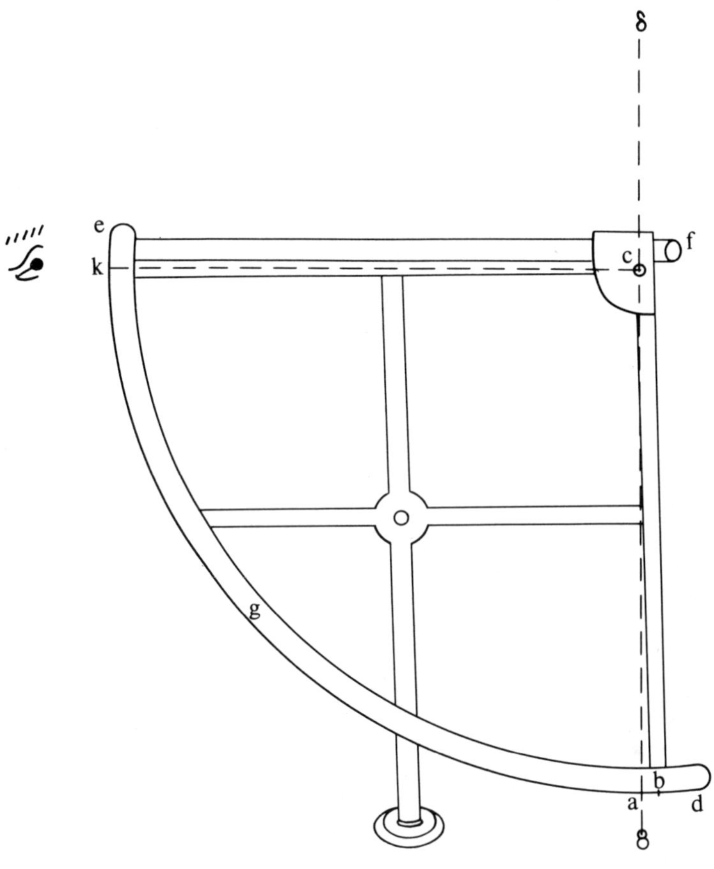

FIG. 5.3

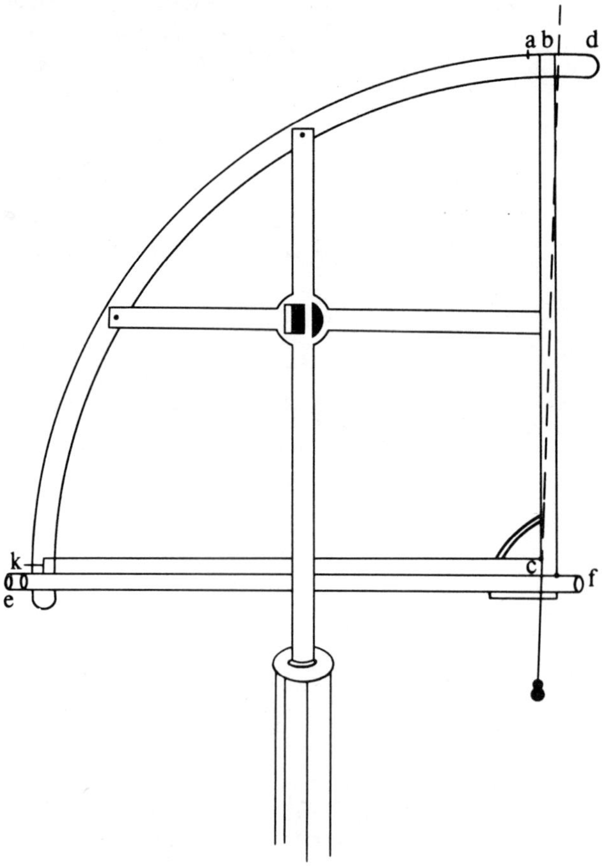

Fig. 5.4

NOTES

5.1. Cf. note 4.17.

5.2. This property is illustrated in the diagram associated with the 138th proposition in Johannes Kepler's *Dioptrice* (Augustae Vindelicorum, 1611), pp. 117–9—a book which Flamsteed claimed never to have studied.

5.3. Two figures on f. 40ᵛ, numbered 5 and 6, follow on in sequence from Lecture 4 and belong to the missing lecture (cf. note 4.17). Flamsteed's numbering convention at this point suggests that there were three more illustrations besides these.

5.4. A similar experiment is described by Descartes (op. cit. [note 4.4]), but Flamsteed obtained his description indirectly from Gascoigne's optical papers (cf. note 4.14).

5.5. Newton's 'experimentum crucis' to demonstrate this fact was described in his Latin letter to Pardies of 11 June 1672, published in *Phil. Trans.* vii, no. 85 (15 July 1672), 5014–18. An English translation of it is given in Cohen, op. cit. (note 4.6), pp. 106–9.

5.6. Op. cit. (note 4.16), pp. 55 ff.

5.7. Newton presented his first reflecting telescope to the Royal Society of London in 1672. An account of it appears in *Phil. Trans.* vii, no. 81 (25 March 1672), 4004–9.

Lecture 6

Wednesday, 26 *October* 1681

The Certeinty of Astronomicall Determinations depends upon these three things, The accuracy of Celestiall observations. and. the skill & diligence of the person that makes them. Hee that has not a Competent understanding in Geometrie, a thorough knowledge of plaine & sphericall Trigonometrie, with the Severall Theoryes of the planets motions, hee that with these is not well acquainted with the Sphere it selfe, & its severall projections & knowes not readily the Constellations, ought not to pretend to the businesse in which all these are required For hee that is not well acquainted with the fixed stars & th[o]roughly instructed in Plane & Sphericall Trigonometrie. shall not be able to deduce the time of ye day or night. nor the place of any appearance from them. Let him know these too yet if hee be ignorant of the Theoryes of the planets motions his other knowledge signifies little, for hee shall necessaryly omit the prime seasons for observations an example of which wee have had not long since in the Work of a very eminent & otherwayes excellently deserveing person,[1] some very ingenuous & curious men & not ill accommodated with the Necessary helpes for observation[2] haveing slipt the principle opportunityes & so ill managed such others as have presented themselves that theire notes serve onely to perplex not to assist the Skillfull Astronomer As I could shew by Examples of Eclipses & some other appearances recorded and committed as a treasure to poster[ity] but that my designe is otherwayes layd at present & that this may more properly be discovered when wee come to treat of Eclipses & what has beene deduced from them but if in the Meane time any one shall desire to see some instances of what I affirme he needs but looke over the observations recorded in Ricciolus his Almagest[3] & Gassendus his booke of observations[4] with the many Papers concerneing the late Comet publisht beyond seas[5] and if hee have skill to judge hee may receave abundant satisfaction:

But yet however skillfull any p[er]son be except hee be accommodated with convenient & fitting instruments, hee cannot put knowledge into practise: nor will his observations be usefull. I told y^{u} in My entrance of these Readeings last Terme that the Ancient

Notes on page 160.

observations are most of them doubtfult and uncerteine tis evident that Ptolemy erred in the Determinations of his equinoxes, since they neither agree with the true lenght of the Yeare stated by himselfe nor yet with those of Hipparchus preceedeing them not 300 yeares. as might be easily proved had it not beene often done hereafter. Wee can not think that that egregious person erred for want of skill. Most therefore conclude that the Armillas hee made use of for this purpose haveing beene placed some Ages before were lapsed from theire first position. tis not improbable. and as likely that the ill Contrivance of those instruments might contribute no little to his error Since tis very evident by compareing of his observations one with another that a person not lesse eminent then hee but much more usefull in his Generation, the Noble Tycho Braher. with larger instruments has often commit[ted] errors in his observations of 3.4 or 5 minutes. And yet the sights through which hee tooke them were much better contrived then those of the Ancients

The reason of which error will be very plaine for by reason the clefts in ye sight next his eye must necessaryly beare some breadth. wee cannot conceave theire widenesse to have beene lesse then 1 minute of his instrument, I rather thinke they were more for I find it not very easy to see a small star through a Cleft no wider then one minute of an instrument of six foot Radius Adde to this the Penumbra which is much complaind of in instruments with plaine sights and it will be easily conceaved how difficult it is for those who employ them to avoid greater errors then wee find the eminent persons of the last century to have committed in their observations.

After ye mention of ye telescope it was the opinion of Ingenuous men that if some way could be found whereby they might be applyed to Quadrants Sextants & such other Instruments as the Noble Tycho had used this difficulty would be removed. but I find not y^t^ it was at all attempted either by Galileus or Scheiner who first applyed optick glasses to ye heavens, the reason I conceave to have beene because theire Perspectives had Concave eye glasses which as I shall shew y^u^ hereafter were not convenient for this purpose

But about some 20 yeares after them, or something more y^n^ 40 since. An ingenuous English gentleman[6] examineing the Nature of opti[c]k glasses more diligently found that if instead of the Concave eye glasse hitherto used hee should apply a Convex eye glasse; the telescope would not onely receave a greater Angle with the heavens but also that hee could clearely perceave when the object passed over a needle thrid or point placed into common Focus & that therefore if such a telescope were placed on ye index of an instrument the diameters & distances of the planets might be thereby exactly measured. his First attempts were to take the Diameter of the Luminaryes & planets in a single tube. his successe therein encouraged him to proceed further; at lenght hee caused a sextant to be made & then began to employ it when our late unnaturall war commenced in which hee became a party & was slaiein in one of ye first battells whereby his invention was left imperfect He[a]ven reserveing the preservation & perfection of it to the more peaceable reigne of o^r^ present Master.

In the meane time his inventions were much talked of by his acquaintance & the reasons of them Frequently discusst but understood by very few. how distances in the heavens might be taken by them was easily understood but how a quadrant or level might be fitted with glasses was not so obvious nor readily apprehended the fault in glasses that they thickest part of poles were not in the Middle hee had showne & those who had either seene or heard of his opticall instruments were acquainted with this quality of theires. but how to remedy it none that I have heard of understood. when Instrum^ts^ with glasses were talked of this quality of glasses was always remembred & the ill effects of it aggravated. Many ingenuous & some famous men declaimed against theire use on instruments & the World was almost persuaded to beleive that all observations made with glasses more doubtfull & uncerteine by reason of this quality as if wee had beene always imposed upon by them and theire had been no remedy nor certeinty in observations made by them.

Especially a famous p[er]son[7] found himselfe p[ar]ticularly touched by my recommendations of them when I had published some few observations or measures taken with glasses & screws placed in theire Focus.[8] Hee was pleased to urge much against theire use. & some of o^r^ freinds were almost p[er]suaded to be of his opinion. I found it necessary therefore to explicate the Nature of Opticke glasses & to Free them from those imputations, on which account I resolved to make this the taske of the last terme. To doo it I was forced to begin with ye Nature of Refraction. I shewed y^u^ the laws of it which were that when a Ray of light passed from a Rare medium into a dense [one] it was refracted towards the perpendicular but from a dense medium into a Rare from it & that the sines of the angles of incidence were proportionable to the sines of the angles refracted. that the proportion betwixt these sines when Raies passed from ye Aire into glasse was nearely as 3 to 2 but more exactly as 300 to 193: & that ye Raies of light passeing through a peece of polisht glasse each of whose sides were parallel to ye other were thence so carryed to ye eye as if they had not beene intercepted by any glasse interposed or with no sensible difference. I left a good p[ar]te of the businesse I intended untouched last terme. I hope now I shall have time more then sufficient to finish what I then began which since it will in a great measure depend on what I then demonstrated I conceave it may be convenient if not necessary to refresh ye memorys with a breife account of what I cheifly noted in my three last lectures.

In which haveing first acquainted y^{u} with the aforementioned law of Refractions I proceeded next to consider what was the way of a Ray of light through a glasse one of whose sides was a plane the other a sphericall superficies or ye whole a segm^{t} of a sphere. & because no writers of opticks had hitherto taken any notice of ye different p[er]formances of such a planoconvex glasse accordeing as its different sides were placed towards the object I tooke occasion to examine theire p^{formance} by a calculus & haveing shewed y^{u} the whole method I concluded from it at last. that if a planoConvex glasse the Radius of whose convexity was 2 inches precise, were placed with the plane side towards the object as in y^{e} 1st Fig [Fig. 6.1] a Ray of light bg. whose course in y^{e} free aire was parallel to y^{e} Axis of y^{e} segment or glasse I.h. & but one tenth of an inch distant from it after its Emersion from y^{e} glasse at $\bar{g}$ would converge with y^{e} said Axis in y^{e} point i at y^{e} distance of 3 inches & 596 millesmes of an inch from y^{e} Pole w. but Another Ray be. whose course in y^{e} free aire parallel to y^{e} Axis was $\frac{1}{2}$ an inch wide from it should after its refraction in e, be so carried that it should converge with the Axis in y^{e} point $\underline{h}$ distant from y^{e} pole $\underline{w}$ onely 3 inches & 327 millesmes of an inch.

So that all the Raies of light which proceedeing from an object infinitely distant fall within halfe an inch of the pole of y^{e} glasse & parallel to its Axis or within a circle of one inch diameter, in this section are collected upon y^{e} said Axis, within y^{e} space of something more y^{n} $\frac{1}{4}$ of an inch, or 16 of the Radius of the section: and all of them lesse distant from y^{e} pole $\underline{w}$ then the Diameter of the sphere in which the glasse was formed. but that which fell upon it neare the pole as y^{e} ray $\overline{ag}$ more remote & y^{e} next y^{e} Diameter at $\underline{i}$. those which fell at a wider distance as $\underline{be}$ nearer at h. so y^{t} if the aperture of y^{e} glasse $\underline{we}$ be an inch & the Radius of the section 2 inches the space ih in which all the parallel raies shall meet the Axis will be $\frac{1}{4}$ of an inch

I next shewed y^{u} that if y^{e} convex side of the glasse were placed against the object. the case were different as also y^{e} manner of y^{e} performance. for then where when the plaine side was next the object the Raies of light suffered but one Refraction & that onely at theire emersion from y^{e} glasse they would now suffer a double first at g and e where they fell in and then at q and ε where they emerged. and that now the Ray $\overline{ag}$ in y^{e} 2^{d} figure [Fig. 6.2] $\frac{1}{10}$ of an inch distant from y^{e} polar ray $\underline{wc}$. of a glasse the Radius of whose convexity $\underline{wc}$ was 2 inches would not converge or meet the Axis as in the former case but more remote for I shewd y^{u} by a plaine & easy tho prolix calculation that if wc were 2 inches wi the distance betwixt the pole $\underline{w}$ and y^{e} place where y^{e} ray ag would converge should be 3 inches & $\frac{476}{1000}$ of an inch but if another ray be halfe an inch wide from y^{e} polar fell on y^{e} glasse parallel to y^{e} Axis it would converge at h. the space wh being onely 3 inches and 411 millesmes or 1000th p[ar]ts of an inch. so y^{t} the space hi in which all the parallel raies not more distant then halfe an inch from y^{e} Axis would converge should be $\frac{65}{1000}$ of an inch whereas when y^{e} plaine side of the glasse was towards the object wee found it $\frac{269}{1000}$ which is neare four times as much.

Hence followed this easy & Naturall Conclusion that planoconvex burneing glasses performe much better and readily when the convex then when y^{e} plaine side is towards the Sun, since all the raies falleing on the same space or within the same aperture are collected in one fourth part of the space when the convex side respects that the employ when the plane side is towards him wherefore they must needs excite a motion more violent & strong proportionably to ye lesse space they are Crouded in, and haveing oft had occasion to make this experiment with a planoconvex that burnes at about 4 inches I can assure y^{u} I have ever found it to answer ye Theory & what I had deduced from ye calculation which was not to be discovered by any other processe.

Haveing thus shewne how ye Raies of light were carried th(o)rough planoconvex glasses when either side was turned towards y^{e} object it naturally followed that I should enquire how they would be collected by a double convex glasse. I proposed one therefore both of whose sides were sections of y^{e} same sphere of 2 inches ra(i)dius & the rayes parallel to y^{e} axis at the same widenesse as before. the nearest $\overline{ag}$ in y^{e} e^{d} figure [Fig. 6.3] $\frac{1}{10}$ and the remoter be $\frac{1}{2}$ of an inch from the Polar raye before theire immersion into y^{e} glasse. the Method of calculation proved more difficult & tedious in this then in y^{e} former cases. but by it at last I shewed that the point $\underline{i}$ where y^{e} ray $\overline{ag}$ converged was distant from $\underline{w}$. The neare pole of the glasse 1 inch & 762 millesmes. but y^{e} point $\underline{h}$ on which the raye bc converged onely 1 inch & 548 millesmes so y^{t} the space hi wherein all the rayes falleing within y^{e} space of $\frac{1}{2}$ inch from y^{e} polar raye would be collected was 213 millesmes of an inch. the depth of which in the plano convex glasse when y^{e} plaine side was towards y^{e} object was 277 of y^{e} same partes but when the convex side was towards it but 70 or scarce halfe so much whence it will follow. that. the single convex glasse shall burne with double ye violence if any combustible matter be to be fired by the Suns raies cast through it, that the double convex wrought in the same sphere does tho at lesse then halfe ye same distance, but the double convex, twice as violently and swift as ye simple convex with ye plaine side towards ye object. if wee may not assigne the duplicate proportion both to ye violence & swiftnesse of theire operration which seemes most rationall, but this I must commit to further experiment & triall not haveing yet had such conveniencys as are requisite in my opinion for this purpose.

It appeares hence that no sphericall glasse however placed will collect all the rayes y^t come from one pointe of a remote object into one point of ye axis, I shewed therefore in my second lecture of ye last terme[9] that tho the sphericall glasse could not yet if wee could forme glasses one of whose superficies should be the section of an Ellipsis or Hyperbola the transverse Diameter of which to the Distance betwixt ye foci had ye same proportion that theire is betwixt ye sines of ye angles of incidence & ye Angles refracted such a glasse would collect all ye uniforme rayes of light into one of the foci. but that by reason of the difficulty of formeing glasses of these figures such glasses were not to be expected[10] I intimated also y^t by reason ye raies of light are either not all of the same forme or differently moved tho wee could as easily give these figure to glasse as demonstrated theire efforts yet would not all the rayes of light parallel to ye Axis be theireby collected into one point on ye axis. and therefore this being onely a matter of speculation hardly possible to be put in practise, & if attained yet likely to prove of little use, I passed from it to shew what was more materiall concerneing the further performances of sphericall optick glasses.

I proposed first a planoconvex glasse the Radius of whose convexity ci in ye 4th figure [Fig. 6.4] should be 82 inches precise. and haveing before shewne y^u where the raies parallel to ye Axis in the Free aire would converge within the glasse. I now thought it necessary to shew y^u how ye oblique or enclined raies such as formed an angle with ye axis at theire Entrance would be conveyed through the glasse & how they would move after theire Emersion to be breife I proved by a strict calculation that the angle doA which ye ray of light formed with ye Axis after its emersion from ye glasse was exactly the same with the Angle PIR which it formed with ye said Axis at its entrance. The point o being taken one third of the thicknesse of ye glasse within it from ye Vertex I

And because a very ingenuous Philosopher[11] has sometimes asserted that the said angle was different if the other side of ye glasse were turned to ye object I proposed the same glasse with ye plaine side of it toward an object & proved that still the angle doA was ye same to lesse then 1 second. which is a quantity altogeather insensible.

I here tooke notice of ye advantage of thin glasses above thick ones. and answered some objections about ye breadth & warpeing of glasses which are usually wrought much broader & thicker then they need to the No little trouble of the Artificer & p[er]haps with some detriment to his worke.

Then I shewd that whatever inclined raies proceeding from ye same point of the object fell upon the sphericall glasse would all converge neare some point in ye polar ray proceedeing from the same point of the object as if ye polar Ray in ye 4th figure [Fig. 6.4] in ye open aire were I R. which haveing entred y^e^ glasse in the Pole I after its Emersion is carried in the line ed and if Rm and qk were two other raies proceeding from y^e^ same point of an object infinitely distant these would meet the polar ray ed in the points d & r the space betwixt which I determined no deeper then $\frac{15}{100}$ of an inch.

I shewed then that if from the point d wee drew y^e^ line dε perpendicular to y^e^ Axis, & if wee supposed the Raies PI & RI to come from the two extremitys of the object, all the raies that came from other points of the said object would be collected into one place on y^e^ polar ray proceedeing from the same point & that here by the picture of the said objected would be projected in ye line dε whereby the picture of the said object would be distinctly painted theire, but inverted. that is those thinges which appeare on ye left hand of others in ye open aire, would heare appeare on the right hand of y^m^ in the picture. the higher part of ye object lowermost &c. by reason because all the raies of light proceedeing from any point of the object are carried into ye distinct base & projected picture theire as if they had onely passed through ye pole So that the brea[d]th of the image or picture painted within the tube subtends the same Angle the distance betwixt ye object glasse and image being ye Radius as the said object itselfe subtends in the open aire the distance betwixt it & ye object glasse being Radius.

Hence I told y^u^ our Ingenuous Countriman Mr Gascoigne derived his Method of measureing small angles in the Heavens such as are subtended by the Diameters of ye luminaries & planets & the distances of small fixed stars by a telescope with a payre of screws placed in ye distinct base which was the same in effect with ye Micrometer wee now use.

In my next Lecture which was ye last of the terme I applyed all my former discourses & shewed the Nature of a telescope of two convex glasses. to doe this it was further necessary to shew the Nature of ye eye.[12] I shewed y^u^ therefore how it was composed & that by the helpe of its severall humors the pictures of all objects which it viewd were projected upon its fundus or tunica retina distinctly & much better then could be by the helpe of any glasse of whatever figure or homogeneous matter that this picture was in an inverted position on the said tunica. & that all objects therefore which were painted inverted here appeared to o^r^ sight erect but those which were painted erect upon it appeared inverted & further that when the raies of light proceedeing from any object fell mixed or confused on the utmost coat or tunica Cornea of ye Eye the picture of the said object would be represented distinct on the tunica retina & the object would be distinctly seene. but when they were distinct on ye tunica Cornea that is when either the object was held close to the Eye or the Eye was placed in ye distinct base of an object glasse it saw nothing of the said object distinctly because all the raies proceedeing from every part of the saide object were collected & throwne confusedly upon the same place in the tunica Retina.

Here I tooke occasion to examine a common opinion & delivered by ye most approved writers of opticks, that the proportion of distinctnesse of objects viewd with ye bare Eye & with a telescope is as the Focall distance of the Eye glasse to ye length of the tube betwixt ye object glasse and the distinct base. and I shewd it was false. for that if it were granted it would follow that any object viewd with a telescope of 18 inches whose eye glasse had its focall lenght 1 inch & an halfe should appeare as distinct as with a telescope of 3 foot the focall length of whose eye glasse is 3 inches for tho theire is the same proportion of 12 to 1. betwixt the focall lengths of both theire object & eye glasses whereas it is & will be evident upon triall that use what eye glasse y^{u} please to an object glasse of 18 inches yet the Moone shall appeare far distincter with a telescope of 3 foot and y^{u} shall see more spots in her face then y^{u} shall ever be able to perceave with ye glasse of 18 inches. I concluded therefore that the true proportion was that of the Focall lenght of the object glasse to ye depth of the Eye which is double in a glasse of 3 foot to what it is in one of 18 inches, & haveing well considered of it since I find no reason to alter my assertion, some objections I have receaved from my friends & some I have made my selfe. which I may have occasion & Leasure enough to answer hereafter.

My whole designe as I tould y^{u} at first was to prove that optick glasses did not impose upon o^{r} senses. then to shew that they might be applyed to instruments & rectified as well as plaine sights. haveing showne & proved the first parte of my taske I came in ye conclusion of my lecture to ye second.[13] To find the true place where the divisions are to be begun upon a quadrant with telescope sights is the onely difficulty in this contrivance. I proved therefore that if haveing fitted ye telescope on ye Instrument with a payre of Crosse thrids. y^{u} placed ye plane of ye Instrument perpendicular to ye horizon getting some fixed object neare ye horizon on ye crosse. & then markeing ye point where the plumb line from ye Center then played if afterwards y^{u} inverted the instrument as in the 6th figure [Fig. 6.6] placeing its plane verticall againe. & then got the same fixed object againe on ye crosse of ye thrids in ye telescope. if y^{u} them move the plumb line to and fro on the limbe till y^{u} find the point b. on which the said line being hung it will play exactly over ye center if from ye center c y^{u} draw a streight line which shall passe exactly in ye middle betwixt these two points d and a as to b. thereon shall be the begining of ye divisions

Thus far I proceeded in my last Lectures of which I have made this long repetition because I intend to prosecute ye businesse of glasses & what I had said before will be requisite to ye understanding of what I have further to adde. I could not but thinke that of the several observables I have here mentiond some might have beene slipt out of ye minds & p[er]haps some p[er]sons might be here present at this who were not at my last lectures to whom an account of what I concluded in each might not be unacceptable. I intend not to repeate anything of what I have already said hereafter on which account I could not but thinke what I have done already necessary & more for yr benefit then my owne. for if these considerations had not prevaild wth me it had beene as easy for me to have given yu fresh matter but to proceed.

I tould yu that ye way of rectifieing an instrument by an horizontall object was sometimes uncerteine by reason yt such sometimes appeare higher then at others accordeing to ye different constitution of the aire which varies sometimes whilest ye are at worke & unperceivd alters whereby ye instrumt is Rendred erroneous. I could give yu an instance of my experience but I Shall wa[i]ve it and in its roome shew yu how to avoid it if yu can onely admit that the p[er]pendicular ray is not refracted which no one ever yet denied, & suppose yt ye p[er]son that is to make ye experimt endued with a Competent stock of patience.

Which when yu are to goe about chuse some star that passes neare ye Zenith in ye night such there will be easily found enow. & haveing calculated the time they are to passe ye meridian dispose ye plane of yr instrument onto it and at that moment so direct it that ye star may passe on ye crosse or move along ye line perpendicular to ye plaine note where ye perpendicular then playes on the limbe. as at a in ye 7th fig. [Fig. 6.7] wherein yu may suppose ye face of the instrumt to have respected ye West. & ye star to have passed to ye south of ye Zenith.

Againe ye next night alter ye position of ye instrument & let ye face of it respect ye East or contrary side of the heavens its plaine being in ye Meridian as in ye 8th figure [Fig. 6.8] and at the time ye Star so passeing over ye Meridian move ye instrument onely in ye said plaine of ye Meridian till ye star be on ye crosse, or at least on yt thrid which is perpendicular to ye instrument neare ye crosse. Note where ye perpendicular playes suppose it be at d in ye s[ai]d figure. The begining of the divisions shall be in the middle betwixt a and d This method is the more tedious of the 2 but the more certeine I have frequently used both the other & this but I prefer this as free from that inconveniency & suspicion of error by ye alteration of the Aire to which ye other is subject. I might further remarque that by useing both these methods togeather it were possible to set of ye arch of a Quadrant on any instrument without compasses but tis obvious to yu I suppose as soone as mentioned I shall not therefore deteine yu with it. nor adde any more at present but that next lecture I intend to shew yu the nature & performances of Concave glasses, in ye same Method yt I have handled ye Convex.

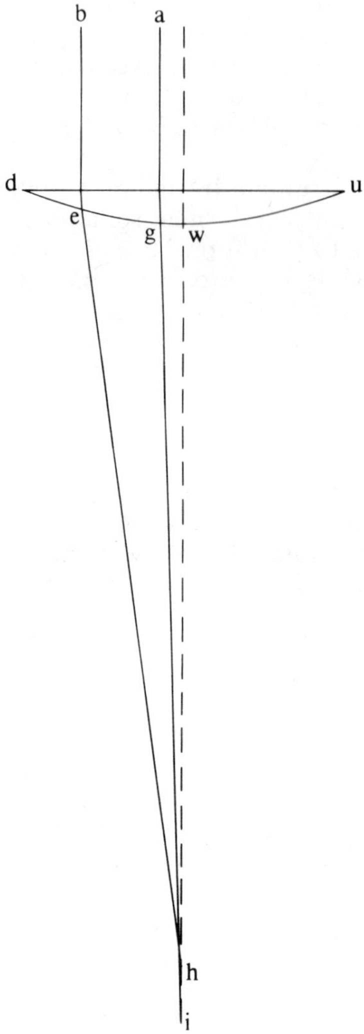

FIG. 6.1

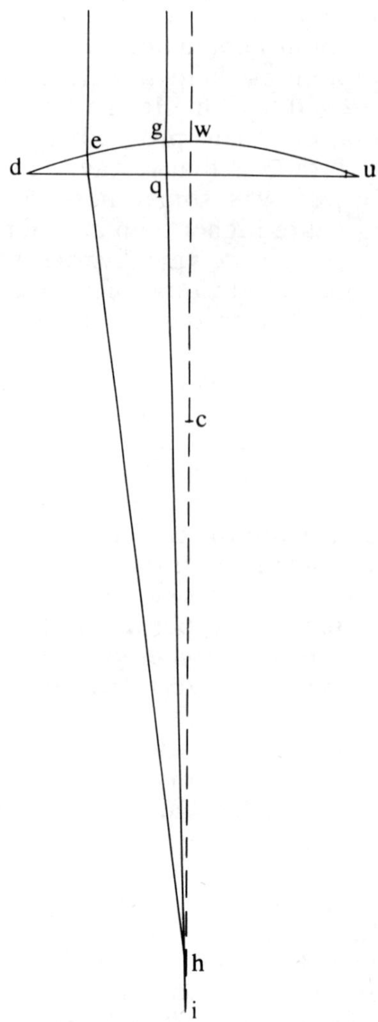

FIG. 6.2

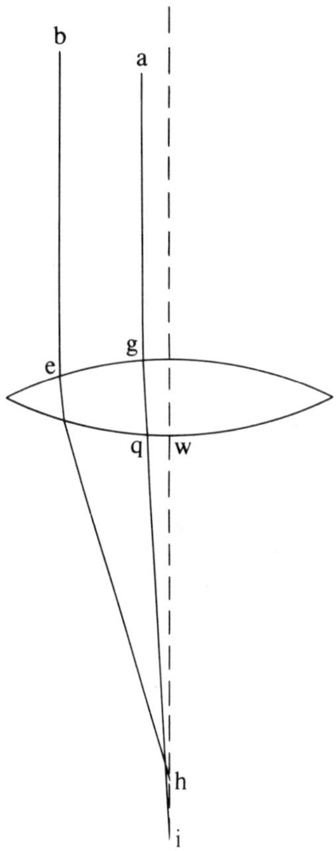

FIG. 6.3

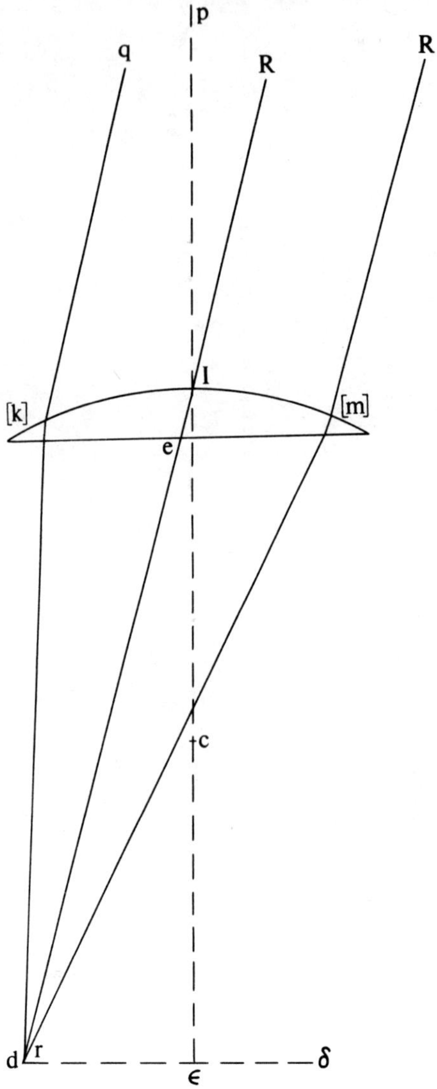

FIG. 6.4

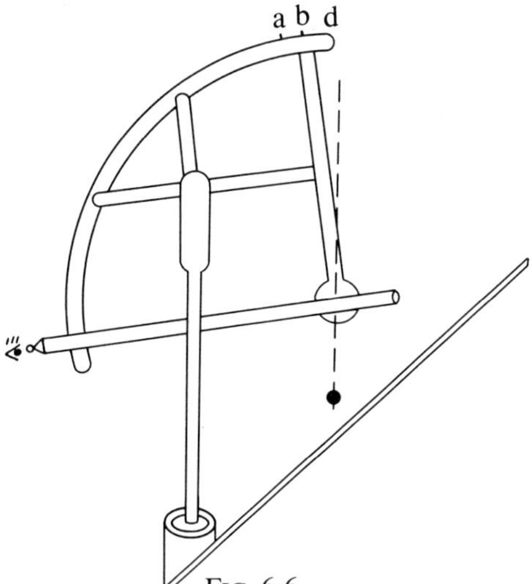

Fig. 6.6

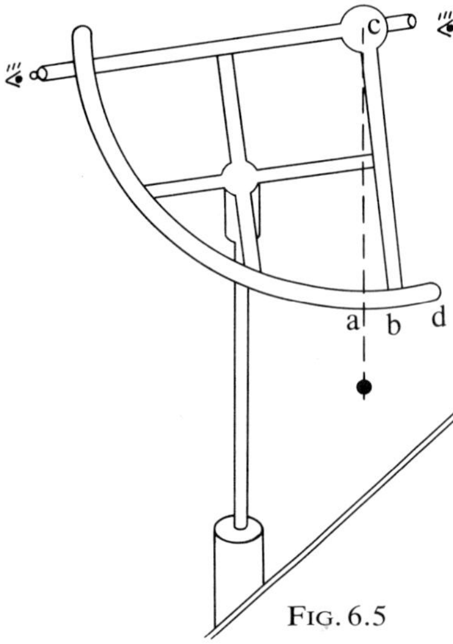

Fig. 6.5

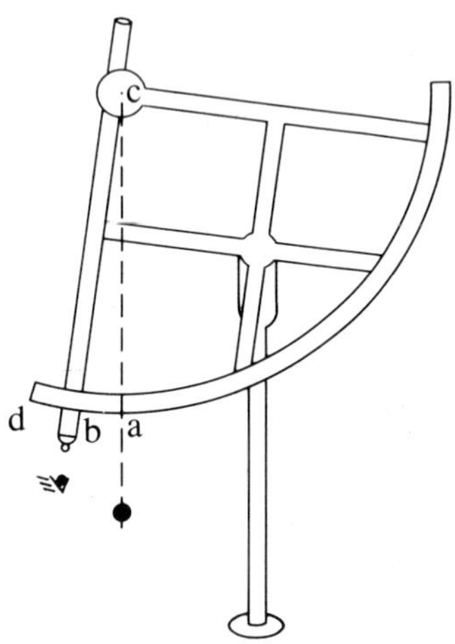

Fig. 6.7

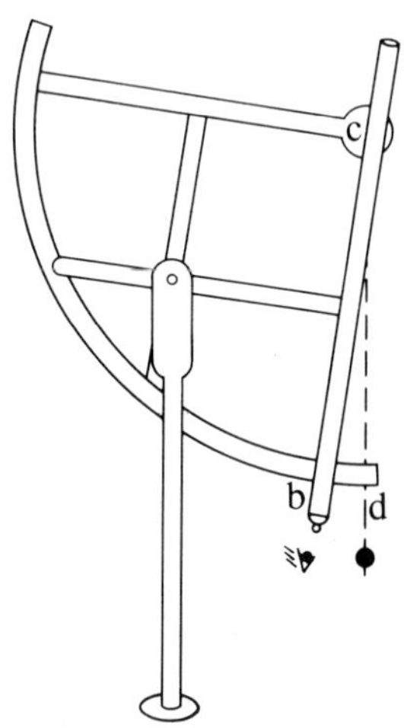

Fig. 6.8

NOTES

6.1. Probably a reference to Riccioli who 'has frequently shewd himselfe' to be an ill-manager of celestial observations (f. 125^{v} cf. p. 225), and whose major work is referred to below (cf. note 6.3).

6.2. Probably those of the Catholic priests at the Jesuit College in Bologna, whose observations are published in Riccioli 1 and 2.

6.3. Giovanni Battista Riccioli, *Almagestum novum*, 2 vols. (Bononiae, 1651)—Riccioli 1 and 2.

6.4. This may be a reference to 'Commentarii de Rebus Caelestibus', in op. cit. (note 3.15), pp. 75–498.

6.5. Cf. note 3.18.

6.6. Flamsteed subsequently identifies this gentleman as William Gascoigne, who joined the Royalist troops at the beginning of the Civil War in England, and was killed at the Battle of Marston Moor in 1644.

6.7. Johannes Hevelius of Danzig. Cf. note 4.2.

6.8. *Phil. Trans.* vi, no. 89 (16 December 1672), 5118–24.

6.9. Namely, the lecture delivered on 15 June 1681. Cf. note 4.17.

6.10. René Descartes was among the first to recognize this physical property of elliptic and hyperbolic curves, and engaged the skilled optician Claude Mydorge to manufacture lenses of these types; but the latter found that the required precision in glass-grinding could not be attained in practice.

6.11. Possibly Pierre Gassendi, following Kepler's 138th proposition in the *Dioptrice* (cf. note 5.2). Gassendi's edition of this latter work appeared together with his *Institutio Astronomica* (Londini, 1653).

6.12. Flamsteed became acquainted with this subject through reading a letter from Gascoigne to Crabtree, dated 8 June [1642], which he partially transcribed in January 1672. Cf. op. cit. (note 4.14), ff. 20^{v}–22^{v}.

6.13. The remainder of this paragraph was added by Flamsteed as a footnote.

Lecture 7

Wednesday, 2 *November* 1681

In my last lecture I gave y^{u} an account of my Readeings last terme concerneing optick glasses & particularly of the nature & p[er]formance of Sphericall glasses. & I shewed y^{u} what was the difference betwixt ye effects of a plano convex when ye plaine side & when the convex was turned towards the object. Thence I proceeded to shew the Nature of a telescope with two convex glasses. & how when they were fitted on a Quadrant or Sextant for celestiall observations such an instrument might be rectified so as that by ye helpe of a perpendicular either playeing loose from ye center after the manner of ye common instruments, or on a fixed point of the limbe the true angle of ye heights of any appearance might be observed & got either by the said loose perpendicular or by a moveable Index if the perpendicular were ordered so as to play on a fixed point of ye limbe Continually. & I shewed how this rectification might be p[er]formed either by an horizontall object or better but with more trouble of attendance by ye fixed stars.

The first telescopes used by Galileo[1] Scheiner[2] Marius[3] & such as made use of them on theire first discovery had onely concave eye glasses. of these I have not yet said any thinge. theire nature I find is rather admired then understood as it ought I shall therefore make it the businesse of this lecture to shew y^{u} first

The Nature of ye sphericall Concave. of which onely the one side is hollow the other plaine. & how the Raies of light are carried through them when either the plane hollow side is turned toward the object, then of the double concave. & that when both Sides are either sections of the same or different spheres. & in doeing of this I shall follow the method I formerly used of calculation for I must & doe affirme that without it it is altogeather impossible to understand either the Nature or the manner of the performances of optick glasses.

I shall consider theire effects first simply, & then I may perhaps have occasion to returne to ye sphericall Convex & to shew, which I did not before, both the manner of its simple p[er]formance. when

Notes on page 173.

an object is viewd through it at any distance and also how the raies are collected by a glasse whose two superficies are the sections of different spheres which in my former lectures was omitted for want of times

I shall then shew y^{u} How the Ellipticall & Hyperbolick concaves would carry the raies of light breifly as I did of the convexes of the same sections, because they are onely speculative. & not likely to be ever brought to practise as I tould y^{u} when I discourst of such convex glasses. & Lastly I shall Shew y^{u} the composition of the telescope with concave eye glasses. the manner of its p[er]formances, why objects viewd through it appeare erect & in theire Naturall position whereas with the convex eye glasses they appeare inverted. I shall shew the place where the measureing thrids are to be fixed in such a glasse & how it may be rectified when applyed either to a Quadrant or a level & all these with Gods assistance & p[er]mission in the same order I have recited.

As in my discourse of convex glasses I began with the planoconvex with ye plaine side towards the object so here I shall begin with ye plano concave with ye same plaine side towards the object. let V in ye first figure [Fig. 7.1] represent such a glasse. & so posited, of which let bg. be the plaine side. cwo the concave. whose center c I call ye center of ye section. & ye semidiameter ow ye Radius of the said section ye line cg perpendicular to ye plaine side its Axis & W the place in which ye Axis intersects ye concave superficies *its pole.* & the Axis produced backward or ye line hc. ye polar ray. The Radius of the Section cw I suppose as formerly 2 inches or 2000 p[ar]tes. & that the sines of the angles of incidence from y^{e} Aire to glasse are in proportion to y^{e} sines of y^{e} sines of y^{e} Refracted angles as 300 to 193.[4]

I will also suppose that two severall raies of light fall upon the said glasse theire course in the free aire haveing been parallel to y^{e} Axis or polar ray hGw but one of them onely $\frac{1}{10}$ of an inch or 1000 the other $\frac{5}{10}$ or 5000 of the like partes wide from it

These both may be represented by y^{e} line abe & for what is said of the one will hold generally of the other mutatis mutandis or onely supposeing the measures of y^{e} lines & Angles allowed

I shall begin with y^{e} Ray which I suppose onely $\frac{1}{10}$ of an inch or $\frac{1}{20}$ of y^{e} radius wide from it which may be represented by y^{e} line ae let be 1000 such partes as ec is 20000 & this distance equall to bg = ex.

It is evident that this ray cannot be refracted at its first entrance into y^{e} glasse at b for that it there falls perpendicular upon y^{e} superficies bg. & is therefore equally enclined every way to it so that it cannot be any wayes inclined in its transit more to one side then another.

But at its emersion from ye glasse it shall because it is then inclined to the plane on which it runs out. to find on what line it emerges first Geometrically.

Because as I shewd y^{u} in my first opticall Lecture the raies of light emergeing from a dense medium to a rare are refracted from ye perpendicular. And that lines drawne from ye center of a Circle or Sphere to ye superficies are perpendicular to that place of the Circle or Sphere on which they fall: From c the Center of ye Section to e ye point on which the ray of light emerges draw ye line ce & produce the Ray ac to d. then shall the angle dec be ye angle of Emersion:

With any convenient widenesse of ye Compasses ed strike ye Arch of a Circle fdm. from d let fall dl perpendicular to ec this shall be the sine of ye angle of incidence to ye same Radius.

Take fk: 300 such parts as dl is 193 betwixt your Compasses: & on the said inch find ye point [k, making fk parallel to dl][6] the other turned about shall onely touch the line ec this shall be the sine of ye refracted angle to ye same Radius & if from ye pointe where ye ray ac emerges y^{u} draw to f the line ef this shall be the line into which the ray ae diverges from ye Axis or polar ray after its egresse. from ye glasse: produce the said line backward till it intersects the polar ray at h it there shews the point whereby the divergency is directed which therefore for ye future I shall call the *pointe of divertion*

When wee enquired formerly how the raies of light were carried thorow a plano convex glasse alike posited wee found that they—all the parallel raies to ye Axis [—] would all converge nearely but not exactly in the same point. wee shall now enquire whether the raies falleing parallel to ye Axis upon ye this glasse will all diverge from ye Axis as if they had proceeded from ye same point h. or not & that in the method wee then used of calculation for the reasons then & since often mentioned.

The processe of which is easy for first from y^{e} point e let fall y^{e} line ex perpendicular to y^{e} Axis in the triangle exc are given ex y^{e} distance of the Ray from y^{e} Axis $\frac{1}{10}$ of an inch or 1000 p[ar]tes such as ec is 2 inches or 20000 p[ar]tes whence the angle ecx = to y^{e} angle of Emersion ced is found 2°.51′.57″$\frac{1}{2}$ then as dl: to fk: or as 193 : 300:: so y^{e} sine of y^{e} angle of incidence dek 2°.51′.57″$\frac{1}{2}$ to y^{e} sine of y^{e} angle refracted fec 4°.27′.27″ from which subtracteing y^{e} angle dec 2°.51′57″$\frac{1}{2}$ the remainder 1° 35′ 29″$\frac{1}{2}$ shall be the angle fed = ehc for the lines ed and hc are parallel:

So then in the triangle ced are knowne the angle hec the complement. of fec to a semicircle 175° : 32′ : 33″ the angle ehc = fed: 1°.35′.29″$\frac{1}{2}$ with y^{e} Radius of the section the side ec 20000 partes whence the side eh will be found 55961 of y^{e} like partes from which takeing cw the Radius of y^{e} remainder wh. 35961 or 3 inches & almost $\frac{6}{10}$ of an inch shall be the distance of the point of divertion h from y^{e} pole of y^{e} concave glasse w.

But if wee suppose ex the widenesse of the parallel ray from y^{e} polar or axis to be halfe an inch or 5000 p[ar]tes. then shall the Angle ecx be 14°.28′.39″. and as 193 : 300 so is y^{e} sine of 14°.28′.39″ to y^{e} sine 22°.52′.03″ the angle cef. refracted. for y^{e} ray passes from the dense medium of glasse to the thinner of aire & therefore runs of from y^e^ glasse in y^e^ line ef:

And if from y^e^ angle fec 22°-52′-03″ wee subtract the angle dec 14°-28′-39″ the residue 8°-23′-24″ is the Angle fed = ehc 8°.23′.29″: which two angles with y^e^ side. ed 20000 being given. in y^e^ triangle ehc the side hc will be found 53266. of the like partes. from which takeing cw 20000 the remain[d]er is the distance of the point of divertion h from y^e^ pole of y^e^ glasse w 33266 p[ar]tes when y^e^ parallel ray falls on the glasse halfe an inche wide from y^e^ Axis

But this distance when y^e^ Ray parallel to the Axis was onely $\frac{1}{10}$ of an inch wide from it was 35961 of the saide p[ar]tes or greater then it is now by 2695. ten thousand[th] partes or something more then $\frac{1}{4}$ of an inch, so y^t^ it is evident the parallel rayes diverge not as if they all proceeded from y^e^ same point h. but those that fall nearest the Axis from a point more remote from y^e^ pole. they that fall wider from a point nearer. of which the reason is evident. for that the sines encrease not proportionally to the arches of the Quadrant but a lesse differences of the Sines answers to ye same difference of degrees as y^u^ remove from ye beginning of the divisions.

It is also evident. that ye point of divertion h. for ye Ray ad is just so far distant from w ye pole of the Concave glasse qwo, as ye point on which the said Ray falleing on a plano convex of the same Radius & placed with the same plane side to y^e^ object would be from y^e^ said pole. for draw y^e^ line qo. & let qiowq represent a plano convex glasse of the same Radius, de. a Ray of light parallel to y^e^ Axis on y^e^ same point. let de be produced to a. ce to μ, the angle $\mu c\delta =$ dec shall be the angle of emersion, with y^e^ Radius $e\mu =$ ef Strike y^e^ arch $\mu\delta\varphi$. then is $\mu\delta =$ me. therefore $\delta\lambda$ y^e^ sine of y^e^ angle $\mu e\delta =$ dl the sine of y^e^ angle dem; and $\kappa\varphi =$ fk. therefore y^e^ $\mu e\varphi =$ fem. and fed $= \delta$eh. so y^t^ y^e^ ray de being refracted in e would run into the line eh and there meet the axis at h. which is the point from whence a ray of light emergeing on the Concave side of the plano convex glasse would be directed after its emersion. this great agreement betwixt a plano convex & a plano concave glasse of the same Radius I could not passe by without remarkeing. whether there will be the same when the other side is turned to ye object I shall enquire & shew y^u^ when I have enformed y^u^ how the rayes of light are carried thorow the same plano concave when the hollow or concave side is placed upon it.

In y^e^ 2^d^ figure [Fig. 7.2] let di represent the plane and bw y^e^ concave side of a plano concave glasse. the Radius of which bc is 2 inches or 20000 partes & let ab be a ray of light $\frac{1}{10}$ of an inch or 1000 of the like partes distant from y^e^ Axis & parallel to it. falleing first on y^e^ concave side of the glasse which is obverted to the objecte.

Produce ab and cb to k and m and with any convenient widenesse of y^e^ compasses strike the arch km. the angle kbm = abc shall be the angle of incidence.

The ray ab. at b passes from a rare to a dense Medium & shall therefore in its passage through the glasse be carried nearer y^e perpendicular ck: to find the line of its passage From m let fall $\overline{mq}$ perpendicular to ck. and because y^e sines of y^e angles of incidence to y^e sines of y^e angles refracted are as 300 to 193 take re 193 such partes betwixt y^e compasses as qm is 300 and therewith find the point e in y^e arch before described so as one point being set on e the other turned about may just touch the line ch the line be shall be that in which the ray of light would move if the Medium of glasse were continued. from b to e.——But because the medium is discontinued at d̲ the said ray moves no further in this line then from b to d̲ on which point it emerges from ye dense medium of glasse to aire & shall therefore be carried from ye perpendicular. From ye point d therefore raise dl perpendicular to di. the plaine side of the glasse; then the angle edl shall be the angle of emersion.

Produce d̲l̲ to n̲ & strike ye arch fnl. from n let fall np perpendicular to dl. it shall be the right sine of the said angle of emersion. take fo. 300 such partes as np is 193 betwixt ye compasses & carreing one foot in ye said Arch. thereby y^u may find the pointe f on which resteing it the other turned about shall just touch the line.dl. if from this point f y^u draw ye line to d. it shall be the divergeing course of the ray after its emersion from ye glasse. & if y^u continue the said line from d backwards to h till it intersect the Axis it shall give y^u the point of divertion h, from which all ye raies falleing on it parallel to ye Axis & emergeing at ye distance id shall be directed to find the place of which point Trigonometrically.

First in y^e triangle bcy are given y^e Radius of the section cb. 20000. with y^e distance of y^e pointe of incidence from y^e axis by. 1000 of the like p[ar]tes hence the Angle of incidence bcy will be found 2°.51′.57″$\frac{1}{2}$ = kbm: Now because y^e ray on its entrance is refracted towards the perpendicular it will further hold: as 300 : 193 : : s[ine of] 2°.51′.57″$\frac{1}{2}$ to y^e sine of 1°.50′-36″. the refracted angle kbe which taken from kbm 2°.51′57″$\frac{1}{2}$ the remainder 1°.01′.21″ shall be the angle dbg = edl the angle of emersion:

Now because y^e ray emerges from a dense Medium to a rare: it will be refracted from y^e perpendicular and as 193 : 300: so y^e sine of y^e Angle edl 1°.01′21″ to y^e sine of y^e refracted angle fdl 1°.35′23″ whose complem[en]t 88.24.37 is y^e angle hdi.

Haveing thus got all the angles formed by. the ray in its passage through the glasse to find the point of divertion h the side di of the triangle dhi must be had to gaine which first I suppose the least thicknesse of y^e glasse w̲i̲ to be $\frac{1}{5}$ of an inch or 200 parts next in the triangle c̲b̲y̲ by y^e side bc given 20000 partes & y^e angle of incidence cby 2°.5$\overline{1'}$.$\overline{57}$″$\frac{1}{2}$ I find the side cy 19975 of the like parts which taken from ci 22000 the residue 2025 shall be the side yi = bg for bg & iy as also y̲b̲ = i̲g̲ are parallel.

Next in y^e triangle dbg with the said bg is given y^e angle dbg

1°-01′.21″ whence the side dg is found 36 of the like partes which added to gi 1000 makes di 1036. such partes as cb is 20000

And now in y^e^ triangle idh are knowne the saide side di 1036 with y^e^ angle hdi 88°.24′.37″ whence the side hi will be found 37249 from which takeing wi the thicknesse of the glasse the remainder 35249 shall be equall to hw the distance of the point of divertion h from w the pole of the glasse. 3 inches and $\frac{525}{1000}$ of an inch.

But if wee suppose that the parallel ray ab. was halfe an inch or 5000 partes = by distance from y^e^ Axis and inquire the distance of the pointe of divertion h from the pole w wee shall find proceedeing after the same manner first the angle of inclination or incidence bcy 14°.28′.39″: the refracted angle kbe 9°.15′.19″ therefor the angle of emersion edl = dbq 5.13.20 for bg and dl are parallel, whence the angle refracted at its emersion fdl 8°.08′.01″: whose complement to 90^d^ shall be the angle hdi 81°.51′.59″

By y^e^ angles of incidence bcy 14°.28′.39″ with y^e^ side bc 20000 given the side cy will be found 19365 which taken from ci leaves yi = bg neare 241 of the like partes & those added to ig 5000 make id 5241 with which and the angle hdi before stated 81°.51′-59″ the side hi will be found 36670 & if from this wee subtract iw 2000 the residue hw 34670 shall be the distance of the said point of divertion h from y^e^ pole of y^e^ glasse w. equall to 3 inches & $\frac{467}{1000}$ of an inch:

Which when y^e^ Ray of light was onely $\frac{1}{10}$ of an inch wide from y^e^ Axis wee found to be 3 inches and $\frac{525}{1000}$ of an inch that is $\frac{58}{1000}$ of an inch or neare $\frac{6}{100}$ greater. the same concave side of the glasse being placed towards y^e^ object

But when the plaine side of the said glasse was towards the object wee found the difference, of the point of divertion for the same 2 ray to be $\frac{269}{1000}$ of an inch that is almost 5 times as much or the fifth part of an inch nearely greater so y^t^ as y^e^ plano convex with the convex side towards y^e^ object p[er]formed best & collected all the parallel rayes nearest into a point so also. hence it is evident that the plano concave with y^e^ concave or hollow side towards ye object disperses them best & nearest from one point of the axis

Wee found that when the plaine side of a plano concave was turned to ye object the point of divertion h was exactly as far from ye pole w as the burneing pointe of a plano convex when ye plaine side was turned on the object. but if ye convex side of a plano convex were turned to ye object supposeing the same radius of the section & distance of ye parallel ray from ye Axis wee shall find y^e^ burneing point a little more removed from y^e^ pole then the point of direction or diverication in y^e^ concave.

For in y^e^ same 2^d^ figure [Fig. 7.2] if y^u^ draw oI p[er]pend[icular] to y^e^ Axis & then suppose oIw to represent a quadrant of the convex glasse its thicknesse wI $\frac{2}{10}$ of an inch & ab the parallel ray at any distance this at its entrance into the glasse shall be refracted into y^e^

line bs & from s accordeing to the law of refraction often mentioned into y^e line sH more removed from y^e pole w then y^e point of divertion. how much wee shall find by y^e calculus fitted to y^e Ray <u>mb</u> whose distance from y^e axis by I suppose $\frac{1}{10}$ of an inch or 1000 p[ar]tes y^e 20th of y^e radius of y^e section

Wee found before y^e angle of incidence bcy to which tbc is equall 2°-51′-57″$\frac{1}{2}$ hence y^e angle cbu in y^e upper parte of y^e figure 1°.50′-36″ zbs 1°.01′.21″ tsx 1°.35′.23″ whose complement to 90° is y^e angle HSI[7] 88°.24′.37″ the Thicknesse of the glasse Iw I supposed formerly $\frac{2}{10}$ of an inch admit the same here then is CF 18000 partes which taken from Cy found before 19975 leaves iy = bz 1975. with which and the angle zbs. the side zs will be found 35$\frac{1}{4}$ which taken from zI[8] 1000 leaves sI[9] 964$\frac{3}{4}$ And now in y^e triangle HSI wee find the said side SI and y^e angle at S knowne whence y^e side HI will be found 34762 to which addeing Iw y^e thicknesse of the glasse the summe shall be <u>Hw</u> y^e distance of y^e burneing point from y^e pole of y^e glasse 36762 or 3 inch & 676 mill which is greater then the distance of y^e point of direction for y^e same ray in a concave of y^e same radius by 1512 partes or something more then $\frac{15}{100}$ of an inch.

Proceedeing in y^e same manner for a Ray halfe an inch wide from y^e Axis I find Hw to be 36112 which is also more then the distance of the point of direction for y^e like ray on y^e concave same sphere by neare $\frac{15}{100}$ of an inch.

How ever it is evident that there is a great relation betwixt y^e concave & convex with theire Sphaericall sides towards the objects for as the one on y^e Axis collects all the parallallel rayes y^t fall lesse y^n $\frac{1}{40}$ of y^e Radius of y^e section within y^e space of 58 1000^s of an inch on y^e Axis or $\frac{1}{32}$ of the Radius of the section so the pointes of direction in y^e concave for all y^e Rayes falleing on y^e glasse parallell to y^e Axis within y^e distance of $\frac{1}{40}$ of the Radius fall within $\frac{1}{33}$ of y^e semidiameter distant from each: but the burneing pointes about $\frac{1}{30}$ of y^e radius of the section more remote from y^e pole then the pointes of direction.

Haveing thus shewne how the rayes are carried thro & from a single concave glasse and from it when either the plaine or Concave side is placed on the object and compared them with the single convex I come next to consider the double concave & how the rayes of light are made to diverge from it

In y^e 3^d figure [Fig. 7.3] let zSwi represent a double concave whose least thicknesse wi is $\frac{1}{5}$ of an inch or 2000 partes the Radius of y^e section ob = ow = ci = cd: as before 2 inches or 20000 partes, and let ab be a ray of light falleing on its superficies parallel to y^e axis $\frac{1}{10}$ an inch wide from y^e point of Immersion to y^e outer center c. draw y^e line ob. which produce to g. as allso to the interior center oc. Continue ab to e with any convenient distance of y^r Compasses strike y^e arch ge & from e let fall el perpendicular to bg. make also fk a sine in y^{e} same arch in such proportion to le as ye sines of ye refracted angle has to ye angle of Incidence. from b to f draw ye line bf. that part of it bd intercepted in the glasse shall be the course of the ray through it

From d to the inner center c draw the line dc then with any convenient distance of y^{r} compasses strike ye Arch rm. make the streight line db intersect the said arch in q. & from thence let fall qn perpendicular to dc make cp a sine. in ye said arch in such proportion to qn as 193 is to 300 and from d to r draw ye streight line dr: this shall be the course of the ray in the free aire after its emersion.

Produce the said ray rd backward till it intersects the Axis it will there give the point of direction h which will be determined trigonometrically thus

First in y^{e} triangle yob are knowne the sides ob and yb whence y^{e} angle boy will be found as in the former example 2°.51′57″$\frac{1}{2}$ and y^{e} side oy as before 19565 which taken from oc 41590 leaves cy 22025 with which & by 1000 the angle bcy will be found 2°.35′.58″$\frac{1}{2}$ which added to y^{e} angle of incidence boy theire complement to 180 degrees will be y^{e} angle obc 174°.32′.04″ whence the side bc will be had—

At b the ray of light passes from a rare medium into a dense & is refracted towards the perpendicular the refracted angle as before gbd is 1°.50′.36″ taken from gbs leaves the angle dbs 1°.01′.21″$\frac{1}{2}$

The angle sbc = bcy is 2°.35′.58″ these added togeather make y^{e} whole angle dbc 3.37.20 with which and the two sides bc and dc before found & stated the angle bdc[10] will be found 176°.00′.23″ whose complement is the angle of emersion qdc 3°.59′.37″.

Now because at d the ray of light emerges from a dense to a rare medium it will hold as 193 to 300 so is y^{e} sine of 3°.59′.37″ to y^{e} sine of 6.12.54 the angle rdc.

Whose complement shall be the angle hdc 173°.47′.06″

The angle cbd was before found 3°.37′.20″, cdq 3°.59′.37″ theire difference 0°.22′.17″ shall be the angle bcd which added to ycb 2.35.58$\frac{1}{2}$ gives the whole angle ycd 2°.58′.15″ hence the angle dhc 3°.14′.58″ & now in this triangle dhc wee have all the angles knowne & y^{e} side cd 20000 whence the side ch will be found 38261 from which takeing cw 22000 what remaines. will be the distance of y^{e} point of direction from y^{e} next pole of the glasse hw 16261 or 1 inch & 626 millesmes

Proceedeing after y^{e} Same manner for y^{e} other ray whose distance from y^{e} Axis I suppose $\frac{1}{2}$ an inch. I find the angle of inclination boy 14°.28′.39″ dbg 9°.15′.19″ therefore dbs 5°.13′.20″, the side oy 19975 therefore cy 22635, bc and the angle bcy 12°.27.23. which added to dbs gives the whole angle dbc 17°.40′.43″. with y^{e} sides bc and dc 20000 and the angle dbc the other angle bdc will be found 159°.23′.31″ therefore qdc 20°.36′.29″ from which takeing dbc the residue Shall be the angle bcd 2°.55′.37″ which added to bcy gives the angle dch 15°.23′.00″

The angle of emersion qdc being 20°.36′.29″ the refracted angle rdc will be found 33°.10′.08″ hence wee have againe all the angles given in y^e triangle hdc whence the distance of y^e pointe of direction h from y^e lower center c will be found 35822 take away cw 22000 & there remaines <u>wh</u> 13822 the distance of y^e point of direction for this ray from y^e pole <u>w</u> Which is lesse then when the ray was but $\frac{1}{10}$ of an [inch] wide from y^e Axis by 2439 partes or neare $\frac{1}{4}$ of an inch So y^t it is evident no Sphericall concave glasse disposes all the parallel rayes as if they had proceeded from one pointe. but the plano concave with y^e howlow side towards the object p[er]formes best; wth y^e plane side it [is] not much worse then y^e double convex.

It is also evident that what ever that ingenuous [person] was who in Dr Barrow^s opticall lecture[11] taught a method of find it both the burneing point & point of direction Geometrically erred enormously since every circle of rayes concentrick to ye Axis has a different point of convergency & direction no one point can be common to all ye parallel rayes I Should have proceeded to have compared this with the double convex but I am abridged of time & therefore must refer the dispatch of what I further promised to my next lecture when I intend God assisteing to Shew from what I have already proved the manner of effects of simple convex & concave glasses & also of ye concave when made the eye glasse of a telescope its place in the tube the place of the thrids the manner of rectifying it with which I intend to conclude this discourse of glasses since I shall thereby prove wee are not imposed upon by them & that instrument[s] furnisht with them are more easily examined & all errors avoided then when they have onely plaine sights. I have broken the ice to farther discoverys which yet because I have not time to presente I Shall not mention. thos that Can abide the paines I have taken in this businesse & will follow it can scarce misse them let them repe the credit I shall thinke it sufficient my selfe that I gave them the occasion:

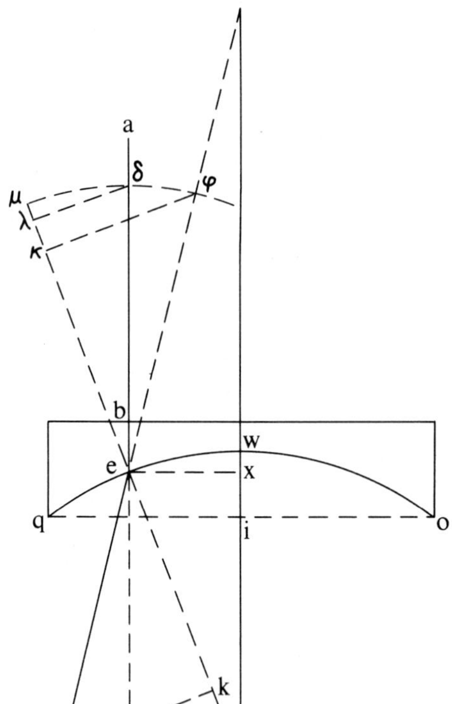

FIG. 7.1

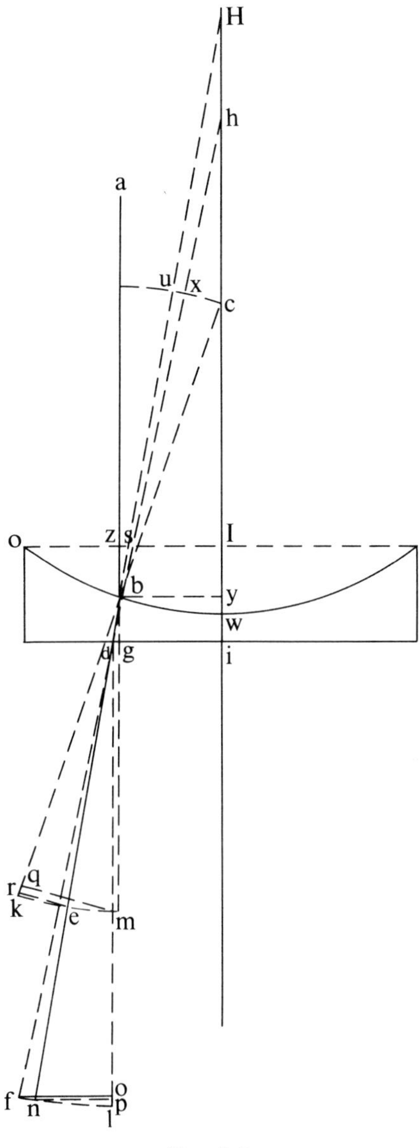

FIG. 7.2

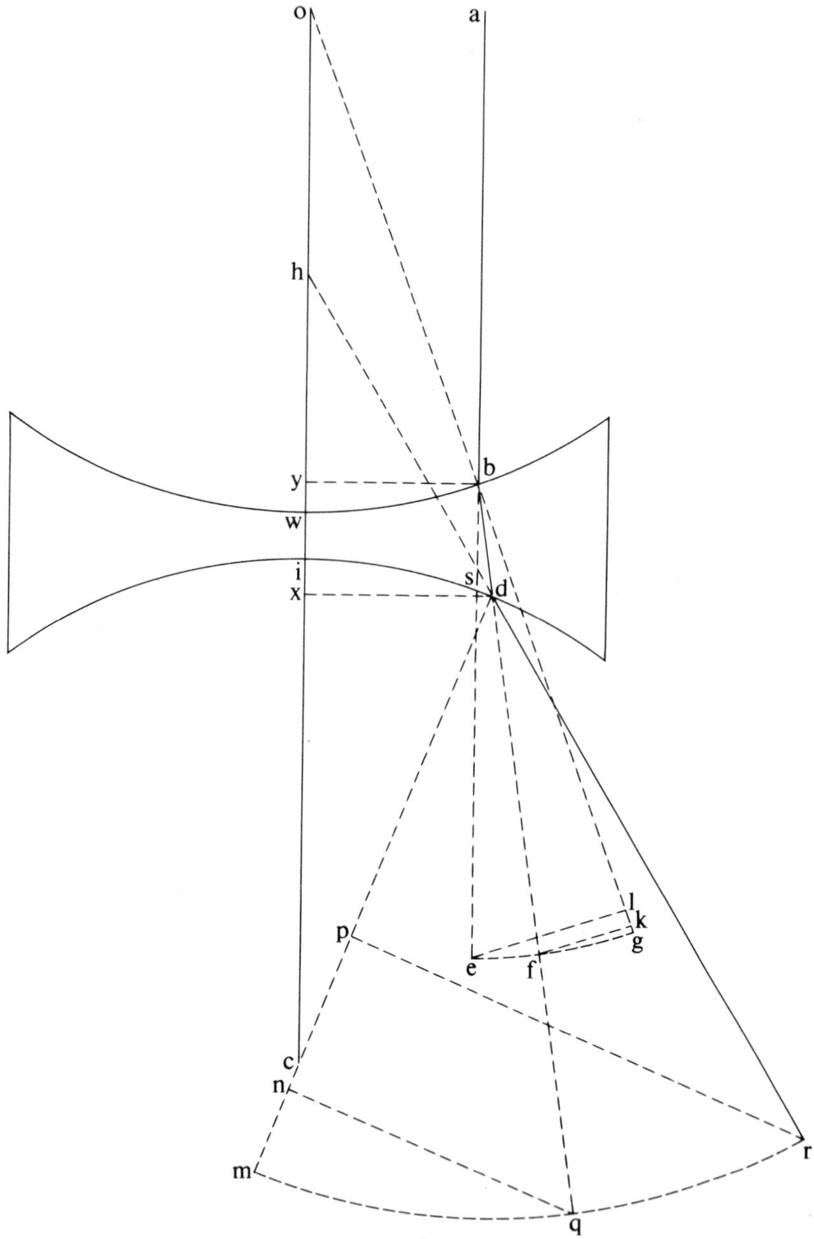

FIG. 7.3

NOTES

7.1. Galileo (op. cit. [note 4.1]) did not discuss the principles of construction of his telescope, although he described the celestial appearances observed with it.

7.2. Father Christopher Scheiner, S.J. (1579–1650), author of the tract *De maculis in sole animadversis* (1612) and of *Rosa Ursina* (Bracciani, 1630), whom Galileo unwisely attacked on the question of priority in the telescopic observation of sunspots.

7.3. Simon Marius (1570–1624), the author of *Tabulae directionum novae* (Heilsbronnae, 1593) and *Mundus Jovialis* (cf. note 24.43), who is usually credited with the discovery of the Andromeda Nebula and who was among the earliest telescopic observers of the motions of Jupiter's satellites.

7.4. This is Newton's result (cf. note 4.7). Flamsteed's previous calculations had actually been based on the approximate ratio $\frac{3}{2}$ used by Descartes (cf. note 4.8) and adopted by Gascoigne.

7.5. Lecture 4 (8 June 1681).

7.6. The words in square brackets have been guessed. A line is missing from the original text.

7.7. Flamsteed wrote SHI.

7.8. Flamsteed wrote i instead of I.

7.9. Flamsteed wrote c instead of s.

7.10. Flamsteed wrote dbc.

7.11. Isaac Barrow, *Lectiones opticae et geometricae* (Londini, 1674).

Lecture 8

Wednesday, 9 *November* 1681

In ye conclusion of my last Lecture. I shewed you how to find ye point of direction in a double concave glasse when the Radius of the sections on each side were ye same. or equall I had not then leasure to compare it with the double convex of the same dimensions which I therefore promised to doe in my next, & shall now performe my promise onely remindeing y^u^ first that as hitherto I suppose alwayes the radius of the Concavity or convexity precise 2 inches So I shall continue it & that I then determined the distance of the point of direction from y^e^ next pole of y^e^ glasse for y^e^ ray Which was parallell to y^e^ polar but one tenth of an inch wide from it to be 16261 such parts as y^e^ Radius of the section was 20000. but for y^e^ parallel ray at the widenesse of $\frac{1}{2}$ an inch that distance onely 13822 of the same partes whence it was evident that no parallell rayes not equally distant from y^e^ Axis had the same point of direction or were carried after theire emersion from y^e^ glasse as it were from y^e^ said point but that for the rayes next the axis the point of direction was more remote from ye pole of the glasse, then when the parallel rayes had theire course at a wider distance from ye polar Ray.

The same I observed by the double convex glasse in my readeings last terme.[1] & shewed y^u^ method whereby I have found repeateing the processe of Calculation since that of two Rayes of light fall upon a double convex glasse the Radij of whose convexity on both sides is exact 2 inches or 20000 p[ar]tes & its thicknesse $\frac{2}{10}$ of an inch, one of them $\frac{1}{10}$ the other $\frac{1}{2}$ an inch wide from the Axis the nearest would converge with it at the distance of 17620 p[ar]tes from y^e^ next pole of the glasse the wider at 16530 of the same partes from y^e^ same pole whence it was evident that the difference of these 1090 or something more y^n^ the tenth of an inch is the space on the Axis wherein all the parallel rayes not more y^n^ $\frac{1}{2}$ an inch wide from it would converge, & that the proportion betwixt the Radius of the section & distance of the burneing place or pointe from y^e^ glasse is as 20 to some little lesse y^n^ 18. but that it may be taken nearely in round numbers nearely as 10 to 9. or something lesse for the wider rayes converge nearer ye lens or

Notes on page 187.

glasse & therefore I have stated the proportion with the least difference or for those rayes that converge at y^{e} greatest distances from y^{e} next part of it

Whereas when wee examined the double concave wee found that y^{e} remotest point of direction for y^{e} ray $\frac{1}{10}$ of an inch was but 16261 partes distant from y^{e} next or inner pole of the glasse which is something lesse then the nearest burneing point in y^{e} double convex. but for y^{e} Ray $\frac{1}{2}$ inch wide it was onely 13.822 of the like partes from y^{e} said inner pole. So y^{t} the distance of the points of direction for y^{e} rayes $\frac{1}{2}$ inch wide from y^{e} Axis is 2439 or neare a quarter of an inch so that tis evident this disposes the rayes neare as ill as the plano concave . . . [for] which this difference was but 2691 of the like partes. & that the distance of the points of convergency in the double convex is scarce halfe so great as the widenesse of the pointes of direction for the same rayes in the double concave. of the same dimensions.

Haveing shewed y^{u} now how the rayes of light are carried through the most simple formes of glasses the plaine or double convexes & concaves I might now proceed to shew y^{u} the composition of ye single telescope composed of a simple convex object glasse & a concave eye glasse but that I esteeme it convenient to dispacth all the businesse of single glasses first & to shew y^{u} How the burneing point is found in ye double convex & the point of direction in the double concave when the Radij of the sections on the different sides are not the same. but of different lengths

In doeing of which I shall take occasion to shew you how y^{u} may know what shall be the focall length of a glasse wrought on its two sides in two different tooles for Concaves Spheres breifly as allso the point of direction for ye like concaves. Afterwards I doubt not but I shall have time enough to shew y^{u} how the[y] performe & how the concave operates in a telescope & may be rectified when placed on an instrument which is the cheife & maine scope of my designe.

In y^{e} first figure [Fig. 8.1] Let zuiw. represent a double concave glasse. the more hollow side wz towards the object. whose Radius wc = wb. I suppose as formerly 2 inches or 20000 partes its least thicknesse wi one fifth part of an inch or 2000 partes but the Radius of the other section oi = od y^{e} quintuple or five times as much as cw. y^{t} is 10 inches or 100000 partes & I will propose it to find the point of direction h. for the ray ab when it is either $\frac{1}{10}$. or $\frac{1}{2}$ of an inch distant from the Axis co.

Let ab. be the ray of light $\frac{1}{10}$ of an inch wide from the axis = yb or 1000 partes to the point of incidence b draw y^{e} perpendicular cb. continueing it to g as also from y^{e} other center o. the line ob. in the triangle ycb. are knowne cb. 20000 & by 1000 whence y^{e} angle of incidence ycb will be found 2°.51′.57″$\frac{1}{2}$ & cy 19975. as in all y^{e} former examples, take cy from co. 122000[3]. so have y^{u} the remain[d]er yo 1025. with which & y^{e} side yb before stated y^{e} angle yob will be found

0°.33′.41″ = obe for be and yo are parallel. y^{u} may also find the Logarithme of y^{e} side bo whose dimensions I give not because for use its logarithme will be found sufficient.

Because at b y^{e} Ray enters y^{e} glasse a dense Medium it will be refracted towards y^{e} perpendicular eg & therefore accordeing to y^{e} laws of refraction it will hold as 300:193,:: so y^{e} sine of y^{e} angle of incidence abc = ebg. 2°.51′–57″$\frac{1}{2}$ to y^{e} sine of y^{e} Refracted angle fbg = dbg 1 – 50 – 36 which subtracted from y^{e} angle of incidence ebg leaves y^{e} refracted angle ebd 1°-01′-21″$\frac{1}{2}$ which added to y^{e} angle obe before found 0°-33′-41″$\frac{1}{2}$ gives the whole angle obd 1°-35′.03″

At d the ray emerges from y^{e} glasse to which point from y^{e} center of y^{e} shallower section o draw y^{e} line od. continueing also the line of its passage through the glasse bd to q.

In the triangle. obd. wee have given the side ob found before, od 100000 & the angle obd determined 1°-35′.03″ whence the angle qdo y^{e} complement of bdo will be found 1°.36′.58″$\frac{1}{2}$ from which takeing the angle obd 1-35-03 the remainder shall be the angle bod 1′55″$\frac{1}{2}$: And this added to y^{e} angle boH gives the whole angle Doh 0°.35′.37″:

Because at d y^{e} ray emerges from a dense to a rare Medium wee must say as 193:300:: so y^{e} sine of the angle of Emersion odq 1°.36′.58″$\frac{1}{2}$: to y^{e} sine of the refracted angle odr 2°-30′-46″ whose complement (if wee protract y^{e} line rd backward till intersecting y^{e} Axis it give y^{e} point of direction h) shall be the angle rdh. in y^{t} triangle 177°.29′-13″ which added to the angle doh before found 0°.35′.37″ the complement of theire summe to 180 degrees Shall be [y^{e}] other angle sought ohd.

hence in that triangle wee have all the angles, & one side od, the Radius of the section 100000 partes knowne, whereby the side oh will be found 130901 from which subtracteing ow. 102000 the remain[d]er 28901 shall be the side Hw. or the distance of the point of direction for this Ray which was onely one tenth of an inch wide from the polar.

As for the Geometricall processe whereby the point of direction h might be found I hope I need not describe it haveing done it sufficiently when I found the like point in ye other concave glasses already handled. if I find I have roome & time I may againe describe it when I come to the meniscus or limular glasse [Fig. 8.5] at present I shall proceede to shew y^{e} place & distance of the said point of direction if supposeing still y^{e} same Radij of the section & thicknesse of the glasse. wee conceave by the widenesse of y^{e} parallel ray to be halfe an inch or 5000 partes.

Which admitted wee shall find cy. 19365 which taken from co 122000 leaves $\overline{\text{yo}}$ 102635. the angle of incidence bcy = ebg being 14°-28′-39″. & therefore accordeing to the laws of refractions fbg 9°-15′-19″ & ebd 5°-13′-20″

2^{d}, in the triangle byo are known yb 5000 and yo 102635 whence y^{e}

side ob will be found with the angle yob = ybe 2°.47′.20″$\frac{1}{2}$ which added to dbe gives the whole angle obd 8°.00′.40″$\frac{1}{2}$

3. in y^e triangle obd are knowne the sides ob and od with y^e said angle obd whence the angle bdo will be found 171°-44′-48″ whose complement 8-14^4-12. is y^e angle of Emersion . . . .

At d the ray emerges from a dense Medium into a rare therefor as 193:300: so is y^e sines of the Angle of Emersion odq. 8°-14′-12″ to y^e sine of the rdo. 12°.51′-43″ whose complement to 180. is the angle odh 167°.08′.17″.

4. The difference of the angle obd 8°.00′.40″$\frac{1}{2}$ & odq 8°.14′-12″ is y^e angle dob 0°.13′.31″$\frac{1}{2}$ which added to yob 2°-47′:20″$\frac{1}{2}$ gives the whole angle yod 3°-00′-52″.

5. So now in y^e triangle doh. wee have attained the two angles at o and d whence the third angle at H will be found 9°-50′-51″. the side do we supposed at first quintuple to cw or 100000 hence ho will be found 130159 from which takeing ow 102000. the residue shall be the distance of [y^e] point of direction for this ray from y^e next pole of y^e glasse w. 28159

which distance when y^e ray of light fel[l] parallel to y^e axis onely $\frac{1}{10}$ of an inch distant from it wee found 28901 that is 742 partes. or neare $\frac{1}{13}$ of an inch greater, so y^t it is evident that tho no two parallel raies have y^e same point of direction yet y^e distances of the severall pointes for all rayes whose angles of incidence are lesse y^n 15 degrees are so small that wee may accept them as spreadeing all nearely from the same point of direction.

It is evident also that the small concavity wrought on the one side ye glasse by a sphere 5 times the magnitude of the section on the opposite side assists it . . . . . but the best way is to place the most concave side towards the object for when wee examined the planoconcave with y^e hollow side towards y^e object the distance of y^e point of direction from y^e pole was found 35252 for y^e nearer ray. which with the double concave for y^e same ray is found 28901 y^t is almost $\frac{64}{100}$ or $\frac{6}{10}$ of an inch lesse.

Hence the distance of the point [of] direction from y^e pole of any glasse whose two sides have beene formed on different tooles may be found: thus. find first what ought to be the distance of the point of direction from y^e pole for y^e lesser radius supposeing the glasse a planoconcave with the hollow side towards the object. this is in proportion to [the] diameter of the sphere by which it is formed nearest as 40 to 35 or as 8 to 7. Say then.

As ye summe of ye semidiameters of the two spheres, to ye semidiameter of the greater, so is the distance of the point of direction for the lesse, to like distance for the glasse when formed in both to trie this proportion by an Example.

The semidiameter of the lesse sphere to y^e greater I supposed as one to 5. theire summe is six, say as 6:5:: so is 35:29: the same distance very nearely y^t was found by y^e calculus: which rule wee shall further confirme by & by:

But if wee suppose another ray αb to fall upon y^e same point b $\frac{1}{10}$ of an inch from the Axis but that y^e inclination of it to y^e axis, is lesse. wee shall find that the point of direction for it will fall nearer the pole still & the ray it selfe after its transmission through y^e glasse will diverge wider from y^e Axis then any of the parallel rayes that fall betwixt it & the said Axis. I shall not need to shew y^u the method of calculation for such an obliq[ue] ray. I shall onely tell y^u tis the same or with very little difference that it is for the parallel ray ab. and that haveing calculated the several angles in its progresse. supposeing the angle of its first inclination αbc to be 1°30′: I find that after its emersion from y^e glasse whereas the ray ab ran into the line dr. this αb diverges wider from y^e Axis in y^e line $\beta\gamma$ so as the point of direction for it shall be 27535 partes distant from y^e pole whereas for y^e parallel ray ab it was 28901 y^t 1366 partes or neare $\frac{14}{100}$ of an inch more.

Let us now consider the simple effects of a Concave glasse & then as applyed to a perspective

In y^e 2^d figure [Fig. 8.2] let O represent a double concave glasse, ag an object to be viewd through it, owiefd the eye viewing the said object, $\bar{p}$ y^e remotest point of direction for y^e said glasse, it is evident that if the breadth of the said object be greater or equall to y^e aperture of y^e eye, the object is not seene by the parallel raies to y^e axis ab and gh for if to b and h the points of incidence wee draw streight lines from y^e point of direction p y^t is bd and hi they will fall without the aperture of ye eye as also will all that fall from ye same points betwixt be and kh. so y^t it is plaine that the object is onely seene by the raies that fall betwixt a and k which subtendeing a lesser angle then kg. the object shall appeare nearly but not exactly as much lesse as ye angle subtended by ek at ye distance ne or nk is less then ye angle subtended by aq at ye distance of na. or ng. for ye rayes proceeding from a and g. & falleing upon the glasse betwixt e and k. or neare them will helpe to project the image om in ye bottom of ye eye

The two raies of light which proceed from the extreme points of ye object, to wit ae and gk divide or spread after they have passed the glasse into ye lines ef:kl: but by theire first refraction are carried nearely parallel to each other into the Christalline humor q. by the helpe of which & ye vitreous uw they are thence so conveyed as that they may project the picture of ye utmost pointes of the said object upon ye bottom of the eye or Tunica retina inverted which is the reason that objects viewd with a simple concave appeare alwayes through it erect, for as I shewd y^u in my first opticall lecture of ye Last terme[5] such objects as are projected inverted on ye bottom of ye eye appeare erect but those which are painted erect, seeme & appeare inverted

But if ye concave be applyed to a telescope object glasse the manner of its performance is much different for let V in ye 3d figure [Fig. 8.3] represent ye object glasse of a telescope which would cast the distinct figure of a remote object on tf. the tunica retina of the eye in this figure. the place of the Concave eye glasse for this perspective Shall be as neare as is possible to ye superficies of ye eye if it be wrought on a small sp[h]ere but more remote if on a larger. now if wee suppose ab in ye said 3d fig a ray of light proceedeing from the one extremity of a remote object, this after its transmission through the glasse shall be carried in ye line bc. to c. where meeting with ye concave superficies of ye eye glasse it is refracted into the line cd. & at its emersion diverges wider from ye axis entring ye eye at e. whilest another ray gh proceeding from ye same point of an object very remote enters ye eye glasse beneath it at k by whose refraction it is made to diverge something wider from ye p[er]pendicular by reason of its greater inclination so that when it enters ye eye at m it is carried by it & made to converge in the same point of ye tunica retina at f with the former neither of them crosseing the opticall Axis by reason of theire great inclination at theire entrance. but projecteing the image inverted as it would have beene cast from the simple object glasse if no eye glasse had beene interposed whence it is that the object appeares to ye eye erect for the Ray ab which wee may imagine to have proceeded from the other extremity of ye object moveing in the line bopwkst projects the other extremity of ye image contrary to its true position above in the eye whereas it was realy the inferior part m in ye aire or heavens in whethersoever of which yu conceave it to have beene placed

It seemes not that theire should be any proper place for the measureing thrids in a telescope with a concave eye glasse by reason that the two rayes ab and gh proceedeing from the same point of the object meet not within ye length of the glasse but in ye bottom of the eye. but if the glasse be a plano convex wrought on a small sphere or former & have its Concave side placed towards the object its plane next the eye a payer of crosse thrids may be placed contiguous to the plane side or rather two lines may be cut on the said plaine side with ye sharpe point of a diamond so as they may crosse each other in the pole of the glasse at right angles which shall performe the desired effect. and observe that part of the object from view. when the raies proceedeing from it passe through the pole of the object glasse upon this place. for all the rayes that proceede from the same point of the object after theire transmission through ye first glasse meet not exactly upon one & the same point of the eye glasse, yet they convene so nearely that in the aperture of the object glasse is not too great nor radius of the concave sections too wide, the thrids or lines upon it will receave the greatest part of them especially of those which fell in nearest the pole & these would meet nearest togeather in the distinct base.

The manner of rectifieing this will be altogeather the same wth that I prescribed for ye Telescope of convex glasses,[6] save that whereas when the Convex eye glasse was used it mattered not whether it was fixed or not. So the thrids were firmely fastened & rendred immoveable, here on the contrary it will be necessary to have the eye glasse firmely fixed that so the thrids may be kept unaltered both dureing and after ye rectification whether made by an horizontall object or ye fixed stars. accordeing as was directed in ye last lecture of the last terme and ye first of this.[7]

I have done with ye Concave glasses, whether simply used or applyed to a convex in a telescope. hitherto I have not said any thing of the convexes whose superficies are the segments of unlike spheres. I shall therefore consider them breifly. and first the double convex.

I shall not need to acquaint y^{u} either with the Geometricall or Trigonometricall processe I have done that already and I feare too largely in some persons opinions I shall therefore onely need to acquaint y^{u} with the result of my labor. & a rule for findeing the focuses of such double convexes whose superficies are the Segments of unequall spheres.

In the 4th figure [Fig. 8.4] let dbg represent such a glasse, The radius of its more convex side cw[8] 2 inches or 20000 p[ar]tes. let be in such proportion to Kd the Radius of y^{e} other as 1 to 5. therefor Kd 100000 or 10 inches. if wee suppose ab a ray of light $\frac{1}{10}$ of an inch wide from y^{e} Axis Kw. the distance of Y the burneing point from w the remoter pole of the glasse will be found. 31127. but if the widenesse of y^{e} said parallel ray from y^{e} Axis be supposed $\frac{5}{10}$ or $\frac{1}{2}$ an inch then shall the distance Yw be 30377: that 81 one thousands of an inch lesse so that is evident this p[er]formes not much worse then y^{e} simple convex with the convex side towards the object but that by how much the side dg is the arch of a lesser circle so much the greater will this difference be

The diameters of the two spheres in which each side of the glasse is wrought being given with the focall distance for the lesser Radius the glasse being supposed a plano convex the compound focall distance may be found after the same Manner as ye point of direction for the like double concave was got before. for in ye same 4th figure it will hold. As ye summe of ye two Radij KC to ye Greater Radius FK:: so is ye focall lenght for ye Lesser Radius Bw to ye compound Focall length wY.

The Meniscus or limular glasse remaines onely to be considered. in the 5th figure [Fig. 8.5] represented by dgb. this if the Radius Ki of the concave side shall be greater then ci the radius of the convexity shall collect the rayes. if lesse disperse them if both be equall all the parallel rayes passe from them as they entred parallel to each other.

Some p[er]sons have conceaved that a Meniscus is the best figure for an object glasse and the reason of this opinion I conceave to have been because the rayes emerged with lesse inclination then in other sections but haveing enquired Trigonometrically in what points they would cut the axis I find that if I suposed Ki. the Radius of the Concave side to be y^e one and halfe of cw y^t is cw 20000 & Ki 30000[9] that ba were a ray of light $\frac{1}{10}$ of an inch wide from y^e axis, the convex side of the glasse respecteing the object the distance of Y the place of the image from I the next pole of the glasse would be 111617 or 11 inches & $\frac{162}{1000}$ but if y^t widenesse were supposed halfe an inch. the space would be 107.240 or 10 inches & $\frac{724}{1000}$ so y^t y^e parallel Ray halfe an inch wide from y^e Axis would not meet but in y^e Space of $\frac{438}{1000}$ or neare halfe an inch upon it,

In either of these cases had I calculated the point in which a ray inclined to y^e Axis falleing upon the point b after its transmission would intersect it. I Should have found y^e place of y^e intersection still nearer the glasse[10] whence wee may easily conceave ye reason why a tube may be produced or contracted from the usuall length & yet ye object appeare distinct for tho it bee seene best & biggest by those rayes which fall in nearest the Axis & therefore are collected widest from it, yet are there enow collected in long glasses some inches nearer whereby it will be seene not much worse. but it is sufficient to have intimated this.

The Rule whereby the focall lenght of the convex side Bw being given (supposeing the other plane.) with the Radij of the spheres cw:Ki whereby they were formed to find the Compound focall lenght is this. As ye difference of the two Radij KC in ye 5th Fig. [Fig. 8.5] to ye Radius of ye concave side KI: so is the focall length for ye convex side Bw: to y^e compound Focall lenght Ym for example. in this glasse last proposed. the Radius of the concave side I supposed 3. inches of the convex 2. theire difference is 1. the focall length of a plano convex of 2 inches or 20000 parts radius, is 36000 it holds. therefore 1:3::36 to 108 or 10 inches $\frac{2}{3}$ neare which wee found it by calculation

I have beene thus prolix upon Telescope glasses that I might shew y^u with what facility & ease it is possible to determine the focall lengths of burneing glasses or object glasses for Telescopes by no more then helpes then plaine Trigonometrie & this I have done because I have not found it handled by any opticall writers as it ought most shuffling ye businesse over as soone as they have entred on it. & soone takeing notice of the different p[er]formances of glasses as they are differently posited by the same method it were easy haveing any convex glasse given & the distance of an object placed in its axis from it to find at what distance from ye glasse it shall appeare most amplyfied & where most distinct. For y^u may remember that tho those be confounded by all writers hither yet I shewed y[u] a vast difference betwixt them & confirmed it by an experiment of which any ordinary p[er]son may be judge.

One thing I have omitted which I promised that was to shew y[u] how the Ellipticall or Hyperbolicall Concaves that had theire diameters to the distance of theire Foci as y[e] sines of y[e] Angles of Incidence to y[e] refracted angles would dispose the parallel rayes of light. I shall cleare my selfe very breifely. the remoter focus of each is y[e] point of direction & all the parallel rayes wherever they fall on the curve shall be carried hence & disperst as if they had proceeded from those points. or Foci[11]

I have done with glasses at present and have no more to say concerneing them at present but that I would advise al[l] those who desire to be throughly acquainted with theire performances & quality to employ his pen as well as his Compasses whereby hee may either avoid or at least excuse himselfe from those difficultys which were otherwayes difficult to be overcome: but DG:[12]

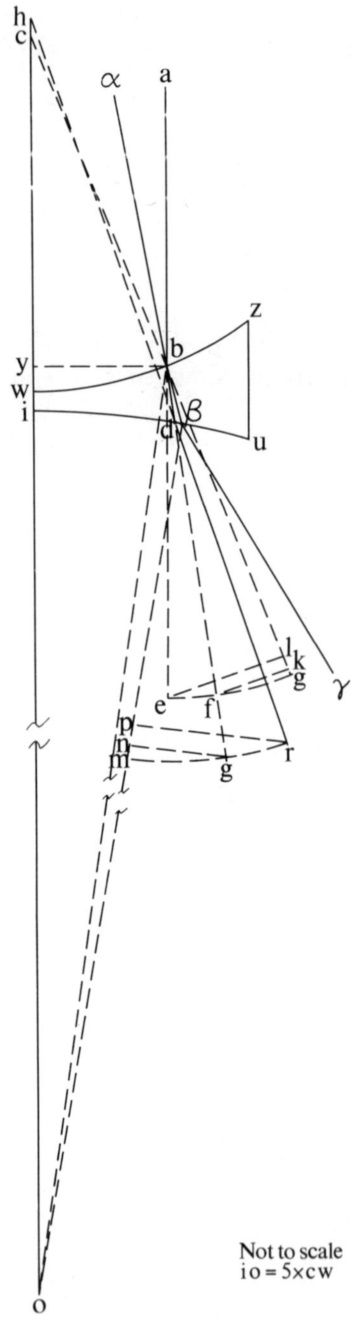

FIG. 8.1

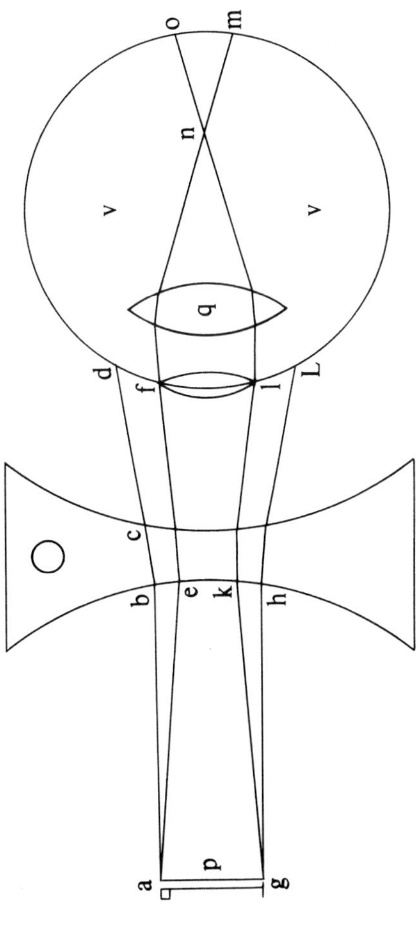

FIG. 8.2

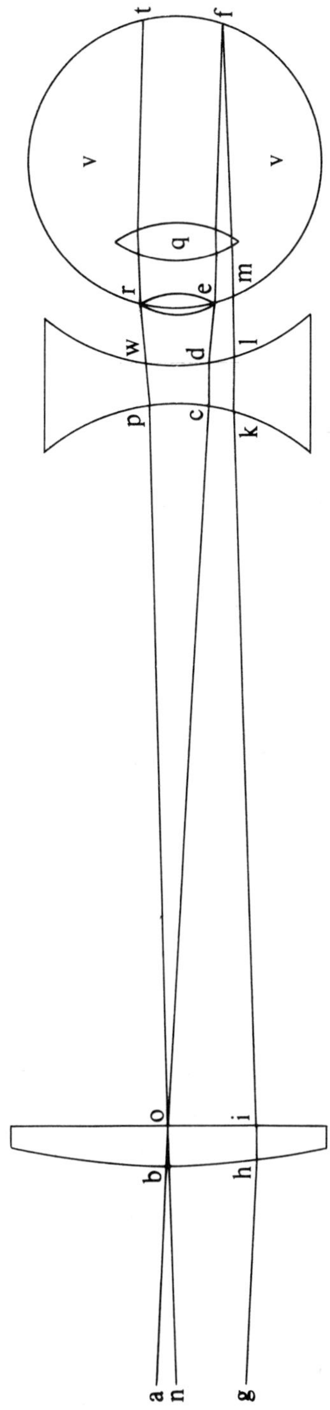

FIG. 8.3

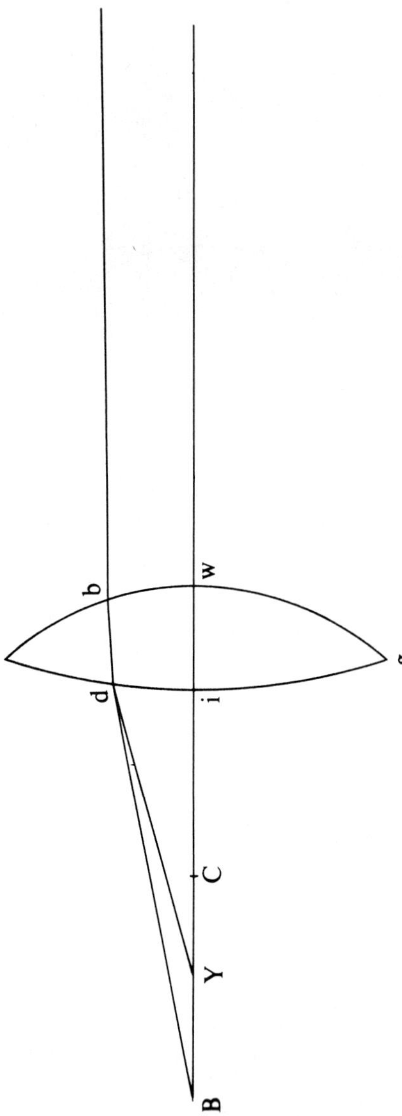

FIG. 8.4

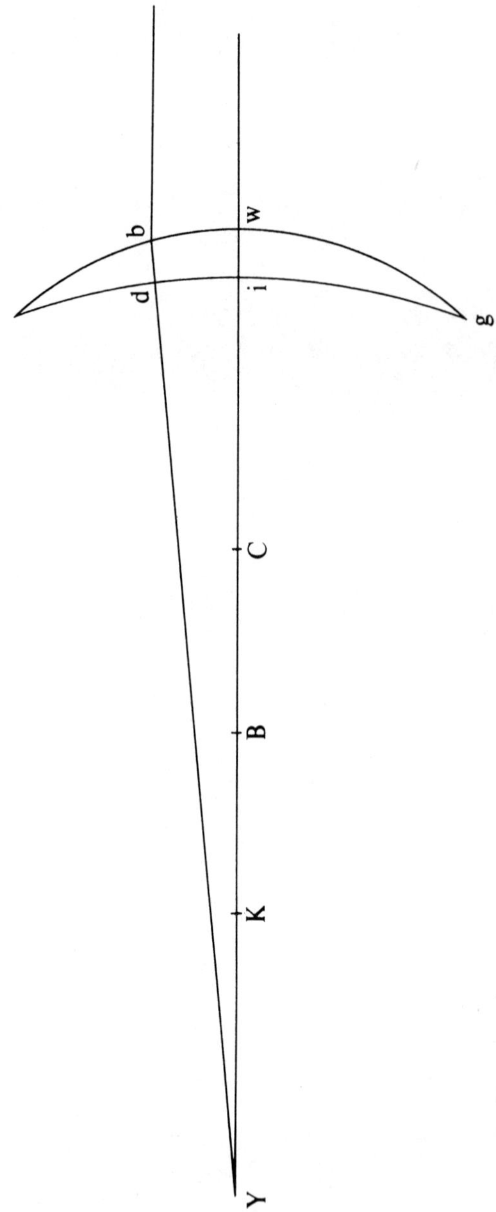

FIG. 8.5

NOTES

8.1. Lecture 4 (8 June 1681).

8.2. Lecture 7 (2 November 1681).

8.3. This figure ought to be 21000.

8.4. Flamsteed here incurs an error, since 180°-171° 44′ 48″ = 8°15′12

8.5. Flamsteed demonstrated this property of the eye in Lecture 5 (22 June 1681), which was in fact the *third* optical lecture of the previous term—bearing in mind that one is missing (cf. note 4.17).

8.6. Lecture 5.

8.7. Lectures 5 and 6 respectively.

8.8. Flamsteed wrote cb instead of cw.

8.9. Flamsteed wrote 300000 by mistake.

8.10. Flamsteed did, in fact, make such a calculation on ff. 81^v and 90^r and obtained a difference of $\frac{4306}{10000}$ inch—namely, $\frac{7}{1000}$ inch nearer the lens.

8.11. Note that Flamsteed had demonstrated the same conclusion for elliptical and hyperbolic *convex* lenses in the missing lecture (cf. note 4.17), as he explicitly states in Lecture 5, f. 48^v.

8.12. Unfortunately, the words on the next (and last?) line of this lecture are missing from the manuscript.

Lecture 9

Wednesday, 16 *November* 1681

In my late lectures I have shewne y^{u} the Nature & Manner of the performances of Optick glasses I have proved that they doe not impose upon our senses, that they may be applyed to Instruments for takeing the Measures of Angles either in the heavens or on Earth, and rectified for that purpose not onely as well but much more easily and exactly then Instrum^{ts} with plaine sights Much more might have beene said as concerneing the Telescope with 3. 4. or 6 glasses. and the colours projected by them on the limbes of objects especially when the Eye glasses are convexes of small spheres or not well placed in theire Cells but those that shall consider them after the manner I used & shewed y^{u}, can not misse of what was further to be said of these. It will be Needlesse to adde more & therefore I shall now returne to my first designe. which was to shew y^{u} by what steps Astronomy has beene improved and arrived at its present state, how the Ancients became first acquainted with ye motions Magnitudes & distances of the Heavenly bodys, & how by compareing their observations with those of this & the last Age wee have attained to the more perfect knowledge of them. In my first Readeing here I performed this in part. as the time would p[er]mit me. I had then but a slender auditory & I can not therefore think that many who are now present heard me. It will be necessary therefore for mee that y^{u} may ye better understand what I have further to say on that subject, to repeate ye summe of that Lecture breifly.

I told y^{u} then first that Kepler in ye first Chapter of his Commentary de Motibus Stellae Martis[1] had shewn how the Ancients by considering the diurnall appearance of ye stars & planets theire accesses to & recesses from each other came to know that the Moone was the swiftest of all the planets, that there were seven in Number & that all the other stars were fixed. but that how they became acquainted with the periods theire Revolutions hee was silent which therefore I undertooke to shew y^{u}. I told y^{u} first that the ancients accounted theire Moneths by the Moone whose Meane returnes to the Sun being p[er]formed 29 dayes and one halfe they allowed twenty nine dayes to one moneth & 30 to the Next and 12 of these Moneths to a yeare.

Notes on page 199.

But that findeing after 3 yeares that theire Spring was not so forward as it had beene 3 yeares before by a Moone. they concluded that theire was one revolution more p[er]formed in three yeares then they had accounted & that therefore once in three yeares they intercaled a moneth where they thought most convenient & that this intercalation was made amongst the Romans by [the Pontifex Maximus,][2] the Jewes by the Sanedrim, with the Greekes at first by the Authority of the Most learned & skillfull amongst them but with the Athenians by theire Auchon for that time being. who called it theire τριετηρισ

This put them upon the inquiry of the time of the Suns revolutions, Who first began it appeares not by any monuments of Antiquity. for all the greate pretences of the Egiptians & Chaldeans wee may reasonably conjecture that ye Jewes began it before them. for by the law of Moses theire offerings were to [be] paid & theire feasts kept at such seasons of the yeare as could not be before determined without a true knowledge of the Suns revolutions. particularly the Feast of Ingathering was to kept at the conclusion of theire harvest & fixed in the Moneth Elul which conteines parte of o^r August & September. Now if they mistimed theire intercalations or made them not of ye due quantity, this moneth might happen before theire fruites were ripe. to obviate this they tooke Notice that the Noone Shadowes of theire houses or any perpendicular body was constantly the same at ye same time of the yeare. & that therefore they might determine the true length of the solar yeare by Counteing the dayes interlapsed betwixt the times in the two Summers when the shadows came to be of the same length. & that heedeing it diligently they found that it was after 12 of theire Lunar Moneths & 11 dayes Nearely So y^t whereas the lunar Yeare consisted onely of 354 dayes the Solar conteined 365.

The Jewes I meane the Ten tribes were first Carried Captives in ye 25th yeare of Nabonasar[3] & from them wee may presume theire Masters the Assirians and the Chaldeans theire Neighbours had the first knowledge of the Solar & Lunar Revolutions For ye first lunar Eclipse wee find recorded to have beene observed by the antients is two yeares after in ye first yeare of Mardekempadi or Meredak Ball Adan. & ye 27th of Nabonasar, the founder of the Babilonian Monarchy & Epocha whose yeares consist of 365 dayes precise whereas the Sacred Story[4] mentions the diall of Ahaz whose reigne accordeing to Helvicus commenced in these & concluded in ye 18th of Nabonasar, whence it appeares that this antientest peece of Mathematicks wee find in any story is at least 7 years older then ye observation of the Eclipse. & it seemes not probable that the Jewes should know how to make a diall without understanding the periods of the Suns revolutions as well as his Course

About 130 years after this Thales the Milesian traveling into ye East learnt of the Chaldeans the true length of the Solar yeare & lunar revolutions Hee is said to have beene the first among the Greekes that predicted an Eclipse of the Sunne & to have wrote a booke de Solsticijs & Equinoctijs, the knowledge of all which wee may most probably conceave him to have atteined by his converse & acquaintance with the Eastern people.

It came now to be noted that ye equall shades hapned twice every year & were all of them twice repeated except at either soltice. Wherefore Meton & Euctemon chose to observe the solstices which the[y] collected wee may presume by the Meridionall Shadows of some few dayes praecedeing & followeing Ptolemy[5] tells us that they determined a solstice on ye 21th day of the moneth Phamenoth in ye Morneing when Abseudes was Praefect of Athens which Scaliger[6] and Petavius[7] make the 27th of June in ye 316 of Nabonassar or 432 yeares before the Incarnation. 152 yeares after which Aristarchus observed another soltice from which observations however difficult it is likely the periods of the Suns revolutions were stated by the Greekes.

Meton placed the begining of theire yeare on the Solstiall New moone as is apparent by Festus Aucenius

Sed primaeva Meton exordia sumsit ab Anno torreret rutilo cum Phoebus sidere Cancrum.[8]

Hee also ordered theire Calendar & Numerus Aureus whereby hee taught them how to find the times of the New & full moones in it through all the yeares of his perriod or ευυεαδεκατηριδ̀[9] tho this was not so absolutely well ordered but that Eudoxus who lived some 70 yeares after him found cause to alter it whose fasti were so well esteemed of by the Greekes that the poet makes Caesar when hee corrected the Calendar to applaud himselfe for it in these words

Nec meus Eudoxi vincetur Fastibus Annus[10]

Euctemon. & Philippus were the first that composed Kalendars amongst ye Greekes which they grounded upon Metons period. These conteined like oe Common Almanacks onely an account of the Moones Age Changes & full with the Feasts depending on them Had it beene attended this period might have beene found out by onely compareing the observed times of lunar Eclipses one with another for they often are repeated on the same day after 19 Yeares. The scarcity of observations affords us not any examples of Eclipses noted of old to have hapned at those Intervalls, but if moderne ones may suffice the curiosity of the present age will furnish us with plenty. The Noble Tycho observed an Eclipse of ye Moone at Uraniburge on ye 18th of October 1594. Longomontanus after 19 yeares. on October ye 18. 1613. another. and on the same day in ye yeare 1632.[11] just 19 yeares after a 3d was observed. Againe Monsr Hevelius observed a lunar Eclipse Octobr 19. 1659 at Dantzick. and at Greenwich 1678 just 19 yeares after another totall Eclipse was observed.[12] So y^t wee cannot conclude it to have beene any thinge but theire ignorance of the exact length of the Solar Yeare and theire Numbring theire Yeares by the Moones that kept the Ancient so long ignorant of this period which if the Eclipses had beene heeded would have beene discovered much Earelyer

After Eudoxus. the next that wee find to have Concerned himselfe in the reformation of the Kalendar & adapteing the lunar to y^e solar Yeares was Calippus who haveing found that the true lenght of the solar Yeare was 365$\frac{1}{4}$ dayes nearely, considered that Metons lunar Cycle if true must not consist of any equall Number of dayes for that in 19 solar Yeares. were 19 of Nabonassars with 4 dayes & $\frac{3}{4}$ of a day. to take away the fraction therefore he quadrupled the period and made it 76 yeares in which space of time accordeing to his account all the feasts would returne and the Phases of the Moone to y^e same day of the yeare. Hitherto wee may reckon it an age of obscurity or as the dawneing of Astronomy. which for want of good records of ancient observations was little improved, but soone after Aristarchus there appeared & florished some diligent & ingenuous persons who considering with themselves that the solstices were very difficult to be determined by reason of the slow alterations of the Noone shades chose rather to observe the aequinoctialls. What Timocharis did in this p[ar]ticular wee have no records of antiquity to informe us. but certeinely his observations were of good use to that great lover of truth & industry Hipparchus who lived about 160 yeares after him and 140 before the Incarnation. & compareing his observations with those of them that went before him concluded that the fixed stars altered theire apparent places & that the true length of the Solar yeare was not as had beene hitherto reputed 365 dayes & $\frac{1}{4}$: but lesse that is $365^d{\cdot}05^h{\cdot}55'\frac{1}{2}$ and that from the time of ye Vernall aequinox to y^e autumnall there were interlapsed 187 dayes praecise but from y^e autumnal forward to the vernall onely $176^d{\cdot}05^h{\cdot}55'\frac{1}{2}$

Hence they argued that since in both these times the suns apparent motion was a semicircle of the Ecliptick, hee Moved slower through the Northerne then through the Southerne halfe of it. and divideing the whole Circle moved in a yeare by the Number of dayes Conteined in it they found his diurnall meane motion something more ye 59′ 08″:

But to find in what places of the Ecliptick hee Moves slowest & swiftest. they were forced to Make use of theire Solsticiall observations of which (not knowing perhaps either how to employ any other or conveniently to attaine them) they were so confident that however difficult they were to be collected by observation yet Ptolemy boldly asserts, *Sed in Solstialibus crede nec nos. nec Archimedem in observatione et calculo ad quartam diei partem errasse.*[13] that is, hee hoped *that neither himselfe nor Archimedes had erred a quarter of a day either in the observations of the Solstices or calculations relateing to them*, tho for the [same reason—] theire difficulty [—] hee rejects the ancienter observations of Meton ... translated by ... & Tome ye 1st p[ar]te ye 1st page 136 are *vetustatis igitur gratia &t.*[14] *that onely for Antiquitys sake the Summer Solstices observed by Meton & Euctemon and Afterwards in the time of Aristarchus might be compared with his. but* Sayes hee *because* y^e *solsticiall observations are very difficult to make, & those delivered by them were not exquisitely observed as Hipparchus also judged wee praetermitte them.* it seemes therefore hee was more confident of his owne then of theires by which hee determined. the time interlapsed betwixt the Suns Entrance into ♈ till hee arrived at ye first point of ♋ to be 94 dayes & 12 houres which obteined his Method of proceedeing to find the pointes of the suns swift & slowest motion is thus

Because all people noted that the Sun came Neare the same stars every yeare at the same time they concluded that hee was carried through ye Ecliptick once in a yeare. & that hee moved equall spaces in equall times in his orbite but because hee stayd longer in the Northerne then in the Southerne signes and in the first quadrant of the Ecliptick then in any of the other. that the Center of this orbite about which his equall motion was made was in the Northerne signes & somewhere in y^e first quadrant Let therefore [Fig. 9.1] ompn represent the Ecliptick o y^e 1st point of ♈, $\underline{m}$ of ♋, p of ♎, therefore op y^e line of y^e aequinoxes and $\underline{mn}$ of the solstices intersecteing it at right angles in c y^e center of the Earth Let e in y^e 1st quadrant of the Ecliptick represent the center of y^e circle duhi in which they supposed the suns equall motion to be performed. connect it with y^e center of y^e Earth by y^e line ce which continue to y^e Ecliptick at $\underline{L}$ through it also draw y^e lines hq. and ui parallel to op and nm.

Now because y^t the Sun Moves over the arch dquhr. from y^e vernall. to autum[n]all aequinox in 187 dayes. his meane motion in that time. 184°.20′ shall bee the measure of that arch. whose halfe 92° 10′[15] shall bee y^e arch. du. but $\overline{uq}$ is 90^d. therefore $\overline{qd}$ 2°-10′ and its right sine 3781 such partes as eq. is 100000 line = $\underline{bd}$ = ef = gc.

From the vernall aequinox to y^e Summer Solstice hee accounted 94^d : 12^h in which time the Suns Meane Motion is 93°-09′ = to y^e arch dl. from which takeing y^e arch du 92-10 the remain[d]er 0°-59′ is y^e arch ul whose right sine 1716 parts. is y^e length of the line lk = ge = cf and now in either of the two similar triangles egc. and cfe wee have the two sides conteineing y^e right angle. whence the distance of the center of y^e orbite from y^e earths center ce will be found 4152 such partes as ea is 100000 this they called the excentricity. A y^e point of the circle in this line produced his Apoge or Apsis whose place or longitude L in the Ecliptick. from y^e 1st point of ♈ is the arch Lo = to y^{e} angle ecf by the aforementioned data hee finds 65° 30′. & thence states y^{e} Apoge . . . Hipparchus 270 years before him. agreeing with him also in his Excentricity which Hipparchus stated y^{e} 24$^{th.}$ parte of the Radius ea.

The Ancients observeing that 60 was a Number which admitted of more aliquot parts then any other lesse y^{n} it, divided not onely the arch of the circumference of a circle subtended by it, but also the radius it selfe. into sixty partes, & theire tables of subtenses was 6 with Cyphers after it. this fashion Ptolemy was taken with. supposeing each subdivision of r[adi]us R to be 60 & states his said Excentricity. 2 parts. 29′. & 30″ his Method is followed by Albategnius or Mahomet of Aracta. who in the yeare of Dijlkarnain or from ye death of Alexander 1194 but of ye Incarnation 883. from like observations states ye place of ye Suns Apoge in ♊ 22°-17′. the Excentricity 3465 such partes. as ce is 100000

I might proceed & shew y^{u} how Arizachel a Spaniard about ye yeare 1065 observeing the difficulty of this Method by reason y^{t} it required ye time of ye Solstice which could not be collected exactly from any such observations as they then could make, contrived another which required onely the Suns places thrice observed at either aequinox & in some place betwixt, but in doeing this I should step too forward & beyond my designe which at present is to cleare ye grounds & fundamentalls of Astronomy & in the roome of those ruionous ones of the ancients to produce better and such as shall stand the test of future ages.

Here therefore I shall stop. The learned Kepler & with him all those who have embraced the true Sustem of the heavens except Landsbergius have all contended that the excentricity ec is not intire but bisected. this they conclude from observations of the places of ye planets, Mars expecially, whose place would erre above a degree from what it often appeares, if the center of the Earths orbite e were removed the quantity of the intire excentricity from ye center of the sun in theire hypothesis. for it is not materiall as for what respects the suns place simply whether wee conceave ye sun or ye earth to be fixed as has beene proved abundantly by able Astronomers.

But least y^{u} should not understand this bisection of the Suns excentricity. I shall explaine it & shew y^{u} the consequences by ye 2^{d} figure [Fig. 9.2] in which ahsa may represent the orbit of the Sun or Earth where y^{u} suppose to move. b its center c y^{e} Suns or Earths or e the point about which the Angles of equall motion are made. as far distant from y^{e} center of the orbite b as is e the center of y^{e} sun or earth on the other side [of] it, cb[16] = be is y^{e} bisected excentricity equall to one halfe of y^{e} entire excentricity ec in y^{e} Ptolemaic forme and the first figure. which if with him wee can admit to be 4152 such partes as ea = cs is 100000 it will follow that ca the suns greatest distance from y^{e} earth shall be 104152 and cs his perige or nearest distance 95848 of y^{e} like partes.

But if wee suppose the same excentricity bisected from ba = bs. 100000 gives ca his Apoge distance 102076. & cs his perige 97924. so that the difference betwixt his least & greatest distances is but halfe, which if wee suppose the Excentricity bisected as if we admit it intire.

Whence it will follow. that since all bodys as globes or spheres such wee conceave those of ye planets, as they approach us nearer subtend bigger angles and therefore appeare greater then when they remove further of[f], so also if the Suns excentricity were entire his apparent diameters will differ twice as much in the nearest & remotest places of his orbite, as when it is bisected. for if wee suppose c in ye 3^d^ figure [Fig. 9.3] to represent ye center of ye earth. bc his perige distance and ye angle dcp his perige ½ diameter. pc his greatest Apoge distance in the bisected Systeme. but ce the same when ye excentricity is entire. in such small angles as ye suns Diameter subtends and at such great distances as his is from us the angle acp subtended by his ½ diameter at y^t^ distance shall be but halfe so much more, then dcp. as it would be if his Apoge distance were ce. for ye angles acp. fĉb : dce ar[e] proportionable to ye distances cp.cb.ce.

This Controversy concerneing the bisection of the excentricity is but of a late date yet of no small moment[17] I shall end[e]avor to decide it & at the same time determine a greater & of no lesse use. about the Suns Diameters

To doe which it will be requisite breifly to informe y^u^ of the severall methods hitherto used, for finding [the sun's angular diameter][18] The Ancient Egiptian Kings caused a fleet horse to be started when ye suns upper limbe first began to appeare above ye horizon. & rode till ye Suns under limbe was ascended to it then compareing the space hee had run with the Stadia or furlongs hee usually coursed in one houre they thence collected how long the time was betwixt the riseing of the Suns upper & under limbes & by converteing it into degre[e]s of a great circle what arch hee subtended. others more skillfull made use of water houreglasses at the times of the aequinoxes & comparein[g] the water fallne from ye upper into the under glasse dureing the times of the Suns riseing. with the liquid descendeing through the same glasse in ye whole day they found them in proportion to each other as 1. to 700 or 750 whence Macrobius collects that they determined ye Suns diameter betwixt 29 & 31 minutes without any regard had to ye obliquity of the Sphere where the observation was made. nor were they acquainty with the Nature of this descent of water or sand through a small foramen & that more rans at first in equall spaces of time then afterwards. Ptolemy made use of a ruler of 4 cubits long or 6 foot with a payre of sights upon it, that next the Eye fixed. next the object moveable & so contrived that at its greater distance it might cover ye Sun with which after hee has acknowledged the difficulty of the observation hee confesses. that to him the Suns semidiameter allwayes subtended the same angle & yet knowing [which] hee makes the diameter of his Apoge Sun 31′⅓ of the Perige 33⅓: theire difference 2 minutes which if hee had not done violence to his Determinations ought to have been 2′-40″

All that succeeded Ptolemy downe to Tycho supposed the Suns Apoge diameter 31 minutes or little different whom wee may therefore conclude rather to have usurped his determinations then to have stated them anew by any observations of theire owne Maestlin excepted who from what annotations of his owne I know not states the same 29′-36″: Tycho made use of severall meanes for this purpose. sometimes hee tooke the difference of the Meridionall heights of the Suns upper and lower limbes but these gave his diameter a minute different from each other, at other times hee applyed a Canalis or Hollow tube of 32 foot long. divided into 1000 partes with a foramen at y[e] remoter end sometimes square at other times round whose diameter was 10 of the like parts or Neare ⅓ of an inch through which the species of the Sun was receaved on a table fixed on the nearer end of the tube, but the foramen being over wide & the observation made as may most probably be supposed in No obscure Roome the limb of the Species was not distinctly seene & so the diameter this way measured was as dubious as by the other. for at y[e] Summer Solstice hee determined it no nearer y[n] betwixt 28′-40″ & 31′-20″: & this I am the more inclined to beleive because I find Mons[r] Hevelius with a Canalis or empty tube of 20. foot & a small foramen in a very darke roome has determined it not much lesse then it is found by a better method I shall mention by & by. With Tycho Scheiner & Riccioli have committed the same errors & with the like event determineing still the Suns Apoge diameter 30′ or at most under 31. but The Method of Gassendus is still worse then these. hee used a strong Ruler of 22 foot long at each end of which hee placed an upright plate at right angles to it hee receaved the shadow of that next y[e] sun upon y[e] other whose brea[d]th being measured halfe y[e] difference betwixt it, & the brea[d]th of the plate from which it was cast (the length of y[e] beame or Ruler being Radius) would be the tangent of the Suns semidiameter but this by reason of the incerteinty of the praecise boundary of the shade agreed worse with itselfe y[n] any of the rest. Ricciolus had another Method hee hung two small lines perpendicular to y[e] Earth in y[e] meridian, & keepeing his eye fixed observed by counteing the vibrations of a pendulum what time the ☉ spent in passeing over them & hee found what arch of any parallel his diameter subtended so by consequence y[e] measure of it, this method was no ill one but hee complaines of its difficulty & determines nothing from it p[er]haps because it agreed not with his predetermination. About this time Vincent Mutas tooke much y[e] better Method at either solstice hee fixed a small telescope upon the Meridionall Sun. Whose species or picture cast through it hee receaved on a Sc[r]ene or table fixed behind it, & then noteing its breadth left y[e] instrument in the same position for some dayes till the suns declination were altered. the better part of his diameter. then hee observed how much higher or lower on his sc[r]ene the species of ye sun past. by knoweing how much the sun altered his declination in the meane time hee knew what arch in the heavens answered this difference, & by compareing it with the whole Diameter its [diameter in y^e^][19] Apoge 31′.18″. in ye Perige 32′.46″: Gabriel de Mouton an ingenuous Frenchman in a peece wrote purposely on this subject[20] affirmes that himselfe often used this same method, but prefers the following, hee also cast the suns species through a telescope on a sc[r]ene & then Counted the vibrations of a short pendulum whilest it passed over a line drawne on it at right angles to ye way of the said species or picture of the sunne, which repeateing oft hee found & determined the Apogeon Diameter 31′45″ the perige 32′.32″ not much different from Matus.

But the best method was that of Mr Gascoigne[21] who tooke his measures by the helpe of a Micrometer or payre of screws placed in ye inner focus of his Telescope whereby hee found the Diameter of ye Apoge sun 31′·40″ of the Perige 32′·50″ with which M^r^ Ts[22] agree. with a telescope of about 14 foot & something better contrived mycrometer[23] at Derby in ye Yeare 1672 I determined the breadth of the Apoge sun 31′.30″ or some little more in his Meane distance on ye 19th of March 32′ 12″ in his Perige 32′ 40″ with which Mr Cassini assures mee the French observations agree exactly.[24] whence it will follow that ye excentricity of the Sun is bisected for let us admit the diameter of the Apoge sun as by the Consent of all observations compared togeather it is 31-40 and ye excentricity intire 3400 p[ar]tes as it is better stated by moderne Authors as in the 1st [25] figure [Fig. 9.2]. the diameters of the luminarys are reciprocally proportionall to theire distances from ye earth it will hold as ye perige Distance cs 966 to ye Apoge. cA 1034. so is Apoge diameter 31-40 to ye perige 33′.54″ which is an whole minute more y^n^ any sound observation will warrant.

But let ye excentricity be bisected as in the under figure . . . such parts as the Apoge distance is 1017 & it will hold as 983 to 1017 so is ye 31′-40″ the Apoge to 32′-46″ the Perige diameter not 6 seconds differing from the observations which is sufficient evidence that the suns Excentricity is bisected as the Copernicans[26] affirme with which haveing done I shall next lecture returne to my Ancients & prosecute the improvement of Astronomy under theire hands, & since, carrying each branch or rather traceing each particular downe to o^r^ owne times.

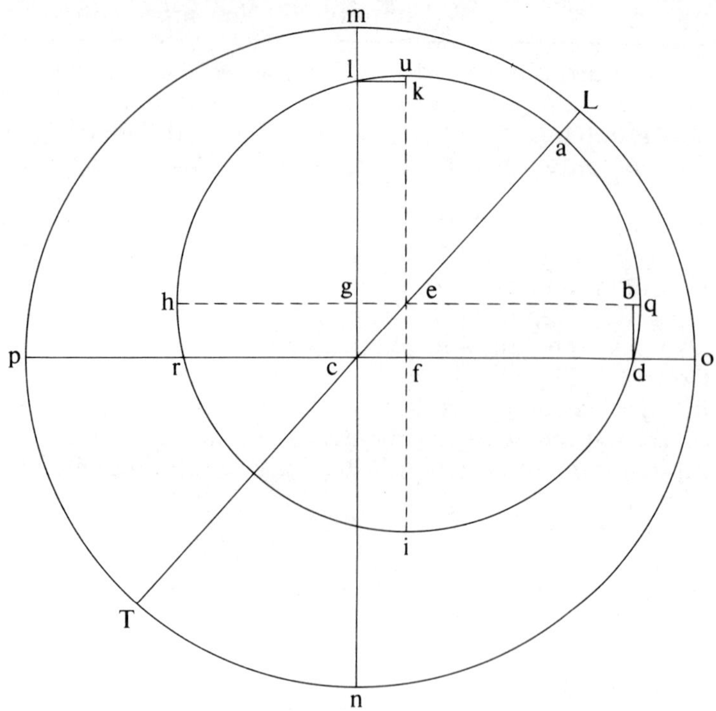

FIG. 9.1

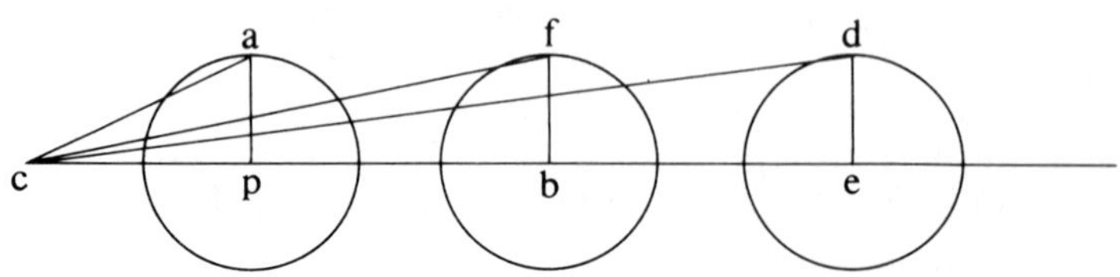

FIG. 9.2

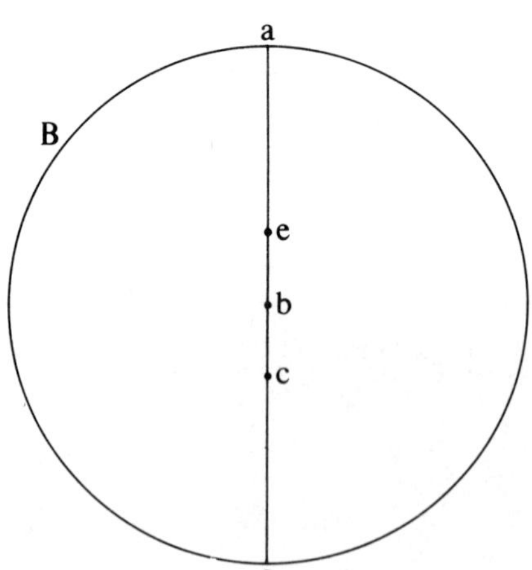

FIG. 9.3

NOTES

9.1. Op. cit. (note 1.3). Flamsteed translated much of this chapter into English in Lecture 28. Cf. note 28.4.

9.2. The words in square brackets are supplied from the text of Lecture 1. A line is missing at this point from the manuscript.

9.3. Cf. note 1.8.

9.4. Cf. note 1.6.

9.5. Cf. note 1.10.

9.6. Cf. note 1.11.

9.7. Cf. note 1.12.

9.8. Cf. note 1.13.

9.9. Cf. note 1.14.

9.10. Cf. note 1.15.

9.11. Cf. note 1.17.

9.12. Cf. note 1.18.

9.13. Riccioli 1, p. 132.

9.14. This is a loose translation of:

> Vetustatis igitur gratia aestiva Solstitia Metonis, & Euctemonis, & postea Aristarchi temporibus observata, his conferenda sunt, quae nostris fuerunt temporibus. Sed quoniam Soltitiorum observationes difficulter discernuntur, & quas illi tradiderunt, non exquisite captae videntur, quod etiam Hipparchi iudicio comprobatur, eas quidem praetermisimus.

Actually in ibid., p. 137.

9.15. Flamsteed accidentally wrote 92° 20′, but used the correct value in his subsequent calculation.

9.16. Flamsteed wrote ce instead of cb.

9.17. It was intimately linked to the question of whether one should adopt Kepler's law of areas or the alternative hypothesis of uniform angular motion about the empty focus of the Earth's elliptical orbit which, in Flamsteed's time, gave an equally satisfactory representation of planetary observations. See E. G. Forbes, 'The Rational Basis of Kepler's Laws', *Journal of the British Astronomical Association*, lxxxii (1971), 33–37.

9.18. The words in square brackets are missing from the original text. The historical survey which follows is from Riccioli 1, pp. 116–20.

9.19. Words are here missing from the original text.

9.20. Gabriel Mouton, *Observationes diametrorum solis et lunae apparentium . . . Huic adjectae est brevis dissertatio de dierum naturalium inaequalitate* (Lugduni, 1670).

9.21. William Gascoigne, in a letter to William Crabtree dated 2 December 1640, op. cit. (note 4.14), ff. 10ᵛ–12ᵛ, enclosed a paper giving an account of some solar diameter measurements that he had made with his micrometer. Cf. op. cit. (note 1.18), pp. 5, 6.

9.22. Richard Towneley, of Towneley in Lancashire, England, with whom Flamsteed had been frequently in correspondence before this time. A summary of the content of seventy letters from Flamsteed to Towneley is given in J. L. E. Dreyer, *Observatory*, xlv (1922), 280–94.

9.23. Flamsteed's modifications to Towneley's micrometer included the substitution of threads, or hairs, for the metal sighting pointers and parallel threads perpendicular to these like a kind of net. Cf. Flamsteed to Oldenburg; Derby, 2 December 1671, Royal Society MS. F. 1. 77.

9.24. Giovanni Domenico (Jean Dominique) Cassini to Flamsteed; Paris, 1 August 1673, states (transl.): 'We agreed with you to the last second [of arc]' in measurements of the sun's angular radius. Royal Society MS. C. 1. 59.

9.25. Flamsteed is referring, in fact, to the second figure, Fig. 9.2.

9.26. Five methods for investigating the eccentricity of the Earths orbit are described in Riccioli 1, pp. 152–6. Cf. note 9.17.

Lecture 10

23 *November* 1681

In my last lecture I shewed you how the Ancients came to find out the manner of the suns motions through the twelve signes its inequalityes & the places where it was slowest & swiftest. I proved by observations of the Suns Diameters of which I shewed y^{u} the severall methods that the excentricity of the Suns or earths orbit admit which y^{u} please to move, was not entire but bisected & togeather determined the quantitys of his least & greatest diameters from late & accurate observation I promised to proceed in the same Method which to mee seemed most Naturall through all the principles of Astronomy. & instead of those ruinous ones of former times to give y^{u} better & such as should abide the test of future Ages. I come therefore now [to] consider the obliquity of this circle how the[y] discovered the quantity of it, & to enquire whether as some p[er]sons conceave it hath beene allways the same & invariable, or whether as others assert it has changed since it was first observed & determined.[1]

Accordeing to the methods I have formerly used I shall shew y^{u} how the first astronomers came to understand that the circle of ye Suns annual motions after more justly & truely attributed to ye earth lay oblique to the Circle of his diurnall motion of which I have not yet said any thing in my former readeings the Method first proposed to my selfe not haveing led me to it

They first noted therefore that all the Stars five excepted rose every night on the same place of the horizon, viewed from theire habitations & the same station that they kept constantly the same distances one from another & came to the South in the same heights, hence they concluded them all to be fixed in the same orbe, but that those stars which now rose or appeard on ye Easterne horizon upon or immediately after sun set about a moneth or two after when the[y] first appeared after the close of the light were some height above ye horizon and 3 or four moneths after that set imediately after the sun hence they conclude that tho the Sun togeather with them all were carried round ye earth riseing & setting in the space of a Naturall day yet the Sun moved round the heavens through these constellations once in a yeare for that alwayes they saw ye same stars setting with him

Notes on pages 209–10.

or riseing as hee set at the same season that they had done ye like ye yeare before.

And tho the stars observed from the same places always rose & set on the same places of the earth yet the Sun altered the places of his riseings & settings as also the Noone Shades were changed sensibly at all times of the yeare twice excepted which times they called theire Solstices the sun then riseing & setting for some days togeather nearely on the same points the Noone Shades also altering little & insensibly.

They noted also that there was a certeine place in the heavens neare which the stars moveing round it described but very small Circles & altered theire heights on the Meridian but little & that all the other stars as they were more removed from this place described larger Circles. & parallel to each other. [they conceaved] them to be carried about the earth in a great orbe which was turned round on an Axis one of whose extremitys was placed in that part of the heavens. & the other under the Earth in a point opposite to it & in a line drawne from that through the earths center which opinion of theires they were the more confirmed in by reason that when they travelled north from theire habitations they lost the sight of those stars which they had seene to rise a little in the south above the Earth in ye night time & that others againe which in theire owne country they had seene rise & set. here on the Contrary never went beneath the horizon: these two points they called the Poles of the heavens, & that great Circle in the heavens which was equally distant from both represented In the figure by ♈ æ ♎ q. the æquator which divided the circle in 360 parts or degrees was 90 distant from each of the poles The height of the Apparent pole they determined by takeing the greatest & least heights of some of those stars which they saw neither rose nor set for the middle betwixt them they ever esteemed the true height not makeing any allowance for refractions which were wholly unknowne in those dayes.

And now layeing all theire motions togeather & considering with themselves that by the riseings & settings of the Stars it was apparent that the Sun moved amongst the Constellations by her returne to ye same stars at the same time of ye yeare it was also evident that hee alwayes kept the same path. which also they were the more easily induced to beleive because the[y] saw the Moone to passe in any moneth neare the same stars shee had passed amongst the moneth before tho with a different face or quantity of light & because each full & quarter Moones was more to the east or amongst stars in consequence of those neare which ye full & quarter moones of the Moneth praecedeing had appeared.

These appearances they found they could not salve nor answer except they supposed the Sun to be carried in a great Circle through the Constellations contrary to his apparent diurnall motion, that is to move from weast to east throug[h] them all once in a yeare & that this circle lay oblique to the Equator & was with it carried round the Earth from east to west once in a day. which being once concluded on it was found that all the appearances of these riseings & settings on severall places when viewd from ye said station the increase & decrease of the length of the Noone Shades & whatever else they had generally noted in the. Motion would be represented. for say they.

Let HæpqaH represent ye meridian pa ye Axis of ye world p ye northerne or Conspicuous pole & end of it a ye Southerne. & let q♈æ♎ be the æquator equally distant from both these poles. about which the diurnall revolutions of the heavens are made. it is evident that all such stars as be betwixt the Northerne pole p. and the northerne horizon Ho shall never set the Circles which they describe by theire motions falleing all of them so much above the said Horizon Ho as theire distances from ye pole are lesse. and of all the rest such as are next nearest the said pole describe the greatest arches and stay ye longest above the Earth ye arch m♋o being proportionally greater then ♈ æ ♎ & that then ndα as those are farther of[f] from ye north pole p or nearer to ye Southerne a

And if wee conceave the Circle in which the Suns annuall motion is performed ♈ ♋ ♎ ♑. to lie oblique to y^e^ æquator ♈æ♎q. & that his motion in it is contrary to ye diurnall revolution of the heavens. & p[er]formed in the space of a yeare all the appearance of the Suns accesse to & recesses from ye fixed stars will be answered togeather with his different meridionall heights & pointes of riseing & setting. for if wee suppose o to be the North. ♈ ye East H ye South & ♎ the west pointes of the horizon. and that at the time when ye Sun rises on ye east point his place is at ♈ in the Ecliptick, & that from the time of this to ye next dayes riseing his progresse in ye Circle of his motion or Ecliptick is from ♈ to e whilest being carried round the heavens by the contrary motion hee has described the Spirall line ♈ æ s q t æ. whereby hee is brought to rise more to the north or nearer the point o. then hee did the day before & continueing the line of his motion to ye meridian at u hee is higher in it by ye quantity of the Arch u æ when at ye aequinoxes is about 24 minutes. but very small when hee comes to ♋ which is about 94 dayes after hee has cut the æquator. by reason that his circle almost touches the parallel of his diurnall Course in ♋o for severall dayes togeather & here they saw the reason of the graduall encrease of the dayes all this time of the Suns passeing from ♑ to ♋ & theire like decrease afterwards. as also why the nights at any season of the yeare were of the same length with ye dayes at the opposite or 6 moneths after. they saw I say the reason of this to be because ye Axis bisected all the parallells of the suns diurnall courses & that by reason of the similar & aequall triangles wsm. αsx. the arch wm of the most northerne parallel intercepted betwixt ye Axis & horizon above ye earth was equall to the Arch of ye most Southerne parallel px alike intercepted betwixt ye horizon & Axis underneath the earth. & therefor[e] the Nocturnall Arch α♑n, in the Southerne tropick equall to ye diurnall arch m♋o in ye Northerne & the day in this equall to ye night in that and because the sun approached every day to the stars at ȳ they set so much sooner after him as his diurnall accesse to them was. but the stars at z in consequence of him on ye contrary rose earelyer or so much the sooner & before him

Who first observed the Obliquity of this circle Ancient authors are not agreed. Ricciolus from Messala tells us that the Indians. by which termes p[er]haps hee meant ye jews had determined it 24 degrees. And Aristarchus of Samos ye same.[2] which is to be understood nearely or not greater for it was the Custome of the Ancients when they could not limit such quantitys precisely by reason of the coursenesse of theire observations to doe it however nearely & within some bounds so that hence wee can onely conclude that they determined it to be not more

Amongst the Greekes Oinopides of Chios is by some said first to have observed it but the time is not recorded. So wee are ignorant whether hee lived before Pythagoras whom some affirme to have knowne it first or not. Laertius affirms that Anaximander ye Milesian who lived about 560 yeares before the Incarnation first set up a gnomon at Sparta for observeing the Solstices and aequinoxes. but if as others assert Thales his Countryman wrote a booke de Solsticijs & aequinoctijs it is most probable that hee who was something older then Anaximander was acquainted with it before him & that hee learnt it in his Travels amongst the Jewes & Easterne people.[3]

The first wee find to have determined it more accurately is Eratosthenes Cyrenensis who makes ye whole distances of ye Tropicks to be 11 such partes as the Circle conteined 83. or 23°.51′.20″ whom both Hipparchus who lived 100 yeares after him. & his close follower Ptolemy agree with Praecisely. His method is described by Ricciolus[4] thus. Eratosthenes sayes hee supposeing that all the Suns rayes proceeded from his body parallel to each other. at ye time of ye summer solstice by a Gnomon placed upright in a Concave hemisphere observed the Suns Meridionall distance from ye vertex of Alexandria 7°.12′. the latitude of Alexandria hee supposed. 31°.03′. from which subtracteing the sayd 7°.12′ the remain[d]er is 23.51′. the obliquity of the Ecliptick. in which determination Riccioli observed a double fault first in that hee accounted all the rayes proceedeing from the Sun to come to us parallel to each other which they doe not for then the sun should have no sensible diameter. Next that hee supposed the Latitude of Alexandria greater by 5 minutes then really it was for Ptolemy who lived after him & whom therefore wee may suppose to have examined & stated it more diligently makes it but 30°.58′ Now since Eratosthenes allowed nothing for the Suns semidiameter if as wee ought wee allow 16 minutes for it the true solsticiall distance of the Suns center from ye vertex of Alexandria will be 7°.28′ which taken from its true latitude 30° 58′ the residue will be the obliquity of the Ecliptick 23°.30′. the same nearely which wee shall find determined by later & more curious observations in or times.

Strabo ye Geographer who lived in ye Reigne of Tyberius Caesar tells us that the lenght of the aequinoctiall shade to ye lenght of the Gnomon was at Byzantium or Constantionople. as 111 to 120: but the Solstitiall Shade to the Gnomon as $41\frac{4}{5}$ to 120. whence Riccioli[5] calculates the Solsti[ti]all height of ye Suns upper limbe at Constantinople 70°.45′.25″: but the aequinoctiall height of the same limbe 47-13-55 whose difference gives the obliquity of the Ecliptick $23°.31\frac{1}{2}'$: tho in his workes elsewhere hee makes it $\frac{1}{15}$ of a limb or roundly 24 degrees.

Ptolemy as I tould yu before allways follows Hypparchus & after him wee find no farther inquiry made in this businesse till the Arabians became acquainted with the Mathematicks. An account of theire observations much hitherto desired I procured from the Learned reader of Astronomy at Oxford & my kind Brother Mr Edward Bernard. with which I shall acquainte yu because I find they have not yet beene published. or at least made common.[6] I hope however this may occasion a further enquiry to be made after them & that some able p[er]son may in time be encouraged to commit those Authors to publick in a language better understood then theire original ye Persian & Arabick.

Al Maimon. an Arabian prince famous beside this for measureing the length of a degree on the Earth assisted sayes Abenezra by many skillfull men observed the obliquity of the Ecliptick 23°-35′. in the yeare of the Hegira 210 of Christ 825 in the manuscript also of Mr Seldens[7] it is affirmed by an uncerteine author that Ischia Ebn Abimansor with severall other learned & ingenuous persons observed the same obliquity & that in ye feilds of Bagdat Beni Musa found also 24°-35′. as Al Nodham has it in Comentaries on ye Astronomicks of Hoseim of Nisibis who addes that this was approved of by many learned men in those dayes & that Alfraganus wholly assented to it

What method these persons tooke & used I am not informed but theire next fellowr. Mahomed Ebn Gaber better knowne by ye name of Albatanius used a long Alhidade or the parallactick Rulers of Ptolemy[8] whereby hee found at Racca or Aracta the least distance of the sun from his vertex in the Summer Solstice 12°-26′. in ye winter 59°-36′ therefore hee states the distance of the tropicks 47°-10′ & the obliquity of ye Ecliptick 23°-35′ Altogeather as the Arabians & theire prince had done before him. Albatanius doubted not to praefer his performances before the bare assertions of Ptolemy. but I feare these rulers were not so exact as they ought to be, a small fault or bending in that whereon his sights were placed might cause a considerable error in ye Suns distance from the vertex. against which for all his confidence this author has said nothing that might secure us. his observation which accordeing to Mr Graves[9] was made in ye yeare of the Hegira 269 of Christ 882. hee died in ye yeare of ye Hegira 317 of Christi 929:

But 20 yeares after the observation of Albatanius Thebit Ebn Corah a Jew whom Ricciolus[10] places in the yeare of Christ 1210. but falsly for it was in the yeare of ye Hegire 289 therefore of Christ 901 found the obliquity but $33\frac{1}{2}$ more yn 23 degrees

At Bagdat in ye yeare of the Hegira 337 of Christ 987 Abul Wafi, Al Buziani & Abu Hamet a Most ingenuous person sayes my author[11] found it scarce 35 minutes above 23 degrees.

Tho Abu Hoseim & the Persian tables of Chrysococca[12] at that time make ye excesse above 23 degrees to be 35 minutes Abul Rihan or Albirunius whom Mr Greaves[13] places in ye yeare of ye Hegira 385. Abulfaragius 463. the first in ye yeare of Christ 995 the latter 1070. with a Quadrant of 15 Cubits Radius observed it 23°-35′. But Abu Gaafar ye Al Chazan with his companion Abulfadius & some others of that age at Edossa observed that the greatest obliquity was not fully 23° . . . but some little lesse

Almeon ye son of Almansor in ye yeare of Christ 1140 states it [23°] 33′.30″.[14] but accordeing to ye Readeings[15] Maestlin & Clavius onely 33′ more yn 23 degrees.

Altho Ismael Abulfeda prince of Hamo in his Manuscript tables retaines it on the Authority of Al Mainon 23°-35′ in ye yeare of the Hegira 711 of Christ 1321.

In ye yeare of Christ 992 of the Hegira 382. in the reigne of Fecrodaul one Abu Mahomet Alcogandi with a Sextant of 15 Cubits. whose limbe was divided into seconds found the obliquity lesse then it had ever beene observed by any of his praedecessors to wit 23°-32′-21″

Hence Aldehamus the Arabian Astronomer affirmes that it was scarce ever observed lesse yn 23° 33′

Tho Arzahell the Spaniard in ye Yeare of Christ 1692 of ye Hegira 1082 proposed it 23°-33′$\frac{1}{2}$

At Maraga a Noble Persian one Chojah Nassir Eddini in ye yeare of Christ 1269 of ye Hegira 668 observed *την λοξοχχν*[16] accuratissime. the obliquity most exactly sayes my Author[17] 23°.30′ which the Commentator upon Astronomick workes of Hoseim Nisabariensis[18] concludes to be the least that anyone hitherto had stated the greatest obliquity of the Ecliptick

And yet Thirty yeares after this Prophatius Judeus in ye yeare of Christ 1303 states it 23°-21′

Ebu Shater of Damascus in a Manuscript of Mr Seldens[19] is said to Affirme that hee corrected the obliquity neglecteing ye Suns parallax which in the horizon hee supposed to be 3 minutes & Concluded his greatest declination 25°-31′

Olagh beigh one of the successors of Tamerlane in ye yeare of Christ 1437 of the Hegira 841 usus Summa cura & maximis Instrumentis says the translator of his workes[20] with large Instruments & great Curiosity observed the Suns greatest declination 23°.30′.17″ Some manuscripts had 27″. but so small a difference is not worth ye Contendeing for.

The Arabians by whom these observations were made thought that from the time of Ptolemy downe to them the obliquity had constantly decreased but very slowly it is no wonder therefore if those that followed Almaimon a prince of theire Country & the Judicious Albategnius were so fearful to depart from theire determinations & that these swayd much with them wee may gather by Abulfeda who lived in later times & could not be ignorant of the accurate determination of Nasir Eddini yet on the Authority of Almaimon retained it 23° 35′.

To which quantity wee find none after them consenting save Albirunius for both Abuhamet & Abugaafar concluded that it was lesse within one hundred yeares after Albatanius, & Thebet ben Chore asserts it not more y^n 23-33½ Scarce 20 yeares him. and Abu Mahomet about ye same times with Albirunius. with a very large instrument determines it but 23°-32′-20″: which is scarce an entire minute more then ye Noble Tycho states it 6 ages after him

After whom Nasir Eddini & Oleg Beigh both find it 23°.30′ whom wee may conclude to have allowed also for the Suns parallax as well as Ebn. Shater they being both Judicious & Skillfull p[er]sons. so y^t compareing all togeather with ye consent of Maimonides wee need Not feare to affirme. that the Obliquity of the Ecliptick which by ye observation of Eratosthenes above 200 yeares before the Incarnation was found 23°.30′ nearely has continued the same and invariable to o^r times. the differences which have beene found in the opinions of different authores & observers proceedeing either from the faults of theire Instruments or thence that some allowed for the suns parallax others not

Theire way of determineing the true obliquity wee may conjecture from Albatanius. who haveing observed the Suns least distance from ye vertex of Racca 12°.26′ his greatest 59° 36′ takes theire difference 47°-10′ for ye distance of the Tropicks & the one halfe of it 23°-35′ for the ☉^s greatest declination.

Which wee may the more probably also conclude for that the acute Regiomontanus who lived in the same . . . not 30 yeares after him determines the obliquity after the same manner but finds it lesse by reason y^t in ye More Northerne Countryes of Germany where hee lived the Suns apparent height in ye Winter Tropick is greater then it ought to be the refractions raiseing & makeing him appeare higher then really hee is which if it be not allowed for causes considerable errors in such conclusions as have been drawne from observations intangled with it. but I shall proceede no further at Present Hereafter I shall have time to examine such observations of the suns greatest declination as will give it more praecisely[21] the times I have passed through were obscure to us by reason of theire remoteness & the barbarity of that people who managed the sciences in them more given to war then arts & to enlarge theire empire then to improve theire soules with generous knowledge wee are arrived at the confines of a more civilized age & a more noble generation, who to make the praises of theire greate Creator greater have left no p[ar]te of Nature no Region of the World no place of the heavens unexamined. to these wee shall passe with ye next opportunity & tho all Mortal[it]y differ from each other in theire opinions of other thinges yet I doubt not but by compareing the observations of Ingenuous men togeather here to prove that they dissent not from each other however They notes have beene wrested & misrepresented by some p[er]verse people who either understood them not aright, or would not manage fairely for feare they would disprove theire prejudiced opinions

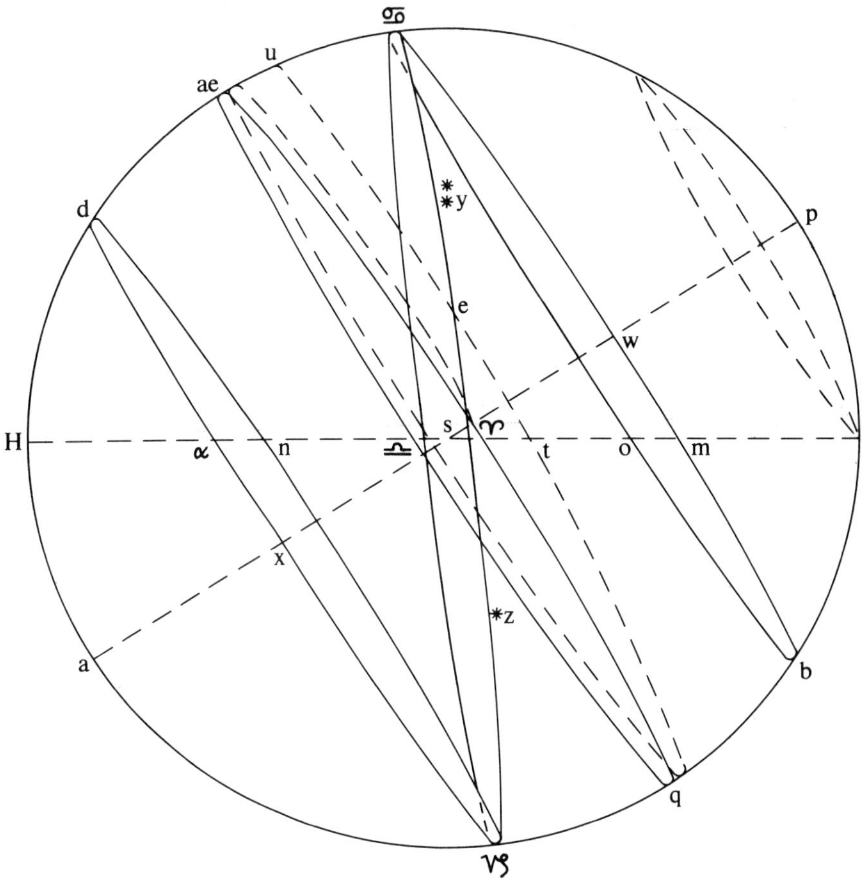

FIG. 10.1

NOTES

10.1. The following passage is crossed out of the text at this point but reintroduced in a modified form on f. 108r,v; cf. p. 204.

Ancient Authors affirme that Oenopides of Chios first observed the obliquity of the Ecliptick. his age I can not certeinely determine. others ascribe this invention to Pythagoras. but if Thales as some affirme wrote a book de Solsticijs & aequinoctijs it is not unlike but hee was acquainted with ye Obliquity of the Ecliptick: Anaximander the Milesian lived about the same time with him yt is 560 yeares before ye incarnation whom Laertius affirmes to have first erected a gnomon at Lacedemon or Sparta for observeing the aequinoxes & solstices. which seemes . . . . for this purpose as being suggested to them by the encrease or Shortning of the length of the shadows of theire houses & trees. but what quantity the [y] determined this obliquity no history that I can meet with has informed us, it was the use of the ancients when by reason of the want of accurate observations they could not praecisely determine such quantitys to doe it however nearely & within some bounds. the Indians stated it 24 degrees as Riccioli informes us from Messala so did

Aristarchus ye Samian [cf. note 10.2] which is to be understood nearely & that it was no more. which considering the rudeness of the observations of the Ancients is as exact as we could expect from that Age.

But that I may not step too far at first and beyond ye apprehensions of such as have not beene acquainted with ye rudiments of Astronomy . . .

10.2. Cf. table entitled 'Declinatio Solis Maxima, seu Obliquitas Eclipticae', in Riccioli 1, p. 162.

10.3. The information in this paragraph can be found, for example, in Part II of the introductory chronicle in Riccioli 1 under: *Oenopides Chius* (p. xli), *Anaximander* (p. xxx), and *Thales Milesius* (p. xlv).

10.4. Riccioli 1, pp. 163–4 is actually summarizing Pliny the Elder's remarks in *Historia Naturalis*, lib. 2, cap. 108.

10.5. Riccioli 1, p. 164.

10.6. The following information concerning ancient Greek, Arabic, and Persian attempts to determine the obliquity of the ecliptic had been communicated to Flamsteed in a Latin letter from Edward Bernard dated 14 August 1681, which was soon to be published in *Phil. Trans.* xiv, no. 163 (20 September 1684), 721–5.

10.7. John Selden (1584–1654) bequeathed all his oriental manuscripts to the Bodleian library, and they comprised part of a collection of over 8,000 volumes sent there from London by the executors of his estate in 1659. A list of Selden's books and manuscripts may be found in Edward Bernard's *Catalogi librorum manuscriptorum angliae et hiberniae* (Oxoniae, 1697), pp. 157–67. Explicit references to those from which the astronomical data presently under discussion were extracted are given by Bernard in op. cit. (note 10.6).

10.8. This instrument, and the errors liable to be incurred in its construction and application, are discussed in Lecture 12, ff. 121^r–122^v, cf. pp. 221–3.

10.9. John Greaves (1602–52), a former Gresham Professor of Astronomy and editor of *Shaw Cholgii Persae Astronomicae quaedam . . . una cum hypothesibus planetarum* (Londini, 1652).

10.10. Cf. note 10.2.

10.11. Cf. note 10.9.

10.12. Ismael Bullialdi, *Astronomia Philolaica* (Parisiis, 1645). Cf. section entitled 'Synopsis Tabularum Astronomicarum Persicarum, ex syntaxi Persarum Georgii Medici Chrysococcae,' pp. 211–29.

10.13. Cf. note 10.9.

10.14. This figure is taken from the table referred to in note 10.2.

10.15. Riccioli 1, p. 160 makes no explicit reference to the value of the obliquity obtained by Michael Maestlin, but cites Clavius's value as 23° 30′ 30″.

10.16. *την λοξοσιν* = the obliquity of the ecliptic.

10.17. Cf. note 10.9.

10.18. Namely, Al Nodham or Noddam, who flourished around 1200 A.D. Op. cit. (note 10.6), pp. 721, 722, 724.

10.19. Cf. note 10.7.

10.20. Namely, Thomas Hyde (1636–1703), Librarian of the Bodleian Library, Oxford, from 1665–1701. He catalogued John Selden's library (op. cit. [note 10.7]) and transcribed several copies of that catalogue. Cf. op. cit. (note 28.13).

10.21. Flamsteed is presumably alluding to the observations of Bernhard Walther and Johannes Werner (cf. Lecture 12), Wilhelm IV, Landgrave of Hesse-Cassel, and Tycho Brahe (cf. Lecture 13).

Lecture 11

1 *February* 1681/2

In my last lecture in this place I began an Enquiry concerneing ye obliquity of the Ecliptick. or the line of Suns course amongst the fixed stars. I had not then leasure to make any great progresse in it I shall therefore continue it now.[1] but because of the time interlapsed since, a great part of what I then told you may have slipt out of y^{r} memories, I shall refresh them with a breife repetition of what I then delivered, & so proceede with what I left & had further to adde on that subject, which may employe the times allotted to my readeings this terme.

[I told y^{u} then] that it was uncerteine whether Oenopides of Chius Pythagoras Thales or Anaximander first observed the obliquity of the Ecliptick amongst ye Greekes; The Greeke Biographer Laertius tells us that Anaximander first set up a Gnomon at Sparta for observeing the aequinoxes & solstices but if as some affirme Thales his Countriman wrote a booke de Solsticijs & aequinoctijs tis most probable that hee who was something older was first acquainted. with it & that hee learnt it in his travells amongst ye Jews & Orientall people.

For Messala an Arabian writers affirmes that the Indians under which name they might be comprehended esteemed ye obliquity more y^{n} 24 degrees about which quantity Oenopides states it whose age & the time hee lived in is for what I can collect unknowne. but y^{t} hee lived about the same time with Pythagoras & Anaximander is guest & therefore his determinations . . . those who for wante of convenient instruments dureing the Infancy of Astronomy could not determine this quantity so accurately as it has beene done by the Curiosity of followeing ages each of which has endeavored to exceede the praecedeing in exactnesse & outdoe them in excellent inventions.

They then thought it sufficient to determine or limit this prime requisite within bounds and accordeing to y^{e} Custome of those times in round Numbers perhaps too they accounted the obliquity not from y^{e} æquator to the suns center but to y^{e} his utmost limbe. which way of reckning would make it by the determinations of his age not lesse y^{n} 23$\frac{3}{4}$ degrees. to which 24 being the next even degree they pitcht upon that, & the rather too, because it was the praecise 15th part of y^{e} Circle or 360 degrees.

Notes on pages 218–20.

And to this opinion I the rather encline because I find the first whose observations have come to our hands counteing on this manner; I meane Eratosthenes of Cyrene in Ægipt, who lived about 230 yeares before o^{r} Savior, or 1900 yeares agone. Ptolemy aequaintes us with his Method, which was thus; in a Concave sp[h]ere hee placed an upright gnomon whereby hee found the distance of the Solsticiall sun from ye vertex of Alexandria to be 7°.12′, whence supposeing the latitude of y^{t} City to be 31°-03″, as Riccioli makes it out, hee collected the distance of y^{e} æquator from ye Ecliptick 23°.51⅓′ or 11 such partes . . . but the same author finds a double fault in this account first. that Eratosthenes obteined the rayes comeing from any part of the Suns body to proceede parallel to each other by reason of his great distance from us which supposition tho it may hold good in all the rayes proceedeing from the same point of ye sun to any place or part of the Earth yet is not to be otherwayes allowed since theire owne observations will make it out that the rayes proceedeing from his center to any point of o^{r} Earth forme an angle of about 16 minutes with the Rayes comeing from his Circumference to ye same point whence it is evident, first that what I before affirmed that the ancient counted the obliquity of the Ecliptick to the remotest point of ye suns periphery, was really true and next that the suns semidiameter which was about 16 min ought to be added to ye observed distance of the Sun from ye vertex which will make the true distance of his center from ye vertex of Alexandria in Eratosthenes his observation 7°.28′ without any regard had to either parallax or refraction which had they beene much greater then wee can suppose them would in that distance be inconsiderable.

The other fault Noted in Eratosthenes by Ricciolus is that hee supposed the latitude of Alexandria to be 31°-03′ which by the later observations of Hipparchus & Ptolemy was onely 30°-58′. & if from this wee subtract the correct zenith distance of Eratosthenes 7°.28′ there will be left but 23°.30′ for ye Obliquity of the Ecliptick which is nearely the same that it has beene found by y^{e} . . . The obliquity of Hipparchus is the same with that of Eratosthenes. 23.51⅓. & Ptolemy who is suspected to have given us little but what hee borrowed from Hipparchus, affirmes the same. we meet not with any who examined it by observation till 1000 yeares at leaste after Eratosthenes. Almaimon The Caliph of Babilon or Bagdat famous for haveing also measured the lenght of a degree on ye Earth assisted by many skillfull men sayes Abenezra. in ye yeare of Christ 828 of ye Hegira 210 observed it 23°-35′. about which time one Ischia Ebn Abimansor. & Beni Musa in distant places are said to have determined its quantity the same. as also Mahomet Ebn Gaber or Albategnius in ye yeare of Christ 882. or 57 yeares after Al Maimon.

But that this was too greate is evident by the observations of Thebit Ebn Core who lived but 20 yeares after Albatanius & before his death noted the obliquity but 23°·33½′ in ye yeare of Christ 901 of ye Hegira 289. as also from the observations of Abuhamet & others who observed at Bagdat in ye yeare 987. that the greatest declination of ye sun was not fully 23°-35′

The Abu l Rihan or Albiruni about ye yeare 1000 pretends to have observed it 23°.35′ with a Quadrant of 15 Cubits Radius

And yet Almeon Ebn Almansor in ye yeare 1140 states it 23°.33½ as Ebn Corah had done before him

Yet betwixt these in ye yeare of Christ 992 of ye Hegira [382. in the reigne of Fecrodaul one Abu Mahomet Alcogandi with a Sextant of 15][2] Cubits whose limbe was divided into seconds observed onely 23°.32′.21″ which is scarce an entire minute more then the Noble Tycho makes it 6 ages after him Note that This praecedes Albirunis determination & from the Circumstances seemes to be the Most exact of any yet given.

Tho wee find Arzahell a Spanish Moore about 100 yeares after or in ye yeare of Christ 1089 makes it still with Thebit & Almeon 23 . 33½

In ye yeare of ye Hegira 668 of Christ 1089 at Maraga a Noble Persian one Chojah Nasir Eddini well knowne by his Geographicall Tables[3] observed the *την λοξοσιυ accuratissime* sayes my Author 23°.30′ which being but 400 yeares after Almaimon and Albategnius suggests to us that this was the most exact & probably the true obliquity of theire times.

For Thirty yeares after wee find Prophatius ye Jew halteing betwixt two opinions & takeing the meane betwixt the old & this. that is 23°.32′

And in ye yeare of Christ 1363 Ebn Shater of Damascus affirmes himselfe to have corrected the obliquity of the Ecliptick not neglecteing the Suns parallax which it should seeme those before him had omitted. but he supposed 3′ in ye horizon & hence determined it 23°.31′

The last of the Moores & Arabians & Easterne people that wee find to have enquired into it is ye famed Gran[d]child [of] Timarleng Oleg Beig who in the yeare of Christ 1437 of the Hegira 841 usus summa cura & Maximis [Instrumentis says the translator of his workes. with][4] large instruments & great Curiosity observed the suns greatest declination 23°.30′.17″ some Manuscripts have 27″ but so small a difference needs not to be heeded.

From what observations these obliquitys of the Zodiac were deduced the Manuscript Paper in which they were transmitted to me[5] sayes not. It onely intimates that the bookes from which they were collected are in being in severall librarys of the university [of Oxford] which if they were consulted by persons skillfull in the Easterne tongues might no doubt afford us not onely this but some other useful observations much wanted in Astronomy

But for theire Instruments & accuracy theire are severall hints in it from which wee may raise faire Conjectures. Albatanius used the Parallactick Rulers of Ptolemy which I may perhaps describe when I come to treat of Bernard Walters observations made with y[e] like Instrument, whence wee may conjecture that the Arabian Caliph. Almaimon who next praeceded him as also Thebit who next followed made use of the same Contrivance. Albatanius sayes his was *longissima Alhidada* therefore the more subject to warpe & prove erroneous, as wee may conclude his observations were, not onely because Thebit with lesse y[n] 20 yeares after him makes y[e] obliquity $1\frac{1}{2}$ minutes lesse, but also for that Abu Mahomet Alcogandi whom my paper calls a most ingenuous person with a sextant of 15 Cubits Radius whose limbe was divided into seconds within lesse y[n] 110 yeares after Albatanius more y[n] the Tychonick observations in the last Century make it. These particulars in this determination tho few, adde much weight to it, & if as it is likely the suns parallax of 3′ in y[e] horizon was involved in this observation, the same when purged from it will not be much more then what wee find in these days.

The observation that seemes strongest for the obliquity of 23°.35′ is that of Albiruni who is said to have observed with a Quadrant of 15 Cubits. but this is onely 3 yeares after Alcogandis which seemes lesse Controvertible & therefore wee may rather suppose that hee asserts it on y[e] Authority of Almaimon & Albatani then from any sound observation of his owne.

To this opinion I am the rather inclined because I find Abulfeday almost 400 yeares after useing the same obliquity that had beene declared by Almaimon & Ebn Gaber or Albatanius when by the Consent of all the skillfull & curious Astronomers of that time it was not more then 23. & 31 or 32 minutes.

Nay lesse for in y[e] same yeare with Arzakell or but 97 after Abu Mahomet Alcogandi. in y[e] yeare of Christ 1089 wee find Chojah Nassir Eddini a Noble Persian to have observed τηυ λοξωσιυ accuratissime The ⊙ s greatest declination most exactly 23°. 30′ altogeather agreeing with that of Oleg Beig made 348 yeares after him, & nearer us, who in y[e] yeare of Christ 1437 determined it with large Instrum[ts] & great Curiosity to be $23°.30'\frac{1}{2}$ scarce sensibly different from Nasir Eddini And the same obliquity was approved by y[e] most Skillfull p[er]sons from the time of Alcogandi to Oleg beig. wee may Collect from Maimonides who lived betwixt them about y[e] yeare 1174. & affirmes that the true greatest declination. of the sun was 23°.30′ nearest. to whom wee may adde Ebn Shater who differs no more from Eddini & oleg beig. then some of those late closet Astronomers who have no meane esteeme in the World doe from others that have taken the most paines & beene the most exact in theire observations as I feare I shall have cause to make appeare before I dismisse this enquiry. but to my businesse

Y[u] may remember that I determined the Suns greatest declination from Eratosthenes his observation duely correct to be 23°.30′ 230 yeare[s] ante Christum, Wee find it here the same by the joynt consent of those who seeme to have beene y[e] most accurate & carefull in theire determinations from y[e] yeare 1089 to 1437 some authors opinions seeme to contradict & make it something more but seeing wee want the Necessary Circumstances whereby they might either have beene confirmd correct or rejected I thinke wee may doe most ingenuously to wave theire authority & that wee may safely pronounce the Inclination of the Suns annual path to the æquator, or the obliquity of the Ecliptick to have beene if not invariable yet the same in the time of Eratosthenes and in these 5 Ages when the Easterne people onely had the honor to be acquainted with & Manage the Caelestiall Sciences.

With which it may not be unpleasant to y[u] to heare how they first became acquainted.[6] the Roman Empire had beene much weakened by its division betwixt the sons of Constantine before y[e] yeare 400 afterwards by the Continuall Incursions of y[e] Goths & Vandalls. about the Yeare 600 Heraclius reignd in y[e] East at Constantinople who being much distressed by y[e] Persian King Chosroes. Calld in the Arabian Saracens to his assistance. by whose helpe becomeing victorious hee dismist them when hee had setled his affaires. not onely without theire pay but with reproaches Mahomet the Author of the Mahumetan sect then was broacheing his opinions amongst this people who wanteing nothing but a head to manage theire rebellion were easily enduced by his wealth & power to accept of him & with him of his new religion. under his Command they seizd on Syria and some part of Ægipt but the totall subjection of it was reserved for Haumar his next successor save one. who with it tooke in all palestine & Mesopotamia. Alexandria in Ægipt gloried in the great schole for sciences till now. these fell into the hands of ye Arabians or Saracens who yet had no leasure at first to consider of theire aequists War & turmoiles afford no leasure for the improvem[t] of Sciences. it was neare 200 yeares after ere the[y] fell upon them & the first notice . . . the learned Scaliger who tells us that hee had seene an ancient version of Ptolemys Almagest from the Greeck into Arabick with this Inscription *Liber hic praecepto Maimonis Regis Arabum qui regnanit in Baldack ab Alhazer filio Josephi filij Maire Arithmetici & Serigo filio Elbe Christiano Anno ducentisimo duodecimo sectae Saracenorum translatus est* that is. this booke at the Command of Sultan Almaimon king or Caliph rather of y[e] Arabians who raigned in Baldack or Babilon was translated by Alhazer y[e] son of Joseph the Sonne of Maire an Arithmetician & Serigus the Sonne of Elbe a Christian in the yeare of the Hegira 212, long before the time of Albategnius as Scaliger theire notes[7] To Almaimon Wee are beholden for the first observation of y[e] obliquity of the Ecliptick after Eratosthenes. as also for y[e] measure of a degree on the Earth. as well as for ye Arabick version of Ptolemy & wee may conjecture that either by his or his successors encouragemts. the Workes of Apollonius, part of whose workes are onely found in the Originall Greeke[8] but entire if I have not beene misinformed in ye Arabick. let it suffice. The Arabians & Saracens became more affectionate to Philosophy & the Mathematicall Sciences then those ever where out of whose hands with them they had wreasted the Empire I meane ye Greekes & Latines. theire law was so well accommoded to ye sensuall appetites of Mankind & they propagated it so vigorously that it was in a short time enterteined all over Africa & ye East & when the Moores invaded Spaine. They carried not it onely but theire Mathematicks along with them as the Arabians and Saracens did to the east. whom by the observations of the obliquity I have already recited wee may understand to have prosecuted these studies earnestly Not onely at Bagdat where they setled a schole, in Persia. & Tartary as the observations of Beni Musa. Nasir Eddini & Oleg beg. make it evident but also at Corduba Toledo & Sevill in Spaine where they had another universite & where Alhazen is said to have studied theire third was Fez which they still retaine & where they are said to have had a large library of Excellent bookes but whether they have beene preserved in theire Many revolutions since. time & opportunitye must enforme us.[9]

But to returne to the Almagest the Greeke Copies of it were so scarce about the yeare 1200 or else so little understood, that ye Emporer Frederick the Second commanded ye version of it to be made from The Arabick Copy. & the learned Kepler tells us[10] that upon examination hee found that both Alfonsus & Copernicus had made use of the Arabick version[11] in theire places of the fixed stars.

The Moores entred & conquered Spaine in ye yeare 724. & from them it is most likely Alphonsus the Compiler of the tables beareing his Names with ye Moores & Jewes imployed by him in that affaire[12] receaved theire Astronomicall knowledge as did the Westerne people neare the same time from the Arabians & Saracens with whom theire Wars about the holy land . . . at last the latines being seized of the Empire of Constantinople. about the yeare 1200. became acquainted with the Grecian Authors And two hundred yeares after some errors being noted in ye Arabick versions of the Almagest crept in by frequent transcriptions. George Trapezuntius sets upon it anew from the originall Greeke & compleates it[13] but not without considerable mistakes & errors which being noted by John Miller commonly & better knowne by the Name of Regiomontanus[14] brought on him the Envy of the sons & heires of Trapezuntius who to revenge the Reflections hee had cast on theire father procured as is reported his end by poison. but this very Ingenuous person had sowed his learneing in so many places & with such successe that the World in a Short while after saw a small Number of Artists prosecuteing what hee had begun with great advantage but I may not passe him by thus hastily the next observations of the obliquity of the Ecliptick wee are indepted to him for. & therefore I Shall examine them.

At Vienna in Austria in yᵉ yeare 1460 Purbachius & Regiomontanus tooke the Suns Meridionall height in yᵉ Summer Solstice 65°.06′ in yᵉ Winter 18°.10′ theire difference 46°.56′ they hence Conclude to be the true distance of yᵉ Tropicks & the halfe of it 23°.28′ the greatest obliquity of the Ecliptick or the Suns Greatest declination not makeing any allowance for the Suns parallax which they esteemed so small as not to be perceived by their Instruments, nor for his refractions,[15] that age & time being wholly Ignorant of theire nature and . . . .

As for yᵉ Suns parallax I have formerly proved in my readeings in this place[16] that it was really so small they might if they had knowne its true quantity have neglected it safely oʳ telescopicall Instruments not p[er]ceiveing it sensible much lesse theires with plaine sights. but for the Refraction it ought not to have beene neglected. I shall not now enter upon any discourse concerneing it but take it up as I have determined it reserveing the proof of the proportion I use to particullar Readeings I intend on this Subject & another occasion.[17]

It will be necessary however to informe yᵘ that by reason oʳ Earth is encompassed with an atmosphere or thicker aire then that of yᵉ Ether or Celestiall matter in which yᵉ bodyes of the planets move & the fixed stars are placed the luminaryes & all the stars above oʳ horizon & not in yᵉ vertex are refracted and appeare higher then really they are. I know Tycho allowes no refractions either to yᵉ sun or stars when they are more yⁿ 20 degrees elevated above the horizon & some others when they are above 45 degrees high. yet both allow the Atmosphere or refracteing Matter to compasse the earth & to be of a considerable thicknesse which granted it will follow that all the Phaenomena are refracted which are not in oʳ Zenith or perpendicular to us which I Shall hereafter make out by experiments. at present I need say no more to the ingenuous but that it seemes a little lesse then absurd to allow yᵉ Cause & yet deny the . . . Consequence.

The Refraction to yᵉ altitude of 65°. is onely 25″ which subtracted from yᵉ observed height 65°.06′ leaves the true or correct height of yᵉ Suns center at Vienna in yᵉ summer solstice 65°-05′-35″

The Refraction answering to 18°10′ of height is 2′.30″ which subtracted from yᵗ observed altitude leave the Suns true altitude in yᵉ Winter solstice 18°-07′-30″

The difference of the two solsticiall heights gives the distance of yᵉ Tropicks 46°-58′-05″. whose halfe shall be true obliquity of the Ecliptick 23°-29′-2½″ which is not much more then a minute different From what it was stated by Oleg beig at Samerkand in yᵉ yeare 1437. just 23 yeares before.

Or altogeather the same if wee consider all Circumstances. Ebne Shater tells us that hee corrected the obliquity of the Ecliptick. I suppose what was observed by others not himselfe, allowing for the Suns parallax, which they had not done, & hee with Ptolemy supposed to be 3 minutes in the horizon Makeing such allowance hee found it 23°.31′. This was in ye yeare 1363. Now wee can not suppose that Oleg Beig who determined it 74 yeares after should be ignorant of this as they thought altogeather Necessary correction. or that hee should omit considering of in his determination which wee have given us without Circumstances. Wee may justly therefore suspect that it is involved in his obliquity. & since none from Al Batani to Tycho supposed the Suns horizontall parallax lesse y^n 3 minutes in his Perigee if wee consider what allowance is necessary for that parallax wee shall find that 50″ are to be subtracted from his obliquity on this account which will make no more y^n 23°-29$\frac{1}{2}$′ for his is 23°.30$\frac{1}{3}$′.

This recited observation of Regiomontanus. Ricciolus in the third booke of his Almagest pag[e] 160 affirmes to have beene made at Nurneburge. but in his Astronom[i]a Reformata[18] not takeing any Notice of his former error hee ascribes it as hee ought to Purbachius & Regiomontanus and tells us it was made at Vienna, nor could it be otherwayes for it gives The Correct latitude of the place 48°-23$\frac{1}{2}$′ which Kepler who was long an inhabitant there attendeing y^e Courtes of y^e 3 successives Emporers. his Masters and so was likely best to knowe it makes 48°-22′ or onely 1$\frac{1}{2}$ minute lesse.

I should now proceede to examine the observations of Regiomontanus his Scholers, & Next successors, Bernard Walter & Werner; but I find I have spent my time I shall therefore refer them to y^e next opportunity & onely adde that it seemes a Miracle to me that so great a person as the famed Alfonsus who about the yeare 1240 & after convocated the most skillfull of the Moores Jewes & Christians to correct his tables & is said to have bestowed 40.000 Royalls of Gold in theire booke instruments & Maintenance should not for all his expence afford us so much as one good observation expecially of the suns greatest declination which was stated differently amongst the different Authors of the Moores Saracens Persian & Arab of those days. but these perhaps have beene lost or falne into ye hands of unskilfull p[er]sons or been cloystered up in Monasterys from whence they expect their freedom of a More ingenuous age & lesse slothfull people then theire present possessors

But enough of this; next lecture I shall proceed directly to Bernard Walter & carry downe ye Enquiry to oure owne home. as time shall p[er]mit[19]

NOTES

11.1. Many of the facts cited in Lecture 10 are again stated on ff. 113^r–116^v, cf. pp. 211–16, but we have not repeated the annotations to that lecture. The two main sources from which Flamsteed's information would appear to have been extracted are op. cit., notes 6.3 and 10.6.

11.2. The words in square brackets, which are missing from the manuscript, have been supplied after a comparison with the corresponding passage in Lecture 10 (f. 110^{v}, cf. p. 206).

11.3. *Binae tabulae geographicae, una Nassir Eddini, altera Ulug Beigi; opera Jo. Gravius publicatae* (Londini, 1648).

11.4. Cf. note 11.2 (f. 111^{r}, cf. p. 207).

11.5. Perhaps that entitled 'Ulug Beghi Tabulae motionum solis', etc., itemized under no. 3143.10 in op. cit. (note 10.7), p. 157. However, on account of the vagueness of the reference, this identification must be regarded as very tentative.

11.6. The following historical survey has apparently been drawn from op. cit. (note 1.11).

11.7. Flamsteed wrote 'Emend Temp pag 394' in the left-hand margin just above this point in the text, but the actual reference to the above-mentioned inscription and to this statement in the edition which he himself possessed (op. cit. [note 1.11]) is p. 215.

11.8. The best Latin edition of that geometer's works published prior to this time is *Apolloni Pergaei Conicorum libri quatuor, ex versione Frederici Commandini* (Bononiae, 1566). Edmond Halley's monumental edition was destined to appear in 1710. Edward Bernard discovered an Arabic version of Apollonius's 'De Rationis Sectione' in the Bodleian Library, and began translating it. Halley completed the translation and published it together with a restoration of another related treatise 'De Spatii Sectione' in 1706.

11.9. G. Delphin, 'Fās, sin Université et l'enseignement supérieure musulman', *Bulletin Soc. Géog. d'Oran*, viii (1907).

11.10. Op. cit. (note 2.16).

11.11. The Arabic *editions*, however, were different. King Alfonsus's Astronomers obtained their star positions via the Toledan Tables of the eleventh-century Cordoban astronomer Arzachel; while Copernicus consulted either or both of the earliest Arabic-Latin translations published at Venice in 1496 and 1515 respectively, before composing the manuscript of his *De revolutionibus orbium coelestium* (Norimbergae, 1543).

11.12. The leader of this group was the Jewish scholar Isaac ibn Sid, whose name is also associated with the invention of various instruments and with a Latin translation of an Arabic work on the quadrant.

11.13. Georgius Trapezuntius, *alias* George of Trebizond (fifteenth century), translated Ptolemy's *Almagest* into Latin from a Greek manuscript in the Vatican Library which was edited by Lucius Gauricus and printed at Venice in 1528. The Greek text itself was first published from a manuscript in Nuremberg (now no longer extant) by Simon Grynaeus in 1538.

11.14. The errors of Trapezuntius are revealed in Regiomontanus's edition of Georg Puerbach's *Epytoma Joānis de mōte regio In almagestū ptolomei* (Venetiis, 1496).

11.15. Internal evidence on f. 120^{r}, cf. p. 218 confirms that this information regarding Puerbach and Regiomontanus was taken from Riccioli 1, p. 160.

11.16. Cf. Lecture 2, where Flamsteed derives a value of 12″ for the solar parallax.

11.17. Cf. Lecture 33 and Lectures 35 to 38 for further discussions of the observations and theory of astronomical refraction which justify the adoption of the particular values of refraction at given altitudes cited by Flamsteed in the text of this lecture.

11.18. Riccioli 3, p. 20.

11.19. On this same page (f. 120^{v}) is written:

> I shewd y^{u} first how the ancients by observeing & noteing the Riseings & settings of the fixed stars might & questionlesse did learne that the ☉ moved

> round the heavens once in a yeare, & by considering the alterations of the lengths of the Meridionall Shades & themselves that the Circle of his Annual Course lay inclined & oblique to his Circles of his dayly revolutions. that the quantity of this obliquity was onely widely guest or stated at first and that it was uncerteine whether Oenipides of Chius &c.

This would appear to be a first (abandoned) attempt to begin Lecture 13.

Lecture 12

8 *February* 1681/2

I told y^{u} in the Conclusion of my last lecture that Without any repetition I should proceed in this to Bernard Walters observations & his contemp[or]arys or Next successors Copernicus. I hope therefore y^{u} expect no account of my last readeing in which I was the more Prolix to excuse it

I shall proceede accordeing to my promise & because both the observations of Regiomontanus his Scolars Walter & Warner, with those of Copernicus, as also the More ancient of Albatani, & probably severall of the Arabian I then recited, were made with ye Parallactick Rulers or Alhidada of Ptolemy I shall give y^{u} first the description of them taken from Snellius his edition of Walters observations printed at Leyden 1618 pag 44 & 45[1]

Prepare sayes hee two Rulers each about 6 foot long of a convenient breadth & thicknesse & let the sides & edges of each be wrought or planed truely parallell, these are represented in ye figure [Fig. 12.1] by AC, and AD, Provide also a 3^{d} Ruler CE of ye same breadth & thicknesse but 9 foot long. haveing fitted ye Rulers, along the flat side of the two shorter AC. AD & in the middle betwixt y^{e} Edges draw y^{e} two streighter lines AC, AD makeing AC. in y^{e} one exactly equall to AD in y^{e} other measure of this length. ac from c, ye center to D in y^{e} 3^{d} or longest ruler on y^{e} edge & divide this space first into 10 partes. eache of which shall thereby be about 7 inches & $\frac{1}{5}$ of y^{e} same measure & each of these into 10. more & divide or conceave of each of these lesser 10 to be divided into 100 partes continueing the divisions beyond D to y^{e} end of y^{e} longer Ruler so shall the whole length CD = CA = AD be divided into 10000 partes.

Let a payre of center pins be fixed into the Ruler FG. at A and C whereby the shortest ruler AD may turne or move on the upper-most, the longer CE on the lowermost center & let the distance of these centerpins y^{t} is the Middle of them be exactly equall to the lenght AD or CD which are equall let a payre of sights pq with small holes in them be fixed on the Ruler AD. so as the line passeing through the said sights or theire small holes may be exactly parallel

Notes on page 230.

to the line first drawne AD. let allso a thin tube of paper or some other light substance be provided & fitted over the said sights so as when the sun is to be observed no light may fall upon ye sight q. but what passes through the foramen of the sight p̄. for otherwayes the faint Species of the sun cast through it on q. will scarce be distinctly visible

In this said thin tube theyre is a hole to be made neare the sight q. through which the species of the sun projected on it might be viewd.

The upwright peece must have a piveot at each end on which it may be turned round as is seene in ye figure [Fig. 12.1].

The rulers were rectified to theire true position by a perpendicular hung where was thought most requisite on ye ruler FG as may be collected not onely from ye Manner of the Contrivance it selfe but also from some speciall notes of Walters relateing to it which wee may have occasion to take notice of by & by.[2]

The Instrument being thus prepared when the Suns Meridionall height was to be observed the[y] turned it by the piveots till the plane of ye 3 Rulers was in the plane of the Meridian & moveing the Ruler AD to and fro till the species of the Sun cast through the foramen or hole p in that sight next the Sun, fell exactly on the middle of the sight q̄ or rather filld a Circle strooke upon it equall to the due breadth of the said species. they brought the longer Ruler CE to touch the extremity & then counted the partes touched by it from the Center c for these were the Chord of the Suns true distance from y^e^ vertex to y^e^ Radius AD = to AC.

But these might be & it is probable were erroneous on a double account & greater then they ought to be considerably. for first. by reason they had neither the use of the compasse in those dayes by which they might direct the plaine of the Instrument nearely into y^e^ true Meridian, nor yet any good time keepers whereby they might find how long after any observation taken in the Morneing before, the Sun would be due South. it was impossible for them to determine it otherwayes then by Conjecture as our Seamen use at present. by takeing the heights of the Sun as hee rises till they find them increase no longer or rather till they begin to diminish & then takeing the greatest for the true. this the ancients might doe but if by mistake they ceased theire notes as sometimes they might before ye Sun arrived at his greatest height then would they have it lesse then they ought. & tis probable sometimes they erred on this account.

But a more certeine cause of theire errors might be the length & slendernesse of the Chord Ruler CE which if it were not carefully provided against might warpe so much as when the Meridionall Sun was to be observed in ye Southerne Tropick would render the Chord of his Zenith distance a quarter of an inch longer then it really was. That is if it either bent side or edgewayes. & so small an error would cause no lesse then 10 minutes fault in the Meridionall height that is would render it so much lesse y[n] it ought to have appeared and would if this had beene provided against.

But in the Summer tropick by reason that the Sight Ruler AD came Nearer to y[e] perpendicular & Consequently cut of[f] a lesse lenght of y[e] Chord Ruler EC. the error Created by its bending would be lesse considerable. At Racca or Aracta where Albatani made his observations. the Zenith distance in y[e] Summer solstice was not more y[n] 12 deg & an halfe so that the length of the chord of this angle would not be more y[n] 16 inches in which short length it is not probable that there could be so much warpeing of the ruler or bending as to cause any sensible error

But in y[e] Winter Soltice the Suns distance from y[e] vertex would be $59\frac{1}{2}$ degrees. the Chord of which being but little more then halfe an inch lesse y[n] the length of the Sight Ruler any warpeing would render the length of the Chord of that distance CD to be greater then it ought and it needed not to be much to render it $\frac{1}{4}$ of an inch too long which as I said before would also render the Suns observed Zenith distance 10 minutes too much & so by Consequence the Widenesse of the tropicks as much too great & the obliquity halfe as much or 5 minutes which if it may be allowed on this Conjecture it will follow that the obliquity of the Ecliptick was 5 minutes lesse then Albatani states it & no more then it was afterwards found by Nasir Eddini & Oleg Beig.

Nor is to be doubted but these same Considerations may serve also to correct the determinations of the Caliph Al Maimon. who praeceded Albatani & whose Methods it is probable this latter followed. So also for those who followed them & used theire Methods or relied on theire Authority.

But enough of those ancients[3] I returne to Bernard Walter. the good agreem[t] of whose observations both w[th] themselves & with others since made w[th] instruments of a different nature & Contriveance p[er]suades us to thinke that hee had provided well against those faults & errors which want of praecaution had caused in those his predecessors.

Haveing turned over his Catalogue of solar observations[4] I find those of the yeares 1478. 1487 & 1503 the fittest for this examination. In that age on the Middle yeare betwixt the leape yeares the sun entred into Cancer betwixt y[e] 12 & 13 of June, into ♑ betwixt y[e] 12 & 13 of December. Now on y[e] 11 as also on the 13th of June 1478 at Nurenburge hee observed the shade of the Suns Zenith distance to be 44890 such partes as the Radius of his Instrument was 100.000 Now because the Chordes of any arch of a Circle are equall to the double sines of halfe these arches. halfeing this Chorde I find it 22445 & entring the table of sines with it, I find y[e] Arch answering to it to be 12°.58′.15″ which doubled gives the suns distance from the vertex 25°.56′.30″

The Refraction in that distance is onely 25″ which added to it makes the true distance of the Suns Center from y^e^ vertex of Nurenburge at the Summer Solstice 25°.56′.55″

Dec y^e^ 11th. Next following or the same yeare hee againe observed the Chorde of the Suns Zenith distance 118.760 of y^e^ same p[ar]tes the halfe of which is 59380 the arch answering to it in the sines doubled 72°-51′$\frac{1}{3}$ the Refraction in that distance 2′.40″ therefore the true distance of the Suns Center from y^e^ vertex of Nurenburge at y^e^ Winter Solstice 72°-54′-00[″]

The difference of the two solsticiall distances is 46°.57′.05″ its halfe 23°.28′.32$\frac{1}{2}$″ the true obliquity of the Ecliptick accordeing to the observation.

I have not Considered the Parallax of the Sun in either of these observations; it being so small was wholly insensible in theire Instruments as I told y^u^ before. for which reason I have also neglected to make some allowance for what the sun wanted of his greatest declination by reason hee was short of the tropicall point at ye Winter observation, that if any thinke it materiall they may allow about 6 or 7″ for it & make the obliquity so much greater which is the most . . . .

Tis Nine yeares after ere I find any more of Walters observations fit for this examination. that is till ye yeare 1487 when June ye 13. hee measured the Chord of the Suns Zenith distance in ye meridian at ye Summer Solstice againe 44890. which gives the arch subtended by it 25°.56′.30″ but Correct by refraction as before 25°.56′.55″ But the Chord of the Meridionall distance of ye sun from ye vertex at ye Winter solstice Dec 13 following 118.790 which gives the Arch 72°.52′.30″ but Correct by Refraction 72°.55′.10″

the difference of these two arches 46°.58′-16″ is the distance of the Tropicks. its halfe 23°.29′-08″ the obliquity of the Ecliptick. accordeing to these latter notes neare halfe a minute bigger then by the former.

I must not omit one note that falls in any way tho not relateing to y^e^ obliquity yet otherwayes of very good use: that is by these observations the latitude of Nurenburg will be found 49°.26$\frac{1}{10}$′: which by the Solsticiall Meridionall distances from y^e^ vertex of y^e^ yeare 1478 is 49°.25$\frac{1}{2}$′

The Next observations I pitcht upon of Walters were made in y^e^ yeare 1503. in which hee often Mentions the Rectification of his instrument by y^e^ Gnomon & otherwayes which suggests to us that he was now more carefull & accurate then formerly p[er]haps upon the discovery of some former errors of his Instrument

By which June ye 12. hee found the Chorde of the Suns Meridionall distance from ye vertex 44846 therefore ye Arch of ye said distance 25°.54′-42″ but Correct by Refraction 25°.55′.07″

Againe Dec 12 hee found the Chord of the ⊙^s^ said distance

118750 therefore the distance it selfe 72°.50′.52″ but correct by refraction 72°.53′.32″

The Arch of y^{e} Meridian therefore intercepted betwixt y^{e} two tropicks 46°.58′.25″ its half the obliquity 23°.29′-12″. and the latitude of the place 49°-24′$\frac{1}{3}$

These 3 yeares observations of Bernard Walter doe sufficiently evince that the obliquity of the Ecliptick was 23°.29′ very nearely in his time. and tho the latitudes of Nurenburge collected from them differ above a minute amongst themselves. I thinke they rather confirme then argue his observations any wayes erroneous & shew that his Rulers kept the same position from one yeares end to another but because of the speciall Notes hee puts upon the last of *rectificato Instrumento*[5] I must esteeme them the Most accurate, and attribute the differences found in the latitudes deduced from his observations to this rectification. for that if the Instrumente remained unaltered after its rectification the obliquity of the Ecliptick collected from the Solar observations of the next two solstices would remaine the same tho ye latitude gathered from them would alter as much as was the amendement of the Instrument.

The next who observed the obliquity is John Werner of Nurenburge: Not long after Bernard Walter who makes it 23°.28′.30″: but from what observations hee determined it wee find not tis probable that being of the same place & time wth Walter hee made use of the same contrivances & Methods & then if wee correct his observations by the allowance of Refractions as wee did Walters the obliquity will be found something more y^{n} 23°.29′

About ye same time wth Werner lived Dominicus Maria[6] Copernicus his Master in Astronomy whose observations are also wanteing from which hee determined it 23°-29′ but it is to be noted that liveing in a more upright sphere then Walter & Werner[7] hee had the suns refractions lesse in ye Winter Solstice & therefore his obliquity tho Corrupted with them yet lesse vitious then his Contempora[r]yes

Riccioli in ye 3^{d} booke of the first volume of his Almagest pag 160.[8] tells us That Copernicus. in the 2 Chap. of his second booke of Revolutions affirmes that the Suns Meridionall height at ye summer solstice at Fraemburge in Prussia was 59°.08′.54″ but in the winter Tropick. 12°.16′.06″ by which hee supposes Copernicus intended the visible. not the heights correct by parallax. & that the obliquity of the Ecliptick hence deduced was 23.28.30 & the height of the pole there 54°-19′-30″: but finding upon further examination that these heights corrected by his refractions would [neither] give the obliquity nor the latitude agreeable to the determinations of Copernicus. not his owne In his Astronomia [reformata][9] . . . the observation to his mind. to redeeme so excellent a person from ye abuses of so ill a manager of Caelestiall observations as Riccioli has frequently shewd himselfe I shall take the observations as layd downe & shew yu that theire is no discohaerence in them when fairely examined

Copernicus used the Parallactick Rulers of Ptolemy as did Walter & Werner in theire observations, the Suns horizontall parallax hee supposed to be 3 minutes in his meane distance. Now because his observed heights are given to odde seconds to which praecisenesse hee never pretended in his observations wee may hence collect that hee had corrected them by the allowance of his parallax. which will certeinely appeare when wee have cleared his observations from it.

In ye height of 59°-9′. his solar parallax will be found 1′-34″ which taken from 59°-08′-54″ his correct leaves 59°.07′.20″ for the height hee had observed more probably true because it falls on ye 3d p[ar]te of a minute

In ye height 12°.16′.06″ the parallax would be 2′-56″ which subtracted from it leaves his height 12°-13′-10″:

And if wee correct the first by the subtraction of 30″. the other of 3′-45″ the Refractions. in the several heights wee shall have the Suns Correct [meridian altitudes][9] . . . at ye winter 12°-09′-25″ whose difference 46°.57′.25″ is the Widenesse of the Tropicks & its halfe 23.28.42 the obliquity of the Ecliptick. agreeable wth Walters.

Hence also the height of the Pole at Fraemburge will be found 54°.21′.53″ which Elias Olaus sent thither by Tycho purposely to observed it found 54°-22′-15″. but Correct by refraction will be 54.21.40 altogeather agreeing with my determination from Copernicus his owne observations

Copernicus was the Person that revived the Pythagorean Systeme of the World & shewd how all the caelestiall appearances might be represented without the helpe of Epicycles or fictitious orbes admitting the Motion of the earth once in yeare in its orbite round the Sun. & in a day about its owne Axis which to represent the appearances of his different meridionall heights hee supposed to keepe the same Inclination to ye plane of ye Ecliptick. it will be convenient to explain this before wee passe any further. which may be done thus.

Conceave the eye placed at any infinite distance in ye plane of the annuall orbe which will then appeare as part of a streight line, let this be represented in ye 2d figure [Fig. 12.2] by ye line ec & let the one of ye circles there described represent the place of ye Earth in the first point of ♋. the other of ♑ whence ye suns visible place is in the opposite signe let pem, the Axis of the earth forme the same constant angle of 66° 31′ pce with ye Axis of ye Ecliptick. So ye arch pe shall be allwayes neare 23½ degrees. let lv represent [the] parallel of London or the path it describes in the Earths diske viewd from an infinite distance as Shee turnes round on her axis whose distance from ye pole p. is 38½ degrees when ye earth is in ♑ of ye sun in ♋ the summe of ye two arches pe the inclination of the Axis 23½, & pv. ye complement of the latitude makes ye suns Meridionall height ve 62° deg but when the sun is in ♋ then theire difference the arch ec 15° deg is the said. height. but when the earth is on eather of the aequinoctiall pointes and all the paths are halfe illuminated then the Complement of the latitude or distance of the Path from the Pole is the Meridionall height of ye sun.

which arches supposeing pe to be the same with the obliquity of the Ecliptick will be the same they would have beene in the old systeme accordeing to which I shewed now they were formed in ye last lecture of the last terme.[10]

Copernicus findeing by the observations of Bernard Walter Regiomontanus & his owne that the obliquity of the Ecliptick was nearely the same to them all but much lesse then that of Albategnius & his lesse then that of Ptolemy Eratosthenes & Hipparchus conclud[e]s [that the direction of the][11] Earths Axis was not the same in all Ages. it seemed to him that allowed of all the Ancients determinations that it was greatest above the time of the Incarnation least at his & neare both those times stationary to salve it therefore he conceaved that it would returne to the same againe after 3434 aegiptian yeares. & 10 dayes. & because hee asserted all the Motions of the heavens to be performed in perfect Circles hee makes use of the following Contrivance. to answer not it onely, but an inaequality he conceived to find in the aequinoxes. the Center of the Circle represented in ye 3^{d} figure [Fig. 12.3] hee supposes placed on the Caelestiall pole of ye AEquator its diameter BE to be 24 minutes. & that the pole of the Earth altered its pointe of direction in this Circle annually 6′.17″.24‴. hence knoweing at what time it was at B & the time interlapsed hee finds the Meane motion the Arch HB or ye Angle HAB with which & the Radius HA Constantly 12 minutes, the Side AG. is easily found Which if ye Anomally counted from B were lesse then 3 signes but more then 9. adde to. but if above ξ & lesse y^{n} 9 subtracted from The Middle obliquity 23°-40′ gave his obliquity for the present time.[12]

But I onely mention this to shew what artifices our praedecessors made use to represent those inequalityes which theire needlesse assent to erroneous observations had forced them to admit I have shewd y^{u} before that neither Eratosthenes his observations nor the Contemporary of Albategnius differ so considerably from those of Walter or Copernicus as that wee need to thinke the Inclination of the Earths Axis or the obliquity of the Ecliptick to have altered. I shall therefore proceed to examine such observations of it as I find following Copernicus next opportunity:

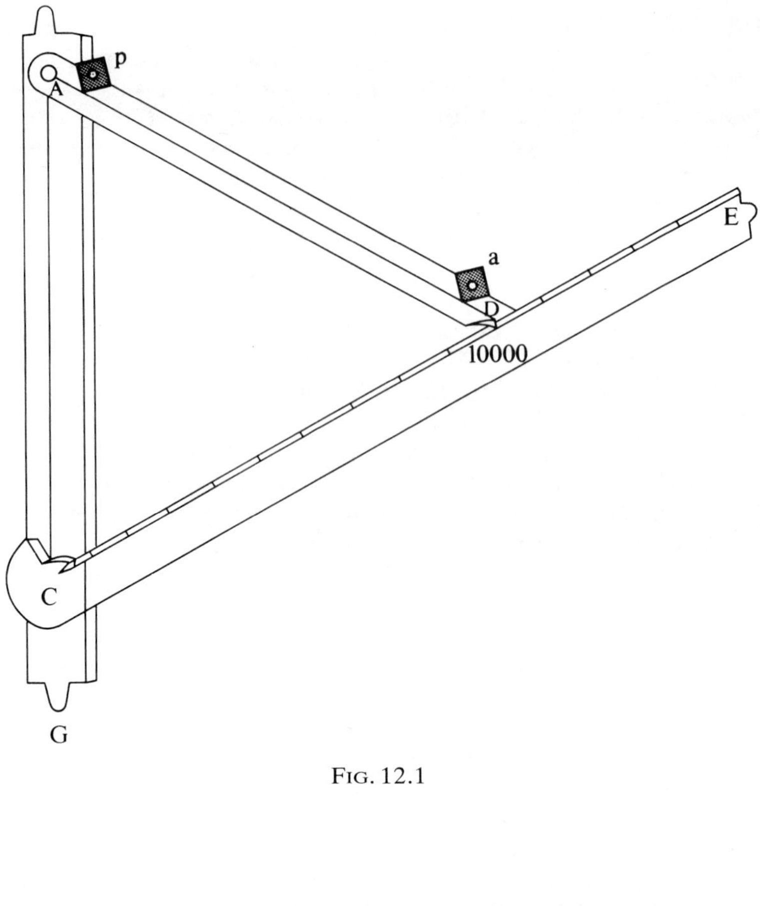

FIG. 12.1

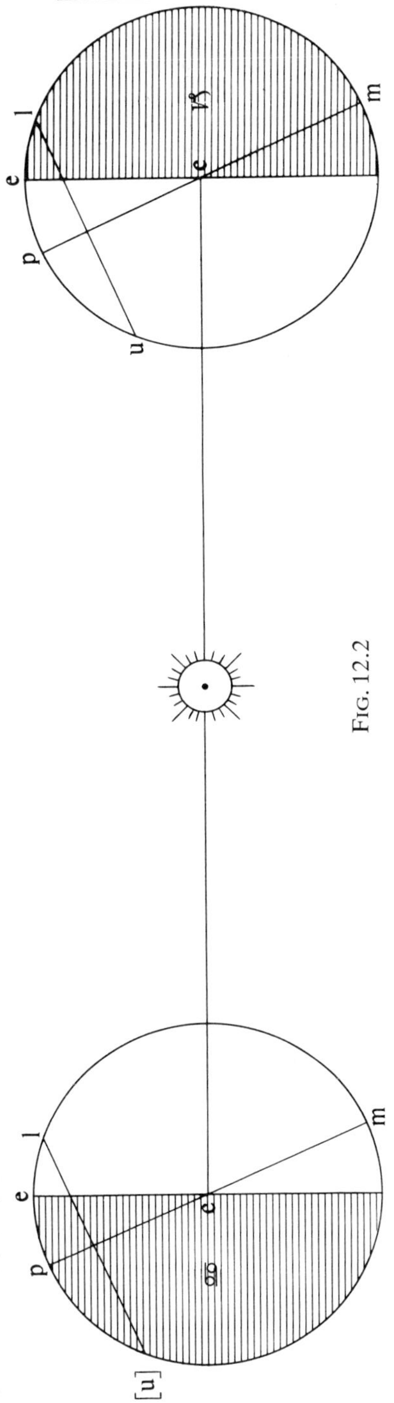

Fig. 12.2

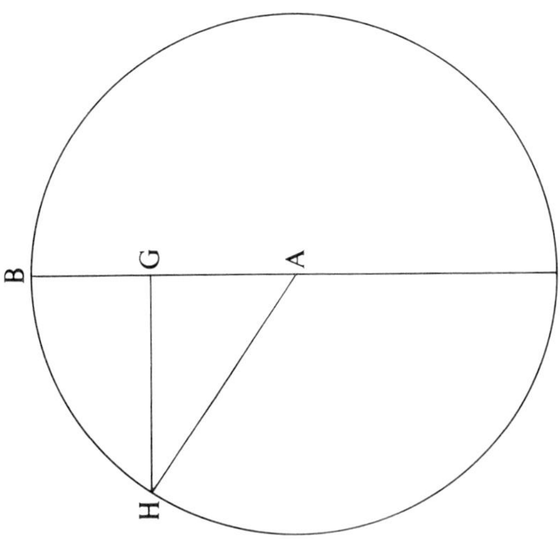

Fig. 12.3

NOTES

12.1. Willebrod Snell (ed.), *Coeli et siderum in eo errantium observationes Hassiacae.. Quibus accesserunt J. Regiomontani et B. Walteri observationes Noribergicae* (Lugduni Batavorum, 1618). Cf. pp. 44, 45.

12.2. The 10-page catalogue of solar observations is appended to ibid. The notes in question are contained on the penultimate page of it.

12.3. Flamsteed is referring here not to the Greeks and Romans but to the above-mentioned and other Arabian and Persian astronomers, whose contributions he has again discussed in Lectures 10 and 11.

12.4. The observations mentioned below are taken from the catalogue appended to op. cit. (note 12.1). Cf. note 12.2.

12.5. Cf. note 12.2.

12.6. Dominicus Maria da Novara (1454–1504) was the Professor of Astronomy at Bologna University with whom Copernicus made his first recorded astronomical observation of a lunar eclipse on 9 March 1497.

12.7. This is another way of expressing the fact that the northern latitude of Bologna (44° 30′) is less than that of Nuremberg (49° 30′).

12.8. Riccioli 1, p. 160.

12.9. Words are missing here from the original manuscript.

12.10. Lecture 10 (23 November 1681).

12.11. Cf. note 12.9. The words in brackets have been guessed.

12.12. Flamsteed's summary of Copernicus's procedure was probably based upon his reading of G. Rheticus, *Narratio Prima* (Basileae, 1566).

Lecture 13

4 *May* 1682

I have informed y[u] formerly that it was my designe & intent in these Readeings to examine the fundamentalls of Astronomy & in the roome of those doubtfull or ruinous ones which have hitherto been made use of by Authors of good repute but suspected by others yet without much amendment before the begining of this Age & the discovery of the Telescope or rather the application of it to Celestiall observations, I say in the roome of these faulty fundamentalls it is my intent to substitute more firme & solid ones & such as may abide the test of future Ages. In order to this worke I began w[th] y[e] suns Diameters & haveing shewd y[u] how faulty the determinations of the Ancients were in this p[ar]ticular I told y[u] what the diameters both of the Apoge & Perige Sun had beene found by undubitable observations. I gave y[u] an account how these were made. & by the Way I decided a Controversy of an old date y[t] is whether the excentricity of the Earths orbite were bisected or Not. & proved that it was.[1]

I had shewed y[u] before how y[e] necessity of true account of the lenght of the solar yeare had caused the Antients to thinke of Methods for findeing out the Motion of the Sun. its unaequalitys & the obliquity of the Ecliptick. in my 3 last Readings[2] I enquired what was their opinions of the obliquity I meane the quantity of it. how it was rudely determined at first about 24 degrees or y[e] 15[th] p[ar]te of the Circle by the ancient Indians or Arabians. after more accurately by Eratosthenes about 230 yeares before the Incarnation. but that hee accounted his obliquity to the utmost Circle of the Sun and that if his observation were duely examined it would give but 30′ minutes above 23 degrees with which the observation of o[r] age nearely agree.

Hipparchus & Ptolemy used the same obliquity with Eratosthenes. & seeing that theire observations were made in y[e] same City of Alexandria with his tis probable the[y] assumed it on his Authority & from his observations for they have given us none of theire owne to Confirme it. I passed therefore from them to y[e]

Notes on page 238.

Arabians. an account of whose observations or rather determinations from them I had lately receaved from an ingenuous freind at Oxford Collected from the Manuscripts of theire libraryes. publick & private.

By these it appeared that the first who after Ptolemy observed the obliquity of the Ecliptick was the Caliph of Bagdat Almaimon about ye yeare of the Hegira 825

The Next after him. Mahomet Ebn Gaber or Albategnius in ye yeare 882 who acquaintes us that his observations were taken with ye Parallactick rulers of Ptolemy but that hee used very long ones & found by them that ye obliquity of the Ecliptick was 35′ above 23 degrees as Almaimon had determined it before him & that probably with ye like Instrum^t^.

But It appeared by the observations of Thebit Ebn Corah made 19 yeares after Albategnius & not unlikely with ye same sort of Instrument that 35 min was too big. for Thebit had it but 33½ & within lesse y^n^ an Age After one Abu Mahomet Alcogandi with a Sextant of 15 Cubits determines it but 23°.32′½ which is but a minute more y^n^ The Noble Tycho found it in The last age admitteing his owne parallax & y^t^ the stars are not refracted when they are more y^n^ 20 degrees high.

But in ye yeare 1089 onely 200 yeares after Albatani a Noble Persian Chojah Nasir Eddini finds it onely 30 min. above 23 degrees. & almost 400 yeares after him Oleg beig the same. whence it was evident that theire predecessors had erred & made it 5 minutes too Much.

Enquireing the Cause of this error I could find none more probable then the ill make of theire Instruments the Parallactick Rulers which I described in my last Readeing. & before I examined Bernard Walters observations which were altogeather made with that Instrument. The Ancients knew well that the longer theire Rulers were so much more accurate would theire observations bee. they therefore made theire Rulers very long but probably tooke not sufficient Care so to fortify them that they might not encline warpe or bend any way. feareing lest by makeing them strong or addeing knobs to them as Tycho & Hevelius have since done in theire Instruments they should render them over Weighty & Consequently Cumbersome & Intractable, for want therefore of such fortifieing of theire instruments if either of theire Rulers bent so much as to make a fault of ¼ of an inch in the said length this would cause tenne Minutes error in theire Zenith distances of the sun at ye Winter Solstice & Consequently would give the obliquity of the Ecliptick at least five minitues too big.

And therefore omitteing the more uncerteine observations of Almaimon Albatani Albiruni & the first of the Arabians I proved that if the obliquitys assigned by theire successors were purged from that faulty parallax which they with Ptolemy had assumed 3 minutes in the horizon the true would be found not sensibly bigger then it has beene determined by the more accurate observations of y[e] two last following Ages

Nay that of the Jewes and Arabians amongst the Curious tho lesse learned in Astronomy as Maimonides who receaved theire opinions from the Tradition of the More Skillfull the obliquity was but 23°-30′:

Oleg Beig the last of the Easterne Astronomers preceded Regiomontanus but about 40 yeares passeing therefore from y[e] Arabian to y[e] European observations I tooke occasion to shew y[u] how those easterne people became first acquainted with y[e] Celestiall Sciences & how from them they were communicated to the Spaniards first afterwards to the Frankes or people warreing in the holy land concerneing which haveing beene formerly sufficiently prolix I shall Make no repetition at present. I shall onely adde which I then omitted that the Moores made not a full conquest of Spaine at theire first Incursion under Musa & Tariq in y[e] yeare of Christ 724. but that they sent severall supplyes into that Country as they carried on theire conquests & their exigences required & that it is more probable the[y] receaved theire Mathematicks by some of theire later rather then by the first Intruders.

But I may not omit to advise those who have occasion to use Alfonsus or Copernicus his Workes to take Notice that both those famous persons made use of the Arabick version of the Almagest & the places of the fixed stars in it which are frequently corrupt as the learned Kepler has observed.

The Arabick determinations wanted the Circumstances of y[e] Suns observed Meridionall heights to confirme them or at least if the Authors gave them they are not yet come o[r] hands. p[er]haps o[r] little Commerce with that people & consequent ignorance of theire language tho wee have many of theire bookes keepes us from understanding them. bee it as it will the Europeans when they receaved these Arts tooke notice of this error & have Carefully conveiyd the Requisite Circumstances of theires whereby posterity may be enabled to judge of them & Correct them as the discoverys of after times should shew reason. & this to theire no small advantage. for otherwise wee must either have concluded the obliquity of theire times to have beene lesse y[n] wee now find it or else have rejected these & all theire other observations as doubtfull & uncerteine

Whereas haveing examined the observations made by Regiomontanus & Walter at Nurenburge of the Suns Meridionall distances from y[e] vertex at either solstice & corrected them by the due allowance of Refractions omitteing the parallax which to the best instruments wee are yet Masters of is almost I had well nigh said altogeather insensible, I shewd you that the greatest declination of the sun was truely 23°.29′. which I also proved from ye Bononian observations of Dominicus Maria. & the Prussian of the famed assertor of Truth the Great Copernicus. in the examination of whose Notes I tooke care to cleare him from ye disingenuous Managem^{t} of Riccioli a p[er]son who by his laborious & methodicall collections had gained very ample commendations. if his wresteing some his inadverteing some & misapplying of other good observations had not detracted from his praises.

Here I concluded my last lecture referreing what remained to be said on this subject to another occasion which since it has pleased providence now to afford us I shall prosecute accordeing to my first designe & intention.

The observations of Bernard Walter both of the sun & planets ended in the yeare 1504: till now Astronomy had beene onely enterteined under private roofes & lived a domestick with ingenuous men in Townes & Cityes. henceforward wee find her enterteined in the Courts of Princes. the necessary charge of Instruments & assistance required for the improvement of this science was found too great for an ordinary estate & fortune. that no part of Ingenuous knowledge might be unimproved it pleased ye great Creator when hee caused the Germans to throw of[f] ye yoke of implicit faith & blind obedience to ye dictates of a pretended infallibility, opening thereby the way to all ingenuous knowledge to fill the hearts of Princes no lesse with ye admiration of his Workes then with the reverence of his word. The Lantgraves of Hesse had receaved Augustane profession of faith & the reformed religion A family eminent then for ingenuous men & able Commanders both in peace & Warre. Philip had gained a great reputation uniteing the Protestant princes in a league at Smalcald for the defence of theire Faith whereby it was susteined against all the opposite endeavors of a Warlike Emperour Charles ye 5th. his Eldest Sonne William dureing his fathers life time began his Caelestiall observations at Cassells and by prosecuteing of those studies procured himselfe a greater fame & much more durable then his father had from all his labors leagues & victoryes, the learned Snellius has transmitted his notes with those of Regiomontanus & Walter in one small volume to posterity[3] they begin with ye yeare 1561. the first Snellius sayes were not so accurate by reason hee had not yet attained to the best contrivance of his Instrument. which were a Couple of Quadrants one greater y^{n} the other but of what Dimensions my Author does not informe us

The Lantgrave therefore. before the yeare 1566. gives the Suns greatest meridionall height 62°.12′ his least on ye 13th of December y^{t} yeare 15°.12′. whence hee collects ye obliquity of ye Ecliptick 23°.30′ and ye height of ye pole 51°.18′:

But by this account it appeares hee made no reckning either of parallax or refraction. and by a Note which I find in his observations 2 yeares after *Deinceps rectificatis Instrumentis*[4] tis evident that his Instruments were hitherto erroneous. I shall therefore wave all before this time and examine those onely which were made afterwards

It being now discovered or within a short while after y^t^ y^e^ Suns Meridionall heights were corrupt with refraction at y^e^ winter solstice. Rothman & Birgius examined the height of y^e^ Pole by y^e^ Meridionall heights of y^e^ pole star. & determined it 51°.19′$\frac{1}{3}$ which before they esteemed neare one minute lesse. the greatest Meridionall height of the solsti[ci]all sun in y^e^ yeare 1587. Rothman affirmes to have beene onely 62°-11′: the refraction in that height is neare halfe a minute. therefore his true height onely 62°-10$\frac{1}{2}$′ & the obliquity 23°.30′ very neare. what wee had found before by y^e^ observations of Regiomontanus Walter & Copernicus but something too big as will be made evident by the followeing & much more certeine observations of the Noble Tycho who flourished at y^e^ same time

It is evident from y^e^ Lantgraves observations that his Mathematicians pretended not to more y^n^ single minutes for none of theire observations are determined nearer then to a single minute. nay it is very probable they attained not to that, for the heights taken by theire different quadrants are sometimes 2 minutes different from each other. what should be the reason of this difference it is in vaine for us to conjecture but wee may hence rationally conclude that his Notes of Meridionall heights are not to be relied on & therefore without examineing any more of them I shall passe from theires to Tychoes who has Carefully informed us both of the Magnitude & fabrick of his Instruments,[5] who had contrived them so as they might shew the parts of a minute & not neglected to acquainte us with all such circumstances as might enable us to judge of the accuracy of his notes to chuse the best & reject the dubious.

This not lesse learned then curious diligent & faithfull p[er]son seemes to have differed from himselfe in delivereing the obliquity of the Ecliptick & the latitude of his Uraniburge in his severall writeings. and at different times for in his Historia Caelestis pag 47. and 48 concludeing his observations of y^e^ yeare 1583. hee states the ⊙^s^ greatest declination 23°.27′. or 10″ more. and the latitude of Hvenna 55°.50′-20″[6] whereas in the observations of following yeares & in his progumnasmata hee makes his latitude 55°.54′.40″. & the suns greatest declination. 23°.31$\frac{1}{2}$′[7]

Yet were his observed meridionall heights of the sun still y^e^ same. it will be therefore requisite to enquire what was the Cause of this difference In doeing of which I shall first shew what was the simple obliquity deducible from his observations as they would have beene managed by the ancients then what it was by allowing of his parallax and afterwards how hee came to enlarge it why hee pitcht upon a faulty Quantity & how his observations if duely considered will give the obliquity not different from what I determined it from Bernard Walters Notes first & has since beene concluded from later & no lesse accurate observations[8]

Our Noble author states the greatest meridionall height of ye sun at ye summer solstice in ye yeare 1583 to be 57-35$\frac{1}{4}$ considering the observations of severall dayes both praecedeing and following but the least at ye winter solstice 10°.39′-30″: Now takeing theire difference after Bernard Walters way for the distance of the tropicks wee shall find it to be 46°-55$\frac{3}{4}$′& therefore its halfe the obliquity of ye Ecliptick 23°-27′-52″: something lesse yn B Walters

But or author supposed the suns horizontall parallax 3 minutes. there his Parallax at ye summer solstice 1′-35″ and the true Correct height of his center 57°.36′.50″ his parallax in ye winter tropick 3 minutes. and his true height 10.42.30: therefore the distance of the Tropicks 46-54$\frac{1}{3}$ the suns greatest declination onely 23°-27′-10″ sensibly lesse then it was found without this unnecessary Correction. the height of the æquator 34°-09′$\frac{2}{3}$ and therefore the poles height 55°.50′.20″:

Which by ye greatest & least heights of the pole star carefully observed in the same moneth of December hee found 55°.54′$\frac{1}{2}$. that is full four minutes greater then hee had collected it from ye greatest & least observed Meridionall heights of the Sun.[9]

This difference I am apt to thinke caused him first to suspect or least to affirme the force of refractions. for if from his greatest meridionall height correct by his parallax 57°.36′.50″: hee subtracted the Complement of the poles height found by ye fixed stars 34°-05′$\frac{1}{2}$ the residue gave him the obliquity of the Ecliptick 23°.31′$\frac{1}{3}$.

But if from ye Complement of ye latitude 34-05$\frac{1}{2}$ he subtracted his least Meridionall height alike corrected 10° 42′$\frac{1}{2}$ there remained but 23°-23′ for the greatest declination of the sun south of ye Æquator. which being lesse by 8$\frac{1}{2}$′ then hee had found it north. hence hee suspects that the rayes of ye sun were refracted in the Winter Tropick and that hee appeared so much higher then hee ought

Henceforward hee became more heedfull of refractions hee seemes to have suspected something of theire effects before for in his Historia Caelestis pag 24. April 17. 1582 delivering the distances of ♀ from ye north horne of the Bull hee Subjoind this note *Hae posteriores minus sunt certae propterea quod ♀ luceret p[er] Radium refractum, erat enim vicina horizonti,*[10] but not cleareley till after. this observation when to ye observations of ye distances of ♀ from ye Sun taken Feb. 3.1584. pag. 74 hee addes *oportet adhibere tam refractionem[,] parallaxin ☉ & ♀ in consilium[,] si ad trutinam sit vocanda operatio.*[11] yet seemes hee not to have instituted any observations on purpose to discover the true quantitys of refractions till y[e] yeare 1585 when hee began at the Solstice to observe the suns meridionall declinations & when hee was at some distance from it with his vast & accurate Armillaes henceforward wee find in his Notes commonly praefixed *pro refractione* & after y[e] observations of Sept. y[e] 10. pag. 111 this especiall remarque. *Has itaque observationes solis idoneo Instrumento feci*[,] *ut constaret ratio ipsius refractionis radiorum.* & a little after hee affirmes the solstiall refractions greater then y[e] Autumnall. *Quemadmodum* (sayes hee,[)] [etiam] Bernardus Waltherus Regiomontani olim discipulus Norimbergae se animadvertisse asserit[12]

But I intend not to discourse further of refractions at this time they will require to be handled apart when I may have an opportunity I shall Informe you more of the history of them & his particular p[er]formance & discoveryes.[13] I shall onely adde now that our Noble Dane haveing assumed a parallax of the sun by much too greate was thence forced to conclude the Refractions likewise greater then the[y] really were by more y[n] as much. I have demonstrated to y[u] in my former readeings both by the lunar Dichotomyes & the distances of ♂ from fixed stars taken when hee was nearest the earth that The Suns Parallax was insensible & the greatest wee could suppose not more y[n] 10″. the Refractions I shall crave leave to assume till I come to reade on that subject when I doubt not to prove theire quantity & proportionall decrease from y[e] Horizon upward to be the same I have hitherto made use of: & now Compareing the severall meridionall heights of y[e] sun observed at y[e] solstices of the yeares 1584 86. 88. 89. 91 & 96 with each other. I find Most of them agreeing to 57°.35′.30″ in which height the true refraction is just halfe a minute. so wee may conclude y[e] true height 57°.35′ praecise.

But the Notes taken at y[e] Winter solstices of the yeares 1590 & 91. it appeares that the least meridionall heighte of the sun was 10°.41′.10″ the refraction in that height is 4′.20″ which taken from y[e] said observed height leaves the true 10°.36′.50″

The difference of the two correct greatest & least heights gives the distance of the Tropicks 46°.58′.10″ its halfe the Suns greatest declination 23.29′.05″ altogeather as I had before determined from the Notes of Bernard Walter Copernicus & Regiomontanus handled after the same manner.

I shall adde one remarke or 2 by the way which will confirme this quantity. that. the middle betwixt the two corrected altitudes gives the height of the Æquino[c]tiall at Uraniburge 34°.04′.55″. therefore the true height of y[e] pole 55°.54′.05″

Which is the same y[t] will be found by Tychoes observed heights of the pole star. in y[e] yeares 1584. 85. 87. and 88. whereby I have collected the simple latitude of his Uraniburge. 55°.54′.40″ which hee also uses himselfe the refraction in that height is 35″ which taken from it leaves the true height of his pole 55°.54′.05″ altogeather the same with what I had collected it from his greatest & least observed Meridionall heights of the sun. & but 10″ lesse y^n it has beene determined by ye French sent from theire observatory some few yeares since to reexamine it.[14]

I shall onely adde more & then conclude. that whereas I used the least meridionall height of the sun at Uraniburge in ye yeare 1583 but 10°.39′-30″ but since 10°.41′.10″. I tooke the first from Tychoes owne determinations that yeare who fairely left them in his notes as hee had observed them whereas I never find it so little afterwards. but as I last stated it in his severall yeares. which makes mee suspect his Instrument was not then perfectly exact. which some notes in his Worke[15] seeme also to intimate but after that yeare hee had all correct and duely examined so y^t wee may rely on his determinations much better then before. which made mee chuse rather his later notes then his former. I have done with Tycho. next lecture I shall confirme this quantity by further experiments & later observations.

NOTES

13.1. Lecture 9 (15 November 1681).

13.2. Lectures 10–12.

13.3. Op. cit. (note 12.1).

13.4. Ibid., p. 4.

13.5. A recent study of these descriptions has been made by Victor E. Thoren, 'New light on Tycho's instruments,' *Journal for the History of Astronomy*, iv (1973), 25–45.

13.6. Op. cit. (note 2.36), pp. 47, 48.

13.7. Op. cit. (note 2.5).

13.8. These are referred to in Lecture 14.

13.9. Cf. op. cit. (note 2.36), pp. 48 and 65, where the values of the latitude are given as 55° 50′ 10″ and 55° 54′ 30″ respectively.

13.10. Ibid., p. 24; but the quote follows observations of 21 April 1582.

13.11. Ibid., p. 74. Flamsteed wrote '... refractiones tam ...' instead of '... refractionem, quam ...'.

13.12. Ibid., p. 111. In the first part of the quote, Flamsteed replaces 'in' by 'solis'; and in the second, he omits 'etiam', replaces 'Wallerus' by 'Waltherus', and transposes the two words 'discipulus olim'.

13.13. Lectures 33, 35–38.

13.14. The results of this expedition are described in Jean Picard, *Voyage d'Uranibourg, ou observations astronomiques faites en Dannemarck* (Paris, 1680).

13.15. Op. cit. (note 2.36).

Lecture 14

11 *May* 1682

In my last lecture I recited breifly what I had collected the obliquity of the Ecliptick to be from ye observations of the Ancients I concluded with ye Tychonick & Made it evident from them that the greatest in his time was not more y^{n} 23°.29′: I shall now proceede where I left of[f] & I doubt not to prove by ye observations of such as have followed him that the same continues unvariable the same very ingenuous Persons have intimated that it is different & alters annually.[1]

Where the Noble Tycho ended o^{r} Countryman Mr Edward Wright An eminent Teacher of Mathematicks in London & well knowne since by his Treatise of ye Correction of the errors of Navigation,[2] began his solar observation that is in ye yeare 1594. the instrument hee used was of Wood something more y^{n} 6 foot Radius. by which on y^{e} 11 & 12 of June the solstitiall dayes of that Age hee noted the suns greatest heights 61°.58′: but on the 12th. of December 1595 his least 14°.58′: which Corrected by y^{e} necessary refraction will give y^{e} greatest obliquity $23.31\frac{1}{3}$ very nearely. which is above two more then it was collected from y^{e} observations of the Noble Dane, the reason of which I cannot attribute to any thing but the Matter of Which his instrument was made. Wood swells & sinkes with every change of weather & more when it is first wrought then afterwards. Wherefore his instrument might alter very Considerably accordeing to the season of the yeare betwixt ye Summer & Winter solstices. I dare not therefore confide in any of his observations findeing his principalls thus erroneous as will be evident not onely by his contemporarys Tychoes but also by those of Lansbergius made some few yeares before him at Tergoes in Zealand[3] who tho hee bee a person infamous for wresteing the observations of others to make them serve his purposes commend his Tables. yet seemes to have dealt faithfully with us in the delivery of his owne, and onely then to have taken the opportunity of wresteing observations to his owne reputation when the Negligence of the observer in not recordeing due circumstances has afforded him an occasion.

Notes on page 246.

In ye yeare 1589 hee observed the suns greatest height at ye summer solstice 61°.58′, at the Winter following 15-02$\frac{1}{2}$: the first being altogeather ye same with that of Mr Wrights corrected by ye allowance of halfe a Minute Refraction. the latter of 3. accordeing as I have hitherto used, gives the distance of the Tropicks 46°.58′ & therefore the obliquity 23°-29′ altogeather ye same yt I collected from Tychoes observations & wholly confirmeing what I Intimated concerneing ye Error of Mr Wrights. Some p[er]sons may perhaps thinke that ye Refractions at Tergoes an Island & foggy place were greater then ors at London & so caused the Sun to appeare higher at ye Winter solstice there then here but to this I have to answer that tho I have observed the refractions in very different constitutions of ye aire, and seasons of the yeare & thereby found it sometimes neare 36′ at other times scarce more yn 26′ minutes in ye horizon, yet in those very different seasons when the sun was so differently refracted in ye horizon I ever found that in 3 degrees of height the refractions came to be still ye same, scarce sensibly more or lesse yn 14 minutes & if they be thus constant so neare the Horizon wee may Well conclude that in all countryes they are the same when the appearance is above 10 deg high. for it cannot be reasonably conceaved that the difference of the constitutions of the aires of any two countrys can make a greater diference of refraction betwixt them, then the different constitution of or aire does with us at distant seasons of ye yeare. but it is not my intent to make a particular discourse of refractions at this time I reserve them to a more proper opportunity & in the meane time I shall assume them as I have hitherto done on a Theory & limitations agreeable to my frequent experience & confirmed by ye agreemt of good observations as I shall hereafter prove.[4]

It is worthy a remarke here that Tycho allowing a faulty parallax but no refraction had stated ye obliquity 23°.31′$\frac{1}{2}$: Mr Wright at first without considering of either onely 30′ above 23 degrees but afterwards heareing of Tychos determinations hee makes the same with him 23-31$\frac{1}{2}$ while Lansberge allowing of Tychoes parallax & some little refraction in the Winter solstice makes only 23°-30′-10″:

Those who since the commence[me]nt of this age have pretended to have observed the obliquity of the Ecliptick most accurately are Albert Lineman Riccioli Cassini Wendelin and Hevelius who all make it something more yn 30 minutes above 23 degrees. that the consent of so many knowne & celebrated Names may not p[er]-suade any one into a false opinion I shall examine theire observa-tions successively as they followed each other in time.[5] whereby it will be evident that the true quantity is not mr. then I have hitherto found it by the accurate observations of others which if I have leasure I may also confirme by my owne.

In ye yeare 1644. June ye 12 or stile at Koningsberg in Prussia.

Albert Lineman observed the suns meridional height 58°.47′$\frac{1}{2}$ the true height therefore Correct by refraction was 58°.47′ praecise.

The latitude of that place hee had observed by y^{e} stars 54°-43′. which being alike correct the true height will be found 54-42$\frac{1}{3}$ & hence the suns greatest declination onely 23°-29′$^{1/3}$ at least an Whole minute lesse then hee esteemed it.[6]

Riccioli in y^{e} 3^{d} booke of his Almagest pag 161 gives us the observed heights of the sun of y^{e} severall solstices of y^{e} yeares 1643.44.45. and 46. which haveing corrected by his parallax twice as big as it really ought to be & wholly omitteing the Consideration of Refractions hee thence Collects the Suns greatest declination 23°:30′ praecise.[7] but

In his *Astronomia reformata* pag. 18. hee ingenuously confesses *utebamur tunc altitudine poli falsa*,[8] that hee then used a false elevation of his pole: how hee corrected it you will find in ye 4th booke of his s[ai]d Ast Ref: pag 246. where hee determines the apparent height of the pole from the observed greatest & least Meridionall heights of the pole star to be onely 44°.30′.10″:[9] which before hee had determined 44°.30′.30″: in that height the refraction is 50″ therefore the true height of the Pole was onely 44°.29′.20″ & of his æquator 45°.30′.40″

On ye 11 of June 1643. with 3 severall Instrum^{ts} hee noted the suns greatest or Meridionall height 69°00′00″ which correct by ye subtraction of 20″ the refraction in that height gave ye Correct Meridionall height 68°59′.40″ from which subtracteing his height of the æquator the greatest declination of ye sun will be found 23°.29′ exactly.

This day at Noone the sun wanted some 20′ of the longitude of ye solsticiall pointe & therefore his declination encreased afterwards about 2″. which is so insensible p[ar]te of a degree in ye best Instruments wee need not regard it.

The next yeare 1644 on ye 10 day of June with ye same instruments hee againe observed the suns Meridionall height 68°.59′.50″: his correct was therefore 68°.59′.30″ but at this time hee was only in 29°.25′ of ♊ so that hee wanted 5″. of his greatest declination. and the Meridionall height for the solstice would bee 68°.59′.35″ therefore the suns greatest declination by this observation is 5″ lesse then 23°-29′

Againe on ye 10th of June 1645 with the same Instruments he gives the suns Meridionall height observed the same it had beene noted by him the same day the yeare praecedeing. the sun was then but in 29°.12′ of ♊ so that hee wanted 10″ of his greatest declination which will thence be collected 23°.29′ exact as by ye observation of ye yeare 1643 June ye 21th.

Lastly in ye yeare following or 1646. on the 11th of June with his greatest quadrant 69 gr. praecies with his middle 69.00.10 the sun wanted now onely 5 minutes of ye longitude of ye solstitiall point in which small arch of the Ecliptick and at that place his declination varies not. So y^t^ wee need not encrease his Meridionall height or declination as in the 2 last examinations on this account. allowing now the same refraction as formerly 20″ the suns correct meridionall height was 68°.59′.40″ or 45″ at most from which subtracting the correct height of y^e^ æquator. 45°.30′.40″ wee have the greatest obliquity 23°.29′ exact or at most but 5″ more ye same as formerly.

About ye same time with Riccioli Gottenfride Wendelin at Folgaquier in Provance. observed the suns nearest Meridionall distance. from ye vertex 20°.24′ the latitude of that place hee had determined from the observed east & greatest heights of the Circumpolar stars to be 43°.54′ whence hee collects the suns greatest declination 23°.30′¼:[10] but if wee make ye necessary allowance of refraction here 20″ in his Solstitiall Zenith distance & 50″ in his height of the Pole wee shall have the suns Zenith distance correct 20°.24′.20″: & his correct latitude 43°.53.10 whose difference is the true obliquity of the Ecliptick by this observation 23°28′.50″ but 10″ less than Ricciolis observations give it

Ricciolis height of the pole was obteined by ye helpe of a large Gnomon made on one of ye Walls of ye Church of St Luke belonging to ye Jesuits Colledge at Bononia. 62 foot high. in the yeare 1655. at Which time Signieur Cassini was alike employd in ye Marquesse of Malvasias house. in ye same towne where by a like contrivance hee the[re] observed the distance of the pole star from ye pole 2°.32′.28″. & the simple latitude of that palace 44°-30′.22″: whence the[y] collected the latitude of ye Church of St. Petronio. 44°.30′.20″: which corrected by ye refraction 50″. gives it[s] true latitude 44°.29′.30″[11]

In this Church in ye yeare 1567. Egnatius Dantes a Personus I had occasion to mention in my readeings last term had contrived a Gnomon.[12] from which the species of the Sun cast on the pavement as it passt a long meridian line shewd his true place. this gave the hint both to Riccioli & Cassini for theire Contrivances the old one being spoild by ye Frequent repayres and alterations of the Roofe & pavem^t^ Signieur Cassini Contrived anew which had these advantages of the former first that it was 71 feet high,[13] which was 5 foot more[14] & then that the Meridian line drawne from ye perpendicular under ye foramen was exact whereas the old deviated not lesse then 9 degrees from ye true Meridian Westward, & that ye pavement upon which it was drawne was truely level which circumstances doe abundantly convince us of the curiosity & diligence of the p[er]son that renewd it & may induce us to believe his observations very accurate

By which on ye 11 of June 1655 hee determined the distance of ye suns center from ye vertex 21°.00′ 04″: which corrected by ye allowance of 20″ refraction makes his true distance 21°-00′-24″: &

this taken from ye latitude 44°.29′.30″ leaves the obliquity of the Ecliptick 23°.29′.06″ agreeing with ye obliquity collected from ye Notes of Riccioli

Againe on ye 20th of June 1658 the observed distance of ye Suns center from ye Vertex was 21°.00′.12″ correct by refraction 21°-00′.31″ which taken from ye latitude 44°.29′½ leaves his present declination 23°.29′.59″ the sun was now in 29°.07′ of ♊ & therefore wanted of his greatest declination 11′. which added to his observed makes the present greatest obliquity of ye Ecliptick 23°.29′.10″ by this observation.

I may not omit to advise those who shall peruse these observations in Riccioli that in his Astronomia Reformata pag 18. ye 2^d^: columme, the observation hee mentions as made ye 21 of June new stile 1658 was really made on ye 22 of June 1655 ye same style.[15] it being exactly the same with it as anyone may p[er]ceave by compareing the place with ye note of that observation pag 26. ye 2^d^ Columme the observation of June ye 21 1658 recorded in the 29 page being noted for dubious & clearely different from this

It was intimated to me formerly & I thinke by Signieur Cassini himselfe that the obliquity of the Ecliptick altered annually to be certeine I wrote to him about it & hee was pleased to informe me in a letter dated November ye 8th 1673. in these Words. *Solis Declinationem Maximam adhuc retineo grad.* 23°.29′ *In Magno Gnomone Bonononiensi apparebat variatio annua secundarum aliquot ut videre est in Ephemeridibus Malvatiae, utrum tamen tribuenda sit differentiae refractionum in Annis diversis an rei ipsi defenire non Ausim.*[16] that is hee reteined the greatest obliquity of the Ecliptick 23°.29′ That in ye Great Gnomon at Bononia there appeared an annuall Change which whether it proceeded from ye different refractions of different yeares or from some reall mutation hee durst not determine but since then I have not heard more of it & therefore conceave it to have proceeded rather from some alteration of ye height of ye pavem[en]t or gnomon by ye swelling or subsideing of the foundation since later observations agree not to any such great change as 6″ p[er] Annum which was ye quantity had beene asserted to mee.

Whilest the Bononian Astronomers were thus employd the learned & accurate Hevelius was carrying on the Like worke at Dantzick with large & curious Instruments of which haveing given us a full account in his Organographia lately published[17] I shall not need to make any description. by these on the 10th of December o^r^ stile in ye yeare 1658 hee noted the Suns least meridionall height at ye Winter Solstice 12°-12′.45″ & found it the same the day following. as also in the yeares 1660. and 61. or within 5.″ an inconsiderable quantity. The greatest hee accordes not so well wth himselfe in. in his severall. yeare. for in 1654 and 55 hee finds 59°-07′.20″- or 25″. in

60. and 61 onely [59°7′5″ or 10″][18] . . . but haveing considered his severall observations and the distance of ye sun from ye solstitiall point at ye time they were made & consequently how much they ought afterwards to encrease. I find the nearest apparent Meridionall height at ye time of ye solstice by his Instruments & observations 59°.07′.20″:

Should we take these observations simply as they lie without any Consideration of Refractions as Bernard Walter did. of old & Tycho in ye yeare 1583. wee should find the obliquity lesse than otherwayes it would be given by ye greatest height compared with ye height of the pole star. for ye least Meridionall height in ye Winter Solstice subtracted from ye greatest in the Summer leaves 46°.54′.40″ for the distance of ye Tropicks therefore the suns greatest declination onely 23°.27′.20″ the height of the Equator 35°.40′ & the latitude or poles height 54°.20′.

But by the greatest height of the pole Star observed by him 56°.55′.25″ on ye 23^d of November the same yeare & the least 51°.50′.25″ on ye 21th of January following the simple height of the pole will be found 54°.22.55″ & therefore the height of the æquator 35.07.05 which taken from ye suns greatest observed height leave the obliquity of the Ecliptick North 23°.30′-15″ neare 3′ greater then was otherwayes found by compareing the two Meridionall heights with each other

But if from this height of the æquator wee take the least solstitiall Meridionall height of the sun as observed 12°.12′-40″: wee shall find the obliquity of the Ecliptick South 23°.24′.25″ neare 6 minutes lesse then it was found by the greatest summer height.

In My last lecture I noted the like incoherence & difference of the obliquitys of the Ecliptick or suns greatest North & Southerne declinations deduced after the same Manner from Tychoes observations by himself. & I told y^u that probably the Consideration of it Caused that greatest person first to find out that theire were refractions in o^r aire & after to thinke of wayes for discovering theire true quantyty & in what proportion they decreased as the appearance ascended from the horizon to the Meridian. the difference betwixt the Northerne & Southerne greatest declination thus investigated from Tychoes observations I found to be above 8 min & should have found it more had his habitation beene More Northerly. for then. the Sun being truely lower in the Meridian would have beene more refracted whereby the apparent distance of the tropicks would have beene found lesse. but the refraction in the height of the pole and greatest height of the Summer Sun altering but little in many degrees change of latitude the greatest Northerne declination would have beene found little different from what it would be determined in latitudes 10 or 20 degrees more Southerly but the Southerne still lesse as the Pole was more raised. & more different from the Northerne why the Simple obliquity of the Ecliptick deduced from Bernard Walters or ye Arabian observations differed so much from Those derived from the later of Copernicus Tychoes & Hevelius his. they lived in lesse latitude & consequently had the sun lesse raised by refraction in the Winter solstice

But to returne to Hevelius. if wee apply o^r^ refractions as to others so also to his observations we shall [find] them altogether coherent & agreeable. the refraction in ye Summer solstice with him is halfe a minute which from 59°.07′.20″ his observed leaves 59°.06′.50″ the true Meridionall height of the Sun in the Summer solstice

And if from 12°.12′.40″ the Meridionall height observed in the Winter solstice wee take [a]way 3′-45″ the refraction. in that height, the residue shall be the Suns true height 12°.08′.55″ therefore the distance of the Tropicks 46°.57′.55″ & the Suns greatest declination 23°.28′.57″: or 23°.29′ exactly the same wee found By Tychoes & Ricciolis observation.

Hence also the height of ye Equator will be found 35°.37′.52″ therefore the poles height 54°.22′-08″ neare which it will be found by his Meridionall heights of the Pole Star.

For by them I collected the simple latitude 54°.22′.55″ the Refraction in which height is 35″. & these subtracted from it leave ye true & correct latitude of his house 54°.22′.20″: but 12″. more then what I collected it from his greatest & least Meridionall heights of the Sun a difference so very small that it Commends both his observations & the[nce] shews the truth of that proportion of refractions I have used which remaines to be further proved hereafter,

The same obliquity of the Ecliptick I found by Compareing ye least distances of ye solstitiall Sun from ye Vertex at Greenwich with ye latitude of the observatory observed by ye polestars greatest and least heights correct by refraction which gave me occasion to enquire for & examine the observations of the ancients. that I might thereby be satisfied whether the obliquity was greater in theire times then now as theire determinations seemed to intimate or whet[h]er they had concluded amisse by reason that they assumed faulty parallaxes & were not acquainted wth ye true nature & proportion & all decrease of refractions as the appearances ascended from the horizon. Nor has my Enquiry been unsucces[s]full it appeareing that could they have managed there observations as they ought to have done they would have found the obliquity altogeather as I have determined it.

The Noble Dane was persuaded that it was greater in the time of Hiparchus then in his age because hee found ye latitudes of Castor & the bright star of the Pleiades lesse by above a quarter of a degree in Hipparchus his Catalouges then his observations gave them as they ought to be if the obliquity of the Ecliptick had decreased being in y^e^

first semicircle of y[e] Ecliptick. but the latitudes of y[e] bright star of y[e] Vultur & y[e] Norther[n]most of the front of y[e] Scorpion diminished from what they are in y[e] ancient Catalogue. but had hee at the same time regarded the latitude[s of] Aldebaran. y[e] Lions heart the Scorpions heart & Syrius hee would have found little reason for his opinion these & severall others rather arguing the obliquity encreased then diminished so that compareing all togeather hee would have beene forced to conclude that the Latitudes & declinations of y[e] fixed stars observed by the ancients were scarce true within y[e] Compasse of a degree & with us that the obliquity of the Eclipticke was invariable.

NOTES

14.1. Riccioli 1, pp. 159–60, cites references to Arzahel, Regiomontanus, Puerbach, Alphonsine, Nonius, Copernicus, Rheticus, Maestlin, Rheinhold, Neander, Moleto, Magini, Snel, Pitiscus, Lansberg, and Lang (in that order) as supporting the idea of a variable eccentricity.

14.2. Flamsteed actually obtained his information from Joseph Moxon's edition of Edward Wright's *Certaine Errors in Navigation* (London, 1657).

14.3. Phillip Lansberg, *Tabulae motuum coelestium perpetuae . . . Item novae et genuinae motuum coelestium theoricae & astronomicarum observationum Thesaurus* (Middelburgi Zelandiae, 1632). The solar observations cited by Flamsteed in the next paragraph are on p. 42 of the last section of this book.

14.4. Lectures 35–38.

14.5. Flamsteed here deleted the phrase 'takeing in also the observation of Gassendus at Marseiles'.

14.6. Riccioli 3, p. 18.

14.7. Riccioli 1, p. 161. The data contained in the table entitled 'Obliquitas Eclipticae ex Nostris Observationibus Bononiae habitis' are the basis of the discussion of Riccioli's results in the following text, f. 142[r,v]; cf. pp. 241–2.

14.8. Cf. note 14.6.

14.9. Riccioli 3, p. 246. This observation was made in the Jesuit College of St. Lucia at Bologna, Italy.

14.10. Riccioli 1, p. 161.

14.11. Riccioli 3, pp. 6, 7.

14.12. A description is given both in Riccioli 3, pp. 5 and 6 and in Riccioli 1, p. 132.

14.13. Riccioli 3, p. 18.

14.14. The words 'then y[e] former' have been deleted at this point.

14.15. Flamsteed actually wrote 'state' instead of 'style'.

14.16. Cassini to Flamsteed, 8 November 1673, in R.S. MSS. C.1. 61.

14.17. Johann Hevelius, *Machina Coelestis pars prior; organographiam*, etc. (Gedani, 1673).

14.18. The missing figures, in square brackets, are the maximum values for the Sun's meridian altitude at 'Solstitium Aestivum' tabulated for the years 1660 and 1661 in Hevelius's *Machina Coelestis pars posterior;* etc., iii (Gedani, 1679), pp. 10 and 12 respectively.

Lecture 15

18 *May* 1682

In my late lectures I have fully proved the obliquity of the Ecliptick to have beene 23 deg. 29′ by the agreeing observations of ye ablest Astronomers of this & the last age, & rendred it more then probable that (whatever the opinions of the ancient seeme to intimate to ye contrary) it was never greater. Those who are but moderately acquainted with Astronomy & understand little more then ye common Rudiments cannot but know how necessary it is to have this prime requisite well limited not can they be ignorant of the different opinions of divers Authorities concerneing it, Which to reconcile it has cost me some paynes. & caused mee to be more prolix then I intended. And now haveing setled this Fundamentall I shall procede to ye superstructure the Doctrine of the Sphere[1] grounded on it which before I enter on, it may be requisite to enquire first what Hypothesis is the most naturall & probable. I have in some of my readeings last yeare toucht upon this subject. but since by reason of the time interlapsed what I then sayd may be slipt out of y^{r}s memorys I shall. resume a part of a discourse I then made[2] & from thence draw such arguments as may convince y^{u} that the Pythagorean susteme is the onely legitimate the rest but suprious.

Thales Milesius was the first amongst the Greekes as history informes us that praedicted the Eclipses of the luminaries for the lunar Eclipses (I meane the duration of such as were Centrall) it was found that the Moones Meane distance from ye Earth was little lesse then 60 of its semidiameters the suns distances they supposed about 5 or 6 times as much as the Moones till Aristarchus the Samian. by observeing the times of the apparent dichotomyes & findeing them but little before the first and after ye second Quadrature proved hence that the suns distance was twenty times as Much as the Moones from the earth. (his booke is extant. & I gave you his demonstration formerly.[3] wherefore I shall not need now to repeate it onely) wee may argue hence. that if the Moones distance be 60 semidiameters of the earth from it the suns shall be 1200. I have formerly proved it to be neare twenty times as much, but that wee

Notes on pages 254–5.

need not stand upon, for it will follow from Aristarchus his distance that it is much more probable that the earth turnes round on its Axis once in 24 houres then that ye Sun with ye whole sphere of ye planets & fixed stars is throwne about it once in that time as the Common & unletterd people imagine

It was found by experiment of the Ancient that about 60 miles measured directly forward on the earths globe answered to one degree of the æquator or Meridian. now one degree of the æquator or 60 miles must passe over the Meridian in 4 minutes of time if the Earth move but if the Sun were supposed to move his . . . be Carried at least 1200 semidiameters of the earth in four minutes of time. which is so stupendious a length that it wholly exceeds humane conception for the swiftest Motion wee perceive on o^{r} earth is that of a bullet from ye mouth of a cannon which yet. is far short of what wee must be forced to allow the revolution of the earth itselfe. for if ye motion of a degree be threescore miles in four minutes of time any point under the æquator moves a fourth part of it or 15 miles in one minute. whereas I doe not know of any experiments that will prove a bullet to be moved a fourth part of that space in the same time. I argue favorably for the antiCopernicans for should I assume the reall measure of a degree as it has beene found by experiment 69 miles and upwards[4] with the Suns distance from ye earth as observation will prove it 20000 semidiameters. the motion of the heavens required by the Ptolemaists would appear exceedingly swifter & therefore as much more improbable

The Consideration of the vast swiftnesse which must be allowed to ye course of the sun & heavens was no doubt the first occasion of that Assertion of Anaximander that ye earth turned round on its prop. hee is the first amongst the ancients that is reported to have asserted any motion of the Earth. but not long after him wee find the Pythagoreans give not onely the diurnall Revolution with him, but the Annuall through the Ecliptick to ye Earth which had beene hitherto allowed for ye Sun. the Reasons for this second assertion were taken From the appearances of the planets. Theire motions sometimes swifte at others slow & sometimes direct at others retrograde & not seldome stationary. all which they found might be represented without ye helpe of any praeposterous or contrary motions in Epicycles as had beene before them imagined supposeing each planet & the earth with them carried round the sun in perfect lines. & in Continually progressive motion. The Manner of salveing such appearances is so well knowne & so common that I hope I need not explicate it by any Figure. it haveing beene done by all the writers of Astronomy that have enterteined this hypothesis fully & at large. Amongst whom it is our unhappinesse that none have undertaken to shew how all the diurnall appearances of the sun and stars are made in this susteme but the reason is evident from the practise of the learned Kepler who haveing shewd us in his Epitome[5] how all the usuall Circles of the Sphere are formed & may be derived from y[e] Pythagorean susteme . . . proceedes to explicate them after the Manner of Ptolemys Followers. the old method would not be easily shaked of by reason it was better accommodated to the then prejudiced judgm[ts] of ManKind & therefore hee retained it not that hee thought it better but more easy to be admitted & understood.

Tis now neare fifteene Moneths. since haveing an opportunity of the presse I tooke occasion to publish a small tract[6] wherein I have endeavored to make it evident that the solution of the diurnall appearances of the sun & stars is not more difficult in the Pythagorean then in the Ptolemaick hypothesis nay much easyer but by y[e] practise of y[e] publisher that booke can come into y[e] hands of few. that the Doctrine of the first part of it may be more common I have resolved to explicate it in My readeings here. but the 2[d] which concernes eclipses I shall not meddle with that haveing beene treated by a very ingenuous freind not long since. in this place,[7]

The propositions of y[e] sphere are most naturally explicated by the Globe it selfe. but that being no portable Instrument projections of it have beene contrived by Ingenuous men which answer & represent what ever is shewed by the Globe not lesse truely & sometimes more exactly. these are of two sorts. Either *Orthographicall* which are described by perpendicular lines falleing from every point of those Circles of the Globe which y[u] would have represented on the plane y[u] would project them upon, or *Stereographicall* which suppose the eye placed on the superficies of the globe & infinite streight lines radiateing from it to each pointe of the Circles to be projected & (continued to y[e] plaine of the projection) one of this sort I made use of it requireing onely y[e] rule & compasses or streight line & Circular in its description for which reason I shall still employ y[t] projection.[8] In that discourse I onely taught how to delineate that projection I used. my leasure not permitting mee nor the businesse requireing more. Since then considering the nature of the lines I had described & More of theire properties then I had then thought how insufferably tedious the learned Clavius has beene on this subject. in his treatise de Astrolabio[9] & how troublesome it would be to a learner to find what hee desired in him. I thought I could not doe better then to resume the businesse of stere[o]graphicall projections & as concisely as I could to demonstrate the properties of the lines I should have occasion to describe in that I intended to make use of this therefore I resolve to doe & that I may not p[er]plex y[u] with such an apparatus of Lemmata as hee has heaped up before his worke. what propositions I shall have occasion to use that have been demonstrated in Euclid or y[e] writers of Conicks[10] I shall onely quote from them such others as I shall find requisate I shall take up & demonstrate as I have occasion in my progresse.

But that I may at the same time shew the reason of y[e] Ptolemaick or Steoflers,[11] & Mr Oughthred[12] stereographicall projections. I shall first shew how all the Circles of Ptolemyes Projections may be derived from ye Pythagorean Systeme

To represent the diurnall appearances. In which the Axis of the earth is supposed to lie alwayes inclined to ye plane of the Ecliptick at the same angle & in what place of the orbite so ever she is posited to keepe allwayes parallel to it selfe. as if in the first figure [Fig. 15.1] the line bac. represent the orbite of the earth viewd by the eye placed at an infinite distance in ye said plane. c. the place of the earths center in ye 1st point of ♑. pcs. its Axis formeing the angle pcg 66°.31′: with bc ye plane of the orbite or aecliptick for I proved formerly in one of my lectures last terme & in ye 2 last of this[13] that the angle pce which the Axis of the earth formed with ye perpendicular to ye Ecliptick was equall to ye Suns greatest declination which by the Consent of the exactest observations I could meet with I also demonstrated to be 23°.29′.

In this position as the earth turnes round on her Axis ps. the sun is successively Verticall to every point of the parallell gf. which is every where 66-31es distant from ye pole p. this therefore wee may Call the Northerne tropick.

Wherever the earth removes the Axis keepes alwayes the same inclination to ye plane of the Ecliptick if therefore wee suppose her to have moved from c to a a quadrant of ye ecliptick, the center a of ye earths diske is now p[er]pendicular under ye sun and as the earth turnes round each point in ye parallel aeq is successively under the sun or has him in its vertex. Now each point in this parallel is 90 degrees from either of the poles p or s. this therefore represents the æquator. But if wee conceave the earth removed a Quadrant further to b. Shee will be then in the first point of ♑. & the pointes on the earth which as shee turnes round passe under the sun or have him in theire vertex will be 23°-29′ more removed from ye north pole or so much distant from ye æquator. towards the south so here wee have the southerne Tropick hi. described on the globe. & the plane of the Ecliptick connecteing the two tropicks thus described on it gbi. representeing ye ecliptick it selfe.

And if through e ye pole of the Ecliptick wee draw the line er parallel to ye Æquator. it shall passe over all those pointes on the earth which can have the pole of ye ecliptick in theire vertex & consequently represent the polar Circle Northerne. the Southerne may be described after the same Manner. the Axis ps. also passeing through the intersection of ye Ecliptick & Æquator shall be the equinoctiall Colure. & the periphery passeing through the solsti[ti]all points the solstitiall [colure.]

So wee have got the principall Circles represented in ye Analemma or orthographicall projection of ye sphere. on ye plane of the Meridian in the Ptolemaic Susteme. it were easy to shew how all the rest must be derived hence. but I leave that to exercise the learner. tis not my intent so much to shew how Nature may be represented by feigned contrivances as how it reall[y] worke[s] all its performances & how they may be clearly apprehended & understood.

For which purpose I shall first show y^u the grounds & reasons of stereographicall projections which being done it will not be difficult to apprehend how any of those circles I made use in the projection I described in my worke may be delineated lastly I shall shew theire uses.) & if in the first part of the taske I impose on my selfe with my pure Astronomicall contemplations. I shall either mix or take up and demonstrate some wholly Geometricall propositions I hope I shall not be thought to deviate from my duty all Astronomicall Conclusions being grounded on Geometricall demonstrations. I hope however the practise of ye Geometry lecturer may excuse mee who thinkes no fault when hee pleases to reade sometimes to divert from the abstracted & v[ery] controvertible rules of pure Geometry to the pleasant conceites of undemonstrable Physicks[14] first then:)

if on any plaine a circle be described as agbf in ye 2^d figure [Fig. 15.2]: & if from a point above it innumerable streight lines be conceaved to be extended to every point in ye periphery of the said Circle those lines shall Include or describe a cone. in which it is demonstrated by Mydorgius in ye 2^d and 3^d propositions of ye first book of his Conicks.[15] that if it be Cut by a plaine either parallel to ye base. or subcontrary that is so as the triangle vse make be like to the whole vba. but placed subcontrary the section shall be a Circle as is also ye section of ye base. whence it will follow That if from ye eye placed in the superficies of the Globe as at b̲ in ye 3^d. figure [Fig. 15.3] wee conceave a line extended through ye Center, bκp. and the Globe be cut by a plane γl or gλ perpendicular to this axis or line bkp. if any circle be described upon the superficies of the globe & from ye eye at b infinite streight lines be produced to each point of ye periphery of this Circle the said lines. conteined to this plane if requisite shall describe a Circle also upon it.

For let gf be the diameter of a Circle described on ye globe. draw y^e lines bg. bf. the triangles bhg and bgf are similar. for y^e angle at b is common to both the Angle gfb. subtended by the arch of y^e periphery gb. is equall to y^e angle hgb: subtended by the equall arch of y^e periphery bλ by y^e 27 of y^e 3^d booke of Euclid[16] therefore also fgb = to y^e angle ghb. for all y^e 3 angles of a triangle are equall to two right angles. the two triangles fgb ghb similar. & the section of the Cone gfb. made by y^e plane gλ is subcontrary to the section of y^e base gf and a Circle as it also is

But y^e plane or section through the Center γ is parallel to gλ therefore the triangles ghb.γ are likewise similar & the lines drawne from the eye at b to every pointe in the Circle whose diameter is gλ describe a Circle also on the plane whose diameter is $\gamma\alpha$ whence it will necessaryly follow that whatever Circles on the Globe shall be described any plane perpendicular to y^e Axis bp by infinite lines radiateing from y^e point b to each pointe in the said Circle the same lines as they passe through the plaine shall describe a Circle on it. and all those lines which are perfect Circles on ye globe shall be truely Circles in all stereographicall projections.

Hence also it will be evident that the distance of any pointe projected in this plane from ye center shall be equall to y^e tangent of halfe its distance from y^e pole. let the place of the point f be required a line drawne from y^e eye at b to f gives it in the plane at α, from y^e center κ draw the line κf. by y^e 20 of y^e 3 of Euclid.[17] the angle αbκ is halfe y^e angle fκp which is measured by y^e arch fp. but $\kappa\alpha$ is y^e tangent of y^e arch αbκ to y^e Radius κb = κp. or the semitangent of y^e arch pf whence it [is] plaine y^t y^e distance $\alpha\kappa$ is y^e tangent of halfe y^e arch fp. as was asserted: Haveing thus shewd y^t all y^e Circles on the globe which lie not in the same plane with y^e eye shall be Circles in this projection & by what pointes they made be bounded I may now apply these easy Theorems to the projection I am to delineate. in which I suppose the eye at d the lower of Southerne pole of the ecliptick. [cf. Fig. 15.1]

As the earth turnes round each point of her super[fi]cies describes a Circle in the contiguous aether London or the parallel of latitude passeing over it is 38 and an halfe from ye Pole of the world[18] this may be represented by ye line lo in ye first figure. to delineate this parallel with ye Ecliptick in a projection on its plane.

Because each point of ye ecliptick in its plane is equally distant from ye earths center & the plane passes through it at right angles to the axis the Ecliptick it selfe shall be a circle whose center shall be the center of the projection. this therefore may be represented by ye Circle ♈ ♋ ♎ ♑ in ye 4th figure [Fig. 15.4].

Through e ye center & pole of the Ecliptick draw y^e [line] ♈ e ♎ and at right angles to it through ye center also ♋ e ♑ of these wee may call the first the aequinoctiall the latter the solstitiall Colure from theire resemblance to the like Circles in ye old projection

The Pole of the World is 23°-29′ distant from ye pole of ye Ecliptick or the Axis of the one is so much enclined to the Axis of the other. count the said 23°-29′ from ♈ towards ♋ in the periphery of the Ecliptick to C lay a ruler over ye opposite point[s] ♎ and C its intersection with ye solstitiall Colure ♋ e ♑. gives the place of the Northerne pole in p.

The parallel of London being 38½ degrees from ye pole. set of[f] 38½ from ye point C on each side it ye periphery of the Ecliptick to ye Capitalls M and X a Ruler layd over these pointes. and ye point ♎ gives two other pointes mx in ye solsti[ti]all Colure through which the parallel of London or the path of its vertex passes.

Now because from what I said before it is proved that this path or Parallel shall be a Circle in this projection wee have found the two extreme pointes in its Diameter. the Middle between them shall be its center. bisect therefore the distance betwixt xm in κ & thereon describe a Circle passeing through the said points x̲ and m̲. this shall represent the said path or parallel. by which all the diurnall appearances of the sun & stars may be represented as theire semidiurnall arches. heights and *Azimuths* but to treat of these at present I should be engaged in a long discourse. I shall therefore here conclude & defer it till another readeing for it will be requisite to shew how some other Circles may first be drawne & divided & to explicate the susteme more fully which I intend to doe in my next God assisteing me

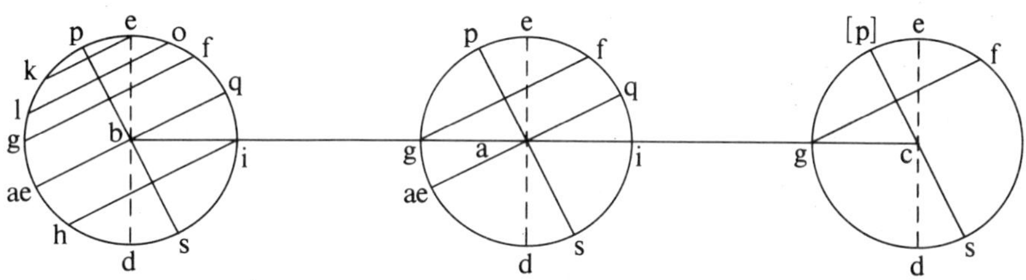

FIG. 15.1

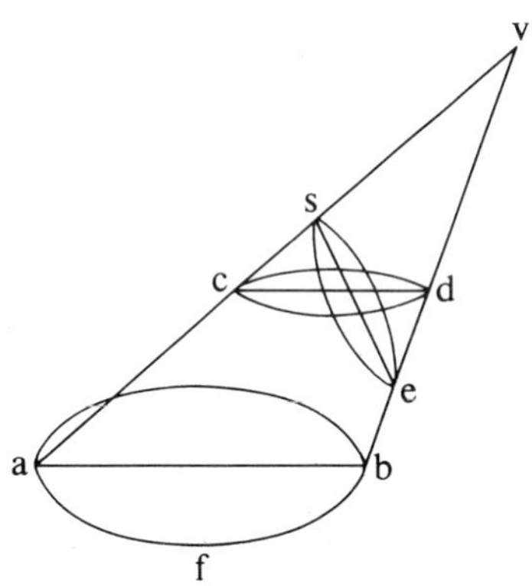

FIG. 15.2

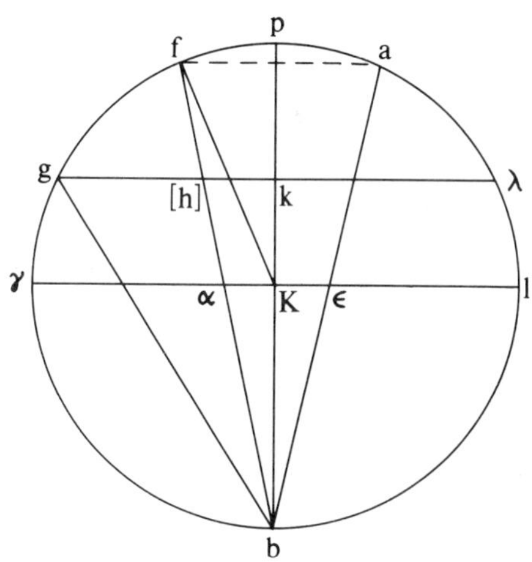

FIG. 15.3

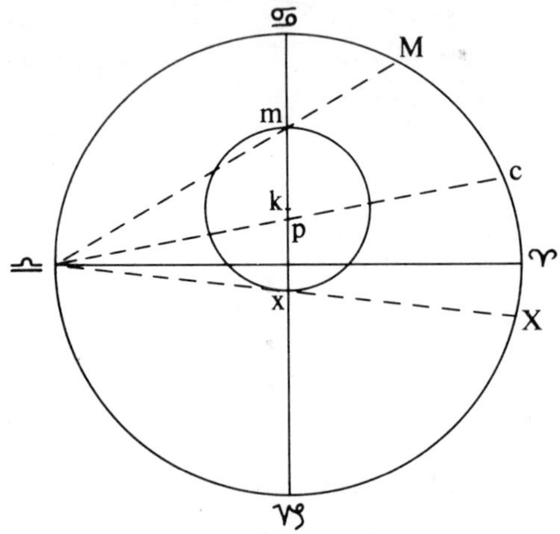

FIG. 15.4

NOTES

15.1. Flamsteed's 75-page tract entitled 'The Doctrine of the Sphere . . .' is published in the first volume of Jonas Moore, *A New Systeme of the Mathematicks* (London, 1681).

15.2. Lecture 1 (27 April 1681) and Lecture 9 (16 November 1681).

15.3. Taken from Riccioli 1, pp. 109, 110. Cf. Lecture 1, ff. 10^{v}, 11^{r,v}, p. 87.

15.4. An account of this experiment was published anonymously by Jean Picard in *Mésure de la Terre* (Paris, 1671). 'A Breviate' of its contents, in English, later appeared in *Phil. Trans.* x, no. 112 (25 March 1675), 261–72.

15.5. Op. cit. (note 2.15).

15.6. Cf. note 15.1.

15.7. Flamsteed may be referring to a lecture, or lectures, delivered by the former Gresham Professor of Astronomy Christopher Wren, who held that office from 1657 until his appointment as Savilian Professor of Astronomy at Oxford in 1660.

15.8. Flamsteed has crossed out the following remark: 'The learned Clavius in his Treatise de Astrolabio [cf. note 15.9.] has discoursed amply of Stereographicall Projections.'

15.9. Christopher Clavius, *Astrolabium* (Romae, 1593).

15.10. Flamsteed possessed a copy of Isaac Barrow's English translation of *Euclide's Elements* (London, 1660), and there is evidence in Lecture 16 that this was the edition which he normally consulted. He also had Barrow's edition of *Archimedis Opera. Apollonii Pergaei conicorum libri III* (Londini, 1675). Another writer on conics to whom he refers explicitly below was Claude Mydorge (cf. note 15.15).

15.11. Johann Steoffler *Elucidatio fabricae ususque Astrolabii* (Parisiis, 1585). Flamsteed could have consulted this particular edition of Stoeffler's popular work in Sir Jonas Moore's library.

15.12. William Oughtred's 'Horologiorum sciotericorum' is contained in his *Clavis Mathematicae*, of which Flamsteed possessed a copy of the fourth (Oxford) edition of 1667.

15.13. Lectures 12, 13 and 14.

15.14. This is a criticism of Robert Hooke, who apparently liked to lecture upon experimental physics whenever the excuse arose.

15.15. Claude Mydorge, *Prodromi catoptricorum et dioptricorum:sive conicorum operis* (Parisiis, 1641), lib. 1, propos. 2, 3, pp. 4–6. Cf. note 6.10. An English translation of Books I and II is appended to [Jonas] *Moor's Arithmetick* (London, 1660); and a full translation, to op. cit (note 15.12).

15.16. 'In equall circles GABC, HDEF, the angles standing upon equall parts of the circumference AC, DF, are equall between themselves, whether they be made at the centres G, H, or at the circumferences, B, E'. Cf. Barrow's 'Euclid' (op. cit. [note 15.10]), p. 67.

15.17. Ibid., p. 63 (Prop. XX): 'In a circle DABC, the angle BDC at the centre is double of the angle BAC at the circumference, when the same arch of the circle BC is the base of the angles.'

15.18. $38\frac{1}{2}$ is the complement of the latitude of London (viz. $51\frac{1}{2}°$) adopted by Flamsteed.

Lecture 16

25 *May* 1682

In my last discourse I began to explaine the diurnall appearances of the sun & Stars & to shew y^{u} how they were p[er]formed by Nature her selfe represented in the Pythagorean hypothesis & because it is requisite for this purpose to understand the several projections of the Sphere. the Orthographicall being easy & not so proper for my purpose I tould you I should make use of ye stereographick. These have beene discoursed of by severall Authors Clavius[1] first but with that prolixity that renders him taedious I had almost said nauseous to a learner my intent was therefore to give y^{u} ye groundes more concise & breifly omitteing all unnecessary propositions. but because pure Mathematicks relish not so well as mixt I thought it might be convenient to refresh y^{u} sometimes with interperseing the uses of such propositions as I should demonstrate: accordeingly I shewd y^{u} how all the Circles of ye Globe made use of in the old Projections might be derived from ye Pythagorean hypothesis onely here was the difference that the Ancient[s] suppose the Globe to represent the heavens & the Circles described on it. such Circles as are really formed on them whereas my discourse. described & fixed them on the earth to which they properly belong & whereon I must desire y^{u} to conceave them delineated so often as I hereafter mentioned them.

I shewd y^{u} then how the pole of the earth might be designed in any stereographick projection the eye being placed in ye southerne pole of the Ecliptick, & also how any parallel of latitude, or as I call in more properly in my published treatise,[2] the path of any vertex might be layd downe in ye said projection. [(] its distance from either of the poles being onely given) before I proceed any further I shall shew y^{u} how the equinoxes with the longest & shortest days are represented in this *projection* In ye fifth figure [Fig. 16.1] ♈♋♎♑. represent the earth on ye 4 Cardinall pointes of ye Ecliptick. with the path of London or its vertex projected in it the eye being supposed in ye southerne pole of ye globe. The Earths Axis I tould y^{u} keepes allwayes parallel to it selfe. this parallel position is represented by y^{e}

Notes on pages 265–6.

arch p̄ē of the solstitiall colure intercepted betwixt ye 2 poles of ye earths globe & ecliptick keeping parallel to its selfe in all positions. now the sun (supposed to be placed neare ye center of this orbite of ye earth) can illuminate but one halfe of the earth at once. So then when shee is placed at ♎ whence shee views the sun on the first point of ♈. onely one halfe of the path or parallel of London is illuminated. & its inhabitants begin to see the sun ariseing when the place or vertex at m passes from the obscure hemisphere of the diske into ye illuminate. & whilest it is transiteing by 1 to B it is day. but when it passes from ye illuminate into ye obscure part of ye haemisphere againe the sun appeares to set & ye night commences. wherefore I call ye line tupemd which seperates the illuminate from ye obscure part of ye globe the Horizon of the diske.

In this position the illuminate part of the Pathe or parallel is equall to ye obscure & therefore the day equall to ye night, & the pole p̄ begins to be illuminated. the line which parts the illuminate hemisphere of the Earth from ye obscure tpemd coincedes with the solsti[ti]all Colure I call it in all places the Horizon of ye diske.

As the Earth removes forwards in her orbite. the Axis p̄ē keepeing as I said before allwayes parallel to it selfe. the pole becomes still more illuminate or enters deeper into the illustrated Diske. when ye earth arrives at ♑ on ye Coelestiall Solstitiall Colure whence shee views ye sun on the first point of ♋ the pde approaches nearest to ye sun. & now the longest dayes are made in all places betwixt ye Equinoctiall & the Northerne pole p̄ for ye horizon of ye diske ted coincedes with the æquinoctiall Colure on the Earth. & that parte of the path of London or its parallel which falls within the illuminate haemisphere is double to that which is included in ye obscure. so that the day is double the length of ye Night at this time at London.

As the Earth removes hence towards ♈ the pole approaches nearer to ye horizon of the diske into which it falls when shee is arrived there: the Horizon now againe coincedes with the solstitiall Colure. & the one halfe of the path as also of all yt can be described on ye Earths globe will be all halfe illuminated so yt ye day will be equall to the night in all places betwixt ye poles. Hence allso the sun appeares on the first point of ♎ & the second æquinox Commences.

As ye earth proceedes from hence to ♑ the North Pole is immerged into ye obscure part of ye diske the horizon againe coincedes with ye æquinoctiall Colure on ye earth & all the Northerne parallels or paths become unequally illuminated that is the obscure parts of them are greater then the light & consequently the nights so much longer yn ye dayes. at ♋ the earth views the sun in the first pointe of ♑ & the second solstice is made. but as the earth proceedes from this place or point the pole againe drawes nearer the horizon of the diske. the northerne pathes becomeing more & more illuminate till she arrives at ♑ the opposite pointe of the Ecliptick. but in the southerne hemisphere of the Earth & to its pole all things happen contrary for as the one pole approaches to the horizon of the diske the other recedes from it So y^{t} all the phainomena of the increase & decrease of ye dayes with ye solstices & æquinoxes may be represented here onely by supposeing the earths continually progressive motion in her orbite & her diurnall about her Axis conceaveing it allwayes parallel to it selfe & preserveing the same constant Inclination to the plane of ye Ecliptick.

And now tis evident that without any helpe of this plane all these appeareances may be as easily & p[er]haps more conveniently represented. for the plane of the orbite eeee. encompasses ye earth round, whose center is every where in it, & therefore a great circle is alwayes described round ye earth. which wee may call y^{e} Ecliptick it selfe. in which the pointes opposite to y^{e} Sun in y^{e} 4 Cardinall positions are each a quadrant distant from y^{e} other. wherefore the earth being placed at ♎ wee may take y^{t} point at i. opposite to y^{e} sun for y^{e} first point of ♈. that at $\bar{t}$ 90 degrees distant from it for y^{e} first point of ♋. at $\bar{o}$ 90 degrees further or y^{e} opposite to ♈ for y^{e} first point of ♎ & y^{t} opposite to ♋ at $\underline{d}$ for y^{e} first point of ♑ for wherever in y^{e} ecliptick the earth is placed the point thus numbred in y^{e} earths periphery shall be in y^{e} line passeing to y^{e} sun & that shewes his place & a line drawne at right angles to it through the Center shall be the horizon of y^{e} diske. onely it must be remembered that the solstitiall & æquinoctiall Colures are alwayes kept the same & unalterable. & the line of Direction $\overline{pe}$ connecteing the two Northerne poles of y^{e} earth or æquinoctiall & Ecliptick still in the same position or parallel to its selfe.

In my last lecture I showd y^{u} how to describe the Path or parallel of any place. I have now shewd y^{u} how the equall or unequall lengths of the dayes or nights accordeing to ye earths different positions in her orbite may be represented by it; It followes that I shew you [how the] diurnall and nocturnall Arches. . . . for so henceforth I may call them) are measured. The revolutions of the Earth about her Axis are equall exceeding that small inaequality ariseing from her unaequall motion in her orbite. & ye unaequall culminations of aequall arches of ye Ecliptick which I have demonstrated some yeares since & which wee need not how consider. So y^{t} aequall arches of ye Equator or equall angles at ye pole are alwayes formed by ye revolutions of the Earth in equall spaces of time. it will be therefore requisite to shew y^{u} first how the Æquator or a Circle equally distant from both the poles of the earth may be described in this projection. how the houre Circles which divide it equally or forme equall angles at ye poles may be drawne as also any lesser Circles, & lastly how they may be divided into equall degrees. for which purpose it will be requisite to premise 2 or 3 Geometricall Theoremes.

Each great Circle on the globe or such an one as divides its superficies into 2 equall parts. Cuts also every other great Circle described on it into 2 equall p[ar]tes and therefore every great Circle to be described in o^r projection shall cut the Ecliptick into 2 equall p[ar]tes and bisect its periphery. I affirme then.

Theor. 3.[3] that ye secant of ye greatest distance of any great circle from ye other great Circle in which the projection is to be drawne shall be the semidiameter of ye great Circle to be projected for In ye 6th figure & for ye 3th Theoreme [Fig. 16.2][4] let ch represent the plane of the great Circle ceg extended & continued, on which the projection is to be drawne. t its northerne & d its southerne pole. let the eye be supposed at d. and let bf be another great Circle to be projected on this plane. from y^e eye at d draw y^e two lines db. and df continueing the latter till it intersect the plane at h. the pointe p shall shew the one intersection & h the other of the said greate Circle & y^e diameter of y^e Circle ceg. produced cg

Because y^e angle bdf is subtended by bf = y^e diameter of y^e said Circle it shall be a right angle. & the point d. shall be in the periphery of a Circle whose diameter shall be the line ph. by 31. e. 3.[5] bisect therefore ph in g̅ & thereon with y^e Radius pg = gh. describe the Circle htp: it shall also passe through the point. d & the line dg shall be = gp = gh. for by y^e 15 def. of. el.[6] all the raies from y^e center to y^e Circumference of a Circle are equall:

The plane of the projection ch I supposed at first to stand at right angles to y^e axis de wherefore the line de bisects y^e triangle pdh into two similar triangles & the Angle pde is equall to ye angle dhe by ye 8. e. 6.[7] the angle egd. ye double of ye angle ehd by 32. e. 1[8] and ye angle edg its complement to a quandrant by Dr Barrows 3^d. Corollary to y^e said proposition.[9]

Now e̅g̅ y^e distance of y^e Center of the projected Circle from e̅ y^e center of y^e projection is the tangent, and gd its semidiameter, y^e secant of y^e Angle edg to y^e Rad[ius] ed.

But by y^e 20. e. 3[10] the angle bet is y^e double of the angle bde = therefore to its complem^t bec = to y^e angle edg the Complem^t of y^e angle egd. y^e double of ghd = edb. whence it follows that the secant of the greatest distance of any great Circle from y^t other Circle on which y^e projection is drawne shall be the semidiameter of the said great Circle in the projection & that the tangent of its said nearest distance set of[f] from y^e center the contrary way shall give y^e center of the projected circle. for pe. y^e semitangent of its greatest distance with eg. y^e tangent of its complement are = to pg = gd y^e secant of its complem^t.

Hence followes the Method of describeing the hourcircles in y^e old stereographicall projections on any plane passeing through y^e poles of y^{e} globe let such a plaine be represented in y^{e} 9th figure [Fig. 16.3] by y^{e} Circle $\bar{p}$[11] $\overline{m}\bar{a}\bar{p}$ which let be the Meridian, p[11]a the Axis, & let it be required to describe an hourcircle 2 houres or 30 degrees distant from the Meridian.

Count 30 degrees from y^{e} Pole p. to b. & therethrough from y^{e} center draw y^{e} line et. from $\underline{p}$ also raise y^{e} perpendicular or draw y^{e} tangent line q[12]t. so is $\bar{q}$[12]$\underline{t}$ y$^{[e]}$ tangent $\underline{te}$ y^{e} secant of 30 degrees.

Take y^{e} length tq[13] betwixt y^{e} Compasses & set it of[f] backwards from y^{e} center in y^{e} æquator from e to g. there shall be the Center of y^{e} hourcircle required:

Take y^{e} line et or y^{e} secant of so many degrees betwixt y^{r} Compasses & setting one foot on $\bar{g}$ with y^{e} other y^{u} may describe the arch of a Circle pha which shall passe through the two poles $\bar{p}$ and $\underline{a}$. and shall therefore be the hourcircle required. this method is generall & will hold for any other required accordeing as was before demonstrated:

But if on y^{e} contrary it were required to draw a lesser Circle or parallel 30 degrees distant from the Pole. then the secant of 30 degrees set of[f] from y^{e} center shall give the center of the parallel & the tangent of 30 degrees its semidiameter. which may be thus demonstrated.

In y^{e} 7th figure. & y^{e} 4th Theor. [Fig. 16.4] let apl represent a semicircle of y^{e} Meridian. $\bar{p}$ y^{e} pole $\overline{po}$ $\frac{1}{2}$ y^{e} Axis pb = pk. y^{e} distance of y^{e} parallel from y^{e} pole. 30 degr. continue $\overline{op}$ and to p draw y^{e} tangent line hpd.

From l. y^{e} supposed place of y^{e} eye. as also from the other extremity of y^{e} æquator, & $\underline{o}$ y^{e} center of the projection. to b & k draw y^{e} streight lines ab; ob, ok lb. lp. lk. continueing this last til it intersect the plane of y^{e} Axis produced in i. and the rest till they cut the tangent line in d.c.e.g.h. the point $\underline{m}$ where m the intersection of op y^{e} plaine and y^{e} line lb is the nearest distance of the projected little Circle from y^{e} center. of the projection. oi its greatest. theire difference $\underline{mi}$ is the semidiameter of the Circle passeing through the points b and k. which is = gd = hc the double tangent of the distance of y^{e} parallel from y^{e} pole or y^{e} arch $\underline{pb}$ = $\underline{pk}$.

For in y^{e} 2 △^{s} dlp, ilp the angles at $\bar{l}$ are equall because the[y] stand on equall arches of y^{e} periphery bp. pk. by y^{e} 27. e. 3.[14] and the side lp. is common. but the angle dpm = ipg is right & the angle opl = gpl. halfe a right angle therefore y^{e} angle ipl = to y^{e} angle dpl. and all the angles & sides of the one △^{e} to y^{e} respective sides & angles of y^{e} other. therefore dp = pi. but the angles at $\underline{d}$ and $\underline{i}$ are also common to y^{e} 2 triangles dpm. ipg & the angle at $\bar{p}$ in both right. therefore mp = pg and the whole line mi = gd.

Againe in y^{e} △^{e} abl. y^{e} angle at b is right because subtended by the diameter so is also y^{e} opposite dbe in y^{e} little △^{e} within y^{e} periphery by y^{e} 15. e.i.[15] and because the angle cbe is = to abo and dbc to obl.

the base de shall be bisected by y^e line ob produced in c. as the base of y^e triangle abl. is bysected equally in o so y^t ce = ed for by y^e 2. e. 6[16] similar Δ^s have three sides proportionall

Further because y^e arch bp = pk the angle okl shall be equall to y^e angle abo therefore the angle gkh = cbe but oc = oh. and ob = ok therefore bc. = kh: but y^e angle bce = ocp = ohp is = khg therefore gh = ce = cd and gd = ch which was to be proved.

But I have before demonstrated that the tangent of any arch with y^e semitangent of its complem^t is equall to y^e secant of y^e said Arch. whence it follows that pc the tangent of y^e arch pb. added to om. the semitangent of its complem^t to a quadrant shall be equall to oi y^e secant of y^e said arch bp. if therefore takeing the lenght of y^e secant oc betwixt y^e Compasses y^u extend it on the Axis from o to q it gives the Center of y^e parallel. & if with y^e distance pc = to y^e tangent of y^e arch pb y^u strike [from q] the arch of a Circle bmk it shall be the parallel y^u define.

Or other wayes if from b either in this or the 9th figure [Fig. 16.3] y^u draw y^e tangent to y^t point bq. where it intersects y^e axis in q. it gives y^u y^e center & bq the semidiameter of the parallel. for the triangles cpo and qbo in ye 7th figure [Fig. 16.4] or etq and ebq [in Fig. 16.3] are both similar and equall:

It will not be unplesant here to remarke the difference betwixt ye Circles that have theire poles in the Diameter, & those that have them in the periphery of the projection. In the first the secants of the angles they make with ye Meridian are their semidiameters & the tangents of theire said distances set of[f] the Contrary way from the Center in ye Diameter give theire Centers. but for the other the secants of theire distances from ye pole set of[f] from ye center in ye Axis give theire centers. & the tangents of theire distances from ye pole are theire semidiameters.

(I have beene thus long in demonstrateing the manner of findeing ye semidiameters & centers of the parallells because it is not done clearely in Clavius[18] but dependant on other propositions which was needelesse.) I shall now give two more easy Theoremes which will be of use for describeing such Circles whose Centers lie in the Diameter of the projection & so Conclude.

If two Circles Intersect each other the angle formed by them at the point of Intersection shall be equall to ye angle comprehended by two lines drawne thence to theire 2 Centers.

Let the 2 Circles be sdf and sgh in the 8th figure [Fig. 16.5]. the Center of the first at a of ye latter at e. from the point of intersection s draw ye line sa and at right angles to it the line bs. this shall be a tangent to ye Circle sdf in y^t point & coincedes with its periphery. For ye smallest arch of a circle wee can imagine is not Materially different from a streight line

From ye same intersection s to e ye Center of the other Circle sgh.

draw ye line se and at right angles to it ye line cs. this shall be a tangent to the same Circle in that point & shall therefore coincede with its periphery for some space. conceave it as small as y^u please. it will be sufficient for my purpose & since the two tangents coincede with the peripheripherys in the point s. it will follow that the sphericall angle g[19]si is equall to ye streight lined angle csb comprehended betwixt ye two tangents cs and bs

But ye angle cbs formed betwixt y^e two tangents is equall to y^e Angle eas.[20] made by y^e meeteing of the two lines es and as. drawne from y^e Center of each Circle to y^e Intersection for y^e angle cse is right by construction. So also is bsa. take away y^e Common angle bse. & there remaines the angle csb = eab.[21] w.w.d[22]

I have one Theoreme still more easy then this to prove which is That if from each end of a streight [line] y^u draw other lines which shall forme the same angle at each end. & from the pointe of theire intersection or Meeteing y^u let fall a p[er]pendicular to the said streight line. it shall bisect the said line into 2 equall p[ar]tes. Let AC. in ye 10th figure [Fig. 16.6]. be the line proposed. from each end of which or ye pointes a̲ and c̲ let the lines ab. and cb. be drawne so as the angles cab. and acb may be equall to each other from the pointe of theire concurrence or Intersection b. let fall the line bp. perpendicular to ac. this shall cut the line ac into 2 equall partes in p̲. for y^e angles pab. pac. are equall by Construction. y^e angles cpb. and apb are right and the side pb. common to both y^e triangles, therefore by y^e 4.e.1[23] the side ap = pc. and the whole line ac bisected equally in p. as was to be proved.

These 7 Theoremes[24] are all that I have found necessary for the descriptions of any Circles of the Sphere Stereographically. had I time I would have demonstrated an 8^[th]. for the division of all such Circles as shall be drawne: but this being somewhat more p[er]plext & long then the rest I shall remit to a future opportunity.

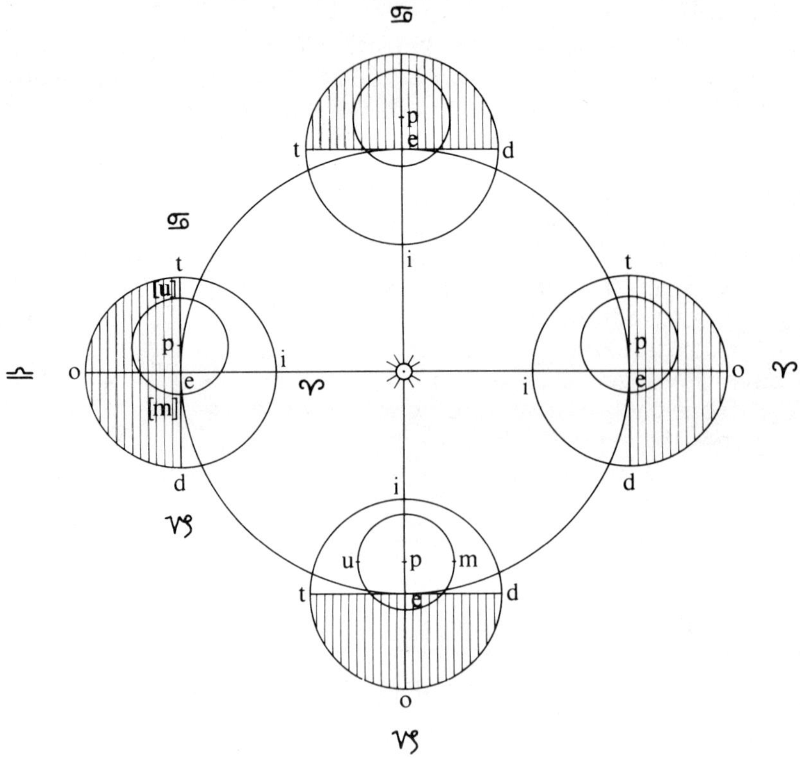

FIG. 16.1

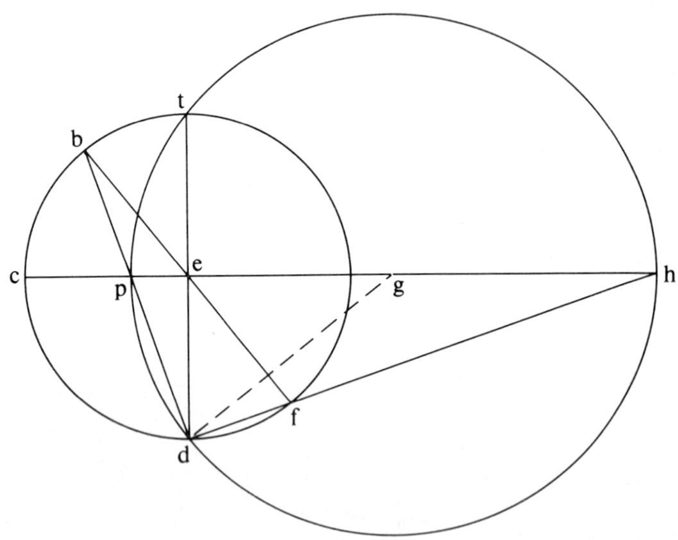

FIG. 16.2

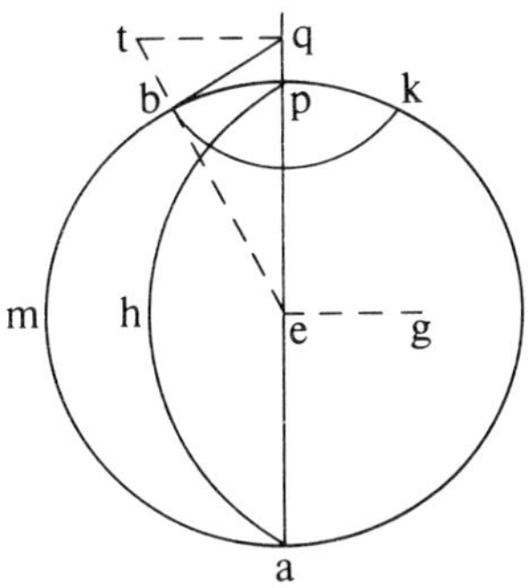

FIG. 16.3

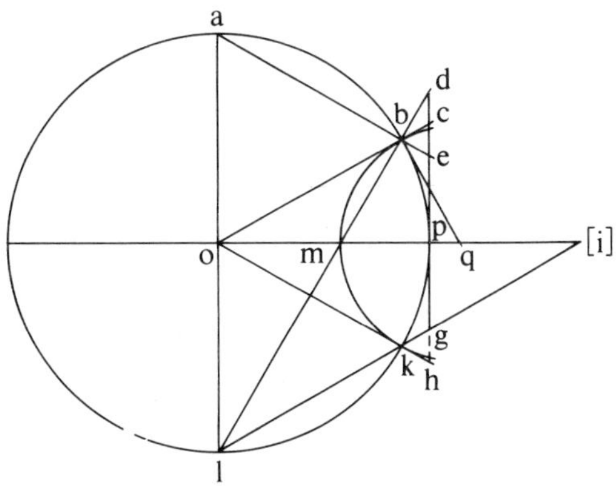

FIG. 16.4

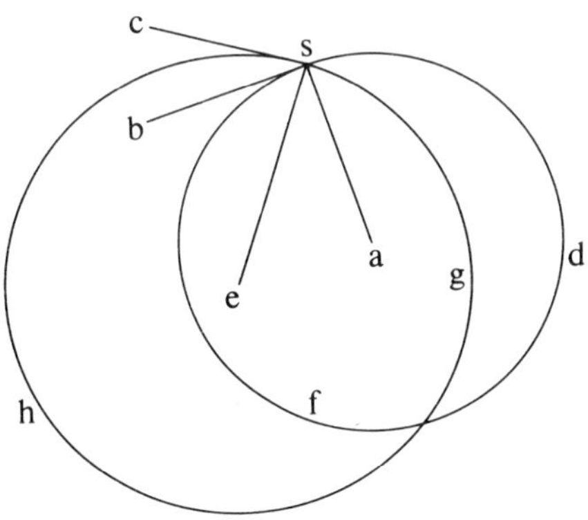

FIG. 16.5

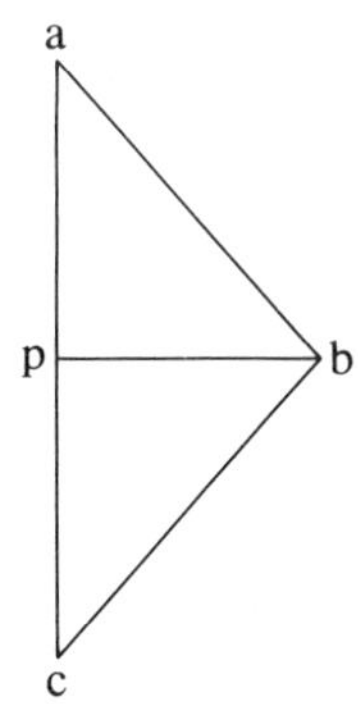

FIG. 16.6

NOTES

16.1. Op. cit. (note 15.9).

16.2. Op. cit. (note 15.1).

16.3. Theorems 1 and 2 correspond to the illustrations Figs. 15.2 and 15.3 respectively.

16.4. The drawing on f. 148r (accompanied by references to three theorems in Euclid) and that on f. 156v (labelled Fig. 6) are identical.

16.5. 31.e.3. means the 31st proposition of the third book of Barrow's 'Euclid' (op. cit. [note 15.10]). Cf. p. 69.

16.6. Barrow's 'Euclid' (ibid.), p. 2. Definition XV: 'A Circle is a plain figure conteined under one line, which is called a Circumference; unto which all lines drawn from one point within the figure, and falling upon the circumference thereof, are equall the one to the other.'

16.7. Ibid., p. 120. Prop. VIII: 'If in a right-angled triangle ABC, from the right-angle BAC there be drawn AD a perpendicular to the base BC; then the triangles about the perpendicular (ADB, ADC) are like both to the whole triangle ABC, and also one to the other.'

16.8. Ibid., p. 25. Prop. XXXII: 'Of any triangle ABC one side BC being drawn out, the outward angle ACD shall be equall to the two inward opposite angles A,B. and the three inward angles of the triangle, A,B, ACB, shall be equall to two right angles.'

16.9. Ibid., p. 26. 'If one angle in a triangle be right, the other two are equall to a right. Likewise, that angle in a triangle which is equall to the other two, is it self a right angle.'

16.10. Ibid., p. 63. Prop. XX: 'In a circle DABC, the angle BDC at the centre is double of the angle BAC at the circumference, when the same arch of the circle BC is the base of the angles.'

16.11. Flamsteed wrote $\overline{qmap}$ instead of $\overline{pmap}$.

16.12. Flamsteed wrote p instead of q three times in this sentence.

16.13. Flamsteed wrote e instead of q.

16.14. Barrow's 'Euclid' (op. cit. [note 15.10]).

16.15. Ibid., p. 16. Prop. XV: 'If two right lines AB, CD cut through one another, then are the two angles which are opposite, viz. CEB, AED, equall one to the other.'

16.16. Ibid., p. 116. Prop. II: 'If to one side BC of a triangle ABC be drawn a parallel right line DE, the same shall cut the sides of the triangle proportionally (AD.BD::AE.EC.) then a right line DE joined at the sections D, E, shall be parallel to the remaining side of the triangle BC.'

16.17. Flamsteed wrote p instead of q.

16.18. Cf. note 15.9.

16.19. Flamsteed wrote ksi instead of gsi.

16.20. Flamsteed wrote eas instead of esa.

16.21. Flamsteed wrote eab instead of esa.

16.22. w.w.d. is Barrow's conventional abbreviation for 'which was demonstrated'.

16.23. Barrow's 'Euclid', p. 10. Prop. IV: 'If two triangles BAC, EDF, have two sides of the one BA.AC equall to two sides of the other ED,DF, each to it's correspondent side (that is, BA = ED, and AC = DE) and have the angle A equall to the angle D conteined under the equall right lines; they shall have the base BC equall to the base EF; and the triangle BAC shall be equall to the triangle EDF; and the remaining angles B, C, shall be equall to the remaining angles E, F, each to each, under which the equall sides are subtended.'

16.24. From the internal evidence in Lectures 15–17, Theorems 1–7 would appear to be associated with: Figs. 15.2, 15.3, 16.2, 16.4, 16.3, 16.5, 16.6 respectively (in that order).

Lecture 17

Wednesday, 28 June 1682

In my last lecture save one I began an explication of y[t] part of the Copernican hypothesis which shews how y[e] diurnall appearances of the sun and stars are formed & represented. I told y[u] these would be best apprehended by the Globe but because that is no portable Instrument Ingenuous men & skillfull in these sciences had contrived Planispheres or Projections whereby all the performances of nature and problemes of the sphere might be represented and answered as easily and sometimes more exactly then by the Globe it selfe. that these projections were of two sorts—either orthographicall which supposed y[e] eye infinitely distant & were designed by perpend[i]culars falleing on a plane passeing through y[e] center of y[e] globe from all the Circles drawne upon it, or stereographicall. which supposed the eye placed any where in y[e] superficies of the Globe and infinite streight lines extended from it to all the said Circles whereby they would be projected on any plaine however placed supposeing onely those lines continued where need was. of these I tould y[u] the Orthographicall were the more difficult to project by reason all those Circles which lay not in the same plane with y[e] eye would be ellipses which require infinite streight lines to designe them. whereas in the Stereographicall they would be Circles of which the great ones y[t] bisect the globe would have theire Centers & semidiameters given, all the rest 2 points in the extremitys of theire diameters whereby all the Necessary lines in this projection would be drawne with the helpe onely of the Rule & Compasses & of this sort of projection for the easinesse of its protraction I designed to make use

The Stereographicall projections I told y[u] had beene treated of by the learned Clavius in his booke de Astrolabio[1] but that his many lemmata & the tediousnesse of his discourses commonly Nauseated his reader: I had long since observed that what ever hee has demonstrated might be proved from a farre lesse apparatus & the whole businesse of stereographicall projections reduced into a much lesse compasse: I informed yu that haveing sometimes since published an explication of that part of the Pythagorean hypothesis that

Notes on pages 274–5.

relates to the diurnall Phainomena of y[e] luminarys stars & planets, which by y[e] practise of y[e] Editor Could come into y[e] hands of few I Made use of Stereographicall projection in it I thought it might not be an unacceptable worke to explaine it here further. but that I might not onely doe what I had done before I intended to mix the fundamentalls of Stereographicall projections with it & to shew not onely the reason of the projection of those lines I had made use of but of such others as are commonly described in Gemma Frisius[2] Steoflers[3] or Mr Oughthreds[4] Projections. I entered on this taske I had imposed on my selfe & in order to it haveing shewd y[u] first how the two tropicks the æquator & parallel of London were formed in the Copernican Theory I next shewd you how the Parallel of London might be projected on the earths diske the eye being placed in ye southerne Pole of the Ecliptick. but to doe this I proposed the two following Theoremes.

The[orem] 1. That if the basis of any Cone be a Circle the subcontrary sections of the same Cone are also Circles. this is demonstrated by all the writers of Conicks to whom therefore I referd y[u] for y[e] proof [of] it

2. The[orem]. The eye being placed in y[e] superficies of the Globe if a streight line were drawne from it through the Center. and a plane placed at right angles to this line, infinite lines passeing from y[e] eye to any Circle described on the globe would also describe a Circle on the said plaine as they passed through it because the plane was a subcontrary section of that Cone which had the given Circle for its base as I then Proved.

Hence it followed that the parallel of any latitude or the path of any vertex being a perfect Circle on the Globe the line of its projection should also be a Circle I showed then how its diameter might be got & how to describe it

In my next lecture[5] I shewed y[u] how the unequall lengths of the nights & days in any place on the Earths globe might be represented by this path of the vertex as I call it in my discourse.[6] to limit theire quantity at all times of the yeare it will be necessary either to have the hourcircles drawne on the Globe or the periphery of this path divided. Either way will serve . . . in which the hourlines are commonly used I chose to show first how they might be described. for which purpose I proposed four other Theoremes to be added to the other two of which therefore y[e] 3[d] in order was

Theor[em] 3. [cf. Fig. 16.2] That y[e] secant of y[e] complement of y[e] greatest distance of any great Circle from y[t] other great Circle whose pole is in y[e] center of y[e] plane. or on which the projection is to be delineated shall be the semidiameter of y[e] great circle to be projected. & the tangent of the said distance set of[f] from ye center of y[e] plane in the Diameter the Contrary way. shall give its Center which I also demonstrated.

From this Theorem I deduced & shewed y^u the Method of describeing ye hourcircles in the Stereographicall projection on ye plaine of ye Meridian. or in the Astrolabe of Gemma frisius.[7] to compleate this the description of the Parallells or lesser circles was wanteing. I demonstrated next a 4th Theor[em] [cf. Fig. 16.4] that ye secant of ye distance of any lesser Circle from its pole. set of[f] from ye center of ye projection, in ye Axis would give its Center. & that the tangent of its said distance was its semidiameter. which I did abstracte & independent from any other Theoreme. this have beene neglected by Clavius.

The 5th Theoreme [cf. Fig. 16.3][8] I proved was that if two Circles Intersect each other the Angle formed by them at the point of Intersection shall be equall to ye angle Comprehended by two lines drawne thence to theire two centers. The 6th or last Theoreme [cf. Fig. 16.6][9] was that if from each end of a given streight line, 2 other lines be drawne formeing equall angles with it. & if from theire Common intersection a perpendicular be let fall on ye given line. it shall bisect. the said given line into two equall p[ar]tes. These 6 theoremes are all I could thinke necessary for the protraction of the Hourcircles Circles of Longitude. Azimuths or parallells or to speake breifly any greater or lesser Circles in stereographicall projections a 7th I intended to have added where any of these Circles might have beene divided but I shall first from these shew how the Hourcircles may be drawne in o^r New Stereographicall projection & when wee shall have occasion to divide any Circle in o^r projection I shall give y^u both it and three severall Methods of doeing it. thence deduced.

I shewd you in my last lecture that the Circle described about ye earth by ye plane of the Ecliptick might be taken for ye Ecliptick it selfe. Conceave it so hereafter and that both it and all the lines I shall describe are formed in a rete or sphere Contiguous to the earth whose axis & Colures are allwayes carried parallell to ye Axis of ye earth & the Colures of ye orbite.

That great Circle which passes through the pole of the Globe & the suns place is the Meridian. & when ever any vertex or place on the globe passes under it tis, then noone at that place. Let us first conceave the sun viewd from ye earth to appeare in the first point of ♋. his place in ye periphery of ye earth or its ecliptick will be at ♋ in y 11th figure [Fig. 17.1] & the solstitiall Colure it selfe ♋ P ♑ shall be the meridian. under which when London passes tis noone there.

All the hourcircles on the Globe meet in its two [poles] . . . y^u to find out the place of ye pole p̄ & to describe the path of the vertex nηxkd.[10] conceave them designed as in the upper Circle of the 11th figure [Fig. 17.1] & let it be required to describe the other hourecircles. but first of all the 6th or that which cuts the Meridian in the Pole at right angles. Now because all the hourcircles must passe thorough both the poles in the projection as they doe through both the Poles on the Globe it will be requisite first to shew the place of the other or southerne pole of the projection.

To find the Northerne count 23½ the distance of the two poles of the globe & ecliptick from ♈ to c. and layeing a ruler over the pointes ♎ and c. it cuts its place in ye solstitiall colure at P.

if from c. through ye center e. wee draw ye line ceπ. this may represent the Axis of ye globe and π its southerne pole. distant from the southerne pole of the Ecliptick & the eye supposed to be placed at ♎ 23½. also. draw the line ♎ πA this by the 3^d Theor[em] shall make a right angle with ye line ♎ Pc for the angle subtended by ye diameter πc is right. & shall give the place of the southerne pole in ye solstitiall Colure produced at A through which all the hourcircles shall passe.

By the same Theor[em]. ye points P♎A shall be in the Periphery of a Circle whose diameter is ye line PA which shall be the 6 a Clock hourcircle: for the Tangent to the Circle A♎P at y^e point P stands at right angles to ♋ P ♑ the tangent of the Meridian whose center is infinately distant & therefore the Circle A♎P. is also at right angles with y^e Meridian or 6 houres distant From w^{ht} I demonstrated [on y^e 3^d Theoreme it is evident that TP = T♎ the semidiameter] of this hourcircle is equall to y^e secant of 66½ degrees the complement of the Arch Px. and eT the distance of its center from y^e center of y^e projection e, to y^e tangent of y^e same arch. the Radius being e ♋

And if Through T y^e center of this hourcircle wee draw the straight line qTw at right angles to the solstitiall Colure ♋T ♑ in this line all the Centers of ye other hourcircles will be found. for. by the 5th Theoreme.[11] any two Circles intersecteing each other forme the same angle at theire intersections that the lines drawne from theire Intersections to theire centers make with each other, & by ye 6th if from each end of a streight line another streight line be drawne so as ye angles formed at each end may be equall to each. other the p[er]pendicular let fall from the point wherein the two streight lines so drawne meet shall bisect the first into two equall p[ar]tes. Now all the hourcircles meeting in each of the poles shall forme the same angles with ye sixth there. that the lines draw[n] from each pole to theire centers make with ye meridian or solstitiall Colure in which wee found ye center of ye 6th & therefore theire centers shall all be found in ye line qTw produced for that bisects ye line PA equally which is so plaine that if I had not before demonstrated it would have beene evident from this very figure.

if therefore it be required to ye strike ye 4th hourcircle From ye Meridian when ye sun is on ye soltitiall point Consider that this is 60 deg[rees] from ye Meridian or 30 from ye sixth.

Make Then TP ye Radius of a tangent line. & takeing from it ye Tangent of 30 deg. set it of[f] in ye line qTw from T on each side to y and. z. set one foot of y^r Compasses on eather of these points suppose z & extendeing the other to either of y^e Poles A or P. strike y^e arch of a Circle AylPs. this shall passe through y^e other pole & be the hourcircle desired for by y^e 5th. y^e Angle ♎ Pl = sP ♈ which it formes with y^e 6 a Clock hourcircle is equall to y^e Angle TPZ = TAZ Which by Construction was 30 degrees.

Hence the Method of describeing either the hourcircles in Mr Oughtreds[12] or the Azimuths in Steoflers[13] projection is given. for in the first if y^e center e of y^e plane be the Zenith then, the tangent of y^e latitude to the Radius of y^e projection e ♑ set of[f] from ye Center in ye Meridian gives T ye Center of ye six a Clock hourcircle. & a line drawne through T shall be that in which the Centers of all the other hourcircles will be found to determine which make TP. the distance of ye Pole from ye said line the Radius of a line of Tangents, the tangents of 15.30.45 to y^t Radius gr &c. set of[f] in ye same line shall give the Centers upon which if y^u set one foot of y^r Compasses Extending the other to ye pole y^u may therethrough strike any hourcircle required. The same Method serves in Steoflers projection onely here the Eye is supposed in the Southerne pole so the Northerne shall be in the Center & the hourcircles will be streight lines because they lie in ye same plaine with ye eye but the Zenith shall be as the Pole in the other & the Azimuths will be Circles whose centers may be found as the Centers of the houreCircles in the other all of them lieing in the line passeing through the Center of the Prime Verticall & being set of[f] by the tangents of theire distances from it. but to returne from theire feigned contrivances to the representations of Nature.

Whereas the Meridian in the Ptolemaick Projection is fixed the Parallells of declination Mutable here ye Contrary the Meridian alters dayly & one parallel or the path of the place in the Earths diske serves to represent all the diurnall appearances all ye yeare long. I have shewed how they are when the earth is on any of the four Cardinall pointes of her orbite. now let us see how they are made when shee is on any intermediall & how the Meridian with the sixth or any other hourcircle is to be drawne.

Admit the Sun to be in ye Middle of 8. or 45 deg. from ye first point of ♈. count 45 deg. from ♈ in the periphery of the Ecliptick to ☉ it there gives y^u the point of the Ecliptick opposite to ye Sun. from which through ye center draw ye line ☉ en. continueing it till it intersect ye tangent line in q̄ this wee may Call the Nonagesimary line for a reason I shall give y^u by & by.[14]

Crosse this line at right angles in ye center e. with ye streight line Her. continueing it likewise till it touch ye tangent line in w this seperates the illuminate p[ar]te of the Globe from ye obscure I therefore Call it the Horizon of the Diske.

Setteing one foot of ye compasses on w extend the other to ye Pole

P. & with that extent describe ye Arch of a Circle P⊙n A this shall cut the Nonagesimary line where it intersects ye Ecliptick & shall be the Meridian to that place of the Sun.

Removeing one point of y^{r} compasses to ye point q. extend the other againe to either of the Poles with this extent y^{u} may describe ye arch of a Circle HPr this shall intersect the former or Meridian at Right angles and shall therefore be the 6 a Clock hourcircle to this place of the sun.

I have shewed you now how the principall Circles are described hence tis obvious that as the earth turnes on her Axis from West to East whilest London moves through the obscure hemisphere from ḡ by ā to ō tis night. but as soone as it passes the horizon of the diske at o. where it is 90 deg. from the sun hee appeares riseing, as it transits the 6 a clock hourcircle at x tis 6 in ye morneing when it arrives at f on the Nonagesimary line the sun is in the Nonagesime or highest point of the Ecliptick whence I gave the line e f ⊙ its denomination. when it passes the Meridian at d tis Noone for then ye sun the vertex & the pole are all in ye same plane. when it arrives at α where it againe passes the 6 a Clock hourcircle tis 6 houres afterNoone & when it comes to ḡ tis 90 deg. from ye Sun and passes out of the illuminate into ye obscure hemisphere againe then hee appeares to set & night commences. I might here shew how other hourcircles with ye parallells of twilight might be inscribed but those I intend to adde as my order & progresse shall require. at present I shall shew how the prime requisites of the sphere are represented hereby & how they may be exactly investigated:

All questions of the sphere suppose the place of appearance knowne. if it be the sun. his is allwayes in the Ecliptick & the requisites are of two sorts. In generall such are. his distance from the North pole of the Globe. his right ascention, and the Angle which the Meridian falleing on his Center makes with ye Ecliptick. or particular and different in different latitudes. such are his amplitudes aescent[ional] differences heights & Azimuth at given houres or the Contrary I shall shew how these are formed & Calculated in theire proper order. & therefore the Generall first.

Betwixt the Meridian ⊙ P. the solstitiall Colure P ♋ & the Ecliptick ♋ ⊙ wee have a triangle formed in which are knowne ⊙ ♋ the Complem[en]t of the suns distance from ye first point of ♈. ♋ P the Circle of Pe the distance of the poles of the Globe & Ecliptick with ye angle at ♋ right to find all the other unknowne p[ar]tes of the triangle of which the angle at P is thet difference of ye Right ascention of the sun and the first point of Cancer. the side P ⊙ is the suns distance from ye Pole, & the Angle P ⊙♋ the Angle of the Meridian & Ecliptick: for ye first by ye demonstrated proporties of sphericall Triangles it will hold R : s: ♋ P : : ct ♋ ⊙ : ct ♋ P ⊙ or As ye Radius. To ye sine of 66°.31′ : ye Comp^{t} of ye distance of ye two Poles

So y[e] tang[t] of y[e] Suns longitude from y[e] Next Aequinoctiall point in this place of y[e] sun 45.

To y[e] Tangent of his right ascention from the said point. 42°.31′½

For y[e] 2[d] Requisite or the suns distance from y[e] Pole. P☉ in y[t] said Triangle. to find it

R : cs ☉♋ : : cs ♋ P : cs ☉ P y[t] is As y[e] Radius To y[e] sine of y[e] suns longitude from y[e] Next aequin[octiall] point so y[e] sine of y[e] Distance of y[e] two poles

To y[e] cosine of the suns distance from y[e] Next. which will therefore be found when y[e] sun is in the middle of ♉ 73°.38′: whose Complement. 16°.22′ is the measure of the Arch Pi. the distance of the Pole from y[e] horizon of the Diske. which is aequivalent to what Wee commonly however improperly call the suns present declination.

The 3d Requisite is the angle ♋☉ P. which the Prime meridian P ☉ formes with y[e] Ecliptick. for the findeing S☉ ♋ : R : : t, ♋ P : t ♋☉P or As the Cosine of y[e] suns longitude from ♈ or ♎ to y[e] Radius

So is y[e] Cotangent of y[e] distance of y[e] two Poles To y[e] tang[t] of y[e] said Angle of the Meridian & Ecliptick: Time will not p[er]mit mee to enter on y[e] Particular requisites of these therefore next opportunity

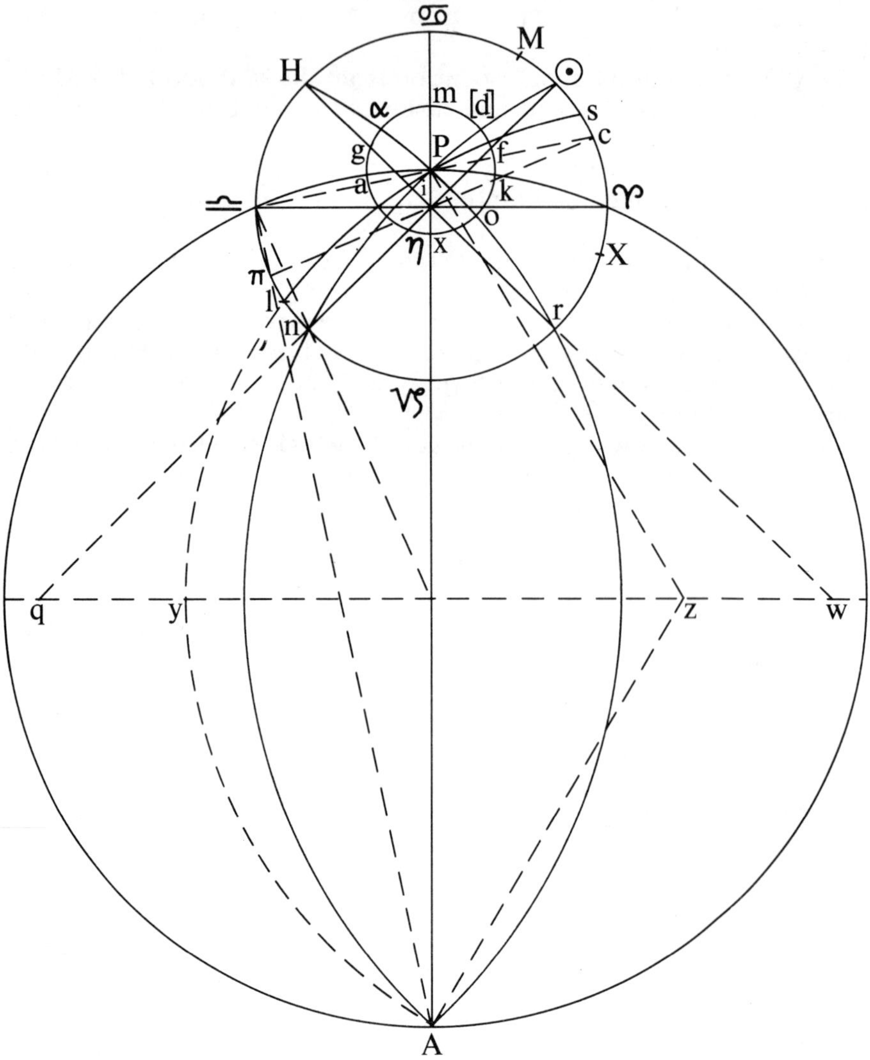

FIG. 17.1

NOTES

17.1. Op. cit. (note 15.9).

17.2. Probably a reference to Gemma Frisius, *De astrolabo catholico liber* (Antverpiae, 1556).

17.3. Op. cit. (note 15.11).

17.4. Op. cit. (note 15.12).

17.5. Lecture 16 (25 May 1682).

17.6. Op. cit. (note 15.1), p. 5.

17.7. Cf. note 17.2. and Fig. 16.3. This is what Flamsteed, in his previous lecture, called Theorem 5 (or Fig. 9). Now he is not numbering it as a theorem.

17.8. Previously denoted as Theorem 6 (or Fig. 8).
17.9. Previously denoted as Theorem 7 (or Fig. 10).
17.10. $m\eta\chi dm$ would seem to make more sense.
17.11. Cf. note 17.8.
17.12. Op. cit. (note 15.12).
17.13. Op. cit. (note 15.11).
17.14. The reason is that the line in question passes through the nonagesimal point.

Lecture 18

Wednesday, 25 *October* 1682[1]

Preface

In my readeings here last terme I taught you the grounds & shewd ye Reasons of the Orthographicall projections of the sphere. & because the usuall or Ptolemaick were onely contrivances to represent the appearances of ye stars & luminaryes I rejected them & shewd a new one concerneing which I have some time since published a small treatise[2] wherein the appearances are answered as they are formed by the Annuall & diurnall revolutions of the earth. I had preceeded to the Generall appearances of the suns Right ascention distance from either of the Poles & such others as are the same in all places of the Earth at the same time the Particular I had mentiond but not Medled with I shall therefore begin now where I ended then

In my last lecture I shewd y^u^ how the generall requisites of the sphere or such as are the same at the same time to all the Inhabitants of the earth were found in the Pythagorean susteme. these were the suns distance from ye North pole of ye globe his Right ascention & ye angle which y^e^. Meridian passeing through his Center formed with ye Ecliptick & required no more then his longitude & the distance of ye pole of ye globe from ye pole of ye Ecliptick to be given. I come now to shew y^u^ how the particular which alter & are different in different latitudes are formed & may be found in the same susteme by the helpe of the same stereographicall projection which I have taught y^u^ how to describe in my former readeings these require the Latitude of ye place & the suns distance from ye pole to be given. Such are his distance from ye vertex at noone. at 6 a Clock & his Azimuth then his distance from ye vertex when hee is on the prime vertical or due East & West & the time of his being so. his Ascentionall difference or the time of his riseing & setting & Amplitudes then; his distance from ye vertex and Azimuth at any given time of the Contrary of which I shall discourse in order and as the time shall permit mee.

First then in determineing the suns Meridionall distance from ye vertex at any time & in any given place. it is to be noted. that whilest his place in the Ecliptick falls in any of ye 6 Northerne signes. or

Notes on page 284.

betwixt ye first points of ♈ & ♎ the Northerne pole hes Continually in ye illuminate diske and his distance from it is lesse y^n a Quadrant by the reflection. but when his place falls in the Southerne Signes or betwixt ♎ & ♈ hee more than a Quadrant from it by the reflection and in all places ye distance of the place or path from ye visible pole subtracted from ye suns distance from the said pole gives his meridionall distance from ye Zenith or vertex For example:

if ye suns place be supposed as formerly at ☉ in ye 12th fig [Fig. 18.1] the meridian described as before was taught will be ☉dP.n. & the sun shall then appeare in the Meridian when London or ye vertex as I call It in my treatise[3] passes the point d where its path cuts ye meridian. the Arch ☉P or the distance of ye sun from ye Pole I before determined 73°.38′: the distance of the pole from y^t Path or Parallel Pd is constantly 38°.30′ which taken from ☉P. leaves the suns Meridionall distance from ye vertex d☉ 35°.08′

But if the suns place were in the other semicircle the Case were otherwise. for then his distance from the pole would exceed a quadrant as much as it wants of one in the opposite point. Conceave him at n in y^e middle of Scorpio. the semicircle Hnr which was before obscure shall here be illuminated. Pn the distance of y^e Sun from y^e visible pole is the Complement of P☉ to a semicircle in this Case therefore 106°.22′:x is the point in which the vertex approaches the sun nearest. Px = Pd = 38.30 taken from Pn leaves nx the suns nearest distance from y^e vertex at Noone 67°.52′

In those places which lie within the Polar Circles at some times of the yeare the sunsets not. but crosses the Meridian under the pole above the horizon. and this case will alwayes happen when the summe of the distances of ye Sun from ye pole & the pole from ye Path is lesse then [a] Quadrant for then that summe it selfe shall be the suns greatest distance from ye vertex in the Meridian & under the visible Pole. but theire difference the least distance from ye vertex above or higher then it.

The two next requisites I proposed to be found are ye distance of ye sun from ye vertex & his Azimuth from ye Meridian at ye hour of six. Now the sun appeares on ye six a Clock hourcircle when ye vertex passes these two pointes in the path where ye Circle strooke at Right angles to the Meridian intersects it these are ye points a and k in ye 12th figure [Fig. 18.1] through which the Suns place & its opposite pointe in the Ecliptick let the Circles na☉, nk☉ be drawne: so have wee a payre of equall triangles P☉k. and P☉a. described in which are knowne P☉ the suns distance from the Pole Pk Pa: the poles distance from y^e Path equall to y^e complement of the latitude. and the Angles at P right. and the side k☉ = ☉a. the suns distance from y^e vertex with y^e Angles Pk☉ or Pa☉ his Azimuth from y^e North Part of y^e Meridian required: to find which

by y^e^ knowne proprieties of Sphericall Triangles it will hold. first for ⊙k y^e^ suns distance from y^e^ Ver[tex]
R : cs P⊙ : : csPk: cs ⊙k:
that is[4]
As y^e^ Radius
To ye sine of ye Reflection. or ⊙ declination
So ye sine of ye Latitude
to ye sine of ye suns height at ye houre of six. which if ye latitude be supposed as before 51-30 and the Reflection 16°.22′ will be found 12°-44′⅓ But to find the Azimuth at 6 or ye angle ⊙kP. it will hold[5]
s, kp: R : : t, k⊙ : t⊙kp : y^t^ is
As ye cosine of ye latitude
To ye Radius
So ye tangent of ye suns distance from ye Pole To ye tangent of his Azimuth from ye North Merid[ian] which the latitude & Reflection being put as before will be found 79°.38′⅓ And because that part of the path intercepted betwixt it and the Pole is the part of the meridian ye Azimuth shall be counted from ye North:

Therefore in what part of the Path soever it happens y^t^ ye Meridian shall be found to cut the line passeing from ye vertex to ye sun at right angles hee appeares due East or West. or as the Ptolemaicks would speake on the prime verticall

Let now two arches of Circles ⊙bn. ⊙ln in ye 13th fig [Fig. 18.2] be so strooke as ye may passe through ye suns place & the opposite point of the Ecliptick & but just touch the Path on each side of it the points of Contingence at b̲ and l̲ shall be the places in the path from whence hourcircles strooke through ye pole shall intersect the Circles passeing from ye zenith to ye sun at right Angles wherefore when ye vertex passes these points b̲ & l̲ the sun appeares due East or West. & now wee have formed another payre of triangles P⊙l. P⊙b in which wee have sufficient given to determine both the Suns distance from the vertex when hee is due East or West & the time of his being so. For in ye said Triangles are given ⊙P ye suns distance from ye pole of ye globe Pb. or Pl the distance of ye pole from ye path. with ye right angles at b̲ and l̲ to find ⊙b or ⊙l ye suns distance from ye vertex when hee is due East or West. and the angles ⊙Pb. or ⊙P.l. measureing the time before or after Noone at which hee is so. to determine ye first or b⊙ it will hold[6]
cs Pb : Rad : : cs P⊙ : cs b⊙ y^t^ is
As ye sine of ye latitude
To y^e^ Radius
So y^e^ sine of y^e^ Reflection.
To y^e^ cosine of y^e^ suns distance from y^e^ vertex which supposeing y^e^ Latitude and reflection as before will hence be found 68°.54′:
But for findeing y^e^ time before or after Noone at which the sun

appeares due east or West, y^t is y^e angle $\odot$Pb = $\odot$Pl: the proportion will be.[7]

ct Pb : R : ct $\odot$P : s $\odot$ Pb.

As y^e Tangent of y^e Latitude
To y^e Radius
So y^e tangent of y^e Reflection.
To y^e sine of y^e hour from y^e 6. or y^e cosine of y^e hour from Noone of this appearance: by which proportion admitteing y^e latitude & reflection as before stated y^e said angle will be found $76°.29'\frac{1}{2}$ which converted into time gives $5^h05'58''$ for the time before or after noone . . . East or West.

The Motion of y^e Earths Rotation about its Axis being as I before have told y^u from West to East the sun then appeares riseing when y^e vertex passes that point of y^e path at o̲ where it cutts y^e horizon of y^e diske and setting when it passes the other Intersection of y^e Path & Horizon at ḡ. Conceave the arches of a payre of hourcircles strooke through the said Intersections of y^e Path & Horizon of the diske. as in y^e 13th figure [Fig. 18.2] as also two other great circles through $\odot$n and the said Point so shall there be two Triangles formed on. each side [of] y^e Meridian y^t is $\odot$Po. oPi. on y^e orientall or ascendeing side. and $\odot$Pg. gPi on y^e occidentall. in either of which wee shall find sufficient data for determineing the time & Azimuth of the suns riseing or setting for in y^e larger triangles $\odot$Pg. $\odot$Po. are given $\odot$P y^e suns distance from y^e pole, Po = Pg the distance of y^e pole from y^e Vertex or Path with y^e sides $\odot$o. $\odot$g Quadrants to find Po$\odot$, or Pg$\odot$ the suns Azimuth in this Case from y^e North riseing or setting. and $\odot$Po or $\odot$Pg the time of either before or after Noone. but here because y^e angles $\odot$Po. $\odot$Pg are obtuse it may be more Convenient to use the lesser triangles in which are given. Pi the Reflection. Po = Pg. y^e Complem^t of y^e Latitude as before and y^e angles on each hand r̲ right to find Pgi or Poi. (the Complements of $\odot$gi & $\odot$oi) the suns amplitudes Riseing or setteing from y^e East or West and iPo = iPg the time of his riseing after or setting before Midnight I shall make use of the orientall of these two triangles in which it will hold:[8]

s, po : R : : s,Pi : s, $\angle$, Poi y^t is
As y^e cosine of y^e latitude
To y^e Radius
So y^e sine of y^e reflection
is to y^e sine of y^e amplitude. which supposeing the latitude & reflection as before stated will be found 26°. 55′: from y^e East or West Northwards. Now to find the angle at P. it will hold[9]

t, Po : R : : t Pi : t $\angle$ oPi. y^t is
As y^e Cotangent of y^e Latitude
To y^e Radius
So is y^e tangent of y^e Reflection.
To y^e sine of y^e Ascentionall difference. which the latitude &

reflection being put as before will be found 21°. 40′: therefore the Angle gPi: 68° 20′ which converted into time make 4^h. 33′. 20″ the time of ye suns riseing which taken from 12 houres gives the time of his setting 7^h. 26′-40″ after Noone.

The suns distance from ye Vertex at any hour of the day when hee is on y[e] æquator is one of the usuall problemes of the sphere. how these are formed in o[r] Projection I have not shewed in my treatise[3] I shall therefore shew it here.

The sun is then on the æquator or in either of the aequinoctiall points when y[e] poles are in the horizon of the diske. at which time the solsti[ti]all Colure shall Coincede with it & be the 6 a Clock hourcircle & the Circle ♈ P ♎ in y[e] 12[th] fig. [Fig. 18.1] the Meridian.

Let it now be proposed to find the suns distance from y[e] vertex at 2^h. 50′ p.m. when hee is on y[e] first point of ♈ ♎

I have shewed y[u] in my former readeings how to find y[e] point T in y[e] solstitiall Colure produced through which if a line qTw be drawne at right angles to it the centers of all the howercircles shall be found in the said line extended infinitely both wayes from T

Divide the said line begining at T both wayes as y[e] tangent of a Circle Whose Radius shall be PT. Convert the time given 2^h.50′ into y[e] p[ar]tes of a Circle the[y] Make 42°-30′. Now because T was y[e] center of y[e] Meridian & the time proposed is afternoone. Count y[e] said 42° 30′ in y[e] divided tangent line from T̲ to w̲. & setting one foot of y[r] Compasses on w̲ extend the other to the pole P. through which strike y[e] hourcircle nP☉, this intersects the Path at d̲ which is the point on which y[e] vertex passes at 2^h.50′ afternoone. Let the Arch of a great Circle be strook through the two aequinoctiall pointes & this point d̲ wee have so a payre of triangles formed in either of which sufficient is given for determineing the probleme. that is from y[e] triangle d♈P. in which are knowne dP the distance of y[e] Pole from y[e] Path. y[e] angle dP♈. y[e] hour from noon & y[e] side P♈, a quadrant. whence the side d♈ the suns distance from y[e] vertex may be found: or else Pdy in [which] are given Pd. y[e] distance of y[e] Pole from y[e] Path y[e] angle dPy y[e] hour from 6. or comple[me]nt of y[e] hour from Noone. & y[e] angle at ȳ right to find y[e] side yd y[e] suns height or complem[en]t of his distance from y[e] vertex which in y[e] former triangle will be got by this proportion[10]

R: s pd : : s ∠ yPd : s yd y[t] is

As y[e] Radius

To y[e] cosine of y[e] latitude.

So is y[e] cosine of y[e] hour from Noone

To y[e] cosine of y[e] suns distance from y[e] vertex. hence in y[e] latitude of 51.30 the time being supposed 2^h. 50′ after noone when ye sun is in y[e] æquator his distance from ye vertex will be found 27° 20′

But if his Azimuth from ye Meridian. at the same time were required this is ye angle Pdy. for ye arch dy is apart of ye south Meridian & the sayd angle may be found by this proportion:[11]

cs Pd: ct∠ y Pd : : R: t∠Pdy : y^{t} is
As ye sine of ye latitude.
To ye tangent of ye hour from Noone
So is ye Radius
To ye tangent of ye Azimuth from ye Meridian which admitteing ye latitude & time as before proposed will be found 35.40.

The next proposition usually is haveing the latitude suns declination. or distance from ye pole given with ye hour of ye day to find his distance from ye vertex.

Let y^{e} ⊙^{s} place be supposed in ye Middle of ♉ & therefore his declination as used in findeing ye Amplitude & Ascentionall difference before. let the latidue [i.e. latitude] be the same but the time 4 houres before Noone or 8 a Clock in ye Morneing.

The Center of ye proper meridian is at w̲ in ye 11th figure [Fig. 17.1]. Count therefore 60 degrees which are equall to 4 houres in the divided tangent line from w̲ towards q̄ the Center of ye 6 a Clock hourcircle these shall terminate upon ȳ which is the center of the hourcircle for 8 in the Morneing when ye Sun is in the middle of ♉. if therefore setting one foot of ye Compasses on ȳ you extend the other to ye pole P. and through it strike the Arch of an hourcircle P8. it shall forme an Angle of 60 degrees with ye Meridian. & the point 8. in which it intersects ye path shall be the place in the path which ye vertex transits at 8 a Clock in ye Morneing or 4 hours before Noone

through this point & ye suns place with ye opposite point let ye Arch of a great Circle be strough. and ye oblique angled triangle P⊙8. are given ⊙P8 the hour angle at the Pole 60^{gr}. P8. the distance of the Pole from ye Path 38°.30′ with P⊙ the distance of the sun from ye Pole. whence by the Second Case of oblique angled Sphericall Triangles[12] the side ⊙8 will be found 58.43½ Which is the suns distance from ye vertex at that time

The Converse of this proposition is of More frequent use in Astronomy. which from ye Suns distance from the Pole. his distance from ye vertex & ye Latitude being given requires the hour of the day or the suns Azimuth at the time & may be answered from the said triangle in which now all the sides are given to find either the Angle at P which is ye hour from Noone or at 8 which is the Azimuth from ye North part of ye Meridian & may be resolved by ye 11th case of oblique angled sphericall Triangles.[13] I doe not here give y^{u} the proportions as formerly there are severall that serve for this purpose & give answers equally true. each of which will be esteemed best accordeing as they are most frequently used: for certeinely every man will distpacth most calculations in the method hee has most beene versed in I meane in Logarithmes for if the Comparison were to be made betwixt the old way of Workeing by Naturall Numbers & the late by Artificiall it will be found that lest the best compendiums be used by the most expert in Naturall numbers the Artificiall will be at least trebly more expeditious.

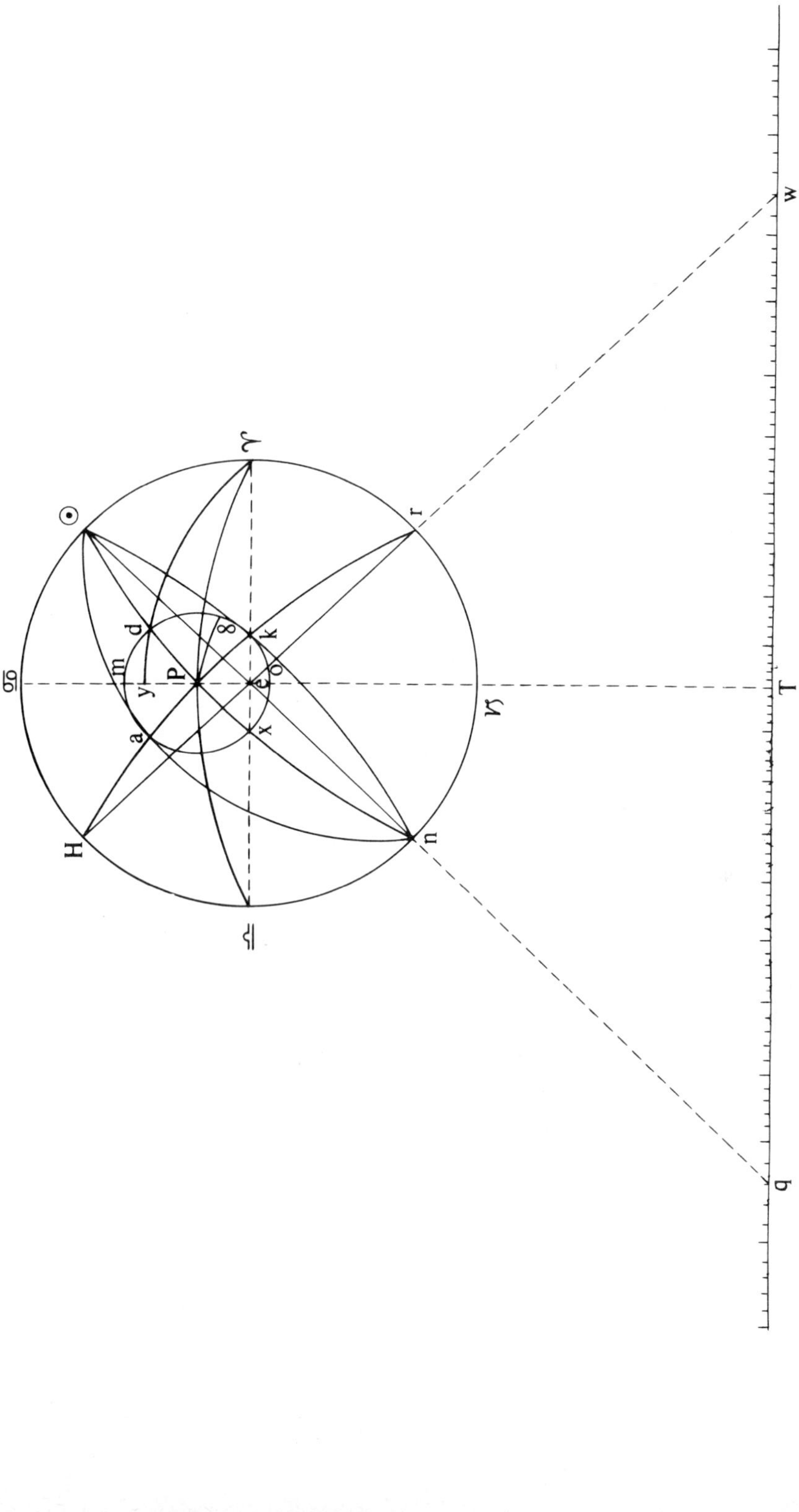

FIG. 18.1

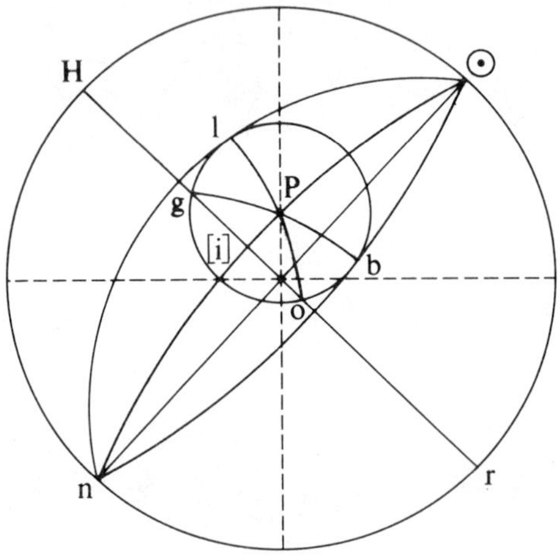

FIG. 18.2

NOTES

18.1. On f. 174r, the folio immediately preceeding this lecture, Flamsteed writes: '[I] was at Gresham Colledge but Dr Wood had dismist ye boyes. so I read not;' from which it may be inferred that this lecture was actually delivered one week later, on 1 November 1682.

18.2. Op. cit. (note 15.1).

18.3. Ibid.

18.4. The convention adopted by Flamsteed for classifying the various types of trigonometrical solutions of right-angled and oblique-angled spherical triangles appears to be that published in John Collins, *The Sector on a Quadrant* (London, 1659), pp. 60–138, of which Flamsteed possessed a personal copy. The particular rule cited here is classified on p. 69 along with five others as belonging to the 15th case of right-angled spherical triangles.

18.5. Ibid., p. 63 (6th case).

18.6. Ibid. (7th case).

18.7. Ibid., p. 65 (10th case).

18.8. Ibid., p. 60 (1st case).

18.9. Ibid. (2nd case).

18.10. Ibid., p. 63 (7th case).

18.11. Ibid., p. 61. (6th case, with the middle terms of the third rule transposed).

18.12. Ibid., p. 77. 'Two Angles together less than a Semicircle with the side between them, alias, the Interjacent side, To find one of the other sides.'

18.13. Ibid., p. 99. 'Three Sides to find an Angle.'

Lecture 19

Wednesday, 8 November 1682

I[n] my former readeings I have shewed you how all such cases & problemes of the sphere are answered by the Copernican susteme & projection as require onely the resolution of right angled sphericall triangles. such were haveing the suns places in ye Ecliptick with the obliquity or inclination of ye Axis to find the Reflection or distance of ye illuminated pole from the horizon of the diske which is answerable to ye declination on the Ptolemaick. with his right ascention. these I called Generall phanomena because they are the same at the same moment to all the Inhabitants of ye Earth. afterward I handled the particular which alter accordeing as the place to which they are demanded is nearer or more remote from either of the Poles such are his amplitudes ascentionall differences heights & azimuths at 6 a clock or on ye prime Verticall as also at any houre when ye Sun is in the æquator. I was proceedeing from these to such as require the resolution of oblique angled sphericall triangles when my distempers obliged me to breake of[f] & remit them to this opportunity.

What I have said of the sun will hold also in the stars when they lie in the Ecliptick. but if they have latitude the case is something different as I may shew you hereafter.

The usuall problemes that require the resolution of oblique angled triangles are. The latitude of any place ye suns declination. & height given to find the hour, of the day his Azimuth & angle of Position. or any 3 of these being given to find the other three.

Or the suns place. & ye latitude being given to find ye pointe of the Ecliptick then culminateing its height. the suns height & the parallactick Angle.

Or the declination & right ascention of any star being given to find its longitude latitude & all the aformentioned requisites. I shall answere the solar first by reason that before I discourse of the stars it will be requisite to shew y^{u} how they come to alter theire Longitudes annually. by the recession of the Earths aequinoctiall pointes & how that recession is formed: how the places are assigned to them in the

Notes on page 291.

Globe & how the place & latitude of any star being given it may be layd downe in ye projection.

In my former readeing I tould y^{u} I would have all the lines I had occasion to make use of to be described or formed in a Rete closely investeing the earths superficies & that the greate circle described round the Globe by ye orbis magnus[1] might properly be called the ecliptick

Let this circle be represented in ye figure before you [Fig. 19.1] by ye periphery ♈ ♋ ♎ ♑ in which let the suns place be supposed at ♋. let <u>kzlm</u> represent the Parallell of London. let the time proposed be 5 houres after noone at which the suns Azimuth & distance from the Vertex is required:

I have shewed y^{u} formerly. that if y line ♋π♑ be drawne through the center it shall represent the solstitiall line & ♎π♈ crosseing it at right angles there in the aequinoctiall Colure.

That if y^{u} count 23°.29′ the distance of two poles of y^{e} earth & globe in y^{e} projection from ♎ to p. a Ruler layd over ♈. y^{e} other end of the aequinoctiall colure & this point p. will give the place of the pole of the globe where it cuts the solstitiall colure at P.

That if from y^{e} point $\bar{p}$ in y^{e} periphery you lay of[f] the distance of y^{e} Pole from y^{e} path or complement of y^{e} latitude on each side to $\bar{x}$ & $\bar{y}$ a Ruler layd over each of those points & ♈. will give y^{e} intersections of the Path or parallel with the solstitiall Colure in k and l. That therefore kl. shall be the diameter of the parallel or path, & that if this line be bisected & on the point of bisection as a center a circle be strooke which shall passe through ye pointes k and l it shall be the parallel or Path it selfe

I likewise proved y^{t} if the tangent of ye complem^{t} of ye arch Pπ. to ye Radius of ye projection, were set of[f] from π on the solstitiall colure produced the contrary way it would give the point T in it Which should be ye center of ye 6 a clock hourcircle when the Sun was on ye 1st pointes of ♋ and ♑.

And that if through this point ye line qTi. were drawne at right angle to the solsti[ti]all colure produced πT. the centers of all the other hourcircles would be found in this line & that the tangent of ye distance of any houre from six a clock in this case. to ye Radius PT. set of[f] in it from T to ye left hand when ye hour was after noone the sun being on the tropick to the right when before Noone would give the center of the said hourcircle.

In o^{r} proposed case the hour is 5 after noone or one houre before 6. to which 15^{gr} are equall. Make PT y^{e} Radius of a tangent line & takeing 15gr from it lay them from T on y^{e} left hand to $\bar{q}$. there shall be [the] center of y^{e} proper hourcircle on which if [with] y^{r} Compasses y^{u} strike an arch through the Pole zPm. it shall be the required hourcircle. & the point z in the Path the place of y^{e} Vertex at y^{t} time

From z to ⊙ or ♋ draw y^{e} streight lin[e] z ⊙ or ♋ which bisect in w hence raise ye perpendicular uw intersecteing the aequinoctiall colure in u now because the centers of all the greate circles passeing through ye two pointes ♋ & ♑ shall be found in the aequinoctiall colure produced when necessary. u in it shall be the center of ye Azimuth z⊙ passeing from ye zenith to the sun at this time

So now wee have formed the oblique angled triangle z⊙P in which zP. is the complement of the latitude or constant distance of ye Pole from ye path P⊙ the distance of the sun from ye Pole. ⊙z. his distance from ye zenith, the Angle ⊙Pz the hour from Noone Pz⊙ the suns Azimuth from the North & z⊙P the angle of Position. any three of which parts being given the rest will be found by the knowne resolutions of ye Cases of oblique angled sphericall triangles.[2] for if as I first proposed the latitude suns declination & height being given. the hour Azimuth & Angle of Position were required then in the said triangle are knowne the side zP ye Complement of the latitude P⊙ ye complem^{t} of y^{e} ⊙^{s} declination & z⊙ the Complement of his height 3 sides to find any of ye other 3 angles or all if required.

If ye latitude hour & declination were given the height Azimuth & angle of Position required. then is the Angle at P ye hour. ⊙P the Complement of the declination zP of ye latitude so here are 2 sides and an angle betwixt them given to find ye 3^{d} side the complem[en]t of ye suns height & the other 2 angles at z his Azimuth & at ⊙ of his position. I give no example for there are severall methods of workeing the Case or resolveing it and little different from each other as to expedition tho I must confesse that where ye angle is given I best like of that of letting fall a perpendicular as being the easiest for a young student who if hee have onely learnt his right angled triangles & understands them th[o]roughly needes no further helpe for the resolution of any case of ye oblique, except when ye 3 sides are given to find an angle or its converse:

Thus I hope I have made it evident that the usuall phainomena of the sphere are as easily represented and the problemes solved in the Pythagorean projection as readily as in the Ptolemaick it remaines that I shew next. how the highest degree of the Ecliptick. its height with the angle y^{t} the verticall Circle passeing by any star are formed and found in the same Theory & projection.

These are of great use especially in eclipses since by them the quantitys of ye parallaxes of the luminary are determined. yet are but very ill represented in the Ptolemaick projections of the sphere but very naturally on this thereby proveing that nature workes the shortest way & that the most ordinary of her p[er]formances shew theire proper effects much more easily then the most ingenuous of humane contrivances can represent them. Hence I thinke too wee may draw one probable Argument for the Verity of the Pythagorean systeme. for since as Nature workes the readiest way. & this representation of her answers all her appearances the most easily and Naturally tis the most likely to be the very truth it selfe & this its p[er]fect figure.

When y[e] 90[th] degree of y[e] Ecliptick is required to y[e] suns right ascention got as was formerly adde the degrees & minutes answering to the time after Noone or the time converted into degrees & minutes. rejecteing a whole circle if the summe exceede one. of this summe if it be lesse then 270[d]. take the complement to 270 thet shall [be] z Pπ. which being knowne by the helpe of it the Meridian zPm be strooke accordeing to the former directions. & consequently the place of the Zenith in y[e] path is given at z.

Through which & the pole of y[e] Ecliptick π. draw y[e] streight line $\pi z\varepsilon$. this cutteing the Ecliptick in ε shewes there the Nonagesme or highest degree of the Ecliptick & the arch πz. measured on the line of semitangents whose radius is equall to the semidiameter of the Projection gives the distance of the pole of y[e] Ecliptick from y[e] Zenith which is equall to the height of y[e] 90. degree.

So Now wee have formed an oblique angled sphaericall triangle zPπ. in which are knowne the side πP the Constant distance of the two poles of the World & Ecliptick 23°.29′: Pz. the Complement of y[e] Latitude & y[e] oblique angle zPπ. the difference of y[e] Right ascention of y[e] Midheaven & the first point of ♑ two sides of an oblique angled sphaericall triangle with y[e] angle included to find the third side πz. the distance of y[e] Zenith from the pole of y[e] Ecliptick & y[e] angle Pπz. = to y[e] arch ♋ε the longitude of y[e] Nonagesime degree from y[e] first point of ♋.

Hence is derived the construction of y[e] Nonagesimary table which shews the 90[d]. of y[e] Ecliptick & its altitude = to y[e] distance of its pole from y[e] Vertex to every degree of y[e] Ecliptick in any latitude proposed. & it may be noted. that the calculation being made to every degree of right ascention from y[e] 1st point of ♋ to y[e] first of ♑ or one semicircle. all the requisites are repeated also & given in y[e] other, for the height of y[e] 90 degree & its longitude from y[e] 1st point of ♑ is the same alwayes when the Right ascention of y[e] Midheaven counted on either side from y[e] 1st point of ♑ is y[e] same.

I have seene some Arabick catalogues of the Right ascentions & declinations of the fixed stars in which the right ascentions are not counted as in ours from y[e] 1st point of ♈ but from y[e] 1st point of ♑. & I beleive it was for this very reason that in the resolution of y[e] triangle. the longitude or angle given is alwayes to be counted from y[e] 1st point of ♋. & the right ascention found by y[e] resolution of y[e] triangle is from y[e] 1st point of ♑ as I shall now make appeare

How the fixed stars come to have theire places and latitudes assigned them on y[e] globe or rather in y[e] Rete encompassing it. I have both shewed in my published treatise & formerly in my readeings here & may perhaps have occasion to discourse further ere long. at present that I may dispatch the business of ye Phainomena at once I suppose there places knowne

The present longitude of Arcturus is ♎ 19°.49′ his latitude North alwayes ye same 30°.57′: this star I proved to be layd downe in ye projection. & his right ascention. declination. semidiurnall arch & c to be found first Mechanically after I shall shew how they may be got by Calculation

Let still ♎π♈ represent the aequinoctiall Colure ♋π♑. the solstitiall, π being the Pole of ye Ecliptick & P of ye world.

Count 19°.49′ ye longitude of Arcturus from ye first point of ♎ in ye periphery or Ecliptick from ♎ to a lay a Ruler over this pointe & center π & by it draw ye streight line asπb this wee may call the line of ye stars longitude. which crosse at right angles through the center with ye line cd

Further set of[f] 30°.57′ the latitude of ye star from in ye periphery from a to e . a Ruler layd over the pointes c and e where it cuts y^{e} line of the stars longitude in s. designes & shews y^{e} stars projected place.

The next thinge required will be the Meridian or great circle passeing through y^{e} star & Pole. to draw this y^{u} may remember. that I demonstrated to y^{u} formerly. that the $\frac{1}{2}$ tangent of any arch with y^{e} tangent of its complem^{t} were equall to the secant of y^{e} same Arch & consequently that the tangent complement of y^{e} distance of any great circle from y^{e} pole of y^{e} projection set of[f] from it in y^{e} line fulleing at right angles on y^{e} said circle would give its center.

Take therefore 30°.57′ or y^{e} arch ae betwixt y^{e} Compasses & transfer it twice from d towards ♈. the second extent will give y^{e} point g in y^{e} periphery. overwhich & c lay a Ruler this cuts y^{e} line of longitude in h. & makes $\overline{\pi h}$ the tangent of y^{e} stars latitude through h draw y^{e} line hi perpendicular to the line of longitude produceing it both wayes when Needfull in this shall the centers of all those great Circles be found which passe through y^{e} place of Arcturus.

I have formerly proved to y^{u} that the centers of all the great circles passeing through the Pole of ye World P. shall fall in the tangent line qT. the Center of the great Circle passeing through the Pole of the World & Arcturus or its Meridian shall therefore fall in the Intersection of ye 2 lines hi and Ti and if y^{u} measure ye length Ti on a tangent line whose Radius shall be equall to TP. it shall give the excesse of its Right ascention above a semicircle. because the center lies on ye right hand ye solsti[ti]all Colure produced & its longitude is in ye 3^{d} Quadrant. The Arch sP is the Complem^{t} of its declination which I shew not now how to measure reserveing that for my next lecture when I shall shew y^{u} how any circle may be divided whether it be great or small to which this p[ar]ticular more properly belongs.

Tho the Horizons of the diske made by ye suns illumination of y^{e}

globe were streight lines in the projection passeing through ye center or pole of ye Ecliptick π. yet for such stars as have latitude the case is different for to them more y^{n} one halfe of ye projected plane is visible if they have north latitude or lesse if they have south & consequently to arcturus haveing both North latitude & declination the horizon of the diske takes in a good space of ye plane beyond the pole as will appeare if it be declineated to doe which. take ye arch ae. betwixt y^{r} compasses & transfer it in ye limbe from d to f a ruler layd over ye point c & f cuts ye line of longitude ab in θ in which point the horizon of the diske shall also cut if for this point is 90 degrees from Arcturus.[3]

Continue the line of longitude ab backwards then take betwixt y^{r} compasses the arch ed the complement of ae & set it of[f] backwards from e to γ. a Ruler layd over ye principall point c. & γ. gives the point α in the line of longitude produced. which shall be the center of the horizon of ye diske.

Extendeing y^{r} compasses from its center α to the points c or d y^{u} may strike the Arch cζd[4] which shall represent the horizon of ye diske. or that line in the projected hemisphere on which the illumination of Arcturus is terminated.

As the earth turnes round on its Axis Arcturus then appeares to rise when the vertex of any place cuts the horizon of the diske this in the path of London is at k. to which if wee conceave Pk. the arch of an hourcircle to be drawne. there will be formed ye sphericall triangle kPz. right angled at z: in which will be knowne Pk. the Complem^{t} of ye latitude Pz ye reflection or declination of ye star with ye right angle at z whence the angle kPz ye Nocturnall arch with Pkz its Amplitude may be found as before they were in the sun And the Nocturnall arch being got the diurnall is knowne for tis but its complem^{t} to 180^{d}. or 12 hours

It will be needlesse for mee to shew how any other of the requisites are found since they are the same with y^{e} solar the horizon of the diske being onely altered. of which I thought fit to say thus much because I had omitted it in my published Treatise.[5]

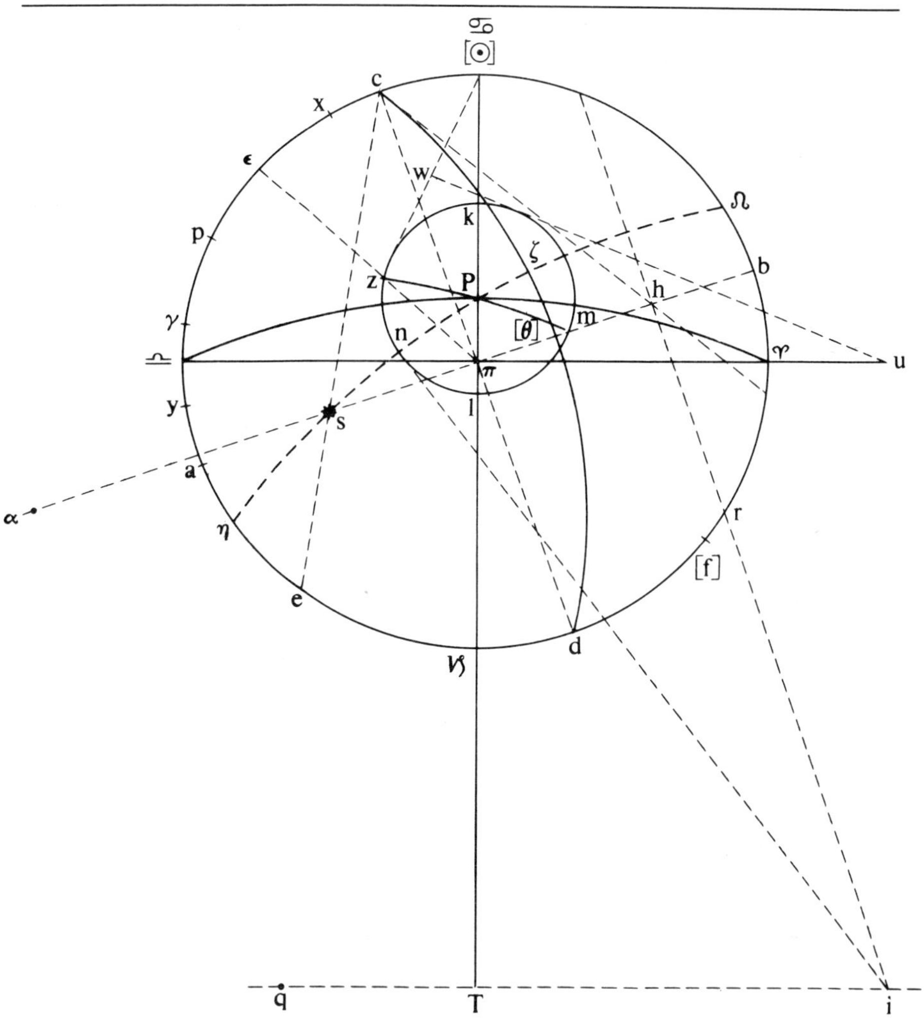

FIG. 19.1

NOTES

19.1. I.e. the celestial sphere.

19.2. These are discussed by Moore in op. cit. (note 15.1), pp. 103–10.

19.3. This latter construction, defining the points f and θ, is not represented in Fig. 19.1.

19.4. Flamsteed wrote cθd instead of cζd.

19.5. Op. cit. (note 15.1).

Lecture 20

15 *November* 1682

In Easter terme last I began my lectures concerneing the stereographicall projections of ye sphere My reason I then tould you was that I *ch*ose this subject because however usefull it had not yet beene cleareIy handled by any Author in o^r Tongue, & that tho the very learned &, in his dayes, much celebrated Clavius had treated amply on it in Latit[u]de, yet his apparatus of Theoremes was so large as might justly fright a young student from lookeing on them at least from peruseing & considering them th[o]roughly. and if hee were master of so much patience as to abide & goe through with this longe labor yet the treatise of Projections[1] dependeing on it was againe so tediously prolix & tempered with fresh propositions occasionally interspersed, that I am confident scarce ever any man since it was wrote has beene of the labor of peruseing it orderly, or if hee has, hee has not thought it worth his while: Considering therefore well with my selfe of this subject I found that to demonstrate what ever hee had tought concerneing projections I needed not make use of more then 8 ordinary propositions or 10 at ye Most. & that the whole subject myght be gone throug[h] with in 2 termes But because the doctrine it selfe would relish but harsh if not applyed I determined to shew the use of every proposition as I tooke it up. And because the ordinary projections of the sphere . . . .
proposed to answer the appearances of Nature I rejected them wholey & shewed you how by the helpe of those propositions Nature herselfe which moves the earth about ye sun & by keepeing its Axis allwayes parallel to it selfe & inclined at the same angle to ye plane of ye Ecliptick. formeing thereby the different length of days & riseings & settings of ye sun at different seasons in ye same or different regions might be represented & that with such facility as all the curious but vaine coments of ingenuous men who could not admit her Motion have endeavor[ed] in vaine to imitate or represent however as I proceeded I also shewed how the Circles of the Ptolemaick projections were to be enscribed in them for though these be no representations of the truth yet may they be derived from such.

Notes on page 299.

I onely layd downe such propositions *then* as *were* necessary for ye delineateing of the Projection & all such circles as are usually required in that or any other but I shewed not how Any circle being drawne *it might* be divided this I conceaved a subject fit to be handled a part & therefore resolved to Manage it last. and in order to it I shall premise one Theoreme[2] From whence the division of any circle in many Stereographicall projection will follow.

If I should give yu the Theoreme I intend to propound in one paragraph its prolixity would render it obscure & . . . .
. . . I shall therefore divide it into three parts & first I affirme. "That if a great circle on ye Globe passe through the poles of any other two great circles described on it, if ye globe be againe cut by a plane set at right angles to ye plane of ye said great circle passeing through ye poles of ye 2 great circles. ye said plane where it cuts ye globe, shall cut of[f] equall [arches] of ye 2 great circles betwixt it selfe & ye other great circle passeing through theire poles,"[3]

As if in ye 15th figure [Fig. 20.1] hof represent ye horizon & æoq ye æquator two great circles on the globe divideing it into two equall partes each. P. ye Northerne pole of ye Globe or æquator n, ye Southerne of ye horizon, or ye Nadir zpqnπh. ye Meridian or any other greate circle passeing through theire two poles p and n let pn be conceaved a plane cutteing ye globe through both ye said poles & standing at rightangles to ye plane of ye Meridian. I say yt ye arches df of ye horizon & gq of ye æquator cut of[f] by this plane betwixt it and ye Meridian which passes through the two poles p and n are equall each to ye other.

For ye distance of every great circle from its pole is allwayes a quadrant. therefor pq. the distance of ye æquator from its pole is a quadrant but for the like reason fn. the distance of the horizon from its southerne or invisible pole is also a quadrant takeing away fq the common arch to them both there remaines the Arch pf = qn: but the angles pod and $\overline{gon}$ are measured by ye equall arches pf & $\overline{qn}$ therefore are they equall to each other and Againe because the sides op and on of the triangle opn are Radij of the same Circle & consequently equall to each other the angles opd and ong & consequently ye triangles shall be also aequall to the other by ye 5 & 6th propositions of ye first booke of Euclids elements. and the side or arch od = to ye side or arch og.

But the arch df is ye complement of od to a quadrant so also gq the Complemt of og to a quadrant now if two arches of Circles be equall to each other theire complemts shall also be equall to each other therefore the Arch df = to ye arch gq. which was to be demonstrated.

Againe I affirme.
"That if with any extent of the compasses two other circles be described on ye globe from each of the said poles through which the great circle passes the afore-mentioned plane shall cut of[f] equall

arches in each of the said circles, (which will be lesser ones) betwixt it selfe & ye great circle aforesaid but in this case there are two varieties either when ye extent of ye compasses is more y^{n} a quadrant of ye globe or lesse I shall demonstrate the proposition in both: that is

First let the extent be more then a quadrant and let the lesser circle described on ye globe from the under pole of ye horizon n be tmxabc and on the other pole p: ursikl. I say y^{t} y^{e} arches ac and il. cut of[f] in y^{e} said lesser circles betwixt y^{e} plane pn and the Meridian pcfqln are equall each to y^{e} other.

I have already proved that the triangle pod is all respects equall to y^{e} triangle nog. therefore y^{e} ∠odp = to y^{e} ∠ogn. but y^{e} lesser circle tc is parallell to y^{e} horizon hf. and ul to y^{e} æquator aeq. therefore y^{e} angle xap = odp and y^{e} angle sin = ogn = xap & the angle pac its complement to a semicircle = to y^{e} ∠nil

But because y^{e} arch nc is = to y^{e} arch pl by construction, takeing away y^{e} common quadrants cf will be equall to ql. & because pf is equall to qn as was proved before: cf substracted from pf leaves the arch pc = qn − lq: therefore the arches ln and pc are equall:

But the angles opf and onq are quadrants and the angles opd and ong are aequall. each of which subtracted from the right angle in which it is included there remaines the angle dpf or apc = to y^{e} angle gnq or inl therefore y^{e} angles apc and inl are aequall.

So then in the two triangles pac and nil. the angle at p in y^{e} one is equall to y^{e} angle at n in y^{e} other the angle at a to that at i. and the arch pc to y^{e} arch nl. therefore the two triangles pac and nil are in all respects equall the angle apc to y^{e} ∠inl which remember pca to ilm and the line or arch ac to ye line or arch i.l. as was asserted.

the 2^{d} case is when ye semidiameter of the circles described on ye globe with y^{r} compasses are lesse y^{n} a quadrant as in ye 16 Figure [Fig. 20.2]. let ye circle described on ye southerne or under pole of ye horizon n be txmabc and with the same distance or Widenesse of y^{r} compasses on ye northerne pole of the globe p. the circle or line usrikl. the arches il and ac intercepted betwixt the plane pidgan, and y^{e} great circles plfcqn passeing through y^{e} two poles p and n are equall.

For as I proved before the two triangles odp. ogn are equall in all respects. adde to each the interjacent triangle odg. then is each part of the triangle odn = to its respective part of y^{e} triangle ogp. therefore y^{e} angle odn = to y^{e} angle ogp: but the circle ul is parallel to y^{e} æquator æq. and tc to y^{e} horizon hf therefore the angle man = odn = ogp = rip and y^{e} angle Pil the compl^{t} of rip to a semicircle = to nac y^{e} complem[en]t of man to a semicircle

The Angles opf and onq are quadrants by construction from which if y^{e} angles opi. and ona [(]which I have also proved equall) be subtracted each from the right angle in which it is included there will remaine the angles ipl and anc equall to each other. but the side pl is equall to ye side nc by hypothesis therefore wee have here in ye 2 triangles pil. nac two angles & one side of the one equall to ye 2 respective angles & a side alike posited of the other therefore the two triangles in all respects equall the angle ipl to ye angle anc. pli to acn. which againe I must entreat you to remember and ye side or arch il to ye side or arch ac which was to be demonstrated.

Lastly I say

"That if the aforementioned plane passeing through the poles of any two parallells be revolved on its axis or the streight line connecteing the said two poles wherever it cuts y^e globe it cuts of[f] equall arches both in the said great circle & the parallells to them equally distant from each of the said poles betwixt it selfe & the great circle which is drawne through the said two poles."

Let $\overline{\text{æq}}$ be the æquator hf the horizon. p the North pole of the æquator. $\underline{\text{n}}$ y^e southerne of y^e horizon ul y^e parallel to the æquator and rc. to y^e horizon as far distant from its pole n as y^e parallel of y^e æquator is from its pole p. as was before supposed both in y^e 15th &16th figures [Figs. 20.1 and 20.2] and let a plane be conceaved to cut the globe in the poles of the two great circles $\bar{\text{p}}$ and $\underline{\text{n}}$. if in y^e 15th figure [Fig. 20.1] wee suppose this plane revolved on its axis pn. and the line of its section on the globe to be pbehkn. I say that the arches hq of the æquator & ef of the horizon as also kl of the parallel to y^e æquator & bc of the parallel to the horizon (which parallells are supposed equally distant the one from ye under Pole of the horizon the other from ye Northerne Pole of y^e æquator) & intercepted betwixt the said plaine on the globe & the great circle passeing through the two sid poles or ye Meridian pcf. qln. are equall. to each other that is the arch ef to hq, and bc to kl: which I thus prove. & first that ef is equall to hq: : :

I have before proved that the angle dpf is equall to the angle gnq. the angle dfp to gqn. as also pf to gn. Now because if a plaine be revolved on its axis its revolution makes equall angles in all places on it the angle dpe shall be equall to y^e angle $\overline{\text{gn}}\underline{\text{h}}$. and y^e $\angle$dpf $-$ dpe $=$ epf shall be equall to y^e angle gnq $-$ gnh $=$ hnq. therefore y^e angles epf and hnq equall to each other. but y^e angle Pfe is equall to y^e angle hqn. & Pf to nq: therefore also y^e line or arch ef $=$ hq. which was to be demonstrated:

And for y^e lesser circles equally distant from theire respective poles let them be as before supposed in either figure [Fig. 20.1 or 20.2]. tmxabc and ursikl. the arche $\underline{\text{bckl}}$ intercepted betwixt the plane & the great circle pfqc shall be also equall. for I before proved the angle apc $=$ $\angle$inl. but the angle bpc $=$ epf is equall to y^e angle hnq $=$ knl therefore the angles knl and bpc are equall. So I like wise shewd before that the angle pcb is $=$ to the angle nlk and y^e arch nl $=$ pc. therefore the whole triangles & each of theire respective p[ar]tes are equall to each other & bc $=$ kl as was to be proved.

the Easiest way to conceave the revolution of this plane upon its axis will be thus. place the globe before y^u set to any latitude. than supposeing y^r eye either in y^e under pole of y^e globe or horizon, let infinite lines be produced from it to all the circles on y^e globe project them on ye plane of ye horizon if y^u supposed y^r eye in ye Nadir or on y^e æquator if in the Southerne pole. conceave ye globe hollow & empty of every thing save the said plane on which the projection is drawne & let an axis be inserted into it from ye eye to the contrary not opposite pole on which imagine ye plane I have so often mentioned to be revolved because the eye is allwayes in this plane turneing on such axis the lines it shall describe on ye plane of the projection shall be allwayes streight. & passeing over equall degrees of that great circle projected on ye plaine which was 90 deg from ye eye shall also divide the other great Circle which was a Quadrant from ye contrary pole or the other through which the inserted axis passes into its answerable but unaequall degrees as if the eye were conceaved at n in either of the figures projecting the æquator æq on ye plane of the horizon hf. the degrees of the horizon so projected shall be every where equall because that circle is in all places equally distant from ye eye. the Mutuall intersection of the plane of the horizon (on which the æquator & other circles are to be projected) with ye plane revolved on ye inserted axis np shall be a streight line allwayes passeing through the pole point d or that wherein the pole is projected, and therefore a ruler layd over the place of the pole in ye projection & the equall divisions or degrees of ye horizon shall cut & divide ye proper degrees in the y^e æquator which shall be unequall in the projection because the circle lies oblique to the eye & projected diverges more from ye axis zn or ye center o in ye quadrant ofqn. then in the opposite zæho.

Or if ye lesser Circle ul parallel to the æquator & every where equally distant from it towards ye Northerne pole being projected on the plane of the Horizon hf were required to be divided. it may easily be done & after the same manner if the Circle tc parallel to the horizon and equall distant from it be drawne and projected on the same plane and divided as y^u would have ye lesser circle ul, the divisions being begun from the Meridian. for ye Plane revolved in ye axis Pn cut equall arches of[f] in both these circles & therefore its plane peirceing ye plane of the horizon cuts also equall arches of[f] in both the lesser circles projected therein & if a ruler representeing the common intersection of these two planes be layd over the Pole P & each of the equall divisions of the Circle tc projected on ye plane Hf it will cut the Answerable Degrees or divisions in ye Circle ul projected on the same plane which yet shall be unequall for the same Reason y^t ye divisions of the æquator projected on it are unaequall as I before proved:

From this method another will result which will be found in many cases much more convenient but it will be best demonstrated when I come to the application of this theoreme & to shew y^{u} how the projected circles are actually divided which by Gods permission shall be next lecture.

What I have proved & demonstrated concerneing the axis or line connecteing the two contrary poles of the Globe P & horizon n the plane revolved on that line as on an Axis will also hold if wee conceave y^{e} two under poles of the globe π & y^{e} same n of y^{e} horizon connected by an Axis & a plane revolved on them For by the same method of demonstration it may be proved that if the plane cut the Globe in the Arch $\beta\varepsilon\gamma$ the arch æε cut of[f] by it in the æquator is equall to y^{e} Arch ef cut of[f] by it in the horizon but with this difference that whereas in the former or. 15 figure [Fig. 20.1] the equall arches were Cut of[f] in the same side of y^{e} aequinoctiall colure & from the same arch of y^{e} Meridian y^{t} is its northerne parte: here they are cut of[f] just contrary the one in the southerne the other in the northerne hemispheres of the globe divided by the solstitiall colure. & the one of the arches æε. in y^{e} æquator from ye southerne, the other ef from the Northerne p[ar]te of the Meridian.

The same difference may be observed in its section of the two parallels to ye horizon & æquator for whereas in ye 15th figure [Fig. 20.1] y^{e} revolved plane pbehkn cut of[f] equall arches in such circles whereof the one fell as much within y^{e} horizon of the projection as y^{e} other did without y^{e} æquator here [in Fig. 20.2] the revolved [plane] kεe cuts not of[f] equall arches in such parallells ul and tc. but the arch $\zeta\beta$ cut of[f] betwixt y^{e} plane & south part of y^{e} Meridian ma parallell declineing south from the æquator is equall to y^{e} arch γc. cut of[f] in a plane equally distant from the horizon towards its southerne or under pole. & betwixt y^{e} said plane & the Northerne part of y^{e} Meridian. for the triangles $\zeta\pi$p and cnγ are equall to each other which I stand not to prove the method of demonstration being the same whereby I proved the arches <u>bc</u> and <u>kl</u> in the 15th figure [Fig. 20.1] to be equall to each other, & therefore both easy & obvious

Now if the eye be supposed at n. projecteing y^{e} severall circles of ye globe ul, æq, ζn, pπ, tc, on ye plane of the great circle hf, the southerne pole of the Globe π shall be projected at ♌ & the plane of ye circle πpeγnπ produced & revolved on the Axis πn shall always passe through the point ♌ in y^{e} plane of y^{e} projection df & the section of this plane with the plane of the projection shall be a streight line running into the point of which shall therefore cut of[f] equall arches in the circles aeq, & hf as also in theire equidistant parallels $\zeta\eta$ and t, but from y^{e} contrary parts of the Meridian & in different hemispheres of y^{e} Globe bisected by the Axis or aequinoctiall Colure pπ. as I shall make appeare by Examples in different Projection next lecture; for which I must desire y^{u} keepe y^{r} figures or leave them here that they may be ready when I shall have occasion to mention them.

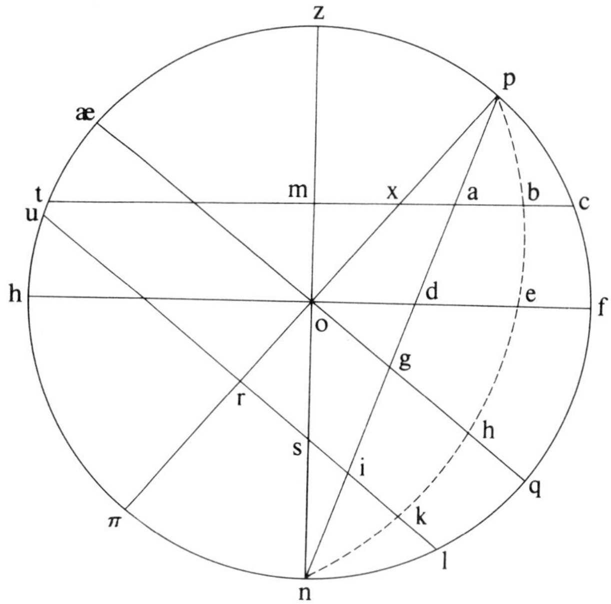

FIG. 20.1

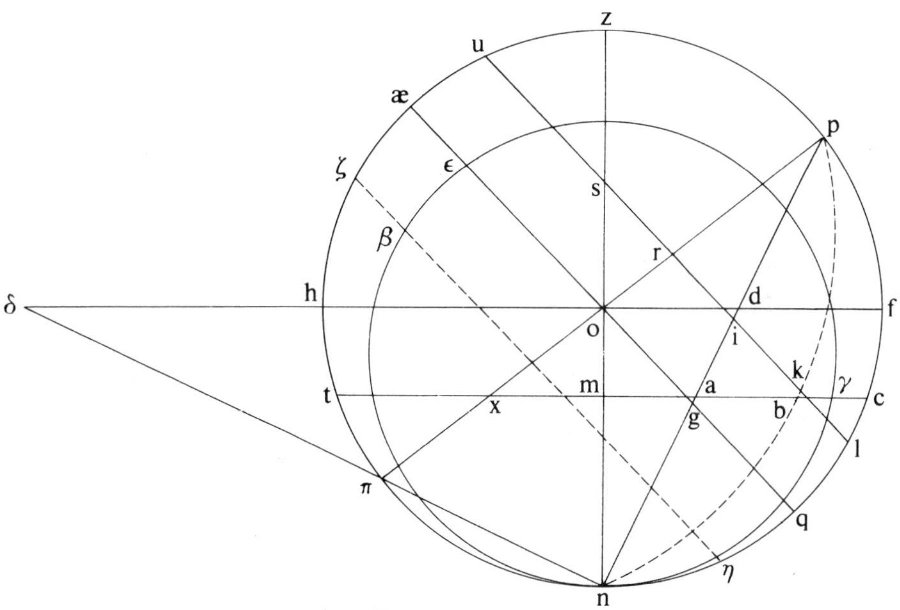

FIG. 20.2

NOTES

20.1. Op. cit. (note 15.9).

20.2. This is actually the 8th theorem which Flamsteed uses. Cf. note 16.24.

20.3. Flamsteed's alternative formulation of this same theorem is quoted in the Introduction (cf. p. 57).

Lecture 21

Wednesday, 22 *November* 1682

In my former readeings here I have shewed the grounds & reasons of y[e] stereographicall projections of the sphere which suppose the eye in y[e] superficies of the globe projecteing all the circles described on it upon a plane placed at right angles to the line produced from y[e] eye through the center of y[e] globe. I also taught y[u] how to delineate a New one in which all the diurnall appearances of the sun & stars are represented as they are naturally formed. in my last lecture I layd downe & proved a Theoreme on which the division of all such circles are grounded. I have said little hitherto to the Ptolemaick projections I shall now shew y[u] breifely how they may be drawne & then how every Circle whether great or small may be divided in them.

I chuse to shew y[u] the method of Projecting the Ptolemaick spheres not for anything extraordinary I have to remarke upon them but because what ever I have to say concerneing them may be conceaved & will hold good in any great Circles & theire parallells projected stereographically on any plane placed as is usuall in those projections. further it is requisite to make a discourse of this nature plane & intelligible that the severall lines to be projected should have some Names appropriated to them & I know none on this account fitter then those that have beene hitherto usually receaved I shall take notice by the way of the great agreem[t] & mutuall correspondence betwixt the projection on the plane of y[e] horizon commonly knowne by the Name of M[r] Ough[t]red[1] and that on the plane of y[e] Equator usually called Steoflers.[2] Which with y[e] division of the severall Circles whether greater or lesser will conclude all I have to say at present on this subject.

Let y[e] eye be conceaved in the superficies of the Globe on y[e] under pole of y[e] horizon n. in ye 15th & 16th figures [Figs. 20.1 and 20.2] projecteing ye meridian zhnf. the Æquator æq. its parallells or the two tropicks ul. ζn with ye axis or equinoctiall Colure Pπ on the plane of the horizon hf. produced every way as shall be thought necessary because the eye lies in the same plane with the Meridian,

Notes on pages 309–10.

this appeares as a streight line & shall be such in the projection.

But because ye horizon is every where equally distant from its pole n in which the eye is placed. this shall appeare a circle and shall be so projected on its pole z so also shall the equator its parallells & all What other circles are conceaved to be described on ye globe except they lie in the same plane with ye eye as I have formerly Fully demonstrated.

Now if from ye eye at n. streight lines be drawne to the intersections of every circle with the Meridian & produced till they intersect the plane of the horizon hf also produced when requisite. the mutuall intersection of these lines will give the pointes in which the projected circles shall intersect the meridian. & because the angles formed at n̲ in the periphery are equall to halfe the angles formed at o the center of the globe by lines extended to ye same pointes of the Meridian, the tangents of halfe the distances of any circle from ye vertex set of[f] on ye horizon from o̲ the same way from ıt that they lie from z. the Zenith shall give theire intersections with ye Meridian in the projection

For Example the two poles p & π being to be projected on ye plane hf. a line drawne from y^e eye at n to p gives the point of the north pole at d & from n to π produced of the southerne in δ the two lines np and nπ formeing a right angle at n because the angle of a semicircle or subtended by the diameter against any point of the periphery is right

And because y^e angle $\overline{\text{onp}}$ is halfe the angle zop. and onπ the halfe of zop. & o̲n̲ a common Radius being the nearest distance of y^e eye from y^e plane, o̲d̲ the distance of y^e pole point from y^e center of the projection shall be the tangent of halfe y^e poles distance from the vertex & oδ the tangent of halfe its complement to a semicircle & se[e]ing all the hourcircles on the Globe passe through the two poles formeing equall angles wth each other there, all the hourcircles in ye projection shall passe through the two poles δ & d̲

if therefore the circles of the Globe be to be projected on ye plane of the horizon. haveing first drawne as in ye 17th figure [Fig. 21.1] the streight line dlfpzuæζhiπ. chuse in it some point z for the center or zenith & on it describe. off cho which may represent the horizon & crosse ye meridian through the center z at Right angles with ye line czo for the prime verticall.

From either intersection of ye horizon & prime verticall c. count the complem^t of ye latitude as from c to d. a streight line dc drawne over this point & ye opposite intersection of ye prime verticall & horizon where it intersects the Meridian in p gives the place of ye visible pole.

And if from ye point. c ye line πo be drawne at right angles to do. this where it intersects the Meridian in π gives the place of ye southerne or invisible pole of the Globe & ye distance pπ the diameter of ye 6 a clock hourcircle which passes through both poles cutteing y^e Meridian at right angles in either of them.

if therefore this distance πp be bisected in T. that point shall be the center of y^e 6 a clock hourcircle. or the aequinoctiall Colure: & zT its distance from the center of the projection the tangent of y^e complement of the arch dc = pz.

For let the line To be drawne. because y^e triangle poπ is right angled at o. and oz perpendicular to pπ divides it into two similar triangles. the angle zπo shall be equall to poz. but because T is the center of the circle passeing through y^e points poπ y^e angle zTo shall be the double of y^e ∠ Tπo therefore equall to y^e arch cd which is the double of the ∠ poz but the angle zoT is y^e complem^t of y^e angle oTz or the arch pz = to it to a quadrant and zT its tangent to y^e Radius zo. whence this generall rule followes, that the tangent of the complement of y^e nearest distance of any great circle from y^e pole of that other great circle on whose plane it is projected is the distance of its center from y^e said plane & because To the semidiameter of ye said Circle is the secant of the said Angle zoT, that the secant of the said Complem^t shall allwayes be the semidiameter of the said Circle to be projected.

To draw y^e æquator in ye said projection because it is a quadrant distant from ye Pole. take df the Complem^t of dc to a quadrant betwixt y^r compasses & transfer it in ye periphery of ye horizon from c to y a line drawne from o to y intersects the meridian at ae in which point the æquator cuts it.

take then the secant of ye arch dc to ye Radius cz. betwixt y^r compasses this set of[f] in the Meridian from ye intersection of ye meridian & æquator at æ gives the point x for the center of y^e æquator on which wth y^t Radius y^u describe the circle cæodg. it shall be the line representeing it in ye projection.

I have formerly shewed y^u that if any number of great circles intersect each other in ye pole of another great circle formeing equall angles with each other at theire intersections they shall also divide that great circle through whose pole they passe as also all lesser circles parallel to it into a like number of equall arches;[3] whence it followes that the hourcircles passeing through ye pole of ye globe & there intersecteing each other at equall angles shall divide the æquator and all its parallels into equall arches

I have also proved to y^u that if two circles great or small intersect each other, two lines drawne from ye point of theire mutuall intersection to each of theire centers forme the same angle there that the two Circles make with each other at ye said point.[4] whence it followd 1. that the centers of all the hourcircles would be found in ye line Tm[5] perpendicular to the meridian & both wayes produced. & that the tangent of the angle which any hourcircle formed with ye 6th to ye Radius Tp set of[f] from ye point T to ye right hand (if the houre was betwixt 6 in ye morneing & noone) otherwayes to the left would give ye center of ye said hourcircle. & the secant of the said angle to ye same Radius would be its semidiameter

Hence arises one way of findeing out the distance of any two pointes assigned on the globe. for first by the doctrine of My last lecture save one of this terme[6] the arch of a given circle passeing through those two pointes as also the line passeing through its pole & cutteing it at right angles in their mutuall intersections may be found through whose Center if a line be drawne perpendicular to the line bisecting both the said arches in it all the centers of the great Circles passeing through the said pole shall be found.

And if the semidiameter of the Great Circle passeing through the pole of that wherein the two given pointes lie be made the radius of a tangent line & the Centers of the 2 great circles passeing through the said two pointes be sought in it, theire distances from the line bisecteing the great circle given & the other passeing through its poles shall be the tangents of two arches whose summe if the centers lie on different sides of the said streight line otherwayes theire difference shall be the distance of the two given. pointes. and as for any pointes lieing in the same great Circle so also for any two lieing in a parallell to it. this Method I must confesse is not the best that can be made use of. I shall propound a better by & by in the meane time I conceave it may deserve an explication by an Example. [cf. Fig. 21.1]

Let the two pointes given be both in the equator at α and θ or in a parallel to it at β and γ. the Meridian dpzπ bisects both the æquator c$\theta\alpha$æo[7] & the aequinoctiall colure cpro which passes through its poles Intersecteing it at right angles at c and o, in which line the Center of the said Colure is found at T & therefore the centers of all the great circles passeing through its pole p. in the line Tn intersecteing y^{e} Meridian at right angles. and In this line. the centers of the great circles passeing through the points $\theta\beta$. $\alpha\gamma$. will be found at b and n. if therefore the lengths Tb. Tn. be measured on a tangent line whose radius is Tp. the difference of theire answerable arches shall be the measure of the arch $\theta\alpha$ in y^{e} equator or great Circle or $\beta\gamma$ its parallel or lesse.

But the great inconveniency of this method is that the tangent line runs out of the plane of the projection in many cases. I therefore have onely mentioned it & now I proceed to another. not so inconvenient.

In my last lecture I proved to y^{u} that if as in the 15 Figure [Fig. 20.1]. the two poles of y^{e} equator $\bar{p}$ and the horizon $\underline{n}$ of any other 2 great circles were connected by the streight line pn. & the globe conceaved to be cut by a plane placed on that line at rig[h]t angles to y^{e} meridian or great circle passeing through the said two poles $\bar{p}$. $\bar{n}$. the arches df and $\overline{gq}$ cut of[f] in y^{e} æquator and horizon or two great circles through whose poles the third or Meridian passes betwixt the plane & the said 3^{d} great Circle would be equall to each other.

2^{d} that the arches ai and ik also cut of[f] in y^{e} lesser circles equally distant from y^{e} contrary poles & betwixt ye plane & Meridian [cf. Fig. 20.1] were also equall, and

3. that if the said plane were revolved on ye Axis pn so that it cut ye globe in any other place. Stil ye arches cf and hq cut of[f] in y^{e} æquator & horizon. as also bc and kl. in ye parallels equally distant from ye contrary poles & intersepted betwixt ye plane & Meridian [cf. Fig. 20.1] would be equall to each other

Whence it followed that seeing the Intersection of the said plaine with ye plaine hf on which the globe is to be stereographically projected would alwayes be a streight line passeing through the pole point, the said line passeing over ye pole of ye globe in ye projection & ye equall degrees of the projected horizon hf,[8] would divide the æquator into its proper but unequall degrees & that if a parallel to ye horizon as wide from without it as the parallel to y^{e} æquator is distant from it inwards or towards its pole were projected on ye said plane of the horizon. a ruler layd over the equal divisions of the parallel projected without y^{e} horizon would cut the answerable unequall degrees of the parallel to y^{e} æquator equally distant within it which to apply to y^{e} projection in y^{e} 17th figure. [Fig. 21.1]

If the visible halfe of y^{e} æquator in this projection c$\theta\alpha$æo were to be divided into its proper degrees. let first the horizon chofc be divided into 360. equall degrees or parts. a ruler layd over the pole & each of these equall degrees of the horizon where it intersects the æquator cuts the answerable degree there. thus if the 30th degree of y^{e} æquator from ye Meridian westwards were required to be given ye 30th degree of ye horizon from ye Meridian westwards is at the point k. & a ruler layd over the pole p̄ and this point k cuts ye equator at α on the intersection of ye 2^{d} hour circle from ye Meridian with it

Or if ye 60th degree were required lieing the same Way this in ye horizon is at e to which point if from ye pole p the line pe be drawne it cuts the 60th degree of the æquator in θ, where also the 4th hourcircle from the Meridian intersects it.

From this method result another I formerly mentiond to y^{u} which in some cases will be much more convenient let ye upper semicircle of ye equator idg. be divided into 180 equall degrees I say a ruler layd over any of these divisions counted from ye meridian on the Contrary side cuts of[f] ye answerable degrees of the equator. that is if it were required to designe the 60th degree of the Equator on ye Westerne side ye meridian. Count 60 degrees in the upper semicircle of y^{e} æquator from α the contrary way to λ a Ruler layd over. this point λ & ye pole p̄. cut the æquator in θ the point required

For let the Circles xλ. λe, ez. be drawne as also through the center of y^{e} æquator & y^{e} line gxi at right angles to the Meridian.as also xo.

Now because xz was Made the tangent of y^{e} poles distance from y^{e}

Zenith to y^e Rad[ius] zo or of y^e arch dc. which is the double of y^e angle poz. the line $\overline{po}$ bisects the angle xoz. & therefore cuts the length zx in such proportion as the sides conteining have: that is. zp. shall be to px. as zo to ox but oz = ze. & ox to xλ therefore in the triangles xλp. poz. it is zp: px : ze.xλ. but by y^e angle λpx = $\angle$ epz. therefore y^e two triangle[s] are similar and each of the angles of the one is = to y^e respective angles of the other therefore the ang pxλ = $\angle$ pze but the angles pzc[9] and pxi which are right being subtracted each from the obtuse angle in which it is included there remaines y^e ang[le] λxi = $\angle$ cze. & therefore y^e arch of the æquator λi is just the same Number of degrees with y^e arch of y^e horizon ce so that if a Ruler be layd over y^e point λ 60 deg. from d & the point p. it will cut the horizon in y^e point e. 60 deg[rees] from h. & the arch æθ cut of[f] in the æquator shall also be 60 deg[rees]

But if it be required to lay of[f] an arch of 30 degrees in a lesser Circle suppose in the tropick of ♑. $\zeta\eta$. then by the demonstration of y^e 15th figure [Fig. 20.1] a circle must be strooke which may fall as much within y^e horizon as the tropick of ♑ lies without the æquator. from which its distance being 23½ [degrees] I count 23½ in y^e periphery of y^e horizon from o to μ & v each way. [cf. Fig. 21.1] a Ruler layd over the point <u>h</u> and $\bar{\mu}$ cut the diameter <u>cπ</u>[10] in <u>s</u> through which I strike a circle on the center z that shall serve for y^e division of the Tropick of ♑.

And layeing againe y^e Ruler over the points h and v a line drawne by it cut the s[ai]d diameter mπ[11] through which a circle strooke on y^e center z shall serve for the division of the Tropick of ♋ that being as much within the equator as this does without the horizon.

Divideing the periphery of y^e Circle strooke through s into 360 equall degrees count 30 of them from <u>m</u> to δ a Ruler layd over this point & the pole $\bar{p}$ cuts the Tropick of ♑ in η the arch $\zeta\eta$ being accordeing to what I demonstrated on y^e 15th figure [Fig. 20.1] 30 degrees of the lesser Circle or Tropick of ♑. with which the hourcircle as it ought coincedes in y^t point.

or if it be required to lay of[f] 30 deg. in y^e Northerne tropick let the periphery of the outer parallel to the horizon be divided into 360 deg. of which count 30 from <u>w</u> to $\bar{\rho}$ a Ruler layd over the point $\bar{\rho}$ & y^e pole $\bar{p}$ cuts y^t Tropick in γ where the hourcircle 30^d from y^e Meridian also intersects it.

Or it may be other wayes found as was the division of y^e æquator. as in y^e 18th figure [Fig. 21.2] if through x the Center of the tropick the diameter kxi be strooke & the upper semicircle iλek being divided into 180 deg y^u count 30 of them from e to λ, for a Ruler layd over ye point λ and p ye pole. where it intersects the opposite point of the periphery in γ cuts of[f] ye arch γu in the tropick 30 deg. as will be easily proved for. let ye circle bwπ be strooke as much without ye horizon as ye tropick lies within y^e

æquator: the summe of the distances of these two Circles from theire Proper poles is a semicircle. theire halfe summe a Quadrant. & theire semidiameters being the halfe tangents of theire true distances from theire proper poles to the Radius of the æquator cz. the one xi[12] shall be the tangent of the complement of the others $\frac{1}{2}$ distance from its pole bz. to y^e^ same Radius but the Radius is a meane proportionall betwixt the tangent of any arch & the tangent of its complement to a quadrant it will hold therefore as xi: cz::cz:bz. & because the triangles xip. & pbz are similar. ix: px::bz:zp: but xλ is equall to xi & zb = zl[13]: & the angle λpx = ∠ lpz therefore the two triangles xλp and plz. are similar & y^e^ ∠ pzl = ∠ pxλ. but the angles pxi.[14] and pzc. included in these are right which abstracteing ther remaines y^e^ ∠ λxi of the one = to y^e^ ∠ bzl of the other therefore. the arches bl and λi. and Consequently their complements to a quadrants eλ & lw conteine an equall number of degrees of theire respective circles & the periphery of the tropick iuγk may be divided either from y^e^ equall degrees of the opposite semicircle iλek: or of y^t^ which lies as much without ye horizon as this lies within ye equator. this demonstration I have given because the learned Clavius has omitted both it & the method in his tedious worke[15]

Or by what I demonstrated on ye 16 figure [Fig. 20.2] either the æquator or the parallells to it may be divided from the southerne pole π[16] of ye 18th figure [Fig. 21.2] layd of[f] as I before directed on the plane of the projection. I must bee breif if it be required otherwayes to lay of[f] 60 deg. in y^e^ æquator from ye meridian or point æ. count 60 degrees in the periphery of the horizon but in the opposite semicircle from f to q. or in the upper semicircle of the æquator continued from d to σ. a ruler layd over the point of the invisible pole in ye projection π & either of these points q or σ cuts of[f] the arch æθ in the æquator 60. degrees or if it were required to lay of[f] 30 deg in the Tropick of ♑ from ζ on the same side. accordeing to what I provd in my last lecture & on ye 16th figure [Fig. 20.2]. if in ye periphery of ye circle πφχg. lieing as much without ye horizon you count 30 deg. in ye opposite semicircle from φ to χ a ruler layd over the pointes π and χ where it intersects ye Tropick of ♑ in η cuts of[f] the arch ζη in it 30 degrees consenteing altogeather with ye 3 other methods of division.

after these Methods may the path of the Vertex be divided in the pythagorean Projections. & haveing proceeded thus far I thinke I have finished. the doctrine of stereographicall projections some small remarkes I might have added but those will easily occur to the Ingenuous who will excuse me for omitting them in this straitnesse of time I have all along in this lecture supposed my more ingenuous hearers to understand some Geometricall easy theoremes, particularly that if any angle of a triangle be bisected equally the line bisecteing it Cuts the opposite side in such proportion as the other containing sides have to each other. which is a Well knowne Theoreme[11] & therefore I conceaved might be admitted without a demonstration especially considering that the Geometry Lecturer[18] has no doubt often demonstrated it in His readeings in this place. till an other opportunity Farewell:

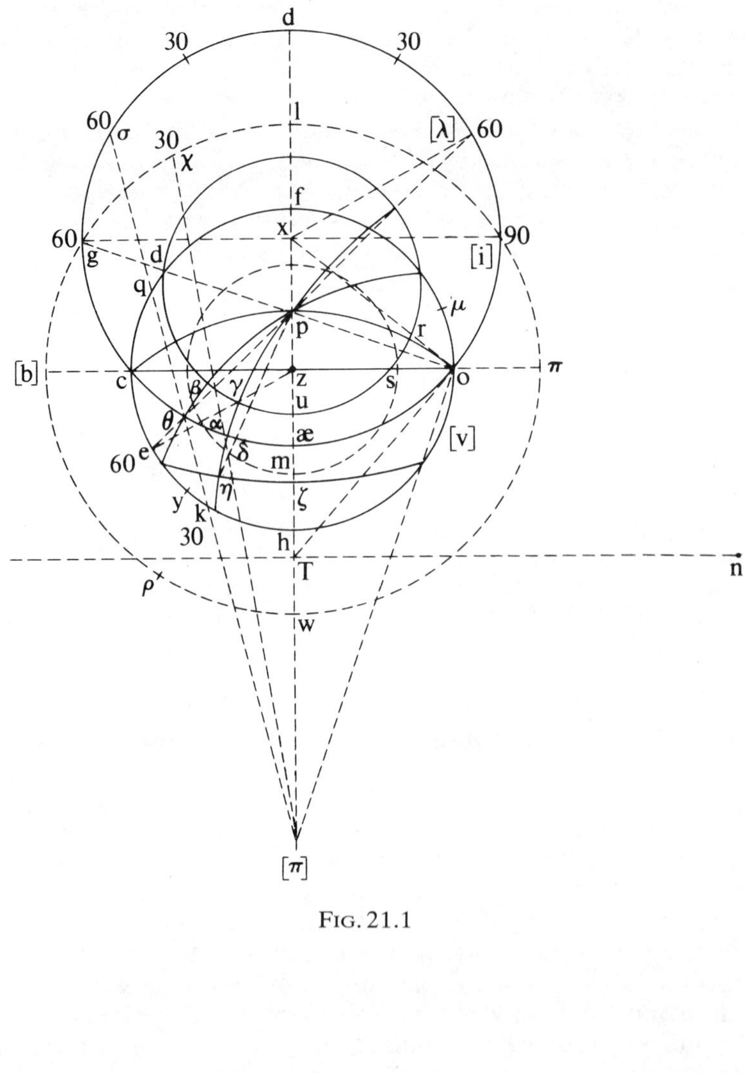

FIG. 21.1

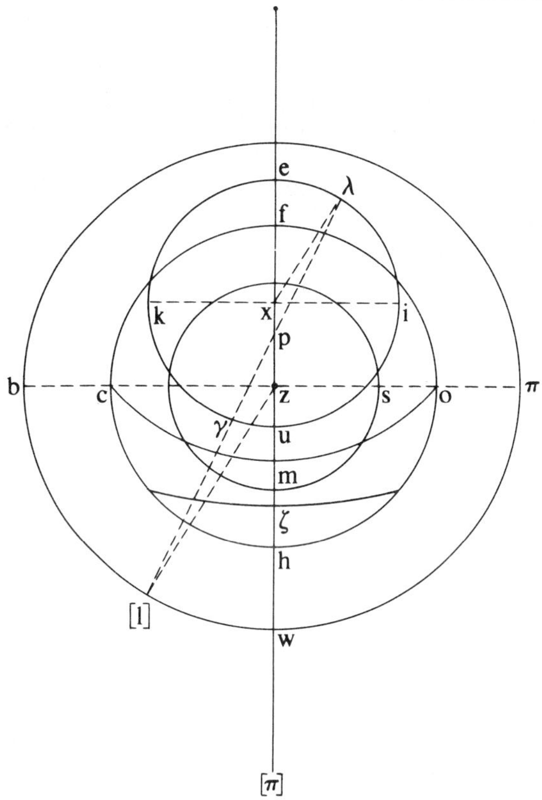

FIG. 21.2

NOTES

21.1. Op. cit. (note 15.12).
21.2. Op. cit. (note 15.11).
21.3. Cf. Lecture 20 for the proof of this theorem.
21.4. Cf. Lecture 16 for the proof of this theorem (viz. Theorem 6 in Flamsteed's *initial* numbering convention, pp. 262–3).
21.5. Flamsteed actually wrote Tn instead of Tm.
21.6. Lecture 19 (8 November 1682).
21.7. Flamsteed actually wrote cθαηo instead of cθαæo.
21.8. Flamsteed wrote hd instead of hf.
21.9. Flamsteed mistakenly wrote pcz.
21.10. Note the ambiguous use of the letter π, which is also used in the text to denote the south pole of this projection (cf. f. 205ᵛ) although omitted from Fig. 21.1 and Fig. 21.2.
21.11. Flamsteed wrote buπ instead of bwπ.
21.12. Flamsteed wrote ci instead of xi.
21.13. The point 'l' has been hitherto undefined, but its position is shown in Fig. 21.1.

21.14. Flamsteed wrote pci instead of pxi.

21.15. Op. cit. (note 15.9).

21.16. Cf. note 21.10.

21.17. That theorem and its converse are to be found in Barrow's 'Euclid' (op. cit. [note 15.10]), p. 117, Prop. III (or 6.e.3. in Flamsteed's notation): 'If an angle BAC of a triangle BAC be bisected, and the right line AD, that bisects the angle, cut the base also; then shall the segments of the base have the same ratio that the other sides of the triangle have, (BD.DC::AB.AC.). And if the segments of the base have the same ratio, that the other sides of the triangle have (BD.DC::AB.AC) then a right line AD drawn from the top A to the section D, shall bisect that angle BAC of the triangle.'

21.18. Robert Hooke (1635–1703) had been appointed Professor of Geometry at Gresham College in 1665.

Lecture 22

24 *January* 1682/3

In my first readeings here neare two yeares since I shewed y^u by what considerations the Contemplative Ancients attained to theire first knowledge of the Celestiall motions, and I thinke I made it evident by good remarkes upon theire observations that the want of a good account of the lenght of the yeare occasioned both theire enquirys into the motions of ye luminarys & planets & at length the restitution of the Catalogue of the fixed stars by Hipparchus whose correction of the Calendar & lunar revolutions being the best of all that had beene attempted before him, seemes to have beene the result of all his other labors & the Marke at which hee leveld & designed them.

Here I shewed how they came first to take notice of ye moones pa[ra]llax or difference of Aspect. by ordinary & course ocular observations at first how it was amended by ye observed durations of lunar Eclipses afterwards whereby her distance from ye Earth was first determined, & hers being got the suns was soone after discovered by considering the Time of her dichotomy or halfe illuminations & how long the[y] praeceded the first or followed the latter quadrature. & compareing some late observations of the dichotomys with each other I made it evident that the suns distance from y^e earth was so large that his parallax was almost insensible as I had formerly also found it by observeing the distance of y^e planet ♂ (neare his Achronicall appearance) from neare fixed stars with each other

This occasioned me to discourse of the best $Instrum^{ts}$ for makeing of observations, & haveing prefered the telescope, its Nature being scarce understood by many I spent y^e lectures of more y^n one terme[1] in Explicateing it & shewed how either Quadrants or Sextants fitted with Telescope sights might be rectified for observation & theire error if any were committed Might be discovered and corrected.

After this I returned to my Astronomicall designe & proved from the observations of severall ages that the greatest obliquity was scarce ever more y^n 23°-29′, & what ever observations seeme to represent it greater were either repugnant to themselves or contradicted by others as old and of no lesse authority. And from y^e

Notes on page 317.

yeare 800 to our time I produced such as made my determination indisputable.

Hence I proceeded to explicate the diurnall Phainomena of the Sphere: or the appearances of the Suns diurnall Course. or rather of the earths diurnall revolutions. for haveing produced some convinceing arguments for the Earths motion I assumed it, & because it had not hitherto beene shewne how the diurnall appearances were made in the Pythagorean Susteme or supposeing the Earth to be moved, I did it for that susteme & made it evident that they were more easily & readily represented in this then in the younger Ptolemaick For I also made it evident that tho it be the opin[i]on of the inconsiderate Common people now & was so formerly that the Earth was unmoveable, yet the more thoughtfull & ingenuous of the ancients asserted the Motion of the earth, & that sustemes of the heavens were by them formed agreeable to it long before there were any made suteable to the grosse opinions of the Vulgar or founded on its fixed stability as were those of Ptolemy & his followers.

To explicate this hypothesis & the diurnall appearances of the sun & stars a Projection of the sphere agreeable to it was requisite. I had published one in a printed treatise on this subject ye yeare before.[2] and now because the doctrine of stereographicall projections at least the grounds of them had never beene treated of in the English tongue. I thought my selfe engaged to explicate them in my way The learned Clavius had wrote a discourse of them entitiled de Astrolabio[3] but with that prolixity & such a long apparatus of Theoremes as might justly deter anyone from undertakeing to peruse him. Considering the subject seriously I found that all his Theoremes might be reduced to nine.[4] at least these few were enow for the whole businesse of stereographicall projections. the grounds of which I therefore taught from them & my new projection I demonstrated also from them after I explained the diurnall appearances from this projection & therewith I concluded last terme.

From ye solar it was my intent to have proceeded to the stellare appearances but I then wanted time, & haveing since thought further of this subject I find it will be necessary to take in severall other Considerations and to examine the opinions of divers philosophers & Astronomers concearning the substance of the stars, theire different magnitudes & colours, theire scintillation, disposition, & rectified Motion to make way to the Theorys of the planets grounded on observations of theire distances from them. to explain which by the agreement of the first lecturers with the honorable trustees the Reader of Astronomy whose place I supply is obliged.[5] of these therefore I shall discourse in order as the occasion will p[er]mit & it will be my endeavor in the roome of the doubtfull & disputable opinions of o^{r} praedecessors to settle such as may be lesse controvertible. and may be admitted without running upon absurdities. which is all in this case I dare p^{r}tend to. For certeinely the soule of man whilest included in the body however it is capable of reasoning & argueing concerneing it selfe & externall objects cannot possibly attaine to the knowledge of ye proper nature of any body as it is nor of the reall causes of those vast differences it notes in ye p[ar]tes of ye great universe. *all the knowledge it receaves is originally conveyd by ye senses. and extends no farther then to distinguish one thinge or body from another by some name accommodated to such peculiar propertys as it distinguishes in different existences* so that really all o^r knowledge is but a Number of Comparisons, but cleare & evident where they are abstract from matter, as when in Mathematicks wee compare lines with lines planes with planes. or solids with solids as ye Curve of a Parabola with a streight line a triangle with a square or parallelogramme. a Cube with a prism a sphere with ye Cilinder or ye like. but obscurer in other matters as when wee compare Gold or Silver with Copper wee distinguish betwixt them by theire Colour fusibility ductility specifick gravity & whereby wee compare them one with another, but the reall differences of the formes of those particles that compose them & the consequent reason of theire cohaesion wee understand not.[6] But then it must be much more difficult to judge or make comparison betwixt the remote heavenly bodys then it is betwixt those of which wee are possest & can touch & handle & other wayes examine. I shall therefore thinke I discharge my selfe well of my undertakeing if first haveing Recited ye opinions of the ancients or best esteemed modernes concerneing the heavenly bodys I shew y^u with what improbabilitys or absurdityes each labors, and afterwards compareing those remote with nearer objects, I shew y^u such resemblances in the one to ye other as may render it probable the appearances in both proceed from ye same cause.[7]

What ye fixed stars really were was a great Controversy amongst the ancients[8] Anaxagoras conceaved them stones drawne up from ye earth by whirlewinds & fired by theire violent motion or circumvolution about ye earth. Zenophanes that they were coles fired by blaseing or burneing cloudes no doubt because hee saw ye Cloudes. oft fringed as it were with fire at sunset. However wee may collect that these persons apprehended not how far the stars were removed from ye earth & to how small a distance the Cloudes are elevated from it else they had never run up on such absurd opinions, Diogenes conceaved the vaulted arch of the heavens to be p[er]forated as pumicestone, are irregularly & that the Celestiall fire above it shineing through the p[er]forations made the appearances of stars. but hee was better at his Morall then Naturall Phylosophy & for that wee may excuse him. but even in those first & as we commonly call them darke times wee find some that by the strength of theire reason had determined nearer truth, if they did not learne the ire opinions from the Jewes or Indians with whom they had conversed in theire travells these were Thales Orpheus Heraclides Pythagoras & Plato of whom Thales Conceaved them burneing Earths, Orpheus Pythagoras & Heraclides taught that every star was a World with its Elements; Plato that they were a mixture of all the Elements, but had most of the fiery in them, whose opinion if admitted with a due latitude of sence will be found not much different from those of the celebrated Philosophers & Mathematicians of o^r age. Brunus Kepler Galileus and De Charte. of whom Riccioli tells us that Brunus in a tract hee wrote *de infinitis & innumerabilibus*[9] affirmes the stars to be so many suns. Kepler in his dissertatio cum Nuncio Sydereo has these words *Altera jucundissima tuae observatio est figurae fixarum radiosae, differentis a planetarum figuris circularibus, quid aliudinde Galileae colligemus; quam fixas lumina sua ab intus emittere, planetus opacos extrin secus pingi, hoc est ut Bruni Verbis iter illas esse solis, hos lunas sive tellures,*[10] y^t is y^r next usefull & pleasant observation is y^t ye bodys of ye planets a[re] p[er]fectly round, the fixed stars radiateing & darting light every way. from which what inference is more naturall then that the stars shine by the emission of theire proper light. the planets by ye suns reflected or in the plane words of Brunus that the stars are so many suns, the planets Moones or Earthes, & Galileus in ye 3^d Dialogue of his Sustema Cosmicum in ye Person of Salviati sayes *vide porro quam concinnae tollatur ex universo praecipitatissimus ille motus 24 horae anno & quomodo stellae fixae quae totidem soles sunt, nostro soli conformes perpetua fruantur quiete:*[11] see further how commodiously that violent motion of ye whole heavens round ye earth once in 24 houres is removed, & how ye fixed stars which are so many suns continue unmoved from their prop[er] plaçes as o^r sun also does. The same opinion is expresly maintained by De Charte in his Principia Philosophiae,[12] & by Borelli in a small tract of his y^t I saw translated into English many yeares agone[13] wherein hee endeavors to prove wth it ye multiplicity of worlds. that I may be breife Generally all those who have admitted the motion of ye earth have likewise asserted this opinion, & none but those who have subjugated themselfes to Aristotles philosophy or the decree of a conclave, since the Invention of the telescope have maintained any other.

Yet such there have been not a few but so unfortunate in theire assertions that seldom any two have subscribed to ye same opinion. Caramuel. conceaves them solid bodys but then hee will have theire light to be onely the suns species reflected from theire planes. this opinion of his is reported by A. M. Schirlæus[14] and refuted because as hee well observes at that great distance whereto they are removed from o^r sun hee appeares no bigger then a lucid point. So y^t very few of his rayes could be reflected from them and considering how many of them must be lost upon theire bodys twould be altogeather improbable that any of those reflected rayes should reach our eye or that they should be seen by them. but then hee delivers as absurd an opinion of his owne or worse. For hee concludes them to be onely the brightnesse of the wholly glorious & lucid Empyreall heaven shineing upon us through the pores or I know not what holes hee conceaves in the next inferior heaven. this opinion is Not much unlike that of Diogenes, & Mastrius[15] thinkes it repugnant as it really is tho the history of the Worlds creation delivered by Moses in ye First Chapter of Genesis. Ricciolus in his Almagest[16] reports severall other opinions of severall men but with all confesses the rejection of them is irkesome to him & not being satisfied with any of them nor dareing positively to maintaine the doctrine of the Copernicans in this point for feare of a Church sensure aforebeares to assert any thinge of theire nature p[ar]ticularly but onely tells us that hee conceaves the great hand that formed those vast bodies made them not to be uselesse whereby wee may collect. that if hee migh[t] safely have done it hee would willingly have asserted the opinions of Brunus Kepler & Galileus & ye rest of ye Copernicans

The opinions of Anaxagoras & Zenophanes that were either stones drawne up from o^r^ earth by the aether or coles fired by blaseing cloudes are so childish and ridiculous that onely to mention is sufficient to refute them, that the aether should attract stones up into it is not probable because if it did so formerly why should it not continue to doe so still. Whi[r]lewinds indeed wee have heard of that teare the Earth sometimes but that the[y] carry not any thing of it into the skyes o^r^ owne experience also tells us what they remove from one place being constantly left in some other, that they should be any matter fired by blaseing cloudes wee must suppose them very neare o^r^ earth for wee see that the hills & mountaines on it superficies often transcend them, the highest cloudes being seldome found to arise so much as 5 miles from its surface but indeed rarely $\frac{1}{10}$ p[ar]te of that distance whereas it appeares by the times the stars continue in o^r^ view that they are infinitely distant & therefore though they blased as Zenophanes supposed yet were it impossibly theire fire should reach those so remote bodys besides as I have have hinted before, it was the figure of the cloudes whose limbes being penetrated by the raies of ye riseing or setting sun represented flame or fire that imposed on Zenophanes but that they were not really kindled by him was evident for that this appearance ceases im-ediately after sunset & seldome continues so long as the Twilight whereas if the Cloudes were really kindled the appearance should in reason continue all night after at least till ye burneing matter were spent it should be more in the night then ye day because there is then no light of the sun which might by its greater force & strength either hide or extinguish it. but I stay too on such idle opinions the next of Diogenes resumed by Caramuel & as wee may reasonably guesse continued downe in ye Peripatetick Scholes to o^r^ times seemes to be the naturall product of that Philosophy which affirmes the planets carried about ye earth in solid orbes & might be rejected as altogeather unreasonable & improbable if the Authority of its first propagators & late assertors did not require us to take some notice of it & not to let it passe without a refutation.

Many of the ancients conceaved that the planets were carried round the earth in solid orbes. amongst the rest Diogenes whose opinion it seemes to have beene that the highest under the empyreall was not onely solid but opake so that the light of the empireall could not be seene by us but through the pervious places of that under it without these solid orbes they conceaved that it was impossible the motions of the planets should be p[er]formed but then wee may enquire how it happens that wee see no such thing for if light can passe through the pervious places of such a solid orbe why shall it Not be reflected from its under solid superficies Why have wee not constantly an arch of light reflected from it envesteing the earth by night by the suns reflected rayes as well as day and light at other times by those transmitted from his body to us. this opinion is therefore altogeather improbable but that there are no such solid orbes are further evident by the observations of comets which if there were must be stopt & impeded in theire motions by them. for it is evident by observation that they have passed through all the limits which the ancients gave to theire solid orbes & commonly the[i]re planes too & therefore if there be any such must either be stopt by them or breake them both which are equally absurd to thinke & hold.

But further the Noble Restorer of Astronomy the Famous Tycho takes notice that if there were such the Orbe of Mars must enterfere & clash with that of the sun,[17] for this planet in his Acronicall appearances or oppositions to ye sun is allwayes considerably nearer to the earth then the sun is. as I shall make it evident hereafter[18] not onely from the observed measures of his diameter but also by ye parallaxes or differences of aspect from o^r earth and ye sun, for y^e ½ diameter of his Epicycle in ye ancient formes is so large that it comes sometimes neare halfe ye Radius of the suns orbe within it towards ye earth in ye Ptolemaick susteme so y^t the bodyes of the sun & ♂ may meet & clash which absurdity some of the late antiCopernicans takeing notice of, have in theire sustemes layd his orbe with that of ♀ & ☿ so as they May have the sun neare the Center And be carried round him & with him round the Earth Particularly Riccioli in his old susteme,[19] & Tycho who for the like reason makes the sun the center of ye orbes of ♃ & ♄. wherein hee has not a few followers.[20] but the severall opinions concerneing the systeme of the World will be more properly & fully handled when I come to discourse of that subject[21] I should now proceed to examine the severall opinions of Thales Orpheus Heraclides Pythagoras & Plato & shew y^u how nearely they agree with & Resemble those [of] Brunus Kepler

Galileo De Charte & Borelli[22] but In so doeing I should engage my selfe in a discourse to[o] prolix for my time & therefore I remit it to my next lecture

NOTES

22.1. Lectures 4–9 inclusive, delivered by Flamsteed during the Trinity and Michaelmas Terms in 1681.

22.2. Op. cit. (note 15.1).

22.3. Op. cit. (note 15.9).

22.4. Cf. notes 16.24 and 20.2. Theorem 9 is probably that enunciated in note 21.17.

22.5. Flamsteed was deputizing for Walter Pope, and was required by a college regulation of 1597 to lecture on the theory of planetary motion. Cf. Lecture 25, f. 233^r, or p. 339.

22.6. Flamsteed is merely restating the philosophical concepts of Plato found in Riccioli 1, 'Praefatio', p. ii.

22.7. The continuation of the text (f. 211^r) was here deleted by Flamsteed.

22.8. The main source of the information which follows on ff. 211^v, 212^{r,v}, (pp. 313–16), was Riccioli 1, lib. 6, 'De Stellis Fixis', pp. 393–6.

22.9. Cited in ibid., p. 394.

22.10. Johannes Kepler, *Dissertatio cum Nuncio Sidereo* (Pragae, 1610), p. 22.

22.11. Galileo Galilei, *Dialogus de Systemate Mundi* (Londini, 1663), p. 456.

22.12. René Descartes, *Principia Philosophiae*, 4th edition (Amstelodami, 1664). Cf. 'Pars Tertia, De Mundo adspectabili', §. 23, p. 56.

22.13. Johann Alphonso Borelli, *De Motionibus a Gravitate Dependentibus* (Regio Julio, 1670), of which a brief English summary is given in *Phil. Trans.* vi, no. 73 (1671), 2210.

22.14. Antonius Maria Schyrlaeus, *Oculus Enoch et Eliae sive radius sidereo-mysticus* (Antverpiae, 1645), p. 177.

22.15. Cf. disput. 2 de Caelo q.2 art. 3, cited in Riccioli 1, p. 394.

22.16. Cf. note 22.8.

22.17. Op. cit. (note 2.5) (1).

22.18. This matter is not, in fact, discussed in any of the subsequent lectures.

22.19. Riccioli 2, pp. 288–9.

22.20. Ibid., pp. 287–8.

22.21. Cf. note 22.18.

22.22. The views of these astronomers are discussed in Riccioli 2, pp. 290–500.

Lecture 23

31 *January* 1682/3

In my last lecture I acquainted y^{u} with the opinions of the ancient & moderne Philosophers concerneing the Nature & substance of ye Fixed stars & examineing those of Anaxagoras Zenophanes & Diogenes of the ancients and Caramuel of the Modernes I shewed y^{u} with what absurdities they labored I come now to the More rationall of Thales Orpheus Heraclides Pythagoras and Plato of whom Thales conceaved them burneing Earths Orpheus Pythagoras & Heraclides taught that every star was a World with its elements, Plato that they were a mixture of all the elements but had most of the fiery in them. whose doctrines taken with some little latitude might be easily reconciled to ye receaved opinions of Brunus. Kepler Galileo De Chart & Borelli, who have al[l] taught that they shine all by theire owne light being so many suns placed at an immense distance from o^{rs} & each other. which that they are will be made probable if wee consider theyre light. figures & that vast distance to which by demonstrations they may be proved to be removed from o^{r} earth.

As for theire light & figure tis much different from y^{t} of the planets or Erratick stars. The discovery of the telescope in ye begining of this century has shewed us this difference & without it wee had wanted those demonstrations of it which are now common & ordinary by it wee see the bodys of the planets are perfectly sphericall & round, and the inferior ♀ & ☿ are liable to all those changes & variety of Phases in theire illuminations y^{t} wee observe in o^{r} Moone theire direct conjunctions with the sun or when they are remotest from o^{r} earth to be fully illuminate in theire greatest elongations but halfe or bisected but as the[y] approach nearer ye Retrograde Conjunction horned. as they ought to doe supposeing they receave theire light from the sun, on the contrary the planet Mars by reason his orbe lies without the orbite of the earth ought alwayes to appeare more then halfe illuminate by reason the earth however posited from him sees more y^{n} one halfe of his illuminated hemisphere & so it really is for the telescopes allwayes shew him more y^{n} halfe illuminated. ♃ & ♄ by reason of

Notes on pages 326–7.

theire great distance from ye earth ought to want so little of appeareing fully illuminated at all times that the defect should scarce be discernable in a long glasse as really it is not. whence it is evident that they all receave theire light from ye sun which is further also proved in ♃ by the shadowes of his satellit[e]s passeing over his body & makeing small Eclipses on it when they are carried in theire orbites betwixt him & ye sun. as also in ♄ by ye shadow of his body on his ring or of it on his body. as for the brisknesse of theire lights the appearance depends upon theire distances from the sun. the light of ☿ & ♀ being strong of ♄ & ♃ weaker & more faint because ye orbes of the first are little distant from ye sun compared with ye orbes of the other. Nor is it to be doubted but that the different matter of which they are composed may contribute much to the vivacity as well as different coloures of theire reflected light. for as in o^r earth theire are an infinite number [&] diversity of soyles. of which some reflect the suns raies or light strongly others weakely, some imbibe & receave it so the bodys of the planets may be such as that some of them may reflect the light briskly others not so. But now if wee come to consider the fixed stars, wee find them with o^r bare eyes like lucid pointes darteing theire raies every wayes, & not much different with glasses save that the ejaculations are less & the better or longer glasses wee apply theire bodyes appeare the bigger lesse hairy & approaching nearer to p[er]fect roundnesse so that it seemes probable that if wee were placed so neare them as that theire bodyes might appeare as large to us as the sun does theire light would strike o^r eyes & affect them as strongly as his now does but of this I shall say more hereafter when I come to consider theire scintillation[1]

The next argument for the fixed stars being either the same or at least of a like nature with o^r sun is drawne from theire real magnitude which can not be proved without some knowledge of theire distances and apparent diameters. As for the apparent diameter Before the invention of the Telescope they were esteemed much more then they have beene really found since for till then there was no way knowne of measureing them but by the estimate of the eye which by reason of their radiation was very Fallacious & uncerteine. Maginus[2] makes y^e diameter of a star of y^e first light 10 minutes in which large quantity none either of the ancients or Modernes agree with him. Tycho more nearely of 2 minutes onely one of y^e 2^d $1'\frac{1}{2}$, of y^e 3^d 1 min. of y^e 4th $\frac{3}{4}$ of a minute of y^e 5th $\frac{1}{2}$ a min of y^e 6th 20″. Kepler before y^e telescope was knowne makes a star of y^e 1st light 3 min over of y^e 2^d onely 2. but afterwards haveing an opportunity to compare them with the planets hee altered his opinion, & made them more Agreeable to the Following measures of Riccioli who in the 6th booke & 9th Chap. of his almagest. Nov. both shewes the method hee used of observeing them & his severall measures. whence hee concludes that the diameter of Syrius is 18″. of Arcturus neare 17″. Aldebaran and Spica ♍s 15″ Regulus & Regel 14 Procyon 12 Aquila 11″. ye Polestar 8″. Algol 7 Propus 6. ye brightest of ye Pleiades 5. Alcor 4″.[3] which yet I must needs confesse I esteeme with ye largest for when in ye yeare 1672 I tooke ye diameters of the planet ☿ & Syrius in ye cleare day with a Micrometer & telescope of neare 14 foot long I found them nearely equall & each 14 seconds.[4] but if either I esteemed Syrius the lesse of ye two

I have proved to y^u formerly In my readeings here both by the observed distances of ♂ in his opposition to the sun & parallaxes thence deduced As also by the observations of the Moones dichotomys that the suns parallax was not more y^n 10 seconds & therefore the diameter of the earth to y^t of ye sun nearest as 1 to 100 & therefore the distance of ye sun from ye earth about 11000 of its diameters[5]

Now that the orbe of the fixed stars is vastly distant from ye Sun will be evident if wee consider that 1st they have alwayes retained the same positions in respect of each other from what part so ever of the earths orbite shee views them for in my former readeing I have brought arguments enew to prove the motion of the Earth to any unprejudiced judgment

Ptolemy himselfe confesses this truth. & allowes the name of fixed to be very properly attributed to these stars on the same account reckning up several ternarys of stars which appeared in the same right line with each other in or before Hipparchus his time & continued in the same in his dayes as they also doe in oures. yet for all this assent the judgment of the Naked eye is so fallacious that little could be attributed to his argum^t if it had not beene found by accurate measures that theire distances have still continued the same & invariable I will not instance In any of Hipparchus his observations they are too Course & seldome nearer truth then within $\frac{1}{3}$ of a degree I must come Nearer home. The Noble Tycho observed in his dayes the distance of y^e bright star of ♈.[6] from the Bulls south eye or Aldebaran 35°.32′. from which neither Hevelius nor my owne measures differ $\frac{1}{3}$ of a minute. but to be more certeine feareing that the distance might be different in distant times of the yeare by reason of the Earths accesse to or recesse from the stars in her orbite I was carefull to repeate it at severall times of the yeare & so to order the time of the observation that tho the refractions were altered they might not Corrupt the measure but I still found it the same or not more y^n $\frac{1}{3}$ of a minute different[7] which is so small a p[ar]te in an instrum^t of 7 foot Radius that tho really sensible yet in this case I thought it of no moment & rather some fault in the instrument for I noted that what difference I obs[er]ved was not alwayes as It ought to be an excesse when the Earths place was betwixt the two fixed stars or defect when shee was removed beyond Quadrature from them but on the Contrary sometimes a defect in ye first Case an excesse in ye second

And yet that it should have altered if the diameter of the earths orbite subtended any sensible angle at the fixed stars, & beene greater when ye earth lay in the line drawne from ye sun to ye middle place betwixt ye fixed stars towards them then when it was in ye opposite p[ar]te of its orbite & ye sun from it in ye same line will be evident from ye first figure [Fig. 23.1] in which let asb. represent the diameter of the orbite of ye earth. passeing through the sun & if produced bisecteing ye difference of the longitude of ye two stars c & d when ye earth is at a the angle caf of theire halfe difference of longitude and consequently halfe theire distance shall be greater than when shee is at b in ye opposite pointe of the orbite & end of the said line. because the angle caf is the outward angle of the triangle bca. therefore equall to both the two angles at c and ∠ cba. & by consequence greater than either of them. but the halfe of any lines & numbers have the same proportion to each other that theire wholes have therefore the whole angle cad proportionably greater y^{n} ye angle cbd.

It is evident from ye figure & might be easily demonstrated if necessary that the greater the angle cad of ye distance of the two stars is so much greater would the difference be betwixt it and ye angle cbd. this caused me to thinke of measureing wider distances at different times of the yeare to compare y^{m} that if there were any I might find it but I still had ye same event. the measures were allwayes the same & if managed with requisite caution to avoyd the effects of refractions insensibly different.

So y^{t} I was forced at length to acquiesce in the opinion or rather subscribe to ye assertion of Copernicus and his Followers that the parallax of the Earths orbite at the Fixed stars was insensible and the angle subtended by it from them not more then a minute.

To comply with o^{r} narrow capacityes I will suppose the semidiameter of ye Earths orbite sb in ye 2^{d} figure [Fig. 23.2] to subtend ye angle bcs. one minute at ye fixed star c. hence still it will follow that the side sc ye distance of the star From ye Sun is above 3000 semidiameters of ye earths orbite bde in ye 2^{d} figure [Fig. 23.2]. or so many times 11000 diameters of the earth which tho it be a stupendious length to those who are used to measure by inches feet yards & miles yet to such as more commonly contemplate the vast & unlimited power & greatnesse of a Deity with ye extent of ye world I am persuaded it wil[l] not appeare so large but that upon good seasons assigned for it they can & readily would subscribe to one much more ample: for to say the stars can not be so far distant from us because the measure is incredible what is it but to limit the workes of a Deity or what can it savor of but a narrow & shallow understanding

The $\frac{1}{2}$ Diameter of Syrius suppose represented by the cx = cf in y^e 2^d figure [Fig. 23.2] let be $\frac{1}{8}$ of a minute[8] = to y^e angle cSx or Sxa. therefore is cx = $\frac{1}{8}$ Sb. & the body of Syrius to a sphere of the same matter whose diameter is equall to that of the earths orbite will be as the Cube of one to y^e Cube of 8. or as 1. to 512 for spheres are in the same proportion as y^e cubes of theire diameters

But the suns diameter is about y^e 220 p[ar]te of y^e diameter of the Earths orbite. therefore his solidity to its as ye cube of 1 to ye Cube of 220 or as 1 to 48400 therefore the solidity of ye Sun to ye solidity of Syrius as 512 to 48400 therefore the body of Syrius exceeds that of the sun 94 times by this reckning which I have purposely cast up with what I advantage I could for these of ye contrary opinion had I beene mindeing to distresse them it were easy to have made ye difference vastly greater without resting any observation

And if I suppose a star to have but one second diameter yet it will be found above double as big as o^r sun. for the proportion of its diameter to y^t of ye Earths orbite will be found as 1 to 60 therefore theire solidytyes as 1 to 21600. but the suns to his orbite is onely as 1 to 48400. therefore the star above twice as big as the Sun.

And now tis evident by this traine of Comparisons and theire naturall consequences that there is not a fixed star visible to ye bare eye but which is equall to if it exceeds not o^r Sun in magnitude And since theire light is proper or rather adherent to theire bodys as o^r suns is, wee have got a couple of as good arguments for theire being suns as the nature of the subject will beare to wit theire greater bodys, & similar light. I shall now compare them further & fetch another argument from theire being such from the extinction of some stars the accession & periodicall appearances of others but I must returne to o^r sun first & enquire what is the reason of that strange appearance of spots in his face observed frequently upon ye first discovery of the telescope,[9] more rarely since.

Tis now above ten yeares since I was first fitted with glasses & instruments necessary for makeing caelestiall observations[10] & tho in all this time I have had frequent occasions to view ye sun especially for this 7 yeares last past yet have I not seene above 2 of them. the French in the meane time whilest I was otherwayes employd observed a third in the yeare 1674.[11] an account of these will intimate to us what ye substance of the suns body may be & consequently if wee find like appearances of the fixed stars what also may be theire Compositions

The first time I saw any spot in the sun was on the 27 of July 1679 whilest I was takeings the suns heights to find ye error of my clocks[12] it was entred then some 5 minutes within his disk and appeared broade in the figure as neare as I could compare it of the sole of a shoe, 2 small indentures one on each side neare one end formed ye appearance of an heele it had some small p[ar]tes neare it not so opake & darke at a little distance as if they had beene broken of[f] from ye maine peece, it continued entire that day ye next & ye following but the forementioned indent widening as if that p[ar]te I called the heele were separateing from ye other. on the 29th after noone they seemed to be connected by a slender neck. but the next morneing when I expected to have found them in two p[ar]tes I saw them planely seperated into 3. which wasted dayly & grew lesse whilest they stayd in ye visible diske of ye sun from which it made its exit on ye fifth of August Mr Halley observed the same appearances and alterations at Oxford so that there could be no doubt but that the changes were really made as I had observed them. Seeing it thus waste I could not expect it should endure so long as to make a second visible returne into ye suns diske however I applyd my glasses to him 14 dayes after but found the spot had disappeard:

In the moneth of October followeing another spot appeared in ye sun but More compact then the former & without any broken p[ar]tes about it as the first had. this was seene here at London on ye 16[13] but not by me till ye 17th day from which till ye 23. I observed it [as] frequently as the serenity of ye heavens would p[er]mit, on ye 9th of November following I found it againe returned into ye suns diske. within which I observed it till ye 20. at which time it seemed to have consistence enough to make a third returne accordeingly on ye 6th of december after I found it come within his visible diske againe but now it seemed much diminished of its former magnitude yet still without any broken p[ar]tes without it. Wee had now much cloudy weather but I had the opportunity to observe it thrice in this transit ye last time was ye 15th of Decem. I hoped it might still returne & make another visible passage over ye suns face but when in ye begining of January I viewed ye sun & sought it I found it gone no trace of it remaineing. this caused me to consider better of the solar spots then I had done before. if they were as clouds & emanations from his body I saw no reason why the latter should not have broken into p[ar]tes as the former or why it should endure three revolutions whereas the first could abide but one and scarce that before it was absorped or dissolved. It seemed therefore most probable to me that as there are vulcanos continueing which cause frequent eruptions of o^{r} earth & throw up mountaines sometimes where formerly there were none or but little eminences, so there might happen in the sun eruptions of his centrall or subcutaneous matter which when they are low though broad are soone worne or melted downe againe by ye menstruum or matter of light encompasseing him. that the first spot I had seene in August was some such thing but the latter of a greater height & density so that it was washed downe gradually & not but after a long time for tho I have met with three observations of some spots in Scheiner[14] which made a second returne into ye sun spot have I never heard of any that made 3 as this did before, and it seemes altogeather improbable to me that any cloudlike matter or slough emitted from his body should be able to continue in that subtle menstruum that envests him 3 moneths. for almost so long I observed it, & tis probable it had beene there some time before I saw it, & Continued some time in his Invisible haemisphere unreduced after it had passed out of his apparent.

But further, it kept the same Constant place in his body; & allowing its revolution as to the former of 25 dayes & $\frac{1}{4}$. With y^e constant inclination of the suns Axis about 7 degrees, I found I could answer all the appearances as nearely as could be desired or expected considering y^e difficulty of the observations, whence wee may collect y^t it adhered to it & had not any thing of y^e nature of a cloude swiming upon it, which might therefore alter its place & be obliged to no constant position within his diske.

The learned Jesuit Scheiner in his Rosa Ursina[14] gives us an account of the Courses of severall spots over ye sun some of which for theire dilutenesse hee calls faculae y^t kept theire due course & distance from ye pole as if the clouds hee supposes to be the spots were fixed, & blased upon ye superficies of the sun, and these often times take up no very small space of his diske. whereas tis much more probable they were onely some lower but broader eruptions of the centrall mater in which the outward & more subtle had soone made it selfe channells & if I may so call them lakes which appeared brighter by reason of the other darke & terrestriall p[ar]tes encompasseing them for light appeares the best & brightest when it shines from, or is compassed with darke objects.

And that the stars are subject to the same accidents is evident from that new one in the Whales neck whereof the excellent Hevelius has given us at severall times a large history first in his ☿ sub sole[15] since in the Philosophicall transactions & letters. to Mr Oldenburge.[16] for the Phaenomena of that star & its disappeareing for about 6 moneths of the time of its revolution can not be attributed to any thing but a thick Cortex or Crust of darke matter covering its land parts for more then $\frac{2}{3}$ of its body otherwayes if wee suppose it halfe darke halfe light it will follow that it ought never to be totally extinguished but at that moment when y^e darke hemisphere is towards us. I might instance also in the new star at y^e swans bill & in her brest[17] but theire Revolutions are slower & not yet so nearly determined. Nor need it be thought strange or the truth of this conception questiond because of the length of the time of its revolution. for tho o^r sun make his but in 25 dayes yet this strange star being in all probability much bigger then hee may require a much longer Revolution. it were not difficult to produce further instances nor is it [difficult to][18] conceave that the greatest part of the sun or his whole superficies might be as easily compassed with spots or faculae as any small p[ar]te of it. so that now wee have a 3^d Comparative Argument to prove the stars are reall suns what remaines to be added I must at present leave to the thoughts of y[e] ingenuous & a further opportunity:

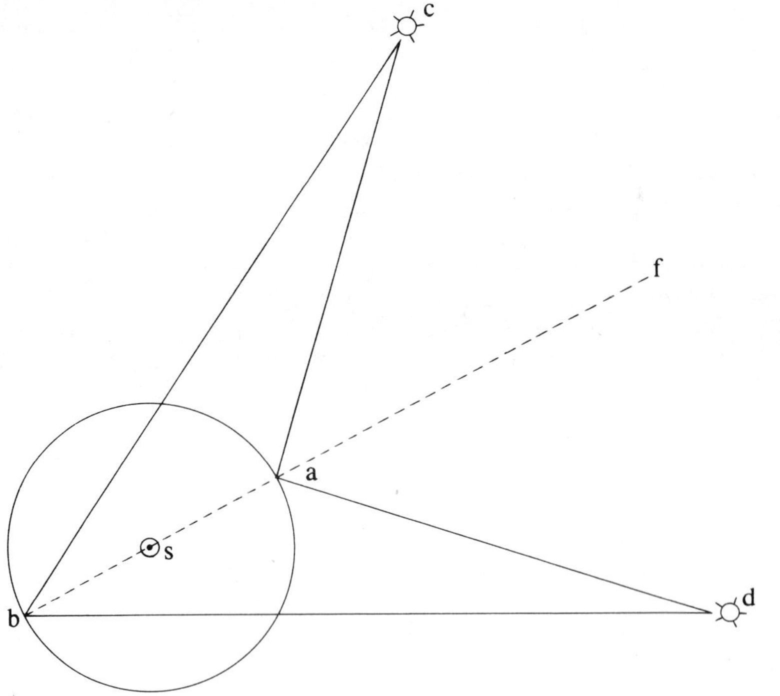

FIG. 23.1

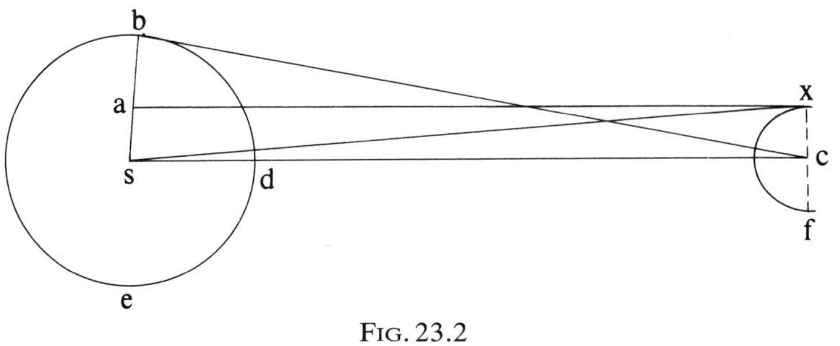

FIG. 23.2

NOTES

23.1. Lecture 27.

23.2. The following figures for the stellar diameters adopted by Giovanni Antonio Magini, Tycho Brahe, and Johannes Kepler, are tabulated in Riccioli 1, p. 424.

23.3. Ibid.

23.4. See the entry for Tuesday, 22 October 1672 in op. cit. (note 1.18), p. 17.

23.5. This figure corresponds to a value of about 9″·3 for the solar parallax, in accordance with the results of the observations made by Flamsteed at Derby on 26 September 1672 (cf. note 2.42) and at Greenwich on 14 January 1678/9 (cf. Lecture 2, f. 24ʳ).

23.6. Viz. α Arietis. The distance of this star from Aldebaran given by Flamsteed is the same as that tabulated in Riccioli 1, p. 426.

23.7. The value of this quantity in op. cit. (note 1.18), part 2, p. 1—obtained from observations made at the Royal Observatory, Greenwich, on Sunday, 29 October 1676—is 35° 32′ 16″.

23.8. This was Flamsteed's own observational result for Sirius's apparent radius. Cf. note 23.4.

23.9. The appearances of Sun-spots first became widely known through the publication of Galileo's *Istoria e dimostrazioni intorno alle macchie solari* (Romae, 1613), although he himself was not the first person to observe them telescopically. Thomas Harriot in England, Simon Marius in Germany, and Father Christopher Scheiner also in Italy each made similar observations independently at about the same time.

23.10. A remark in Flamsteed's letter to Collins on 20 March 1670/1, published in S. P. Rigaud's *Correspondence of Scientific Men*, ii (Oxford, 1841), p. 110, indicates that his first telescopic observations may have been made that same evening.

23.11. A description and diagram of a Sun-spot observed in February and March 1678 are contained in *Journal des Sçavans de l'an 1678* (Amsterdam, 1679), 93–95.

23.12. This is an allusion to the use of the method of double (or equal) altitudes for finding the time of apparent noon by the clock. Full details of Flamsteed's observations of this first Sun-spot appear in op. cit. (note 1.18), ff. 361ʳ–362ᵛ.

23.13. Flamsteed's observations on this second Sun-spot appear in ibid., f. 362ᵛ, pp. 363–6.

23.14. Flamsteed possessed a copy of Scheiner's *Rosa Ursina* (cf. note 7.2) to which he refers at the beginning of f. 223ᵛ.

23.15. Johannes Hevelius, *Mercurius in sole visus Gedani, anno ... 1661: etc.* (Gedani, 1662).

23.16. Hevelius's letter to Bullialdus of 15 March 1667 regarding the nova in the Whale's neck is published in *Phil. Trans.* ii, no. 25 (6 May 1667), 460. A letter to Oldenburg of 2 January 1667 containing an account of his further observations of this and two other novae (cf. note 23.17) is in ibid. xii, no. 134 (23 April 1677), 853–8.

23.17. Hevelius's observations on these two novae are reported in: *Phil. Trans.* i, no. 19 (19 November 1666), 349; ibid., no. 21 (21 January 1666/7), 372; ibid. vi, no. 73 (17 July 1671), 2197–8; ibid. xii, no. 134 (23 April 1677), 853–8.

23.18. For some inexplicable reason, the words 'difficult to' were actually deleted by Flamsteed.

Lecture 24

7 February 1682/3

In my first Lecture this terme[1] I shewed y^{u} by Instances that all o^{r} knowledge was but comparative. & that wee could not understand the nature of Caelestiall bodys but by compareing those remote objects wth such as we are conversant with at lesser distances. Accordeing to this principle I consider the fixed stars & by compareing them with our sun I made it probable that they were like substances from theire greater magnitudes, similar light, & the periodicall appearances, & disappeareings; I instanced in the New stars in the Whales neck & the Swans brest, for which I gave y^{u} such Reasons as the subject would admit & since sometimes wee can proceed no better in ye search of truth then by rejecteing absurd opinions in o^{r} Comparisons, I shewed the absurd[it]y of the old doctrines of Anaxagoras & Zenophanes & Diogenes concerneing y^{e} ★^{s} in my first lecture[2] & the proba[bi]lity of the stars being Suns in my next[3] from ye aforementioned arguments.

I made it also probable. that ye Nucleus of the Suns body was not, as it has. hitherto beene esteemed, of the same matter with ye outward apparent matter of light but of an opake & dense body, which by some accidents in it like ye Volcanos of o^{r} earth was throwne up. higher then his lucid sup[er]ficies which made ye appearance of spots in his face till it was wasted away or washed downe by that subtle & active menstruum that envests him

And if the inner substance of the sun be such as the Phenomena of the spots intimates it is why may wee not subscribe or assent to ye opinion of Orpheus Thales Heraclides & Pythagoras who conceaved them to be Worlds with all theire elements. for certeinly the different duration of them intimates that the severall spots are of different compositions and as in o^{r} earth the soyle & mineralls are various in Regions not far seperated from each other so allso in the inner substance of the sun there may be as great a variety, but this is not to be supposed without a different mixture of the Elements that compose it, so that takeing the World, as they did, for a substance composed of theire severall elements, & the light of the sun (which

Notes on pages 336–7.

is the fountaine or originall cause of ye etheriall heate) for fire it followes that the sun it selfe & so every star May be a World with all its elements as Plato also thought;[4] it may seeme strange that I should intimate that the sun himselfe is not fire, but of this I may have occasion to say more hereafter. at present to confirme my assertion concernein[g] ye substance of y^e^ ★^s^ it is requisite that I give an account of the New Stars that appeared in Cassiopeas Chaire 110 yeares agone.[5] in ye Swans breast Anno 1600.[6] in Serpentarius 1605[7] In the Girdle of Andromeda.[8] & at ye Swans bill of late[9] whose appearances at first sight may seeme repugant to this opinion but if duely considered will be found to Confirme it.

And that wee may the better doe this it will be necessary to take ye account of them from those authors who have lived in the times they first appeared in & wrote purposely concerneing them.

The New Star in Cassiopeas Chaire. was very carefully & Curiously observed by The Noble Tycho.[10] who first saw it on ye 11th of November 1572 as he returned from his Chymical laboratory in the Night time. but it had appeared & beene seene before. For Wolfgang Schuler[11] had seene and observed it at Witteberge in Germany on the 6th of November betwixt which & ye 2^d^. precedeing it first appeared for on the 2^d^ at Night Munosius[12] a learned Spaniard professor of the Hebrew tongue & Mathematicks in ye University of Valentia sheweing the stars of the severall Constellations to his scholars tooke no notice of any such thing as the New Star in Cassiopea whereas had it then beene visible hee could not have passed by so remarkeable an appearance I shall give y^u^ an account of it in ye Tychoes Words but o^r^ owne language. When it first appeared sayes hee it was bigger. then either Syrius or ye bright star of ye harpe Nay then ♃ tho hee was at that time in ☍ to ye sun and consequently his diameter greater then at any other time this yeare. So y^t^ it was neare as big as ♀ & visible, to sharpe sighted people, in the day time. In December it was diminished & now onely equall to ♃ but yet bigger then either Syrius or the harpe star, as it continued in January 1573 tho Now lesse then Jove In February & March it appeared nearely equall to these stars. in Aprill & May of the bignesse of stars of the 2^d^ light in June July & August of the 3^d^; in Sept. Octob. & Novemb. of ye 4th. in December & ye Following January 1574 scarce bigger y^n^ stars of ye 5th light. in February of the sixth & in ye moneth of March it became so small that it was now no longer conspicuous. So y^t^ it continued for above 16 moneths gradually decreaseing but Not so fast at first as afterward Its colour was a briske & shineing white at its first appeareing after more yellowish. in March reddish like Mars or the Bulls eye. in May it became of a dull pale Whitenesse like Saturne. which it kept till it disappeared but continually growing lesse intense & Weaker. it[s] place he determined then in ♉ 6°.54′ with 53°.45′ North latitude.

The apparent Diameter of this New Star Tycho makes $3\frac{1}{2}$ from the invariablenesse of its distances from the adjacent fixed stars hee concludes it to be placed amongst them, & then supposeing theire distances from our earth to be onely 14000 of its semidiameters hee concludes that its diameter was to o^{r} earths as 15 to 2. & therefore y^{e} star bigger then o^{r} globe 360 times & to conteine the sun or to be to its solidity as $2\frac{2}{3}$ is to 1.

I should be needlessely prolix if I should goe about to give y^{u} the severall opinions of all those that wrote discourses concerneing this new appearance which Tycho himselfe has done in his Progymnasmata[13] let it suffice therefore that the most ingenuous Mathematicians of those dayes concluded with Tycho that this star had no parallax, that therefore it was as far removed from o^{r} earth as the fixed [stars] that in the other circumstances of its magnitude & colour they all agree with him. Claromontius indeed his & the learned Keplers professt & subtle antagonist endeavors to overthrow his opinion & to prove from his observations compared with the Courser ones of others that this appearance was lower y^{n} ye Moone but so unhappily that Ricciolus his profest freind is forced to yeeld that his Arguments prove not what hee intends,[14] tho in favor of him hee will needs say the same of those hee opposes. the reason of the appearance I shall not discusse at present it will be better to take up & consider it when I have given y^{u} an account of other later appearances

Of those the next appeared in ye yeare 1600 in ye Swans breast and was first taken notice of by William Janson better knowne by ye Name of Blaue. this by the Consent of those that viewd it says Riccioli[15] was of the third light. or equall in magnitude to that in ye brest in ye bill. in either of the pinions of the two wings or in ye extremity of the southerne Wing all which are reckned of equall bignesse. in which place Riccioli[16] affirmes it to have continued to ye yeare 1621 till which time himselfe with the fathers at Parma observed it, which both Blancanus[17] affirmes & Licetus.[18] Gloriosus[19] affirmed that it continued in ye yeare 1624 when hee wrote his booke of Comets, Argolus[20] that it was still visible in ye yeare 1629. but whether it diminished in its magnitude or not none of them say any thinge. at Prag[u]e the Disciples of the Noble Tycho who died not Long before tooke its distance with one of his sextants from Scheat in Pegasus 35°.49′ From ye pole star 52°.00′[21] whence they determined its true place in 16°.18′ of ♒ its latitude North 55°.30′ or 32′[22]

What yeare it first disappeared is not knowne. Claromontius doubts whether it was not gone before ye yeare 21 but Blancanus & Licetus on theire owne experience affirme it then visible. So does Argolus to ye yeare 1629 Ricciolus who with Blancanus had viewd it at Parma in ye yeare 1621. affirmes that from ye yeare 1640 to the yeare

1650 certeinly it was not seene When it became next againe visible I find not but it appeares by one of the learned Hevelius his letters to ye R. S. [i.e. Royal Society] printed in Mr Oldenburges transactions Nov: 1666.[23] that it was visible in November ye yeare 1658, & that hee then measured its distance from Scheat 35°-51′-20″: not sensible lesse then Tychoes disciples had found it 58 yeares before. it appeares there that it continued so visible till ye yeare 1662. from which time till the date of his letter hee sayes it had continued almost altogeather hid. but that then it was againe become visible & hee had measured its distance the same from ye same star of Pegasus, that it continued visible from the Moneth of September this yeare til ye Moneth of August 1670 is probable because the same excellent p[er]son in a letter of his to Mr Oldenburge printed also in the Transactions[24] mentioning another new star beneath the Swans head tells him it was different from this & considerably distant from its place: but of what light it was hee sayes not onely that it was of ye 7th when hee first observed it whether it since it hath disappeared againe I know not but this I am certeine of that I observed it on ye 7th of October 1680 something more then 2 yeares agone it being then one of ye least of ye stars of ye 6th light. its distance from Scheat was 35°.51′:[25] not sensibly lesse yn what Hevelius states it which makes mee apt to thinke it did not disappeare before that time nor doe I thinke that since the yeare 1666 it has encreased for then so oft as I have viewed that constellation I should have certeinely taken notice of it. but its smalnesse was the reason I did not regard it, however I am satisfied it was then visible, & haveing beene seene formerly as large as big stars of the 3d light it may be ranked with the New appearances.

The next new star appeared in ye yeare 1604 in the begining of October.[26] of this the acute kepler has wrote a learned treatise,[27] A Pomeranian preist[28] affirmes it appeared on the 27th of September because on that day the Ephemerides of Maginus[29] & Origanus[30] place the ☌ of ♃ & ♂ as if such a Congresse could create this appearance, others[31] that it first shone on ye 8 of October New Stile, but Kepler[32] affirmes that it appeared not till ye 9th himselfe & others haveing viewd the heavens both on ye 27th of September foregoing & also on ye 7th 8th & 9th of the following October & taken no Notice of any such remarkeable object as certeinely they must have done had it then appeared p[ar]ticularly his Master Mestlin had observed on the 9th at night at Tubing[en] & seene no such thinge but on the 10th which was Sunday night one Jo: Brunovsky when ye clouds were suddenly dispersed after sunset saw it in ye place where ♂ & ♃ had beene in Conjunction on ye 12 Maginus saw it at Bononia & Roeslin in Alsatia, on ye 13 Fabritius viewd it, on ye 14th Maestlin. on ye 16 Schuler on ye 17th Kepler himselfe first viewd it all of them concurreing in theire relations that it was biggest at its first appearance.[33] Some judged it bigger then ♀ but hee judged it certeinely lesse yet greater then either ♃ ♄ or ♂ who were all them neare & to whom therefore it might easily be compared. So that hee esteemed it altogeather such another as that which Tycho had observed in Cassiope Chaire in ye yeare 1572 It continued all the moneth of October of ye same biggnesse. on ye 16th of November it set Heliacally or with ye sun at Prag[u]e: on the 3^d of Jan 1605 it rose Heliacally or before him still bigger then ye Scorpions heart but lesse y^n Arcturus on ye 20th of March it was lesse y^n ♄ & not much bigger then the stars of ye 3^d light in ye knees of Ophiuchus April 21. it equalled ye bright star in ye right knee of Ophiuchus on ye 13th of September lesse y^n y^t in ye legge on ye 8th of October it could scarce be compared wth it by reason of the light of ye twilight but was sufficiently well distinguished from it by its greater & stronger radiation

Its light. was bright & vivid its scintillation greater then in any of the other fixed stars. Santutius calls it the biggest of them & affirmes it bigger then Syrius Kepler tells us *Coruscatione seu vibratione tam rapida*[,] *ut negaverint quidam sibi dum viverent unquam quicquam in caelo visum esse aequae pernici motu ut ipsi loquebantur*[34] y^t is its radiation or vibrations were so quick that some p[er]sons to use theire owne expression affirmed they never saw any thinge in the heavens move so quick that is glitter & shine so briskly. some p[er]sons hee sayes compared it to a torch light moved With ye winds sometimes extendeing & imediately contracteing its flame. but I sayes hee who cheifly regarded ye Colours of Iris in it rather compared it to a diamond cut into many sides and as it is turned round reflecteing the rayes of the sun on the eyes of the beholder in different manners. As for its Coleur[s] Maestlin agrees wth mee. that they changed every moment being now yellow presently Flamelike. by & by purple or red. but most commonly white. when it was elevated above ye vapors of the Horizon.[35] As for its figure it was plainely round without any perwigge of Rayes or tayle extended any way from it its distances from ye fixed stars continued the same all the time of its appearance. being from ye head of serpentarius by ye severall observations of Kepler Tengnagell Byrgius & Fabritius 34.00 From Aquila by Birgius 45°-43′. & from ye next knee of Ophiuchus 7°-47′. Kepler determined its place in 17°-40′ ♐. its latitude 1°.56′. Bor.[36] & adds this remarke to his observations. that all the planets applyed to it before it was extinguished.[37]

The next New star after this appeared in the Whales brest & was first noted there by John Phocillides an ingenuous Hollander well knowne by his Examen of Landspergius his Astronomy.[38] of this the Learned Hevelius has wrote a tract hee calls Historiola Mirae Stellae[39] wherein hee transcribes some paragraphs out of a booke of his concerneing it hee tells then that hee first saw it in ye month of December 1638 that it then appeared bigger then the stars of ye 3^d light such as those in ye mouth & cheeke of ye whale or ye knot of y^e

line joyneing ye two fishes are but sensibly lesse yn yt in ye jaw or ye bright star of ♈ are. that it disappeared the next year 1639 that it appeared & hee saw it first againe after much Cloudy weather on ye 7th of December. yt yeare. by an Epistle of Fullenius to one Other it appeares that it was visible againe but not before ye 23d of September 1642 that it appeared not on ye 23 of September 1644: in ye yeare 1647 Dr. Joachin Jungius the Hambororow professor of Astronomy saw & observed it againe. as also in following yeare 1648 but from thence to the yeare 1659 it was not heeded nor p[er]haps sought after but from that time to this the Learned Hevelius has given us a Journall of its appearances in his sayd Historiola. & letters after it was published sent to Mr Oldenburge & printed in ye Philosophicall Transactions.[40] wherein in yt of June 1666 Nom. 21 hee gives us an account of a booke entituled Ismaelis Bullialdi Monita duo ad Astronomos.[41] whereof the first [advice] is concerneing this very star. From Hevelius his observations it appeares that it seldome is seene longer then 5 moneths at once, & that from the time it begins to be of ye 6th. till haveing passed all its degrees of increase & decrease it becomes of the same light againe is usually about 4 moneths. its returnes from its greatest appearance to ye same againe. hee makes 333 dayes. or 11 Kalendar moneths. so yt In every of its revolutions tis above 6 moneths invisible. his period I can not esteeme altogeather exact for that accordeing to it the New Star ought not to have appeared before ye 24th of November in ye yeare 1677 whereas I saw it but smaller then any of ye adjacent scarce bigger then the 7th. light on ye 7th of that moneth. So yt it seemes a day may be abated every yeare & wee ma[ke] its revolution to ye same phasis increaseing or decreaseing to be onely 332 dayes. accordeing to which reckning it ought to have disappeared this yeare past in ye beginning of November & not to returne to its visibility againe till May day next.

As for the Nebulosa ad Cingulum Andromedae Bullialdus tells us that it was seene & taken Notice of by an Anonymouse observer above 150 yeares before ye publication of his booke[42] that Simon Marius saw it in the Yeare 1612. whose plane & full description of it I find in ye preface to his Mundus Jovialis[43] which I thus translate "Since ye 15 of December I have found & seene such a Wonderfull fixed star as the like can scarce be found any where besides in ye heavens, tis neare ye 3d & Norther[n]most in ye girdle of Andromeda, lookes to the Naked eye like a small Cloud but with ye p[er]spective there are no small stars to bee seene in it as in ye Nebulosa of ♋ and others, but onely whitish raies which the Nearer they are to the center appeare the brighter. in ye center is a dull pale light. the diameter of this appearance is about ¼ of a degree if yu view the light of a candle placed behind a transparent horne at a great distance yu see an appearance like it tis not much unlike the Comet which Tycho Brahe observed in ye yeare 1586.[44] so far hee. Bullialdus addes that in ye moneth of November in y^t yeare 1665 when hee wrote those Tracts[45] it appeared much lessened & very obscure whereas but two yeares before it had shone out very bright & cleare: I have not often sought for it but I remember that When I observed it on ye 29th of December 1680 it shone againe very cleare. nor doe I remember that I have mist it at any time since when I have viewd this Constellation however hence it appeares that tho it has no superficiall light as ye Sun & other stars have yet with them it has the vicissitudes of it appeareings & disappeareings & may therefore be numbred with the New appearances of the Stars in ye Swan Cassiopea Serpentarius & ye Whale of Which the Mathematicall Philosopher ought to give an account.

The Nebulosa in ye sword of orion is alike appearance save that it has not beene observed to encrease & decrease in its light hitherto by any one Galileo haveing discovered y^t ye Milky way or y^t large Nebulosa which invests the heavens was onely an aggregate of very small stars placed neare each other & the like of the cloudy place in ♋ seemes to have concluded that all the Nebulosae in ye heavens and this as one of them were alike and accordingly hee gave us a figure of it in his Nuncius Syderius,[46] but hee may be excused for so doeing because the telescope was then but a new invention which attained but lately to its p[er]fection if it be yet arrived at it and Hugenus who gives a different account of it in his Systema Saturnium[47] had much better glasses. his words concerneing are to this effect. "One thing more I have noted in ye fixed stars worth observeing & of which none that I know of hitherto hath taken Notice. neither indeed can it be duely observed but with such large Telescopes (his was a good one of 22 feet) There are 3 stars placed in the Sword of Orion by Astronomers Viewing the middlemost of these in the yeare 1656. in ye roome of one twelve stars p^r sented themselves to my view of which the three y^t are contiguous with 4 other shone as it were from beyond a cloude, it was then very cleare & the rest of ye unilluminated heavens appeareing p[er]fectly darke this looked like as if there were some open breath through which y^u had a prospect into a bright or illuminated place, I have seene the same since contriveing to this day without any alteration so that wee may probably conceave it hath beene there of old & so is like to continue. whatever it bee I have seene nothing like it amongst the fixed stars"[48] to which I must needs assent for as often as I viewd the same place I have found the same appearance unchanged and there is this remarkeable in it that whereas the Nebulosa in Andromedas Girdle, tho not round, yet seemes enclineing to that figure this y^t on ye contrary has nothing of roundnesse in it but is of a figure p[er]fectly irregular. Besides these there has shone more lately another New star beneath ye Swans bill of Which the accounts given in ye Transactions are too long to be collected & repeated at this time. hereafter if I have another opportunity I may abridge them to y^u [49] & then shew y[u] how agreeable these Phaenomana are with y[e] spots in o[r] sun & how they make it probable that the stars are like substances which if the Necessary Circumstances in the appeareings & disappeareings of the New stars had not beene more then I expected to have met with I had done at this time, by these however What I have suggested will be better confirmed then was to be at first expect[ed] but I must remit this taske to a further opportunity.

NOTES

24.1. Lecture 22 was the first lecture given by Flamsteed during the Hilary Term.

24.2. Ibid.

24.3. Lecture 23.

24.4. Cf. note 22.6.

24.5. Cf. 'De nova stella anni 1572', op. cit. (note 2.5) (1), 822 pp.

24.6. Riccioli 2, p. 166.

24.7. Johannes Kepler, *De Stella Nova in pede Serpentarii . . . Trigono Igneo* (Pragae, 1606).

24.8. See the Preface to Simon Marius, *Mundus Iovialis anno 1609 detectus* ([Norimbergae,] 1614). Cf. note 7.3.

24.9. The first notice of this nova occurs in a Latin extract of the letter from Hevelius to Oldenburg, Danzig, 17/27 August 1670, published in the *Phil. Trans.* v, no. 65 (14 November 1670), 2087–91.

24.10. Op. cit. (note 2.5) (1). This aspect of Brahe's work is discussed in Riccioli 2, pp. 133–66.

24.11. Riccioli 2, p. 136, referring to Brahe (ibid.).

24.12. Hieronymus Munosius's observations are referred to in Riccioli 2, p. 135 (quoting Brahe, ibid.).

24.13. Brahe, ibid. References to other writings on this subject are given in Riccioli 2, pp. 164–5.

24.14. Riccioli 2, p. 163. The discussion of Scipio Claramontius's method and results conclude with the words (transl.) 'he has transgressed the limitations of his own demonstrations'.

24.15. Ibid., p. 166.

24.16. Ibid., p. 168.

24.17. Josephus Blancanus, *Sphaera Mundi, seu cosmographia demonstrativa* (Bononiae, 1620), pp. 349–50. Riccioli 2, p. 168, cites a reference from the later (Mutinae, 1653) edition of Blancanus's book.

24.18. Fortunius Licetus, *De novis astris et cometis libb. sex* (Venetiis, 1623, 1622). Cf. Riccioli, 2, p. 168.

24.19. Ioannis Camillus Glorioso, *Responsio . . . ad controversias de cometis Peripateticas* (Venetiis, 1626). Cf. Riccioli 2, p. 168.

24.20. Andreas Argoli, *Pandosion Sphaericum*, 2nd ed. (Patavii, 1653), p. 289. Cf. Riccioli 2, p. 166.

24.21. The respective distances of the nova from Scheat Pegasi and Polaris are tabulated in Riccioli 2, p. 166.

24.22. Riccioli 2, p. 167.

24.23. *Phil. Trans.* i, no. 19 (19 November 1666), 349.

24.24. Cf. note 24.9.

24.25. Op. cit. (note 1.18), p. 70. The observation was made at 7.49 p.m.

24.26. The observations are discussed in Riccioli 2, pp. 168–73.

24.27. Op. cit. (note 24.7).

24.28. Georgius Spate (alias Herlicio). Cf. Riccioli 2, p. 168.

24.29. Giovanni Antonio Magini, *Ephemerides coelestium motuum . . . ad annos XL, ad anno Domini 1581 usque ad annum 1620* (Venetiis, 1582).

24.30. Davidus Origanus, *Novae motuum coelestium Ephemerides Brandenburgicae, Annorum LX, incipientes ab anno 1595, & desinentes in annum 1655 . . .* (Francofurti cis Viadrum, 1609).

24.31. E.g. Johann Brunowsky of Prague, Helisaeus Roeslinus in Alsace, and David Fabricius in East Friesia, whose names are all cited in Riccioli 2, p. 168.

24.32. Op. cit. (note 24.7).

24.33. The above-mentioned observers and the following observations are referred to by Riccioli 2, p. 168.

24.34. Quoted in Riccioli 2, p. 169.

24.35. This brief description is a loose translation of the following sentence in op. cit. (note 24.7), p. 5 cited in Riccioli 2, p. 169, 'De Coloribus Maestlinus plane mecum, eos ad momenta singula variari, ex flava mox croceam, e vestigio purpuream & rufam, ut plurimum candidam videri, ubi ex vaporibus paulo altius elevetur.'

24.36. The figures cited here are taken from Riccioli 2, p. 169.

24.37. Riccioli 2, p. 169, remarks: 'Memorabile autem est, quod advertit Keplerus, hanc stellam, antequam extingueretur, aditam fuisse a Planetis omnibus ipsisq. coniunctam saltem platice.'

24.38. Johannes Phocylides, *alias* Johann P. Holwarda, *Dissertatio astronomica sc. de eclipsi lunari 1638; de astronomia Lansbergiana et de novis phenomenis sive stellis* (Franek., 1640).

24.39. Johann Hevelius's 'Historiola novae illius ac mirae stellae in collo Ceti' was appended to op. cit. (note 23.15). Note that this tract, and not Riccioli (op. cit. [note 6.3]), is the source of the information contained in this paragraph.

24.40. Cf. note 23.16.

24.41. *Phil. Trans.* no. 21 was actually published on 21 January 1666/7. The account in question is on pp. 381–3 of this issue.

24.42. That is, about 1515, or almost a century before the invention of the telescope.

24.43. Cf. note 24.8.

24.44. Ibid. [p. 5].

24.45. *Ismaeli Bullialdi Monita duo ad Astronomos* (Parisiis, 1667). Cf. note 24.41.

24.46. Galileo Galilei, *Sidereus Nuncius* (Venetiis, 1610), on the back side of two unnumbered pages interposed between pp. 16 and 17.

24.47. Christian Huygens, *Systema Saturnium* (Hagae-Comitis, 1659).

24.48. Translated from ibid., p. 8 and lines 1–6 on p. 9.

24.49. This is done in the following lecture (Lecture 25).

Lecture 25

2 *May* 1683

Endeing the last terme I was forced to breake of[f] in ye middle of an historicall account of some unusuall appearances of the fixed stars I intend to proceede & compleate it wherefore it will be necessary to give y^u a breife account of my last readeings that so you may ye better apprehend the scope of my designe & the Evidence of my conclusion.

The Astronomicall Lecturer is obliged to explaine the Theorys of ye Planets by a mutuall stipulation with ye Honorable trustees of ye colledge made 86 yeares agone: when it was first setled.[1] before I understood this I had happily falne upon subjects not disagreable with this stipulation. The Theorys of the planets motions being derived from observations made with instruments contrived for this purpose I had shewed what sort were the best & haveing explained the fundamentalls of dioptricks I also taught how glasses applyed to Quadrants or sextants might be rectified for observation. then because the obliquity of the Ecliptick was one of the prime requisites in Astronomy I proved both its quantity & that it was invariable by examineing the best of the ancient & Moderne observations. hence I proceeded to explicate the doctrine of the sphere, & shewed y^u how all the diurnall phainomena of the sun and stars were formed & represented in ye ancient Pythagorean susteme of the World Which supposes the suns fixed place in the Center of o^r susteme but the earth & all the other planets to be moved about him in stated periods of time which is most consonant to reason had beene revived about 200 yeares agone by Copernicus. demonstrated more lately by Kepler and as most agreable to experience approved of by the best reputed Astronomers of o^r Age. which I could not find to have beene before p[er]formed by any that had enterteined this subject

The true longitudes of ye sun were obteined by Compareing his Meridionall heights observed at different times of ye yeare with one another how the Ancients arrived at ye knowledge both of his inequalitys & the lunar [ones] & by what steps they proceeded I

Notes on pages 345–6.

have formerly shewne it will not therefore bee needfull to make any relation of these. The places of all the other planets were gathered or determined from theire observed distances from the Fixed stars. whose places had beene rectified by Hipparchus from theire observed distances from the Moone. whose susteme they thought next the suns they had the perfect knowledge of the experience since has informed us that the inequalitys of her motion are as little if not less knowne then any of the other planets

The Fixed Stars are then the basis of Astronomy all that are skillfull in it haveing ever made use of theire places & the distances of the other planets from them for determineing theire motions. & inequalities. It was therefore necessary before wee proceeded to the Theoryes to enquire what were the opinions of the Ancients concerneing theire Nature & that little Motion which long tracts of time & ye continued observations of distant Ages seemes to have discovered in them.

With this inquiry I began last terme. and first with the opinions of ye Ancients concerneing theire Nature or substance. of whom I told y^{u} that Anaxagoras taught they were stones drawne up from ye earth by whir[l]ew[i]nds & fired by theire violent motion or Circumvolution. about ye Earth. Zenophanes that they were coles fixed by blaseing or burneing clouds. Diogenes that the arched vault of the heavens was perforated as pumice stones. & that the Caelestiall fire shineing through the perforations Made the appearance of stars Thales that they were burneing earths Orpheus Pythagoras & Heraclides that every star was a World with its elements. with whom our Neotericks Brunus Kepler Galileo & de Charte differ not much. they asserteing the stars to be suns and a Composition of all ye elements.

The opinions of Anaxagoras Zenophanes & Diogenes were easily refuted. for it was not proba[b]le that such vast bodyes as the first must suppose them could be lifted up from o^{r} earth by whirle winds of old since wee see nothinge like it p[er]formed by them in o^{r} dayes. if they were onely the appearance of the Celestiall light through the p[er]forations of the vaulted arch of the heavens then it would follow that the sayd vaulted arch should reflect ye light of the sun at all times and Wee should have no night Nor could the Comets which have peirced althose Places the ancients assigned to ye solid orbes of the planets have passed to & fro so easily & equably as they have beene observed to doe through the heavens these opinions therefore I passed by as absurd. and proceeded to enquire from ye obvious appearances of the fixed stars what more probably might be then theire Nature & substance.

The remotenesse of the Celestiall bodyes from us p[er]mits us not proceede otherwayes in this inquiry thence compareing the more remote with such as are demonstrably nearer & these againe with such wherewith wee are dayly conversant Can touch & handle them.

All our knowledge is indeed but a traine of comparisons tho in some subjects more cleare and evident then in others but most of all in Mathematics & things abstract from Matter thus wee can easily distinguish betwixt lines & lines as ye Curve of a Circle & a Parabala by theire propertys. in like manner betwixt ye planes of a triangle & a square the solids of a prisme & a Cone. but in Materiall objects as Gold & Copper wee can distinguish & compare them no otherwayes then by theire fusibility ductility & specifick gravity color &c. the More reall differences betwixt the particles that compound them shunning the comparison of o^{r} senses by reason of theire Minutenesse or the grossnesse of o^{r} organs. Seeing then wee can not attaine to the perfect knowledge of those bodyes with which we are dayly conversant nor of theire Reall differences it cannot be expected that wee should accurately determine what the remote Caelestiall bodyes are, or what Reall differences there are betwixt them otherwayes or mor then a traine of Comparisons can direct us. I tooke therefore this comparative method[2] to enquire the Nature of the fixed stars & haveing proved theire vast distances from y^{e} [Earth] . . . also confirmed I proved thence that theire bodys were vastly greate generally more then the suns whose bulke in all probability is a million of times bigger then our earths. Againe from theire briske & vivid light not unlike o^{r} suns. the fixednesse of theire places. & theire long duration in the same distances one from another I argued theire being as hee is the centrall fires or Rather Magneticall powers of theire severall vortices. for it may be very much questioned whether o^{r} sun be really fire or not but that hee attracts all the planets & repells them from accordeing to the position & respects of theire Axes to his I thinke is lesse to be doubted

Since the Invention of Optick glasses it has beene discovered that o^{r} sun turnes round upon his owne Axis carrying round with him by this rotation such spots as were formerly frequently found and seene in his face of late yeares more rarely these observations suggest to us a solution of some rare Phenomena of Nature before this Invention not taken notice of I meane the disappeareing of some fixed stars formerly noted for bright & large ones, the appearance of others extinction & stated returnes. An account of these I knew would be very gratefull to this Auditory & Instructive in my present designe. I therefore gave y^{u} an history of such New Stars as have been the Most remarkeable and taken notice of by the ingenuous observers of the past and present Century. I began with ye New star that shone in Cassiopea & was taken Notice of first[3] on ye 6th of November 1572. betwixt which and ye 2^{d} day of the same Moneth it had first appeared this at its first appearance was bigger then ♃ neerly equall to ♀ & so continued for above a moneth decreaseing gradually till ye Moneth of March 1573 when it disappeared. being when first seene of a bright resplendent whitenesse. in ye March following Reddish like ye planet ♂ or ye Bulls eye. in May following of a dull pale Whitenesse which colour it kept till it became invisible to ye Naked eye.

The next of these strange appearances was first taken Notice of William Janson alias Blaue. the Dutch Globe maker who had beene a domestick of the Noble Tychoes & if I am not misinformed the best contriver & former of his instruments. In the yeare 1600 in the Swans breast being then of equall magnitude with those in the pinions of the wings by its distances from ye pole star & scheat the disciples of Tycho then lately deceased determined its longitude in 16°.18′ of ♒. its latitude North 55°.30′ in which place by the joynt Consent of Riccioli, Licetus Gloriosus, Blancini & Argolus it continued visible till ye yeare 1629 When it first disappeared is not noted by any that I can finde onely Riccioli[4] tells us that from ye yeare 1640 to 1650 certeinely it was not seene. But in November 1658 it was by the Learned Hevelius who measured its distance from scheat & found it the same that Tychoes disciples had formerly determined it hence it continued visible till the yeare 1662. then agane almost continually hid till September 1666[5] from which time till August 1670 it was againe visible when it disappeared I cannot learne but certeine I am that I saw it on ye 7th of October 1680 when I observed the stars of this Constellation & that I measured its distance from scheat 35°.51′ nearely the same with what Tychoes disciples & Hevelius had found it before me. Since I have not taken notice of it but o^r Next cleare nights will show us whether it have encreased or is disappeared since.

In the yeare 1604 another new star appeared in the Moneth of October on ye 10th day 17°.40′ of ♐ with 1-56 North latitude neare which place there had beene a conjunction of ♄ & ♃ not many dayes before. this was observed in distant places by Reostlin Maginus Fabritius Maestlin & Kepler all agree[6] that it was biggest at its first appearance & bigger then either ♄♃ or ♂ who were all neare it, it continued in ye same place and at ye same distance from all the Circumjacent fixed stars all the time of its appearance continually decreaseing in Magnitude till ye moneth of October the yeare following at which time by reason of the light of the twilight it Could not be compared with ye other fixed stars neare it tho its radiation was still strong enought to render it visible afterwards the sun approacheing it, it became invisible & as the star in Cassiopeas Chaire was not since seene.

The next new star after this was observed in the Whales Brest in ye Moneth of December 1638 by John Phocillides an ingenuous Hollander & well knowne by his Examen Astronomia Landsbergianae[7] since then by Fullenius & Jungius but much More exactly by Hevelius who has wrote a tract of it which he entitles Historiola Mirae Stellae.[8] in it he gives us an account of the severall observations of the forementioned learned persons from theire tracts &

letters to which hee addes his owne. And the learned Bullialdus in his Monita duo ad Astronomos[9] haveing considered both Hevelius & the others observations collects that this star turnes about its owne Center once in 33 days that the intervall of time betwixt its begining to appeare equall to stars of ye 6th light & its endeing to doe so is about 120 dayes that its greatest appearance continues neare 15 dayes. but compareing Hevelius his observations with my owne since made I find his period one day too long. & concluded. that its first appearance equall to the stars of ye sixth light this yeare would happen on ye 1st of May. but at this time it is hid under ye beames of the sun so that in our horizon it will scarce be visible till the latter end of June when it emerges from his light & rises visibly before nearely equall to stars of ye 3^d^ light, & to disappeare in ye ende of August or begining of September at longest.

Simon Marius in ye preface to his Mundus Jovialis[10] gives us an account of a New Nebulosa which hee had observed with his glasse in ye yeare 1612 to the North of ye 3 stars in ye Girdle of Andromeda, I gave you it in English. but Bullialdus tells us that it had beene taken notice of by an anonymous observer neare 100 yeares before tho neither Tycho nor his disciples Kepler and Longomontanus make no mention of it in theire Catalogues, this is a bright & lucid place in ye heavens but has no stars in it as the Cloudy place in ♋ and the Milky way a like but vastly large resemblance of it seeme to have. Bullialdus tells us that in ye Moneth of November 1665 when hee wrote his tracts[11] it appeared much lessened & shone but dully whereas two yeares before it had shone out out very bright & cleare. On the 29th of December 1680 I saw & observed it my selfe. & takeing its distances from severall fixed stars.[12] A large comet then appeareing whose head I could compare to nothing better then to this Nebulosa.[13] hence it appeares that it is subject to its vicissitudes as well as the other New appearances but. what they are time onely & continued observations can informe us

The last strange appearance I mentioned was that of the Nebulosa in ye sword of Orion first taken notice of by Galileo, but better since described by the ingenuous inventor of ye Pendulum Clock & wat[c]h Mr Huygens.[14] who tells us that 7 small stars shine in that place as it were through a bright cloud, & that all the rest of the Heavens appeareing darke this place being viewd with a telescope it lookes as if there were some open breath through which y^u^ had a prospect into a bright place. but it is Cheifly remarkeable that whereas the Nebulosa ad Cingulum Andromedae appeares, tho not p[er]fectly round, yet inclineing to roundnesse; this on ye Contrary appeares altogeather irregular. but all wayes in the same forme as I find by my observation made since compared with his & without any sensible alteration either of its light or figure.

I come now to ye history of the New star at ye Swans bill first taken notice of by Monsr Hevelius in June & July 1670 of April . . . whereof hee gave an account to Mr Oldenburge in a letter Aug 17. 1670.[15] by his observations of it taken July 25 yt yeare hee concludes its place in 1°.52′ of ♒ its latitude 47°.25′.20″ North. and because that Neither Hipparc[h]us Ptolemy Tycho nor his scholars have it in their Catalogues nor yet Bayerus in his Mapps tho hee have severall small stars omitted by them, he concludes that it neither was seene by those ancients nor in Tychoes Time

yet by Bayer in ye yeare 1603 when hee Wrote his Uranometria.[16] Hevelius himselfe had observed all the stars in that Constellation in ye yeares 1559. 60. & 61 & then tooke no notice of it tho hee noted severall small ones neare that place. whence hee also infers that it was not then Conspicuous. but whether it appeared before this time or not hee sayes nothinge

But by an extract of 2 letters sent from Paris & published togeather with this[17] it appeares that it was seene on ye 20 of Decem. 1669 by one father Antheline a Carthusian & that it was then of the 3d light as ye Learned Hevelius also makes it. & by other letters from Paris[18] that the same Carthusian on ye 20 our 10 of June 1670 had seene it againe more brightly then before being at least of ye 3d light that the 3d of July following it continued still of the same Magnitude but its light sensibly fainter, that if afterward decreased dayly so yt about theire middle of August it became Invisible by reason of its exceeding smallnesse. here theire observations fall in with those of the excellent Hevelius who observed it whilest decreaseing on ye 25 of July and thence deduced its longitudes & latitudes as I gave them before

In March ye next yeare Monsr Antelme got sight of her againe theire 17th day at night shee was now nearely of the 4th Magnitude. which being signified to the Academy at Paris they lookt out & met with her in the same place they had found her the yeare before. Aprill ye 3 Signieur Cassini found her bigger then ye 3d light stars so yt now shee was encreaseing, but lesse yn ye star in ye Swans Bill but the night following nearely equall to it and much more radiant of this biggnesse shee continued till the 30 of this Moneth. haveing beene observed in the Meane-time by Monsr Hevelius whilest her light was greatest on ye 29 day at night. whose remarkes agree very well at this time with ye French.

The first 6 dayes of May shee decreased not but was still rather greater then the stars in ye Swans bill but afterwards shee decreased fast so that on ye 15 of May shee was lesse yn ye stars of ye 3d light. in ye Middle of July shee was onely equall to ye stars of ye 5th light in ye begining of August of ye 6th, August ye 17 shee was scarce any longer visible, & on ye 25. shee could not be found by ye Learned Hevelius to whose register[19] wee are beholden for all the notes of her Magnitude since ye begining of May.

In ye following yeare 1672 Monsr Hevelius sayes that on ye 29th of

March it was scarce equall to ye sixth light stars & yet hee tells us after on ye 9th of August *nusquam hoc anno fulsit*[20] it shone not at all this yeare. if so hee grants the first Note a mistake. & such I am persuaded it was. but hee certeinely affirmes that from this time March. 1672 to January 1677 it continued invisible

Compareing these last remarkes of Hevelius & the french togeather it is evident that this star was visible neare if not wholly 6 moneths. from its first to its last Appearance for Anthelme saw it of ye 4th light on ye 17th of March whence tis probable it had first appeared a Moneth before or in ye Middle of February. Hevelius to ye 25th of August Following

The French make several fine observations upon it as that its revolution to the same appearances is in tenne moneths whereas that in ye Whales Neck makes its returnes in 11. whereas if it had made a[ny] it should have Made it greatest appearance visible to us every yeare. yet Hevelius who was never wanteing in Circumspection Could not find it from that time to the yeare 1676. since I have sometimes observed the stars in that Constellation & when I purposely sought it in Sept. & Octob. 1680 at what time I made the stars of that Constellation my businesse I could not find the least appearance of it. Nor have I since tho I have purposely viewd that constellation oft in quest of it, so that its revolution must either be of a longer date which is improbable or else, the accidentall Cause of y^t appearance in ye body of the star is ceased which as most likely I rather chuse to suppose & shall endeavor to give y^u a reason for it in my Next lecture

To these Phainomena I may adde what was remarked by ye Ingenuous Montanari,[21] some time since that severall remarkeable stars are missing

the heavens which were noted by the Ancients particularly two in ye Poope of ye ship marked with β & γ in Bayerus his maps.[22] & called bright or resplendent ones by Ptolemy, & that in other parts of the heavens other stars seeme to have changed theire Magnitudes particularly in ye Constellation of Andromeda & in ♒ there is one marked with σ in Bayerus his maps[23] & placed in Tychos Catalogues[24] which I could never find & therefore conclude to have disappeard since his time. but of these enough. I shall endeavor next lecture to shew that these appearances confirme the opinions of Brunus Kepler & Galileo with most of o^r eminent Astronomers since that they are really suns & this argument will be all the subject will beare. & therefore in my opinion unanswerable. &c.

NOTES

25.1. Cf. Introduction, pp. 3–4, for further details.

25.2. Flamsteed is here adopting the approach advocated in the preface to Riccioli 1, which is inspired by Book 7 of Plato's *Republic*.

25.3. We have deleted the word 'betwixt' from the text here, since Flamsteed begins the next page (f. 235^{v}) with an additional phrase and then reintroduces it.
25.4. Riccioli 2, p. 166.
25.5. Cf. note 24.23.
25.6. These astronomers and their observations are referred to in Riccioli 2, p. 168.
25.7. Cf. note 24.38.
25.8. Cf. note 24.39.
25.9. Cf. notes 24.41 and 24.45.
25.10. Cf. note 24.8.
25.11. Op. cit. (note 24.45).
25.12. These observations are not recorded in op. cit. (note 1.18).
25.13. Lecture 3 (11 May 1681). Flamsteed's observations of this comet between 10 December 1680 and 5 February 1681 are contained in op. cit. (note 1.18), pp. 104–7.
25.14. Op. cit. (note 24.47), pp. 8, 9.
25.15. Op. cit. (note 24.9).
25.16. Op. cit. (note 3.17).
25.17. Op. cit. (note 24.9), p. 2092. The letters are dated 5 and 19 July 1670 [N.S.] respectively.
25.18. Perhaps this is a reference to the English translation of the account in the Journal des Sçavans printed at Paris on 22 June 1671, which appeared in *Phil. Trans.* vi, no. 73 (17 July 1671), 2198–202.
25.19. This nova in the Swan's bill was actually detected by Hevelius for the first time on 29 September 1670. His notes on it are in op. cit. (note 14.18), pp. 538, 542, 560, 561, 563–9, 595.
25.20. These words actually occur after the entry for 11 August 1682 in ibid., p. 606.
25.21. Geminiano Montanari, *Cometes Bononiae observatus anno 1664 et 1665 astronomico-physica dissertatio* (Bononiae, 1665).
25.22. Op. cit. (note 26.16). Cf. 'Tabula Quadragesima. Navis'.
25.23. Ibid. Cf. 'Tabula Trigesima Secunda. Aquarius'.
25.24. Op. cit. (note 2.36).

Lecture 26

9 *May* 1683

I finished the historicall account of the strange appearances of the fixed stars last lecture I come now to the Philosophicall & to shew how by applyeing Mathematicall rules to phusicall observations, & compareing the Phenomena of these remote appearances with some Nearer us, wee may arrive at the reason of them, the knowledge of those great workes of the Almighty hand, which the divine Psalmist tells us *give him continuall praises.*[1]

For I esteeme it the duty of the Astronomer, as I have often told y^{u}, by applyeing the demonstrations of Mathematicks to naturall philosophy, to give an account of those thinges in the heavens, which being too far removed from o^{r} senses, cannot be exactly discovered by them; but may be sufficiently understood by this mixture of knowledge, & the right application of o^{r} reason, if wee endeavor as wee ought to avoid absurd comparisons, & conclusions contrary to Mathematicall demonstration.

My designe has been to prove the assertion of Brunus Kepler Galiles Decharte, & other Mathematicall Philosop[h]ers of the last & present age, that the fixed stars are really suns, I shewd y^{u} that proofes of things of this nature could onely be made by compareing the appearances of the one with ye other. & that from theire light like ye suns, & theire magnitudes considerably greater, it is probable they were altogeather such as hee is

I have also shewd y^{u} by considering the Phenomena of the solar spots, that that bright object is not as some have contended a ball of fire, but rather a shell, or (if ye Novelty of the terme will permit) y^{u} to receave a liquid sea of light continually moved upon a solid globe from which the spots are onely eruptions caused by accidentall alterations within his body & cast on high through this sea of lucid Matter which by its extreme subtelty & violent motion weares melts or washes them downe againe some in a little time, others in longer.

That from the returnes of these spots to the same pointes of ye heavens or starry sphere found to be made alwayes in 25 dayes 6 houres, it is evident the globe or solid parte of the sun turnes round

Notes on page 355.

upon its owne Axis in that space of time; tho by reason the earth runs forward neare 25 degrees in the mean time, the Returnes or revolutions to ye same longitude within the sun are not accomplished but in 27 dayes & 4 houres[2]

If now I shall shew yu the like revolutions of the fixed stars about theire owne Centers, & that from or Solar spots such appearances may happen in him as have done in them, that is hee may be totally covered as some of them have beene I account I have discharged my selfe of my undertakeing For hereby I shall adde two as good arguments more as the nature of the subject will beare for theire being suns in the sence I asserted them, & hee that in such remote bodyes requires better proofe must have lost his discretion & become ridiculous

And that they doe make the like revolutions about theire owne centers, is evident first from ye New Star in ye Whales breast, which is found and observed to make ye Returnes of its appearances of ye same magnitude in 11 of or Calendar moneths nearely, or more precisely in 332 dayes: to continue visible for about 120 of them or neare 4 moneths, but invisible for 7. its greatest appearance continueing neare 15 dayes. its phases betwixt being greater or lesse accordeing as they are nearer to the greatest appearance or further from it; the same thinge may be argued from ye New star under ye Swans bill, for it appeared neare 6 moneths of the tenne it made its revolution in, as I shewed yu in ye historicall account I gave you of this stars appeareing in my last lecture and of that in ye Whales breast in ye last of the last terme[3] from ye severall observations of Phocillides Fullenius Jungius Hevelius & my owne compared togeather.

If the reason of this appearance in the New star in ye Swans breast be enquired into, it will be evident that it cannot be because the one hemispher[e] of it is bright the other darke, for then it should cease to be visible onely one moment in ye Whole tenne moneths of its revolution as if in the first figure [Fig. 26.1] wee suppose ye outward figured Circle to represent the star and a. 11. 1. 2.3.4. g c d to be its illuminate hemisphere next the eye & the other the obscure removed from it conceaveing the sphere turned round upon its axis & pole c the said hemisphere is no sooner removed or turned out of the view of the eye & the obscure 4, 5, 6, 7, 8, 9, 10, a. succeeded in its place: but the point g, which terminates the light & darke is arrived at a, & it begins againe to be visible on that part next. a. so that here it could scar[c]e ever be twelve houres invisible, tho its light were exceedeing small at the begining of its increaseing appearances when ye illuminate hemisphere enters into the visible periphery of the star at a, or when it is going forth of ye view at g.

It is therefore absurd to thinke or affirme that one halfe of the body is illuminated, the other obscure. since that it would follow ye star must be continually visible or within an insensible intervall of time. if any one shall object that the small part of y[e] illuminate haemisphere is not visible to y[e] naked eye for some time after it comes first into y[e] visible diske or before it leaves it such may doe well to consider the phases of the moone neare her conjunction with y[e] sun, or of ♀ in y[e] like position, both which are never the lesse then observable tho exceedeing small portions of theire spheres are illuminated, as also the Eclipse of ♃[s] satellit[e]s which diminish sensibly & some considerable time before they are totally immerged in his shadow so y[t] it is very probable th[ey] are visible till the illuminate parts of theire periphery are as small as the Moones in her first phasis

Some other reason must therefore be thought of for this appearance. What then if wee suppose that onely some small Arch of the periphery or rather a round or irregular space of it is illustrated. it will follow still that the star shall be seene for above one halfe y[e] time of its revolution, considerably more or lesse accordeing to y[e] latitude or breadth of the shineing space; for if the shineing part were no more then a visible point & the arch a.b.1.2.3.4. g. the visible disk of y[e] star from y[e] Earth. the sayd lucid point in its periphery shall appeare to y[e] earths inhabitants as soone as it touches the visible diske in a, & shall so continue whilest it is moveing through the visible periphery from a to b to g where it passes into the invisible or obscure the hemisphere; but if wee conceave the shineing part to have breadth then, it will appeare proportionably longer as y[e] breadth is greater plaine adde the brea[d]th, which y[u] may suppose ab. in y[e] first periphery to y[e] semicircle g. 6.7.8.9.10. a; the time of its visibility, to the time it lies hid, is as the Arch gab is to the arch bg. that is if wee suppose ab y[e] breadth of y[e] shineing spot to be one twelveth part of y[e] periphery the time of its appeareing to the time it lies hid shall be as 7 to 5. but no such thing was observed in this; For on the contrary it appeared 4 moneths but lies hid 7. & therefore this appeareance proceedes from some different cause.

Since therefore neither one hemisphere of the star illuminate and the other obscure, nor a lucid point or broad spot in y[e] superficies will answer this appearance. tis evident that the cause cannot consist in either of these conceptions they laboring equally with absurdities but however wee must suppose some part of the body obscure. another light let us trie how it will happen if wee conceave the darke parte to have some thickenesse & the light to be included

Because the star makes it returnes to the same appearance in 11 moneths let the periphery of the first figure [Fig. 26.1] be divided into 11 partes so as each may represent the space it revolves in one moneth

Conceave y[e] line ch drawne from y[e] center of the star to y[e] earth to be alwayes the same for the earths orbite subtendeing no sensible angle at ye star. this may very well be admitted alwayes the same and invariable. crosse this line with another adcg through ye center c then may ye semicircle abg represent one halfe of ye diske of ye star visible from o^{r} earth.

Because the star appeares onely foure moneths of the 11 in which it makes its revolution, let the aforementioned division be so ordered that ye 2^{d} may lie in the line ch where it intersects the periphery of the star. or let be begun from that point & continued round thence so then the first appeareance of the star shall be when the middle of its light is at or against ye point marked 11. the last at y^{t} marked 4.

I conceave the outward part of the planet to be a solid shel[l] of a darke sub[s]tance which by reason of its vast distance from o^{r} sun or the adjacent fixed stars receaves very little light from them & consequently is not seene by the light reflected from it as are o^{r} planets. The Inward substance to be a ball of lucid matter which may be seene through a broken or open place in this shell

Conceave the semidiameter of this open place represented by 11. a equall to 11. b in each and all the figures by the underextremity b. let a line ebd be drawne parallel to hc and within the outward circle another dikl on ye center c so as it may just touch the line de if the aperture of the broken place be very wide as in ye 2^{d} fig [Fig. 26.2] or to the intersection of the said line de, and another drawne from ye center c to ye other side of ye hole a if it be narrower then the Radius of ye said inner circle. as in ye 1st 3^{d} and 4th figures [Figs. 26.1, 26.3, 26.4 respectively] this shall represent the globe of light included in the outward shell of dense or opake matter & the larger wee suppose it the lesse shall the aperture or widenesse of the open broken part in the shell be by reason that the laterall visuall ray de departes the further from the Centrall ch. but then it will follow that the star should be but small & could scarce ever encrease to the magnitude of the 2^{d} light stars as this has nearely beene observed. and besides tis not probable that the included globe of light should so much exceede the conteining shel[l] of opake matter as it will if wee conceave ye one beares to the other as in ye 3^{d}. figure [Fig. 26.3] & further the violent motion and agitation of the of[f] Centrall matter would easily breake more holes in the includeing Cortex. tis not probable therefore that this figure should be the Naturall likenesse of the appeareance.

But if wee conceave the outward shell thicker & conteined globe of subtle matter lesse as in the 2^{d} figure [Fig. 26.2] then it will be difficult to apprehend how so little Central matter should forme this appearance, besides the shineing part df will Not be much broader then in the 3^{d} figure [Fig. 26.3] & still one of the former objections takes place that is the appearance of light will be too small for a 2^{d}

light star. except wee conceave the bulke of the star as bigge as Syrius or ye harpe which considering how few stars there are of this magnitude is not probable.

Tis therefore most likely that the Corticall or dense superficiall matter the aperture or broken place in it & the included globe of light are to each other as in ye 1st. 4th and 5th figures [Figs. 26.1, 26.4, 26.5 respectively]. in which I have represented the positions of it at ye first or smallest increaseing, last or least of ye decreaseing, and middle or greatest appearances. for if the Middle of the Aperture be at 11 in ye 1st figure [Fig. 26.1] the line eb. proceeding from ye Earth & toucheing ye under or nearest side of ye hole if continued reaches the upper & inner side of it & the Contiguous globe of light or subtle matter whose raies emitted to o^r Earth forme the appearance of the star. & this is the first begining of the star to appeare. for if wee conceave the point b never so little more removed from ye line hc. the ray eb proceedeing from o^r earth & passeing by the point b parallel to hc will fall somewhere between d and a. nearer a as the remove is greater upon ye opake broken part of ye Corticall substance & from of[f] or *besides* the globe of light which therefore is hid from ye eye & star is wholly invisible

But now as ye star turne round on its Axis the point b approacheing nearer the Centrall ray hc Aperture or hole ab opens to the Eye & the laterall visuall ray eb passeing by the next side of ye hole & determineing ye breadth of ye lucid appearance. passes more & more dayly upon the lucid included Globe. of subtle Matter fdikl & the star increases in its visible Magnitude till nearly the center of the said open place traverse ye line or centrall ray ch or be arrived to ye point 2 where it lies wholly open to ye eye. as in the 5th figure [Fig. 26.5] & this is the greatest appearance.

Which by reason the centrall visuall ray hc falls not upon either of the sides of the open place but betwixt the space ab for some few dayes before and after that it cuts the middle, must continue for some dayes stationary as I may say and unaltered.

But as the center of the aperture recedes from this place of the centrall visuall ray the space ab. narrows again & is visibly contracted the star lessening proportionably till ye point in ye middle of the said open space be arrived at ye point 4 in the Periphery where the visuall ray ae in ye 4th figure [Fig. 26.4] falls upon & is intercepted by the side a of the open space & onely touches the included globe of light which is here barely visible but no further seene by reason that as the revolution continues forward accordeing to ye succession of the figures the laterall visuall raie ae falls first upon the broken darke side of ye aperture a.b. and afterwards on the obscure shell if selfe.

The length of the revolution of the star about its Axis may raise some scruple about this hypothesis. Those who are used to measure spaces by theire owne lenghts & time by theire owne ages when they consider that the time allowed for this revolution is 332 times as long as the earths & 13 times as the suns. may thinke improbable that this star should require so long a time to make one revolution in but if they would togeather consider to what vast distances Mathematicall demonstrations have removed those bodyes & how vastly large & bulky they make them theire Wonder would imediately cease especially if they weigh with themselves those long periods of Revolutions of the superior planets about ye sun. which caused the Ingenuous thoughtfull & subtle Kepler to conclude that hee had a Revolution upon his Axis before any other seemes to have dreamt of it[4] indeed hee was mistaken in his period for hee judged that when any new Invention should helpe us to this discovery wee should find the Revolution made in 3 dayes or therabouts which by the helpe of the telescope soone after ye publication of his Commentaries de Motibus Stellae Martis[5] was found to be made in No lesse y^n 25 dayes which by the way May Intimate to us the shallownesse of humane understanding & how infinitely short it falls of the divine Wisdom but to returne to my first subject & conclude it breifely to thinke or assert that the revolutions of the fixed stars can not require longer time then that of o^r sun to limit ye power of the great Architect who is unconfined by time or Measure & to come to o^r selves wee may as well argue that the Antidiluvian patriarchs lived not so long as the divine history informes us because none in o^r dayes attaine to theire Ages

The solution I have proposed of this strange appearance is so naturall that I doubt not but it will be satisfactory also to those that have or shall duely consider it but there remaines one p[ar]ticular more in the history which I have not touched that is that tho it was not commonly taken notice of before the Yeare 1638. & that it is not in Tychoes Catalogues. Nor in Keplers so y^t it is not probable that it was seene by Tycho or his disciples, yet is it in Bayerus his Charte of this Constellation which was published in ye yeare 1603[6] so that wee cannot thinke it to have appeared whilest Tycho was observeing these stars for then hee had not missed one so remarkeable & yet it must have appeared soone after or Bayer Could not have placed it in his figure. how a New star should Come to appeare is then the Question which will best be resolved by considering how those which have appeared formerly in ye poope of the Ship & else where in the heavens come to have disappeared Nor can this be well done without returneing to o^r sun & remindeing you of what I formerly sayd Concerneing his spots.

That is that I conceaved the light encompasseing him to be of a liquid substance of no great depth. which by its subtility & violent motion might weare it selfe a passage into some part of the Earthlike Globe I conceaved placed under it whereby, the inward sulphureous parts might by fixed & a violent Eruption Caused in some adjacent part of his body whence the matter cast out might forme a vast prominence or Mountaine in his superficies, that the liquid sea of light being not very deepe this eruption might be cast much above it whereby a spot would appeare to us in his superficies, of longer or lesse duration accordeing as the Matter Cast out were more or lesse & of a more or lesse tenacious substance for as theire are infinite varietys of Matter in o^{r} Globe of earth it is but reasonable to conceave ye like in ye sun. or any other of the planets where there are solid globes of such vast bignesse.

The subtle Matter envesteing the sun soone frets & weares these downe againe leveling them equall with his superficies. but in the meane time wee must conceave the liquid sea of light encompasseing him to be lessened in his superficies by so much as was receaved within his central globe thus the superficiall matter may by continuall agitation worke it selfe wholly into this centrall globe whereby that Glorious body may at last become darke and scarce visible but as the star I have now given an account of for some part of its revolution when the place or orifice where the Matter entred shall be obverted to ye view I say not that it will be so but that it may & that thus *pereunte mundo* the sun may lose his light

And since I have proved from ye appearances of the New star in the Whales breast that it has a shell of an opacke substance encompasseing a centrall light why may wee not conceave it to have hapned in this Manner that it was of old a Glorious fixed star or rather Sun appeareing so at a vast distance, that it had its luminous substance on the outside which has worne it selfe a passage & place in the Middle of its body after the Manner I have described. some stars seeme to have decreased in theire Magnitudes assigned to them in the Ancient & Tychoes Catalogues it is easy to answer them in this hypothesis. as also that of ye star in ye Swans breast and bill and Cassiopeas Chaire, for wee may readily conceave that the orifice through which the luminous matter wrought it selfe from ye superficies into ye body of the star may by severall accidents be closed. that fresh Eruptions may be Caused through which the lucid matter might be conspicuous againe. for some time, so long as it continued open, & that in the star of the Whale tis so wide it cannot be easily filled up againe or closed.

De Chart[7] conceaves that all the Planets may have beene suns or fixed stars haveing originally centrall globes of luminous matter in them these may be also admitted & the appearances still solved I conceave both true that some have this subtle matter centrall others superficiall & that from hence wee may fetch the solution of one of the Greatest Secrets of Nature, of which God spareing mee an opportunity I have more to say hereafter.

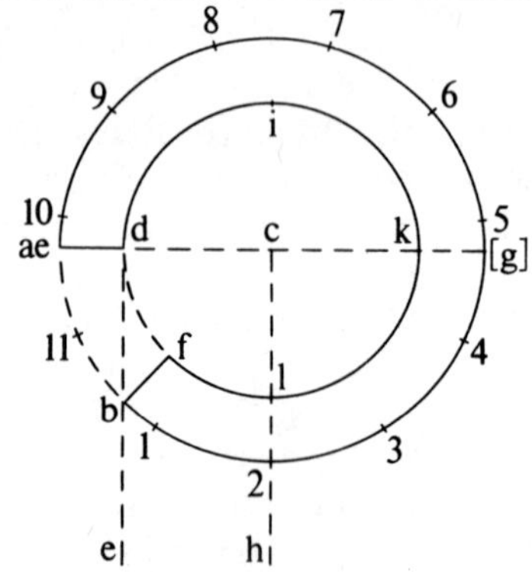

FIG. 26.1

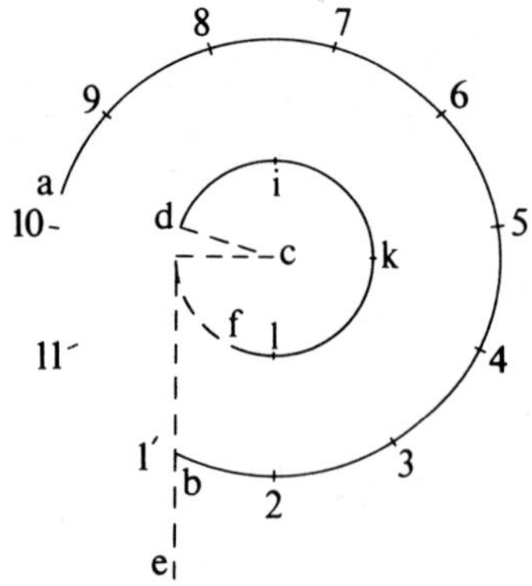

FIG. 26.2

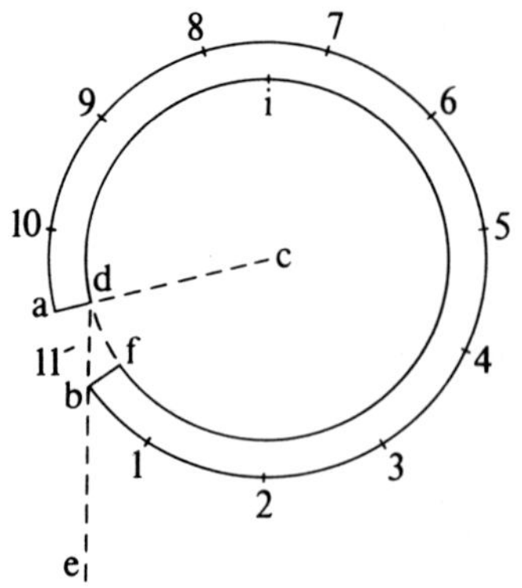

FIG. 26.3

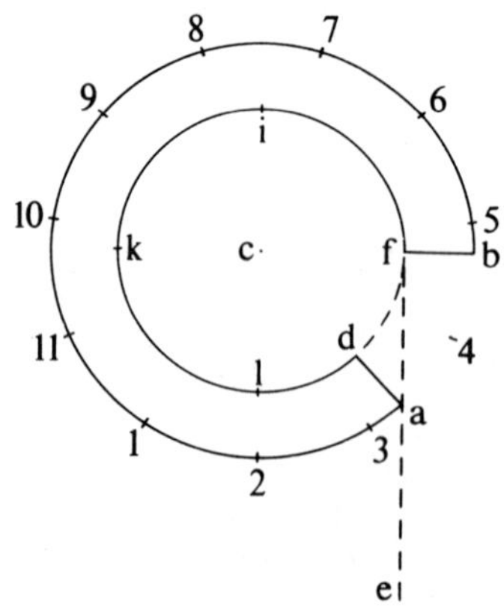

FIG. 26.4

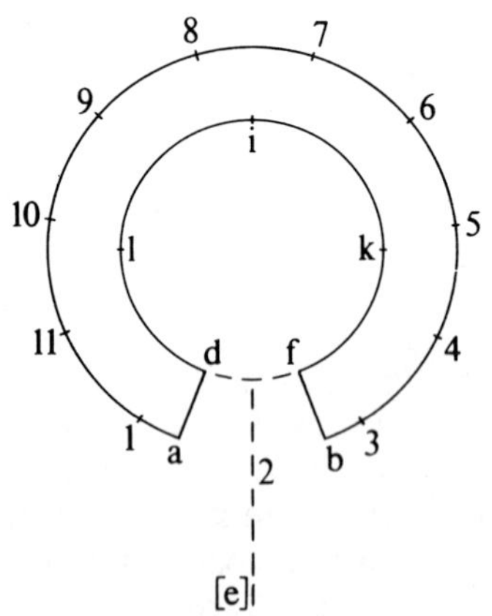

FIG. 26.5

NOTES

26.1. *Psalmorum Davidis paraphrasis poetica Georgii Buchanani Scoti* (Herbornae Nassoviorum, 1624) was one of several books in Flamsteed's library from which these words may have been taken or translated.

26.2. This is the *synodic* period of the Sun's axial rotation; 25d 6h being the *sidereal* period.

26.3. Lecture 24 (7 February 1682/3).

26.4. Op. cit. (note 24.7).

26.5. Op. cit. (note 1.3).

26.6. Op. cit. (note 25.16). the nova in question is that in the Whale's breast which was observed by Hevelius in December 1638 (cf. op. cit. [notes 23.16 and 24.39]). Flamsteed thought that if this had been the same nova witnessed between the time of Tycho's death and the publication of Bayer's charts (op. cit. [note 25.16])—that is, *circa* 1602 or roughly 36 years earlier—perhaps it was the same as that reappearing in 1682, or 36 years later.

26.7. René Descartes, *Principia Philosophiae*, 3rd edition (Amstelodami, 1656) was the source of Flamsteed's knowledge of that author's cosmological ideas.

Lecture 27

16 *May* 1683

In my Last lecture. I shewd y^{u} from the appearances of ye New star in ye Whales breast that it turned about on its owne Center that the outward part of it was a shell of dense matter, the inner a luminous substance shineing through a broken & open place in that shell which as often as by ye revolution of ye star on its owne Axis it was brought into ye view of the Eye made the appearance of a star small when first visible. then encreaseing for about 2 moneths when it arrived at ye bignesse of a star betwixt ye 2^{d} & 3^{d} light of which magnitude it continues neare a forth night afterwards continually and gradually decreaseing for almost 2 moneths longer when ye orifice by the stars revolution being removed from ye Eye it becomes againe invisible & so continues for 7 moneths after, I told y^{u} also by the by, that the time of its greatest appearance this yeare would be in the begining of July next, and that it would disappeare in the begining of September the period of its returnes being 132 dayes.

Enquireing how this subcorticall light might come there I shewd y^{u} from ye appearances of the Solar spots that it was probable the luminous matter envesteing him might find passages into ye centrall substance whereby It might gradually worke it selfe receptacles in that the spots appeareing on his superficies were onely eruptions of the included matter cast up to a very considerable height in his superficies & through the ambient luminous fluid that hence, the exteriour coate of luminous . . .
and that probably in time ye exterior part may all become darke so that onely once in 27 dayes a light place may be seene in him where the fluid had entered the globe it invested before. & that in all proba[bi]lity the appearance of the New Star was caused by some accidents

For theire are some of the bright fixed stars missing which wee cannot easily conceave are removed from theire places, but wee may readily that they are extinguished in them after the Manner that I have supposed the sun may be darkened pereunte Mundo, others seeme to have theire light diminished this hypothesis may answer it

Notes on pages 363–4.

as also it suggests to us whence the Comets arise but haveing spoken of them formerly in some lectures I read the terme after the appearance of the last great one I shall say no more of it at present least I involve myselve into a subject which I designed not to meddle with at present & have onely occasionally touched upon.

Yet that I may not be misapprehended I must crave leave to adde that tho I doe not thinke that every sun or fixed star upon the reception of the outward luminous Matter whatever it be into its earthlike globe placed under it, does lose its place or to use the Cartesian termes has its vortex absorpd & become erratic yet I grant that sometimes it so happens they may & I am very apt to beleive that some of the Comets wee have sometimes seene are of this sort[1] & . . .

Hence also it will be easy to conceave how those New stars in Cassiopeas Chaire Serpentarius, & two in Swan have hapned to appeare shine bright & disappeare againe in undetermined tracts of time, for the active lucid matter which once encompassed them being receaved or rather haveing made it selfe away into the globe of solid Matter may there after some time meet with sulphureous matter which may cause ye shel[l] it has formed about it to breake & in theire returne so to fall as to cover all the orifice through which the outside matter had entered & beene once admitted: so that the body shall no longer bee seene afterwards then while that fuel burnes & blazes. this I conceave to have beene ye case with ye star in Cassiopeas Chaire & Serpentarius, in others the place throug[h] which the eruption was made may fall in gradually So I conceave it to have hapned with ye star in ye Swans bill. in some it may if large, continue open so I conceave it & have shewd it to be in that of the Whales breast. p[er]haps in some there may be more then one open place in ye shell else how should that in the Swans breast be still visible for I have seene it within a very few nights but exceedeing small scarce equall to stars of ye sixth light.

There are two phenomena still left to Consider & of which I have given no solution that in ye North part of Andromedas Girdle, & another like it in ye Sword of Orion

Though that in Andromedas Girdle be not a perfectly round appeareance yet its figure approaches something to roundnesse when viewd with a good glasse. there is however nothing like a star in it but rather an hazy misty brightnesse. here I conceave some remarkeable star to have beene seated formerly, but the outside luminous matter being removed into ye middle of ye broken masse shines t[h]rough a great number of small foramina or holes it has made in it by frequent eruptions. Bullialdus[2] informes us that tho the Noble Tycho has taken no notice of this appearance yet it had beene remarked above 100 yeares before him by an anonymous observer tis not in Bayer[3] yet have I seldome mist it when I viewd that Constellation in o^{r} Cleare darke nights hence tis evident that it has appeared sometimes clearely at others not. which wee may conceave to [have] happened according as the Matter through which ye Nuclesis shone had more or fewer open places in it or as they were more or lesse obstructed. I once thought the substance whatever it be in that place might be opake. & the light reflected from it onely what it receaved from ye adjacent fixed stars but then the alteration of its brightnesse allowes not this conception which requires it to be allwayes the same, & therefore I am forced to Reject that opinion

As for the other bright place in Orions Sword the same hypothesis will answer it onely some persons may scruple to admit it by reason of the irregularity of the figure. But sure the Almighty hand is not obliged to observe the Circumstances of such formes as wee esteeme regular in his Workes, for what appeares irregular & out of order to us may be excellently disposed if wee were well acquainted with ye reason of such disposition, but I have this reason also for applying my hypothesis to this appearance. there appeare in it as the Excellent Hugenius informes us[4] seven small bright fixed [stars] shone within its compasse all as it were through it & 3 of them contiguous to each other. Now since wee cannot conceave the Matter in that Place pellucid, all the other Caelestiall bodyes being opake. as o^{r} planets are proved by theire eclipses & illumination o^{r} sun by his spots. it is not conceaveable that the stars as they appeare to doe should really shine through this hazy appearance. further wee find that the great hand which ordered ye universe has placed o^{r} sun or the fixed star of o^{r} vortex at a vast distance from any other yet is hee considerably smaller then many of the fixed stars which wee may therefore reasonably conclude to be placed at greater distances from each other then hee is even from ye next him if so then the shineing matter in ye Sword of Orion must be vastly removed beyond ye small stars appeareing in it & consequently can reflect little light from them to o^{r} eyes what wee see must be innate & proper to it emitted from it selfe. & therefore most like sent from its interiour parts through severall spiracula or holes formed thick in its outermost superficies. The vicissitudes of ye appeareing & disappeareing of severall fixed stars in stated periods of time shewes us therefore that as our sun so they have motions or make revolutions about theire owne Axes & Centers, and further tis probable that the darke broken shells of Matter includeing the luminous shineing substance was formerly all Centrall but by the violent motion of ye luminous subtle matter includeing it had foramina wrought in it whereby the outside luminous substance wrought it selfe into ye grosser Centra 11 matter & by forceing continuall eruptions as the superficiall Matter does in o^{r} sun caused first large spots upon them afterwards weareing theire Cavity wider it wholly left the superficies & became all Centrall as I shewd y^{u} in tracts of time it may probably happen in o^{r} sun & Now I thinke it can no longer be doubted but that they are as hee is the Centrall suns of severall vortices & that as o^{r}s so they may have sustemes of planets moveing round them & performeing theire periods in spaces of time proportionable to the periods of the Revolutions of theire Centrall suns, theire Magnetique powers. & theire owne. proper Magnitudes & Consistencys.

This hint may suggest unto us the originall of Comets which being vast bodys of matter wee can not easyly conceave how they should otherwayes be generated but admitting them to be the planets of some vortex whose centrall sun is thus extinct, wee may easily conceave that tho his attractive power be in a great measure or totally destroyed by the subtle superficiall light leaveing the superficies & retireing to ye center of the star through orifices made in the globe once placed under it yet the planets runneing of may retaine so much of theires as may serve to incline them towards the Centrall sun of any other vortex when they fall into it Neare the plane whereing the attractive faculty of that sun or star most powerfully operates & exerts it selfe but haveing spoken more fully to this in a lecture I read formerly concerneing the great Comet that appeared 30 moneths agone[5] I shall not say any more concerneing it at this time

But onely that if as de Charte[6] conceaves & probably it is those bodyes which have not the luminous active matter in theire super[fi]cies such are our earth & the planetts of o^{r} systeme shall have it in theire Centers as these strange fixed stars have, then wee may p[er]haps fetch the Doctrine of Magnetisme[7] or attraction from ye Centrall subtle matter in ye planets & superficiall in o^{r} sun much better then by the helpe of those p[ar]ticulae stratae which de Charte has Coyned in his principles[8] to serve this purpose whereby tho hee shewes how it happens that the loadestone may attract from yet hee has not explained to us how the sun attracts & repells the planets retaineing them still in the same due distances from him in theire proper places of theire orbites which in my opinion ought to have beene his first consideration. for except wee allow some Magnetike virtue in ye sun it will be impossible to conceave why they continue to move in ovall figures about him & doe not rather run of[f] in streight lines or indeede why they should respect him at all or have theire Motion directed round him

Haveing thus shewed y^{u} what sort of body the fixed stars probably are I come next to consider theire Colour & scintillation or twinkeling.

Theire colour in theire Riseing & settings is observed continually more dull & reddish then when they are any considerable height above it, the reason is becau[s]e in and neare the horizon theire rayes are intercepted and broken by the vapors of ye atmosphere or grosse exhalations from o^{r} earth whereby as many of those which come directly from ye star towards the Eye are intercepted & diverted. whereby theire naturall colour, is rendered more obscure & dull then it is afterwards when the stars are raised the some small height above these vapors

But besides this accidentall there is a reall difference in theire colours ariseing from some innate principle as the different shapes of the particles of the fluid substance envesting them for it is not reasonable to conclude that all fires or fluid lucid matters shoud be composed of the same substance any more then that there should be more then one peculiar substance in o^{r} earth, or in ye Centrall globes of the stars themselves.

Concerneing theire scintillation or twinkeling Riccioli reckons up six severall opinions of learned & Ingenuous Men.[9] The Aristotelian's who taught that objects were seene or (to speake theire sence in proper words) rather arrested & brought into ye eye by rayes emitted from it, Conceaved that the eye trembled & danced at ye view of remote objects & that therefore the sun seemed not to vibrate but onely neare ye horizon because of ye elongation of the sight from him there I speake theire owne words for they with some of o^{r} vulgar conceaved the sun more remote upon the horizon then when hee was elevated from it, that the Planets twinkeled not for they also were concluded neare us so y^{t} the eye by its owne strength could reach them but ye stars vibrated by reason of theire great distance. but this the Ingenuous men of o^{r} Neighbour age found but a trick of words contrived to hide theire ignorance in Naturall Causes & amuse the commonly people with subtle termes y^{t} onely seemed to signifie much because they had no sence & were not at all understood. This opinion therefore as knowledge encreased was soone rejected & lookd upon as ridiculous so y^{t} Copernicus argued the vast distance of the fixed stars compared with that of the planets from theire scintillation. for sayes hee, Quod enim a supremo Errantium ♄°, ad fixarum spheram adhuc plurimum Intersit scintillantia illarum lumina demonstrant[10] y^{t} is. the scintillation or twinkeling light of fixed stars proves that there is a vast space betwixt the highest of the planets ♄ & the fixed stars And the Noble Tycho writeing Concerneing the New Star in Cassiopea's Chaire more expresly scintillatio in super eximia, & praeceteris Caeli luminibus Coruscans, omnium apertissime ostendebat, illam inter affixarum stellarum supremam vastissimamque regionem locum sibi delegisse omniumque planetarum gyrationes nedum humillimum elementarem Mundum exuperasse[11] y^{t} is Moreover the Wonderfull scintillation or sparkeling of ye star beyond any other of the Caelestiall lights does most evidently shew, that its place was given it beyond o^{r} low elementary regions amongst the highest of the fixed stars, And in another place, hee sayes Citra omnem tergiversationem &c[12] y^{t} is, Without any controversy wee are obliged to confesse, because this star continually twinkeled, that it was placed beyond all the orbes of the planets amongst the remotest fixed stars

The Noble restorer of Astronomy shewes no reason for this scintillation in the Fixed stars but Vitellio[13] Scheiner[14] & Blancanus[15] Enquireing the Cause ascribe it to the Medium through which the Rayes or Species of the fixed stars are transmitted to us, Scheiner is very expresse *Stellarum scintillatio est vinca et sola specierum in oculum délapsarum Intercisio facta ab irrequieta vaporum varie affectorum intercisione*[16] the scintillation of the fixed stars, is onely the breakeing of theire raies, by the continuall agitation of the Vapors of o^r atmosphere, the others[17] adde that the stars vibrate sparkle or twinkle by reason that the Raies comeing through that vast distance betwixt them and our earth have not sufficient force to breake through the vapors of o^r Atmosphere or impresse theire motion directly on them, but the planets being placed much nearer us tho theire light be onely borrowed & reflected have. that therefore the[y] sparkell not all nor sun or Moone except when they are very neare the horizon where theire raies are broken by the grosse vapors emitted from o^r Earth. for then indeed the sun seemes to undulate so does the Moone too but the planets sparkle a little yet not so violent as the most sedate of the large fixed stars, that the scintillation of the stars and the undulation of the suns limbe neare the horizon is made by ye moyst vapors adjacent to o^r earth & breathed from it seemes to me confirmed by My dayly experience for haveing some 12 yeares agone occasion to trie a good telescope of 14 foot on a land object placed about 300 yards from mee in moist Weather[18] I found that when y^e sun shone strongly on the object which I had so placed that my station might be betwixt the sun and it on purpose to have it well illumined the object appeared to danse & undulate so briskely that I could not distinguis[h]e black markes made in white paper $\frac{1}{8}$ of an inch broade & at above 2 inches distance from each other upon it, tho when the sun Ceased to shine the Clouds interposeing I saw them very distinct & the object undulated very little; which suggested to mee that the motion of the interjacent aire was onely caused by the agitation the suns raies strikeing upon it put it into whereby it interrupted the light reflected from ye object. since then I have contunually found the same appearance in horizon objects viewd when the aire is moyst or inclineing to raine; so that tis not to be doubted but that the interruption of the raies of light in o^r Atmosphere is the true reason Not onely of the undulation of the limbes of the luminaryes neare the horizon but also of the scintillation of the fixed stars.

Of which yet the ingenuous Kepler will have some parts to be attributed to theire rotations about theire Axes,[19] but these I have shewed to be so slow by the observations of ye stars in ye Whales breast & Swans breast & bill that I doubt not but this Concept is out of dores at least No otherwayes esteemed then as one of the bold strokes of that great wit & judgment.

As for theire disposition some of the Ancients conceaved them all placed in the same sphere. but this opinion soone was exploded & it was more truely concluded that theire different Magnitudes proceede not so much from any reall differences in theire bulke as from theire different distances whereby the more remote appeared smaller then the greater by reason they subtended lesser angles at o^r eye as for theire order the ancients who supposed them all placed in the same sphere conceaved it wholly irregular. but sayes an ingenuous author if wee "barely consider them as they appeare they seeme rather scattered too & fro by accident in the heavens then placed there by any determined advice & forecast seeing that they are not disposed at equall distances or in any regular figures like the flowers in o^r gardens," But this sayes hee is an Argument of some greater wisdome and forecast for those thinges which to humane wisdome appeare irregular may to the divine be very regular & ordinate.[20] and de Charte in Epistle to Mersennus the 57 in order of his second booke expressely Quamquam temere admodum in Caelo &t[21] y^t is Tho they seeme scattered there & there without designe in the heavens yet I doubt not but theire is a very regular & determinate order observed in theire disposition, the knowledge of which is the key & groundworke of the greatest & most p[er]fect knowledge that man can attaine to of materiall beings, for that by the helpe of it the different formes of all terrene bodyes may be found from theire first Causes, but without it wee are forced to Collect them theire effects. which assertion I esteeme as that I before mentioned of Keplers onely a bold stroke of that great mans fancy for wee find not any thinge either before or after relateing to it in any of his workes so that I am apt to beleive Whatever hee then thought in the heate of his New Notion hee found it not to answer his expectations in the prosecution. however since wee can scarce conceave the Wide expansum of the heaven's bounded wee may safely conclude the stars innumerated & in that vast space the great mind which observed a Wonderfull order in all its workes has not disposed them disorderly. but for theire substances or Natures scintillations and order this may suffice theire apparent motions distances & position I shall consider when God offers us another opportunity which may it be as the Founder designed[22] ever usefull & happy.

NOTES

27.1. Cf. Flamsteed's views on the cause of cometary motion in Lecture 3, ff. 33^r–34^v, or pp. 114–15.

27.2. Op. cit. (note 24.45).

27.3. Op. cit. (note 25.16).
27.4. Op. cit. (note 24.47), p. 8.
27.5. Lecture 3 (11 May 1681).
27.6. Op. cit. (note 26.7).
27.7. Introduced by William Gilbert and applied by Kepler and others to explain the motions of bodies in the solar system. Cf. note 3.22.
27.8. Op. cit. (note 26.7).
27.9. Riccioli 1, pp. 396–7.
27.10. Quoted in Riccioli 1, p. 396 out of Copernicus, op. cit. (note 11.11) (lib. 1, cap. 10).
27.11. Quoted in Riccioli 1, p. 396 out of Brahe, op. cit. (note 2.5), p. 401.
27.12. Cf. note 27.11. The complete sentence was excerpted by Riccioli from Brahe's text, and translated here by Flamsteed.
27.13. Op. cit. (note 33.6), pp. 449–50. 'Scintillatio accidit semper omnibus stellis fixis propter divaricationē formae in loco imaginis ex motu subiecti corporis accidentem.'
27.14. Christopher Scheiner, *Disquisiitones Mathematicae, de controversiis et novitatibus astronomicis* (Ingolstadii, 1614), pp. 68–72.
27.15. Blancanus, op. cit. (note 24.17), p. 342; or p. 181 in the 1653 edition.
27.16. Quoted in Riccioli 1, p. 396 out of Scheiner, op. cit. (note 27.14), p. 70. The last word ought to be 'intercursatione' and not 'intercisione'.
27.17. Riccioli 1, p. 396, further refers to the opinions of Giordano Bruno, Franciscus Aguilonius, Raphael Aversa, Tycho Brahe, Kepler, Bartholomaeus Mastrius, and Scaliger.
27.18. Flamsteed had made this experiment early in 1671 for the purpose of testing the quality and optimal arrangement of the objective and eye-glass which he had obtained through John Collins about three months previously. Cf. Flamsteed to Collins; Derby, 30 January 1671 in op. cit. (note 23.10), p. 106.
27.19. Riccioli 1, p. 397, referring to Kepler, op. cit. (note 24.7), p. 95.
27.20. We have unfortunately been unable to trace the source of this quotation.
27.21. René Descartes, *Epistolae, partim at auctore latino sermone conscriptae, partem ex gallico translatae, etc. pars secunda* (Amstelodami, 1668), p. 233.
27.22. The founder of Gresham College, Sir Thomas Gresham (1519–79).

Lecture 28

13 *June* 1683

In my readeings here last terme.[1] I discoursed Concerneing the Nature substance light & scintillation of the fixed stars. theire remotenesse from o^{r} habitation the earth p[er]mits not any absolute proofe what they are but onely comparative. I considered the Nature of o^{r} sun & suggested to y^{u} what in all probability was the reason of his spotts I shewed y^{u} that tis likely the shineing matter is but a shallow fluid encompasseing a globe of more solid matter that this fluid has wrought it selfe passages into the globe it envests & receptacles there, that tis probable in time the whole ambient fluid may leave the superficies of this globe, & become centrall so that onely a small lucid p[ar]te may remaine visible (through the passage at which it entered) for some small time of his revolution or it may be totally covered & the sun extinct from Whence the destruction of the rest of ye planets of o^{r} susteme may follow.

I gave y^{u} an account of some strange fixed star whose appeareings & disappeareings were periodicall of which inquireing the cause I could assigne no other then that which I had intimated to o^{r} sun, that is they had formerly theire lucid matter with the rest of the fixed stars in theire superficies. but that by like accidents spotts were at first made or eruptions cast up in it till such time as the whole superficiall matter had receded into & buried it selfe in the more substantiall globe it once invested, that where the passage through which it was receaved is large, & so not easily filld up againe the star appeares for some time of every revolution as y^{t} does in ye whales breast, but that it often happens, tis totally closed up & so probably twas in ye New Star of Cassiopeas Chaire, Serpentarius the Swans bill, some in ye poope of ye Ship & other places of ye heavens. I shewed y^{u} also from the little or no parallax observed in the fixed stars that theire distances were vastly large they were not to be determined by humane skill, & haveing proposed them as little as I reasonably could & stated theire apparent diameters I proved that scarce any of those visible to ye bare eye were lesse y^{n} our sun & summeing up my severall Comparisons that they were such as hee is the Created Centrall motive & attractive powers of severall vortices

Notes on pages 371–2.

I intimated by ye way the probability of De Chartes opinion[2] that all the globes of o^{r} susteme may have centrall subtle matter in them such as investes the more solid bodyes of the fixed stars. & that the doctrine of Magnetisme may p[er]haps be derived hence much better then from those particulae striatae which that great man has Coinded for this purpose. I shewd y^{u} also whence theire scintillations proceeded, & what was the opinions of some very ingenuous men concerneing theire disposition in ye heavens with which I concluded last terme.

The method I proposed first to my selfe requires that I should next shew y^{u} how theire places in longitude & latitude came first to be assigned to them, how the longitudes alter annually & from what cause; Theire latitudes have been held invariable ever since they were determined by most ingenuous men but because some p[er]sons of no small repute have asserted that they have changed[3] I shall also enquire whether they have or not & togeather settle the Annuall motion of the fixed stars or rather recesse of the Earths equinoctiall pointes.

Before wee shew y^{u} how theire place in longitude & latitude came to be assigned them it may be convenient to run back a little further & shew y^{u} what were the opinions of ancients concerneing the motions of the Heavens which I thinke cannot be better done then in the expresse words of ye subtle Kepler.[4] I translate them thus. "Before y^{t} men had yet distinguished betwixt the first & secondary motions of ye heavens viewing the sun Moone & stars they tooke notice that theire diurnall courses were circles nearely but such that one seemed alwayes connected to ye next after the manner of ye thrids wound upon a clew, that the Circles described in ye sphere were most Commonly lesser sometimes but seldome great ones cutting the æquator AB in C parte of them declineing to the South, part to the North of it, they observed also that theire was a difference in ye swiftnesse of them in this theire diurnall & apparent motions. The fixed stars they concluded to be the swiftest of all others. For when any night they had seene the Moone setting upon ye horizon at f within a star h, the same fixed star haveing described its diurnall Course by 1. k. & comeing to the horizon at i they saw the moone now a good distance above it in g, & that shee set not of some time after, that the sun in like manner approached dayly to these fixed stars which soone after his setting had beene seene any night in the Meridian, but the Moone was slower to theire Imagination then hee because that when at any time shee had set within a at f next night shee set after him and was above the horizon at g when hee touched it, the like they observed in the planets of all which they esteemed the Moone the slowest because these alterations were the most p[er]ceptible in her motion, & therefore when the Pythagoreans distributed theire Musicall Notes in the heavens they attributed the Weig[h]tyest or y[e] πατεη to the Moone whose motion seemed the slowest to them. hence can also the termes so often used by the ancient Astronomers προηγούμενοζ. ὑπολειπτιχός[5] in antecedence or consequence. those stars which set first being accounted in antecedence of those which came next after them to y[e] horizon, or these latter in consequence of those that set before them.

This (sayes hee) is that picure of Astronomy, which consists in no explication of the Cause but in the bare dull & deluded experience of o[r] sight, and differing continually from it selfe can neather be explained by schemes nor Numbers & therefore cannot be conveyed downe to posterity because it differs from it selfe continually no one spirall revolution being exactly equall to y[e] next or any other in length or time, and yet this is that which some p[er]sons after two thousand yeares paines & experience have endeavord with all theire skill learneing & knowledge to restore, succesfully procureing themselves to be admired & esteemed by the Vulgar. whom yet the more skillfull know onely to have trifled & if they arrogate to themselves the Names of Philosophers to be rationally Mad

But in processe of time the Astronomer understood that two simple motions were here mixt or confounded to wit y[e] primary & secondary or the common & proper whence that connected series of windeing motions had necessarily followed, & now they found the diurnall revolutions being seperated & withdrawne from the other that the fixed stars were not the swiftest and y[e] Moone the slowest as had beene hitherto Imagined but on y[e] contrary the Moone the swiftest & the stars either the slowest or unmoved.[6] This was a great step towards the understanding of the simplicity of the planets motions for now in y[e] roome of infinite spiralls succeedeing & connected to each other there was left onely almost single Circles & one common motion of the planets contrary to the Motion of y[e] Universe supposeing the earth to stand still, but admitteing with Aristarchus that the Earths globe turned round, the same way all from West to East. so far the Ingenuous Kepler.

It being thus found by the different times of the ariseings & settings of the fixed stars that the suns apparent motion was round the heavens & the Next thinge to be enquired was neare which of them his annuall Course was directed.

In my former & first readeings I shewd you that the necessity of an exact account of y[e] yeare was the occasion of the Ancients enquiry into the motions of the luminary. that the first contrivance for this purpose was the gnomon which tho history tells was first set up at Lacedemon by Anaximander yet it is probable that it was earelyer used by the Jewes or Indians under which name the Greekes concluded all those people that lived from y[e] Mediterranean sea Eastwards.

That hereby first the obliquity of the line of the suns annuall Course to ye diurnall revolutions was determined almost 24 degrees. the line of his Annuall Course they called the Ecliptick the greatest of the diurnall revolutions or ye Circle bisecting the heavens equally distant from both ye poles *æquator* & the pointes wherein these great Circles intersected each other ye equinoctiall pointes. the great Circle connecteing them & passeing through the two poles of ye Mundane sphere the aequinoctiall Colure.

The Gnomons were Instruments onely proper for solar observations not for the stars. to place these the aequinoctiall & other Armillas were continued. where both the suns & theire declinations or distances from the æquator were first determined. But the fixed stars not being visible by day or observable from ye sun on that account. some other meanes was to be thought of for determineing how far they were from one at any time. The Moones motion was the next after the suns that they regarded the Ancient moneths being lunar, & by the meanes of her Eclipses or passages in opposition to him through the shadow this was next understood the Inaequalitys in it determined & tables made for correcteing her motion. the Moones distance from ye sun was observeable by day from ye stars by night. compareing these distances with theire declinations it was not very difficult then to determine the places of such fixed stars which had beene observed from the Moone the same day that shee had beene observed from the sun, & that this was theire method seemes further confirmed by the accounts Menelaus & Ptolemy have left us of the Moones transits over fixed stars whereby they doe not pretend as wee now to correct her place but supposeing that truely given by theire tables the thence derive the place of the fixed stars or Pretend to shew those of theire Tables allready determined to bee confirmed by them.

And indeed it is evident that theire knowledge of ye Moones motion was much the best next the suns they had of any of the other planets, the meane motions aequations. Motions of ye Node and Apoge. being much nearer the determinations of or best & latest Astronomers then they are in any of ye other 5 primary planets

Which causes mee to thinke that they had much more ancient observations of her then any come to or hands, & that by theire Armillas they often observed her from ye sun in all positions & places of her orbite. But theire instruments either not admitting exact divisions or being ill contrived, or p[er]haps theire faulty determination of ye solar inequalitys or all togeather caused the moones places found by them to be very erroneous.

I have beene enformed very lately that, amongst some ancient mansucript Arabick bookes brought over hither. there are some that treate of the ancient greekes Instruments & compareing them with the Arabian shew they were much inferior both in contrivance & division,[7] these if translated & made publick might give us some light into the observations of the Ancients & shew us how far they are to be relied on.

At present wee can onely judge by compareing theire latitudes of the fixed stars, collected from theire observed declinations, & y[e] Tychonick corrected latitudes with y[e] Noble Tycho whereby it will be evident that they seldome came nearer then 20 minutes, tho a mixture of y[e] false latitudes & obliquity of y[e] Ecliptick, at least 20 minutes too greate, has caused a suspicion of much greater errors. as I shall more fully shew y[u] hereafter for now time to returne to o[r] first designe to shew what is the motion of the Fixed stars or rather first how it came to [be] discovered & limited after which I shall endeavor to shew y[u] the naturall & genuine cause of that seemeing motion which y[e] experience of many ages has discovered in them & why theire longitudes are altered theire latitudes remaining invariable. I shall not here concerne my selfe with theire faulty method of determineing the places of the stars by theire observed distances from y[e] moone or occultations that I shall have occasion to be larger upon hereafter at present I am cheifly concerned with theire declinations & latitudes of which I shall take the ancient as the observers give them it being impossible to correct them on any considerations except it be the error of Manuscripts or printed editions of bookes from them which have beene so well canvassed that tis not probable our diligence should herein gaine any thinge upon y[e] labors of o[r] predecessors[8]

The first observations of the fixed stars places & the oldest are those of Tymocharis made at Alexandria about 300 yeares before y[e] Incarnation. hee observed the latitude of y[e] star following y[e] brightest of y[e] pleiades to the Eastward 3°.40′ in y[e] 465 yeare of Nabonassar which Agrippa in Bythnia in y[e] 840 yeare of Nabonassar 375 yeares after him determined the same. Againe Tymocharis had stated the latitude of Spica ♍[s] 2°.00′, Menelaus at Rome in y[e] first yeare of Trajon & 845 of Nabonassar. makes it altogeather y[e] same. Further Tymocharis makes y[e] latitude of y[e] Norther[n]most in y[e] front or breast of y[e] Scorpion 1°.12′ North. Meneleaus found it altogeather agreeable. whence Ptolemy concludes the latitudes of the fixed stars immutable in all ages but theire longitudes to alter neare one degree in one hundred yeares for the bright star of y[e] Pleiades aforementioned was placed by Tymocharis in ♈ 29.00 by Agrippa neare 400 yeares after him in ♉ 3°.00′: that in the Scorpions forehead by Tymocharis in 2.00 of ♏ by Menelaus in 6. after neare 400 yeares. so course were the observations of the ancient & so far from y[e] truth as will appeare by takeing in some other collaterall observations of theires & compareing them with o[r] owne which allow not a degrees motion to more then 72 yeares.

Tymocharis observed the declination of Spica ♍[s] in his time 1°.24′ north. Hyparchus 145 yeares before o[r] Saviour 0°.36′ north also Ptolemy at neare the same space of time after him 0°.30′ south of the æquator. and it was by Compareing these declinations with ye observed distances of this star from ye sun that Ptolemy & those before him found it to ye south of the line his motion describes in the heavens neare 2 gr but its distance from ye vernall equinox encreased & larger in ye latter then in ye former observations

Hipparchus was the first who tooke notice of this alteration of ye declination of ye fixed stars. & that theire latitudes continued neverthelesse unmoveable if wee may rely upon ye account of Ptolemy who tells us in his Almagest Hipparchus quoque circa Zodiaci polos fieri hunc motum consensisse videtur &t.[9] yt is Hipparchus also seemes to assert that this motion of ye fixed stars is made upon ye poles of ye Ecliptick for in his treatise concerneing the translation of ye solstitiall & aequinoctiall points[10] hee gathers from Tymocharis & his observations that Spica ♍s Keepes allwayes the same distance not from ye Equinoctiall Circle but from the Ecliptick & that both of old & late times it was 2 gr. distant to the south of that great Circle which passes through the Middle of the signes. wherefore in his tract concerneing ye length of the Year hee[11] supposes this motion to be made about the poles of the Zodiack Yet was hee doubtfull as himselfe confesses, because hee neither thought the observations taken in ye Time of Tymocharis sufficiently accurate, they being but rudely made, nor yet the time interlapsed betwixt Tymocharis & him sufficient for this determination to which Ptolemy[12] Imediately subjoynes But wee haveing found it by observations made at wider intervalls of time & in allmost all the fixed stars, doe affirme that this motion is made about ye poles of ye Zodiack, for haveing observed theire latitudes from ye Ecliptick or ye great Circle described on ye poles of ye Zodiack wee have found them allwayes the same & equall to those which were gathered & delivered to us by Hipparchus, which hee confirmes by the fore[me]nt[ion]ed observations of ye Pleia[de]s. Spica & yt Norther[n]most in the front of ye Scorpion.

The latitudes of ye fixed stars determined by Hipparchus & Ptolemy were retained by theire successors for many ages after. I find not when they were first altered. but tis probable that when about the yeare 830 Almaimon the Caliph of Babilon had Caused the obliquity of ye ecliptick to be observed & found it lesse then was stated by Hipparchus & Ptolemy by above one quarter of a degree & his next successors that his determination was neare tho something bigger then the truth. they began from this time to suspect the latitudes of ye fixed stars stated by theire predecessors, for when they found that theire & positions remained the same & theire latitudes not much alter, tis proba[b]le that they might attribute the small variation noted in them rather to the little change they had found in the obliquity of the Ecliptick then any thing else. I have not Met with any ancient Catalogues besides that of ωleg beig[13] which differs from the Ptolemaick. Copernicus who was later then hee retaines his latitudes[14] so does Clavius[15] to both whom tis probable this worke of ye gran[d]child of Timurleng was unknowne or if they had it, the great authority of Hipparchus & Ptolemy & the obvious Faults of Olegs Catalogues no lesse visible then those in Hipparchus & Ptolemys might cause them to refraine the use of the Junior till they were better satisfied of its accuracy & authority

The differences betwixt the determinations of Hipparchus Ptolemy & the Following Arabians had Caused some Ingenuous to conceipt that the fixed stars moved unequally forward sometimes slower at others swifter, & to suggest severall contrivances for salveing this inequality.[16] I shall hereafter enquire what reason they had for this opinion but first it will be requisite that I should shew y^u the faults of ye Ancient methods of observeing, & what better Methods have beene used by the Noble Tycho & his successors from whose determinations when wee have corrected y^e true longitude of the fixed stars wee may determine Annuall motion. which therefore with Gods Assistance shall be done next lecture.

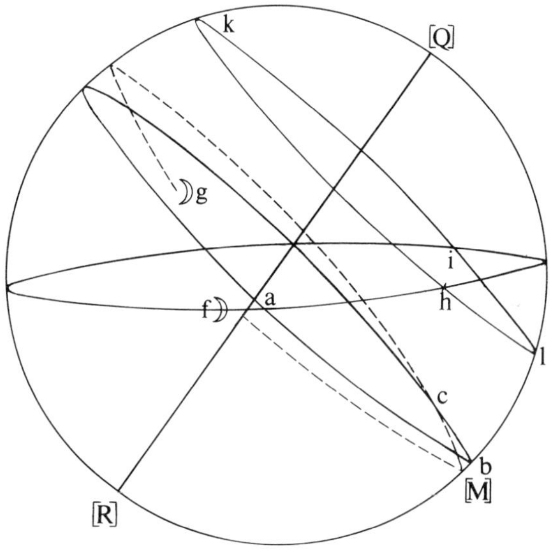

FIG. 28.1

NOTES

28.1. Lectures 22–27.

28.2. Op. cit. (note 26.7).

28.3. E.g. Pomponius Gauricus and Tycho Brahe. Cf. Riccioli 1, pp. 441–2.

28.4. The long quotation which follows is translated from the first chapter of op. cit. (note 1.3), pp. 1–3. Cf. note 28.6.

28.5. Flamsteed's own spelling of these Greek words was incorrect.

28.6. Flamsteed omits Kepler's brief description of the diagram in op. cit. (note 1.3), p. 2 reproduced in this text as Fig. 28.1.

28.7. Cf. notes 31.2 and 31.3.

28.8. Flamsteed's main source for the following information would seem to have been Riccioli 1, pp. 440–8.

28.9. The sentence is quoted in full in Riccioli 1, p. 440, from a Latin edition of Ptolemy's *Almagest* (lib. 7, cap. 3).

28.10. Hipparchus's treatise 'On the Change of the Solstices and Equinoxes', as well as those 'On the length of the Year' and 'On Intercalation of Months and Days' are mentioned by Ptolemy, *Almagest*, cf. lib. 3, cap. 1, 145 and lib. 7, cap. 2, 12 in Karl Manitius, *Ptolemaeus Handbuch der Astronomie* ([Lipsiae,] 1912–13).

28.11. This comment on Hipparchus's view regarding precession was translated by Flamsteed from Riccioli 3, p. 201.

28.12. Translated from the Latin sentence in Ptolemy (op. cit. [note 28.9]), quoted in Riccioli 1, pp. 440–1.

28.13. Thomas Hyde (ed.), [*Persian Title*] *sive tabulae long. ac lat. stellarum fixarum ex observatione Ulugh Beighi, Tamerlanis Magni nepotis* ... etc., 2 parts (Oxonii, 1665).

28.14. Flamsteed does not appear to have possessed a copy of Copernicus's major work (op. cit. [note 11.11]). He probably got his information from his copy of Rheticus's summary of it (op. cit. [note 12.12]).

28.15. Op. cit. (note 15.9).

28.16. Cf. note 30.18.

Lecture 29

20 *June* 1683

In my last lecture I showed y^u by what methods the ancients enquired the places of the fixed stars. of these I tould y^u that the first who give us any tolerable observations was Tymocharis hee lived about 300 yeares before ye Incarnation, the next was Hipparchus about 150 yeares after him Ptolemy at neare the like distance after o^r Saviors time was the last, who seemes onely to have transcribed the Catalouge of Hipparchus, without any considerable addition to him

I promised y^u in the Conclusion of my last lecture to shew y^u the errors of theire methods which upon serious consideration I find to have beene divided betwixt theire instruments not so well contrived as the[y] ought to have beene, & theire processes.

As for theire Instruments they were of two sorts the Parallactick rulers which I have described & shewed the faults of in my former readeings' here.[1] with these they tooke the Meridionall heights of ye stars whence they gathered theire declinations, but if these were large as they ought to be when the observers desired to be accurate theyre lenght if it was not duely provided against would cause them to swerve & render the zenith distance some minutes different from what it really ought to have beene.

For theire Armillaes the Instrument whereby they determined theire Right ascentions Tycho Brahe in ye First booke of his Progumnasmata page 104.[2] makes almost the same objections against them I had made against the Parallactick Rulers. I shall give y^u his sence in English. The Armilla's which Hipparchus & Ptolemy most confided in Tho doubtlesse they were made of solid mettall curiously & magnificently at ye charge of y^e Ægiptian kings, yet however they were contrived they could scarcely answer the end for which they were intended. for if they were so large that y^c could admit the divisions of single minits theire owne weight & bignesse would render them intractable, or if they were smaller that theire weight & bignesse Might cause no incumbrance; theire smallnesse would render the divisions too close & consequently not easily to be distinguished, or if they were continued & made of a moderate size that so they might admit divisions sufficiently exact & yet not be over cumbersome by

Notes on pages 379–80.

reason of theire weight yet consisteing of Six Circles all which were not turned about on the same poles & that which measured ye latitudes being revolved within ye Inward Periphery of ye Zodiack Circle, they could not be moved freely. or if these observations of ye latitudes were to be taken by ye helpe of a ruler with sights on it or a slit in ye middle of ye Axis one p[ar]te of ye instrument would sometimes incumber ye other, & ye disagree[m]ente of ye severall p[ar]tes would hinder them from answer[ing] theire expectations praecisely: so far Tycho[2] to which hee addes some further inconveniencys of the Armillas but not very Materiall which seemes to intimate that the all hee had objected might probably be true yet hee was something too severe on the Ancients. a greate pte of whose errors may be attributed to other causes which they wanted the experience of some ages & such helpes as onely this last has afforded but neither diligence will nor subtility to avoid besides it is to be suspected that when his owne Instruments come to be considered they will not be found wholly free from those faults he objects to his Ancients.

But to come nearer to theire methods. they supposed that they had attained a true Theory of the Sun, and then by measureing his distances from ye moone by day & hers from ye stars by night they collected how much the suns place was in consequence or antecedence of y^e stars whence his being given the stars was also knowne. in which processe some things were absolutely faulty others dubious which they supposed accurately knowne. as first they supposed the obliquity of y^e Ecliptick 23°.51$\frac{1}{3}$ which neither theire owne nor y^e observations of following times would make more then 23°.30′ as I have fully proved in my former lectures here[3] this p[er]haps was in p[ar]te the Cause why they determined y^e greatest equations of y^e suns orbite 2°.23′ which is at least 27 min more then later observations warrant it compare these 2 togeather it is evident they might erre a whole degree or more in ye true place of the sun not mindeing a small error they committed in supposeing the true length of ye yeare some few minutes more then it really was.

Adde to these that they assumed the horizontall p[ar]allax of the sun much larger then it really is, & the Moones not as it ought to be

But an error not lesse materiall was committed in theire places of ye Moone in whose sustemes they totally omitted the variation or Reflection an inequality dependeing accordeing to Tycho & those since him on her true distance from y^e ☉ & not discovered or introduced by any before him.

Theire remaines another inconvenience to be mentioned which not onely was incumbent on those ancient but on all o^r predecessors after them from Hipparchus downe to John Muller or Regiomantanus, & that is ye difficulty of allowing for ye Parallaxes of the Moone in Longitude & latitude in theire observations which tho now it bee much more easy (then it was possible for them) to calculate yet is esteemed the most laborious in ye Worke of ye Astronomers in Hipparchus his time they had no tables of Naturall sines or tangents but onely of substenses or Chordes so that ye labor of resolveing these cases of Sphaericall triangles which were requisite for determineing the parallaxes was much more then afterwards when the tables of sines & Naturall tangents were Made besides theire Radius was allwayes sexagenary even till Regiomontanus his time in whose workes[4] I find a Canon of Naturall sines in which the Radius is 600000, & yet when it was afterwards reduced to ye Centesimall. the labor of Calculation was 10 times as much as it is found since ye Artificiall sines & tangents or Logarithmicall were invented in the begining of this Century[5] which allmost p[er]suades mee to thinke that they little considered y^e ☽^s parallaxes when they derived the places of ye fixed stars from theire observed distances of ye Moone from them in ye Night . . . & the sun by day. or her transits over them or their occultations of them by her.

The Noble Tycho when hee set upon the restitution of ye places of ye fixed stars considering well the inconveniencys of this method of the ancients & the difficulty of applying it thought out another; hee considered with himselfe that the planet ♀ neare her greatest elongations from ye sun was visible to ye naked eye in the day time for some moneths togeather & that her distances from ye sun might then be exactly measured by his large & costly Instruments as also from the stars by Night whereby the suns place being supposed given the places both of ye planet & the star might be easily attained the Meridionall heights both of ye sun planet & star being observed by his large quadrants theire distances with his sextants, this was his first method of observeing for in his Historia Caelestis I find onely distances heights & Azimuths taken, in ye yeares 1582 & 83.[6]

But as in the Invention of Navigation men first rowed in small boates along ye bankes of rivers afterward adventured to crosse them then by the seashore but were long ere they durst adventure into the deepe or before the[y] enlarged theire vessells or added sayles to them, so has it hapned to the improvers of Astronomy the gnomones were the first contrivances of the ancients then the parallactick Rulers after this they attempted, to imitate the Circles of the sphere in theire Armillas. these were the Alexandrian Instruments. whose inconveniencys being noted by the Noble Dane, hee first set up large quadrants & sextants, I meane in his owne observatory[7] for the Hassian Lantgrave.[8] had taken notice of ye advantage of these contrivances by reason of theire simplicity before him & had caused some to be made of solid metall, but o^r great man [Tycho], gave them theire best improvem^ts & haveing afterwards found that the Armillas would give the declination of ye stars in all places of the sphere which his Quadrants onely shewed in ye Meridian, hee contrived how to avoid the inconveniencys in the structure of the ancient Alexandrian Instruments & Now w[it]h his hee could determine both the declinations of the stars & the differences of right asscention betwixt the sun & ye planet ♀ by day, betwixt her & the stars by night whereby hee could either confirme or correct ye observations made with his sextant & quadrant & had ye right ascentions both of ye planet & stars given him supposeing the suns knowne & allowing for a small variation in ye planets caused by her Motion betwixt ye time When hee first observed her from ye sun & afterwards from ye star if shee were occidentall or ye Contrary when orientall.

This method of the Noble Tycho was excellent & shewed him to be a man of solid judgment & quick apprehension, but there were severall faults committed by him in the practice & application of it which I shall mention not so much to reproach his Memory, far be it from mee, as to shew how observations are to be corrected & amended if any one shall thinke fit to examine them over againe, for his diligence & forecast was admirable his sincerity undoubted, nor did hee erre through pure but inevitable Ignorance & for want of such helpes as this last age & a few yeares after his decease afforded us so that I reckon I shall rather commend or extoll him then otherwayes whilest I recount his errors & shew how they may be corrected.

Of these the first was in his Solar Theory. The Earths distance from the sun hee determined, not as it ought to have beene done, from celestiall observations either of the Moones Dichotomys or ye parallaxes of the planet Mars. indeed it appeares not that hee was acquainted with the first, but the latter hee had thought of & caused observations to be made purposely for this triall but when they came accordeing to his order to be calculated. that so it might appeare what parallax they would give, his idle Amanuenses had onely calculated what the parallax of ye planet ought to be in those observations according to ye Copernicans, not what it really was; & the product of this calculation they put upon him as the result of the observations duely examined, but for want of a good contrivance whereby hee might be sure of the accuracy of his observations hee found them not all exactly agree in y^e same measures & determination, & therefore contented himselfe in a Rabinicall conceipt of Offusius[9] whereby hee had concluded y^e $\odot^s$ distance from y^e earth 576 diameters of the Earth. this differing not much from y^e Copernican determinations y^t had beene put upon him by his slothfull Calculators in y^e Roome of y^e result of his observations.

From this distance it followed that the suns horizontall parallax was neare 3 minutes, & the Nearest distance of ♀ from y^e earth in the Perigee conjunction being little more y^n $\frac{1}{4}$ of y^e earths from y^e sun that the greatest horizontall pa[ra]llax of ♀ was neare 4 times as big as y^e suns. or about 10 min. Tycho determined the distances of ♀ from y^e earth accordeing to the Copernican Hypothesis which must necessaryly be erroneous

For both by the Moones Dichotomys & the parallax of ♂ observed in his Acronicall appearances it is evident as I have formerly proved here[10] that the suns horizontall pa[ra]llax is not more y^n 10″ or the sixth p[ar]te of a minute whence it will follow that the parallax of ♀ is never more then 4 times as much or 40″.

And whereas Tycho never allowed but much more then this, it will necessaryly followe, that his determinations were faulty first on this account

Next this a greater error was by it carried into his Solar Theory, for admitteing the horizontall p[ar]allax 3′, hee corrected his Meridionall heights by a proportionall allowance of it, hence raiseing & correcteing his Solar Theory hee derives the greatest aequation[11] 7 or 8 minutes too big, and Consequently the suns true place as much erroneous; so that was also propagated & spread into his determinations of the true places of ye planets & stars.

These seeme cheifly to affect theire longitudes but if wee attend wee shall find they had theire effects on the latitudes too, for too greate a parallax supposed rendered the obliquity of the Ecliptick bigger then it ought to be

To which his omission of Refractions when the stars observed were above 20 degrees high, added & made it $2\frac{1}{2}$′ bigger then really it was, & consequently the stars in Cancer & Capricorne had theire latitudes as much too great or too little stated by Tycho, lesse aMisse nearer the aequinoctiall pointes, but no where true except upon them. Nay scarce there considering that the pole star it selfe was refracted near 1 minute, & consequently his true Equator so much higher then hee esteemed it

For Tycho esteemed the refractions of ye stars when elevated more y^n 20 degrees above the horizon either nothing or at most insensible in his Instruments; & so indeed they were; for it is evident both by compareing his observations of the same distance in heavens wth each other, & also his measured diameters of the Moone; that they seldome agreed nearer then within 3 minutes which is more y^n ye refraction in ye height of 20 degrees

But since the aire in which the refractions are made & whereby they are caused envests ye earth round as well over o^r heads as before or behind us, it is certeine the refractions cease not from the horizontall to ye perpendicular ray, & therefore allowing his Measures to be exact, both in his distances & Meridionall heights, they ought to be corrected by an allowance for this refraction; which how greate it is & what its effects, the Rules of Geometry doe easily shew & teach us. but I shall not at present discourse of ye nature properpties & effects of Refractions. these require to be handled simply by themselves, & if occasion be I may treat amply of them hereafter.[12]

O[u]r Noble observer himselfe well understood there was some sensible fault committed in his calculations, or rather ye application of his observations; & therefore haveing laide all the right ascentions of the bright star of ♈ collected by severall processes & differeing sensibly from each other togeather, hee takes the geometicall meane betwixt as the most probable Right ascention of the star, whereas if his misapprehended Corrections had beene omitted, & the suns place correct they would have agreed much better then hee has found them to doe.

This method of his has brought his places of the fixed stars so neare the truth, that tho there was a controversy betwixt him & the Lantgrave of Hesse concerneing them,[13] & some hasty people reckned a difference of 6 minutes betwixt them, yet the more judicious have ever since given ye advantage on Tychoes side, & made it evident, that the difference was not as it was esteemed but really a difference in the yeares to which the places were fitted; the Lantgraves being accommodated to ye yeare 1593, Tychoes to 1600 compleate. so that allowing both true, there ought to be about 6 minutes difference betwixt them.

But it is further remarkeable that the latitude stated both by the Lantgrave & the Noble Dane differ so inconsiderably from those of Hipparchus,[14] sometimes in ye excesse at others in ye defect. & that too so irregularly that the difference cannot be attributed to the supposed decrease of ye obliquity of ye ecliptick (which himselfe was ready to suggest), that wee may hence conclude however theire longitudes are altered theire latitudes have continued allwayes the same & invariable. & that those differences which are found betwixt Tycho & the ancients, ought to be attributed to ye perfection & accuracy of his instruments above theires, & his knowledge of the effects of refractions, of which they were wholly ignorant.

Seeing therefore the Ancients have left us nothing of theire observations of the fixed stars but onely theire places deduced from them, and some few declinations; takeing those for true, & the Tychonick latitudes I shall hereafter enquire what were the true places of the fixed stars in theire times. which compared with Tychoes will shew us the true recesse of ye earths equinoctiall pointes & how much ye longitudes of ye fixed stars are annually encreased. but this Must be the Worke of the Next opportunity

At Present I have onely time to remarke that the Method Tycho Makes use of for rectifieing the places of the fixed stars was really the onely proper one for correcteing the Solar Theory or finding his True longitude from them, whereby the sydereall place of the Aphelion & the true quantity of the Excentricity might have beene much better determined then from his Meridionall heights

In which an error of onely 24″ seconds would cause a fault of a whole minute in the suns place derived from them. & that it was almost impossible to determine nearer is plaine from ye severall observations of the same Meridionall height of ye sun taken with severall Instruments, which sometimes & not rarely are found to differ a minute or more from each other whence an error of neare three minutes would be carried into the suns place on this account, not Considering what further fault was committed by leaveing out all consideration of refractions when y[e] sun or stars were above 20 degrees high & allowing an unwarrantable parallax which togeather as I told y[u] before have made his places of the sun sometimes 8 minutes false.

It will now be enquired how these effects of parallaxes & refractions may be avoyded. I answer tis impossible to avoyd them altogeather but that nevertheless the times of y[e] necessary observations may be so chosen as that the effects of both refractions & parallaxes may become either wholly Insensible or if sensible the alteration made by them may be easily & readily discovered & limited thus. Let the time of the observation be ordered that both y[e] ☉ & planet ♀ or ♃ (for either of them may be observed from y[e] sun by day) may be in equall heighte with him & in the night observation let such stars be pitched upon may at y[e] time of y[e] observation be in equall heights with y[e] planets from which they are to be measured, for in this case: the verticall circles neare y[e] horizon & in distances lesse then 60 degrees being little inclined to each other the dilation of y[e] true distance by parallax even in y[e] Moone becomes very small & in y[e] other planets wholly insensible And the contraction by refraction neare y[e] horizon small also but in higher altitudes which those that doe not well consider will wonder at greater, yet never more y[n] a whole minute & the true quantity easily determined from y[e] Tables of refractions tho p[er]haps a little faulty. so that one caution seemes to me to rectifie all that was faulty in Tychoes methods, & therefore wherever I could I have alwayes carefully observed it; tis difficult I confesse to practice in a right sphere but no[t] so in o[r]s & p[er]haps there too there may be a convenience equivalent to what wee enjoy in o[r]s if duely Considered

The Onely thing wanteing here will be an equinoctiall pointe which may easily be found by observeing 2 times neare y[e] aequinoxes when y[e] sun comes to have the same declination, for tables framed of y[e] ☉[s] sidereall place deduced from y[e] aforementioned observations will give y[e] arch of y[e] Earths motions betwixt them, whose excesse above or defect from an intire semicircle halved is the true distance of the Earth from either aequinoctiall pointe at y[e] 2 observed times when y[e] sun had y[e] same declination. but of this more hereafter God Willing when wee shall have stated the true motions of y[e] aequinoctiall points. & this with his p[er]mission shall be the businesse of o[r] next opportunity

NOTES

29.1. Lecture 12 (8 February 1681/2).
29.2. Op. cit. (note 2.5), p. 104.
29.3. Lecture 10 (23 November 1681) and Lecture 11 (1 February 1681/2).

29.4. Flamsteed possessed a copy of A. Shato's edition of Regiomontanus's *Tabulae directionum ... Acceserunt tabulae ... per E. Reinholdum supputatea* (Witebergae, 1606).

29.5. John Napier, Baron of Merchiston, is usually credited with the discovery of natural logarithms, which he published in his *Mirifici Logarithmorum Canonis descriptio* (Edinburgi, 1614). The Gresham Professor of Geometry Henry Briggs was largely responsible for the introduction of common logarithms only ten years later. It was not long before Henry Gellibrand, Adrian Vlacq, William Oughtred, and others produced tables of logarithmic functions. Cf. notes 39.6–39.10.

29.6. Op. cit. (note 2.36), pp. 1–72.

29.7. 'Uraniborg' in Herritzvad Castle, on the island of Hven, in the Danish Sound.

29.8. William IV, Landgrave of Hesse-Cassel (1532–92), a pioneer in the art of astronomical observation. Cf. note 12.1.

29.9. Joannes Francus Offusius, *De divina astrorum facultate, in larvatam astrologiam* (Parisiis, 1570).

29.10. Lecture 2 (4 May 1681).

29.11. Details regarding Tycho's contributions to astronomy are contained in the excellent biography of J. L. E. Dreyer, *Tycho Brahe* (Edinburgh, 1890).

29.12. Lectures 33, 35–38.

29.13. This is one of the themes discussed in *Tychonis Brahe Dani, Epistolarum Astronomicarum libri* (Francofurti, 1610).

29.14. This remark is probably based upon the comparison of the longitudes and latitudes of ten bright stars presented in tabular form in Riccioli 1, p. 440.

Lecture 30

7 November 1683

In my last lecture I shewed you what errors the Ancients committed in determineing the true places of the fixed stars, I tould y^{u} theire faults were to be divided betwixt theire Instruments & theire Processes; what I sayd concerneing the Alexandrian Armillaes may in some measure be Applyed to Bernard Walters,[1] & what I discourst of Ptolemyes parallactick rulers will be also applicable to ye Arabian;[2] who tho they boasted of greater accuracy then theire predecessors, will yet be found Not to have reached the utmost exactnesse as will be very readily concluded if y^{u} remember what I discourst formerly concerneing the Arabian obiquity of the Ecliptick, of which I gave y^{u} many observations, & some very considerable ones, whereby I proved it to have remained invariable to o^{r} sence in all Ages.[3]

I told y^{u}, that tho Tychoes instruments were not so well accomodated for observation as they might have beene had the use of optick glasses beene knowne in his dayes, yet hee had used so much Caution & foresight in determineing the places of the Fixed stars, by divers observations, & different processes with divers instrum^{ts}, that by compareing them togeather, & electeing the meane amongst his different results, hee had pitcht near[e] on the true; & that the difference noted betwixt his and the Lantgravian Catalogue,[4] was no reall thinge, but a mistake in those who Compareing them togeather regarded not the intervall of 7 yeares distance which they were fitted; & that this being allowd theire agreem^{t} would be found as good as could be expected.

My businesses is now to determine the Annuall motion of the fixed stars or rather recesse of the Earths aequinoctiall points, whereby the apparent change of theire places is caused; for seeing they keepe theire distances unchanged from each other, & theire positions allwayes the same, I see No reason why wee should allow them any motion of theire owne, or conceave them, as the Ancients did, all fixed like Nayles in ye Rimme of a Chart wheele & all moved togeather as they did

For this determination I shall take Tychoes places of the fixed stars, or rather some of theire declinations observed by him as the

Notes on pages 387–8.

Basis or limit on the one side, For tho I esteeme not his places nor observations so accurate as they ought to be, yet Considering what course ones I must fix upon For ye other limit or bound of my examen I cannot but conclude them exact enough for my purpose; tho hereafter I may perhaps compare some of the best of the last or praecedeing century with my owne, whereby it will be evident, that tho the large intervall of time betwixt ye ancient & moderne observations render ye errors in ye Annuall apparent progresses of ye stars & planets very small, yet by compareing exact observations tho made at lesser intervalls wee may attaine to as great & p[er]haps better certeinty in o^{r} determinations.

But it will be most convenient to returne to our Ancients first & brushing the dust, if I may so say. from the laborious & usefull monuments, to see what certeinty wee can collect from them. The face of Antiquity is venerable, but truth much more valuable & precious, This I shall seeke & maintaine, but with a due respect to ye other

The Catalogue of ye fixed stars composed by Hipparchus was transmitted downe to us by Ptolemy in his Almagest, with little alteration, certeinely not much if any thing amended: Hipparchus flourished about 130 yeares before the Incarnation, Ptolemys tables are fitted to ye begining of ye Reigne of ye Emporour Antoninus, of Christ 137 or 138 From Ptolemy to Albategnius wee find no notice taken of the fixed stars, at least no observations of them. but now the learned Arabians becomeing acquainted with ye Mathematician Mahomet Al Batani or Albategnius in ye yeare of Nabonassar 1627 of Christ 880, determines ye place of ye Lions heart in 14 gr of ♌; which hee tells us Ptolemy had observed in 2°-10′ of ♌ in y^{t} 886th yeares of Nabonassar. whence hee concludes the fixed stars to have moved forward 11°.50′ in 747 yeares interlapsed betwixt them & so much hee addes allwayes to ye Ptolemaick longitudes to make theire true places in his age; not altering theire latitudes at all, because hee esteemed them invariable, & exactly determined by Ptolemy: but here wee must note that tho the Copy of Al Batani had ye place of ye star onely in ♌ 2°-10′, yet y^{t} which Riccioli used had 2°.30′. or 20′ greater longitude.[5]

57 yeares after al Batani, Ebenensophim[6] an Arabian determined ye place of ye same star in 15°-12′ of ♌; and 180 yeares after him or in ye yeare of o^{r} Lord 1115. the Persian mathematicians restored the places of ye fixed stars:[7] The Alphonsine tables[8] are made to ye yeare 1256. Rabbi Levi[9] to 1336. & in ye yeare 1364 wee have another verification of an anonymous Author:[10] in ye yeare 1440 wee have Cardinall Cusanus;[11] All these seeme onely to have added a certeine number of degrees to Ptolemyes longitudes, which how they found wee shall shew y^{u} imediately but to have kept his latitudes invariable. as After them Copernicus[12] & Clavius[13] both did of Whom more by & by.

Here wee must take in the famous Catalougue of Ulug Beigh,[14] ye gran[d]child of Timur Leng. rectified to ye yeare of Grace 1437 of ye Hegira 841 wherein the Author pretends to have rectified theire places by his owne observations made with a very large Quadrant, what other instruments hee had I find not, but I am apt to thinke hee had no other for tho I find every where thiere latitudes altered from ye Ptolemaick Numbers, yet theire differences of longitude are the same with Ptolemys or not much better sometimes worse; which p[er]suades mee to thinke that hee altered them onely some little to make them answer his observed declinations, For in the Ptolemaick Catalogue the difference of longitude betwixt ye bright star of ♈ & ye Bulls south eye is 31°.40′ in Ulug Beigs 31°-48′ the true in Tychoes 32.07 greater then Ptolemys by 27′ minu[tes] then Ulug Beighs by 19. Againe Betwixt the bulls south eye & the lions heart Ptolemy has the differe[nce] of Longitud[e] 79°.50′. Ulug Beige 79°-42′ Tycho 80°-03′ so that here Ulug Beig erres more y^{n} Ptolemy by 8 minutes at least. it were easy to adde more, where there were greater differences but the Catalogue is easy to be had & these may be sufficient to shew that Ulugs Catalogue is not much Preferable to Ptolemys

And that Ulug Beig made use of no other Instruments but his Quadrant for this rectification I am the rather persuaded because I find that all those who went before him Hipparchus & Ptolemy excepted used onely the like method: Timochares 295 yeares before ye Incarnation observed that Spica ♍^{s} was 1°.24′ north of y^{e} æquator; Hipparchus 128 yeares before Christ or 167 yeares after him finds its declination onely 36′ north. Ptolemy in ye yeare 137 of Christ or 265 after Hipparchus determines it 0.30′ south of the æquator the latitude of ye star Ptolemy finds ye same with Hipparchus his determination, or so neare y^{t} hee well concludes it to have remained both in this & all the rest invariable and y^{n} proceedeing upon this supposition hee inquires how much this change of declination would alter y^{e} ★^{s} longitude. & concludes that it had moved a degree nearely in 100 yeares As Hipparchus had done before him by compareing his observations with those of Tymocharis. but Hipparchus ingenuously acknowledges that Tymocharis his observations were both too Course & too neare him for this determination, & therefore doubts & is not at all Positive in his opinion. The same objections o^{r} Moderne Artists make to ye observations of both Hipparchus & Ptolemy; that tho they are remote enough as to time, yet they are too rude. wee may learne however by them How far wee may confide in ye ancient observations. & p[er]haps wee shall find at last that the Alexandrian Armillas with which Hipparchus observed, were slipt from theire pristine position in Ptolemys time, which may be the reason of the difference betwixt his & Hipparchus his observed aequinoxes remarked by all those Who have made Tables of the suns Motion, & agree to be nearely a whole day erroneous[15] But to come nearer to our ancients, & make some estimate how neare theire determinations were to truth; I shall Compare some of Ptolemys differences of longitudes betwixt bright fixed stars with the Tychonick as also theire latitudes with his which will give us the information wee want, after I shall consider what observations of theires are the fittest for determineing how much the stars are found in consequences of theire ancient true place, & what is the annuall recesse of the Earths aequinoctiall pointes, whether they have receded equally or not, & what may probably be the Cause of this Variation

The longitude of y^{e} first bright star of ♈ which in the Ptolemaick catalogues is y^{e} lucida informis supra Caput is placed by him in 10°.40′ of ♈. by Tycho in ♉ 2°.06′. Aldebaran or y^{e} Bulls south eye in ♉ 12°-20′. but Tycho has it ♊ 4°-12′$\frac{1}{2}$ hence the Ptolemaick difference of longitude is 31°-40′ y^{e} Tychonick 32°-06′$\frac{1}{2}$ exceedeing y^{e} ancient 26$\frac{1}{2}$′.[16]

The Ptolemaick difference of Longitude betwixt the Bull's eye & y^{e} lions heart is 79°-50′. The Tychonick 80°-04′ exceedeing it 14 min

Betwixt y^{e} Lions heart & Spica ♍^{s}. the Ptolemaick arch of longitude 54.50. Tychonick 53-59. y^{t} is 51 min lesse:

Betwixt Spica ♍^{s} & y^{e} Scorpions heart ye said difference is 45°-40′ by Ptolemy, Tycho makes it 45-57 y^{t} is 17 min greater.

From y^{e} Scorpions heart to y^{e} bright star of the Vulture Ptolemys intervall of longitude is 51-10 Tychos 51-56 is 46 min greater.

From y^{e} bright star of y^{e} Vulture to Markab in Pegasus y^{e} Intervall in Ptolemis tables is 52°-50′ in Tycho's but 51°.47′$\frac{1}{2}$ which is lesse y^{n} y^{e} ancient by 1°-02′$\frac{1}{2}$.

And y^{e} difference of longitude betwixt Markab and y^{e} bright star of ♈ by Ptolemy is 44°.00′. which Tycho finds 44°-10′. exceedeing y^{e} ancients 10′.

Tis remarkable in this Collation. that the summe of the Tychonick differences of longitude make up as they really ought an intire circle or 360. to which I have found both ye Lantgravian & my owne alwayes agree very well. Tycho makes use of such trialls to prove ye accuracy of his instruments. But the Ptolemaick too to my wonder doe the same thinge, for haveing added them up togeather I find them to make a circle exactly, tho some of them, particularly the difference betwixt Regulus & Spica is lesse y^{n} the Tychonick by 51 min. betwixt Aquila and Markab 1°.02′ & so much faulty; but the excesses in the rest make up the defects in these, & recompense the error.

Whence it is evident that there is a fault of a degree comitted in ye ancient observations, which yet I can scarce thinke to have beene made all at once by theire instruments,[17] but rather by accumulation of errors the same way; which how great they were p[er]haps wee may better learne by compareing theire observed & determined latitudes with Tychoes.

The true latitude of ye bright star of ♈ is in Tychoes Catalogues 9°-57′ North in ye ancient 10°-30′ but in some copies onely 10°-00′ agreeing nearely with ye Tychonick

The latitude of ye Bulls eye in ye Tychonick Catalogue is 5.31, in Ptolemys 5°-10′, wanteing a third part of a degree of ye truth.

Tychos latitude of the Lions heart is 0.26½ North. the Ptolemaick 0-10. or 16′ lesse y^n it ought to be

Spica ♍^s has 1-59 south latitude in Tychoes Tables but 2.00 in Ptolemys but in some Copys 2-10 exceedeing the truth, if wee take nearest onely one minute or at most but 11′.

The distance of the Scorpions heart from ye Ecliptick is in Tychoes Catalogues 4°.27′ south; in Ptolemyes onely 4 degrees. wanteing 27′ of what it ought to be

The bright star of ye Vulture is 29°-21′½ distant from ye Ecliptick Northwards in Tychoes tables. but 29° in ye Ancient. or 11 minutes lesse than truth.

Markab or ye Wing of Pegasus has 19.25 North latitude in Tychoes workes but 19°-40′ in Ptolemys exceedeing ye due measure 15 minutes.

The greatest difference here betwixt the old late true tables never exceedes 27′ & is that onely in the Scorpions heart. but in ye rest takeing the Meane amongst them is about 15 minutes. whence wee may collect that the observations of the Ancients I meane of theire Meridionall heights were not more accurate then within ¼ of a degree I am apt to beleive the Armillaes the[y] used were small & tho p[er]ha[p]s they were divided into sixths of a degree. yet either by reason of theire ill contrivance or intractability yet this was the nearest they came to truth. tis remarkable that neither theire longitudes nor latitudes are assigned nearer then to ye sixth p[ar]te of a degree & tis not unusuall especially with the ancients to give theire calculations finer & to lesser p[ar]tes than they could make observation as I may perhaps hereafter have occasion to shew y^u.

But an old objection must here be considered which possest a many of o^r praedecessors & may still vex those Who will maintaine the stability of the Earth, & the Motion of the Fixed stars. From ye Arabian observations of the obliquity of the Ecliptick it was found lesse y^n Eratosthenes Hipparchus & Ptolemy had stated it & from then downewards to Tycho Astronomy travelling from ye Southerne effeminate to ye Northerne Climates, that the observations of it simply considered (for they knew little of Refractions before Tycho certeinely they understood not how to allow for them) would still give it lesse then it had beene determined by the Artists of ye precedeing ages. So y^t they concluded the obliquity of the Ecliptick to have altered & decreased gradually & both Kepler & Bullialdus tho they beleived no such thing as well as Copernicus & his followers have given us Contrivances to salve it.[18] Admit this & then it seemed but reasonable to Conclude, that the Ecliptick or line of ye suns Annuall Course being altered, the latitudes of those stars which lay in ye first semicircle of ye Ecliptick betwixt ♈ forward to ♎ north of it or from ♎ forward to ♈ south of it should be encreased the Ecliptick line recedeing from them but on the Contrary, those which had south latitude in ye first semicircle or North in ye second should have theire latitudes diminished, ye Ecliptick inclineing & approacheing to them. the greatest alteration in ye Tropicall stars or neare the first pointes of ♋ & ♑ being equall to ye decrease or change of ye obliquity. but lesse continually as they were removed in longitude towards either aequinox. & theire insensible

Riccioli in ye 6th booke of his Almagestum Novum pag 442 gives us a Table of Tychoes conteineing ye longitude of 11 fixed stars corrected by ye allowance of what hee found requisite to reduce them to the times of Tymocharis Hipparchus & Ptolemy with theire latitudes to theire severall ages corrected by theire declinations then observed.[19] whence hee would infer that theire latitudes have altered betwixt theire times & ours, proportionably to the alteration of the obliquity of the Ecliptick by him supposed, But Riccioli observed not that hee uses the obliquity of the Ecliptick $23°$-$51'\frac{1}{2}$ as Ptolemy employed it, which being granted him, & that the observed declinations were true his consequence necessarily followes, whereas if hee had taken the obliquity of the Ecliptick, as his observations make it, & theire declinations hee would have found no such thing but rather that the latitudes were invariable as I shall shew y^{u} from ye ancient latitudes I have lately recited compared with ye Moderne.

The first star of ♈ haveing its latitude North in the first semicircle of the Ecliptick if the obliquity decrease as is supposed, ought to have its latitude augmented whereas tis rather diminished, Tychoes being some little lesse y^{n} Ptolemys.

The Bulls eye haveing south latitude in the same semicircle ought to have its distance from y^{e} Ecliptick diminished, whereas Tycho has it 5.31 at least 20 minutes greater y^{n} Ptolemy.

The bright star of y^{e} Vulture lieing neare y^{e} solst[it]iall Colure with North latitude ought to have its distance diminished whereas tis rather encreased $11\frac{1}{2}$ Tychoes latitude being so much bigger y^{n} Ptolemys.

These are all bright & remarkeable fixed stars in stateing whose places wee need not doubt but the Ancients used as much diligence as in any of the rest, of which therefore tho some seeme to favor the opinion of the Mutability of theire latitudes, yet these so apparently contradicting it, wee may hence reasonably conclude that as the obliquity of the Ecliptick so they have continued alwayes invariable & unchanged; what differences there seemes to be betwixt the ancient intervalls, either of longitude or latitudes, haveing beene caused either by the smallnesse ill divisions or intractability of theire instruments & not by any reall change of them amongst themselves or recesse of the ecliptick from some, & approach to others of them.

For besides what I have already urged it may further be noted, that all those fixed stars which the ancients observed to lie in the same streight line with one another, of which Riccioli reckons up at least 20 ternarys,[20] remaine in the same position to this day whence wee have got another Argument to prove the stability of the fixed stars. Since therefore they keepe allwayes the same fixed places in the heavens the same intervalls of longitude. the same latitudes from ye ecliptick invariable. theire longitudes from ye equinoxes onely altering some little anually, & thereby Changeing theire Right Ascentions & declinations; wee can have no better way of findeing out the quantity of this annuall alteration then Admitting theire latitudes as observed in o^r^ age, theire declinations as noted of old by Tymocharis, Hipparchus, or Ptolemy & of late by Tycho & o^r^ selves, to enquire what alterations of place the change of declination requires, which will be no difficult probleme as y^u^ will find by my next lecture & then the intervall of longitude divided by ye yeares interlapsed betwixt ye observations will give ye Annuall alteration of theire longitude

But all ye stars are not alike fit for this examen the dayly declinations of y^e^ ☉ change fastest neare the aequinoctiall pointes, slowly neare ye Tropicks, or the Colures so in ye stars those change theire declinations fastest which lie nearest the Equinoctiall Colures slowest y^t^ are neare the solstitiall wee ought therefore to pitch on such stars for this examination whose longitudes were not much in antecedence or consequence of the aequinoctiall colures, at ye times when ye ancients observed; & the brightest are the most proper for o^r^ purpose because they were the easiliest seene & therefore in all likelyhood the Most exactly observed. Such are the bright star of ♈ Andromedas head. ye Virgins spike & Arcturus. of these 4 I like Spica ♍^s^ the best because it is so remarkeably bright, has small latitude from ye Ecliptick, & was made use of for this very purpose both by Hipparchus Ptolemy Albategnius and all those other who since them have endeavored to determine the Annuall recesse of ye aequinoctiall pointes or motion of the fixed stars with this therefore I shall begin my examen God assisteing next lecture.

NOTES

30.1. Lecture 12.
30.2. Lecture 11.
30.3. Lecture 10.
30.4. Op. cit. (note 12.1).

30.5. Flamsteed is here consulting Riccioli 3, pp. 215–17, which is the source of the information contained in the following paragraph.

30.6. This name ought to be spelt 'Ebennesophim'. Cf. Riccioli 3, p. 215.

30.7. Op. cit. (note 10.13), p. 226.

30.8. The *editio princeps* of 1483 is entitled 'Alfontij regis castellae illustrissimi coelestiū motuū tabulae', etc.

30.9. Rabbi Levi, *Defensionum Dei*, lib. 5, cap. 62.

30.10. The table in question is printed in op. cit. (note 3.15), pp. 534, 535, as an appendix to the second of three epistles to Wendelinus.

30.11. See the star catalogue appended to Nicolas of Cusa's 'De docta ignorantia' (1440) at the end of the Basel, 1565, edition of his *Opera.*

30.12. Nicolas Copernicus (1473–1543). Cf. op. cit. (note 11.11).

30.13. Christopher Clavius, *In Sphaeram* (Romae, 1607). Cf. the star catalogue on pp. 177–206 and the tables of the Sun's longitude on pp. 320–7.

30.14. Op. cit. (note 28.13).

30.15. A much more convincing explanation of the origin of this error was given by the German astronomer Tobias Mayer (1723–62) in a letter to Leonhard Euler dated 22 August 1753; an English translation of which is given in Eric G. Forbes, *The Euler-Mayer Correspondence 1751–1755* (London, 1971), pp. 73–76.

30.16. Flamsteed is probably comparing the values for the celestial longitudes of the two stars in question tabulated in Riccioli 3, p. 211 (from Ptolemy's star-catalogue) and p. 239 (from Tycho's table). The data used for the other longitude and latitude comparisons which follow on ff. 275^{v} and $276^{r,v}$, or pp. 384–5, are also to be found in this same source.

30.17. Cf. note 30.15.

30.18. The fictitious phenomena of trepidation of the equinoxes and the motion of the eighth crystalline sphere, to which Flamsteed is here referring, are discussed in Riccioli 1, pp. 448–56.

30.19. Riccioli 1, pp. 441–2.

30.20. Riccioli 3, p. 203.

Lecture 31

14 *November* 1683

Haveing shewne you in my last lecture that the fixed stars keepe both theire distances differences of longitude & theire true latitudes invariable. & that therefore the best way of determineing how much they have changed theire apparent longitudes betwixt ye Ancients & us would be by compareing some of theire observed declinations with ours I come now to put that method into practise. I told y^u^ there were four remarkeable fixed stars very proper for this examen the bright star of ♈. Andremedas head Arcturus & Spica ♍^s^ because they lie not far from ye aequinoctiall Colure. & therefore alter theire declinations faster by much then those y^t^ are more remote. but of these I find that Spica onely has had its declination observed of old by Tymocharis Hippa[rchus] & Ptolemy delivered downe to us; wee must therefore content o^r^ selves with what wee can learne by examineing the ancient observations of this star[1] and p[er]haps that may be sufficient and satisfactory

Tymocharis in ye 454th yeare of Nabonassar or 295 before ye incarnation observed ye declination of Spica ♍^s^ 1°.24′ North. its place. ♍ 22°.20′. but hee esteemed the Quantity of ye Eclipticks greatest declination too big with Eratosthenes & ye latitude of ye star p[er]haps 11 minutes greater then it is since found & really was in his time. let us accept it as given by Tycho 1°.59′ south then.

Let now in ye figure [Fig. 31.1] ♋♎♑♈ represent the Ecliptick e its pole. p ye pole of ye world. ♋pe♑ the solstitiall Colure s ye place of Spica ♍^s^ ps. ye complement of ye stars declination observed by Tymocharis 1°.24′ North. therefore ps 88.36. se its distance from y^e^ pole of y^e^ Ecliptick 91°.59′ in y^e^ sphericall triangle pse are knowne the side pe the constant distance of y^e^ two poles 23°.29′ with y^e^ other two sides before mentioned ps and es. whence the angle pes the longitude of y^e^ star from y^e^ Tropick of ♋ will be found 81°-46′$\frac{1}{4}$ or from y^e^ aequinoctiall point of ♈ in ♍ 21°-46′$\frac{1}{4}$

But in o^r^ observation of Hipparchus from his data the said Angle will be found 83°37′$\frac{3}{4}$: & therefore y^e^ stars true place ♍ 23°.37′$\frac{1}{4}$

And applying y^e^ declination as observed by Ptolemy. tis found 86°-10′$\frac{1}{3}$. therefore its correct longitude ♍ 26. 10$\frac{1}{3}$

Notes on pages 396–7.

Which Tycho in ye yeare 1600 compleate places in ♎ 18-16 The[re]fore from Timocharis to Tycho the stars longitude is encreased 26°.30′ or 95400 seconds which divided by 1895 the yeares interlapsed betwixt Tycho & Tymocharis give ye Annuall Motion 50″ & almost $\frac{1}{3}$ p[ar]te of a second.

The time interlapsed betwixt Hipparchus and Tycho is 1728 yeares. the arch of ye stars longitude encreased in yt space 23°.38′ or 88680 seconds which divided by the Intervall of yeares 1728 gives the Annuall Motion 51 seconds & $\frac{1}{3}$ p[ar]te of of a second or 20 thirds

Ptolemys observation is elder then Tychoes determined place by 1464 yeares, & the place of ye star deduced from it 22°-00′ in 79200 seconds in antecedence of it which divided by ye yeares interlapsed betwixt them or 1464 leaves ye annuall recesse 54″ & about 06 thirds.

here wee find a difference of 4″ in ye annuall motion of ye ★★s which propagated & extended from Tycho back to Tymocharis would make his place of the star 2°.07′ faulty or Hipparchus his more yn 1$\frac{1}{2}$ degrees.

To know upon which of these to pitch wee must consider the Authority of theire Authors. Hipparchus professes that hee esteemed the observations of Tymocharis to be but Course ones & that therefore hee durst not rely upon them. this intimates that hee was more curious & therefore certeine of his owne. Ptolemy also gives Hipparchus a large character that hee was an equall lover of truth and Industry. Pliny is Hiperbolicall in his Commendations. & indeed his observations of the aequinoxes prove him to have beene both a good observer & faithfull relator of his observations. the disagreemt of Ptolemys on the Contrary lay a blemish upon his which seeme far inferior to those of Hipparchus. on his therefore I shall pitch as the Nearest to the truth. And admitting it I shall enquire what the place of this star really was both in ye times of Tymocharis & Ptolemy and what was its reall declination where by which it will appeare that the error of theire observations & instruments was as I have before stated it about $\frac{1}{3}$ of a degree.

Supposeing therefore ye Annuall recesse of ye earths aequinoctiall points 51″-20‴. in 1895 interlapsed betwixt Tymocharis & Tycho it will be 27°-01′ which taken from 18°-16′ of ♎ The Tychonick place of ye star leaves ♍ 21°-15′ its place in the observation of Tymocharis which by his observation seemes to be ♍ 21-46 or 31 minutes forwarder. further with its place ♍ 21°-15′ & true latitude 1.59 south seekeing its declination I find it 1°.39′ which Tymocharis gives onely 1.24 that is 15 min lesse yn it really was & if wee may trust Hipparchus his observation, so much erroneous.

The apparent recesse of ye aequinoctiall pointes allowing at ye same rate for 1464 yeares interlapsed betwixt Ptolemy & Tycho is

20°.52′½ which Taken away from 18.16 of ♎ the true place of ye star in Tychoes time leaves

♍ 27°23′½ its true place in Ptolemys. whose observed declination places it in ♍ 26°-10′ or 1°. 13′ lesse y^n it ought to be. & if with ♍ 27°-23′½ longitude & 1°-59′ I enquire its its declination it will be found 0-47′ south which is 17 minutes more y^n Ptolemys faulty observation makes it. who therefore seemes to have erred as much in its declination south as Tymocharis did in makeing it lesse north then it ought to be.

& Considering that both theire Measures seeme faulty in ye defect, it seemes suggested to mee that they Considered not the structure of theire Armillaes. in which if wee suppose ye latitude of the fixed æquator just halfe a degree & that Tymocharis counted his declination North from ye upper edge of ye Circle. Ptolemys his southerne from ye under or southerne limbe it will follow y^t both theire Measures were about ¼ of a degree too small & restoring this or augmenting each of them ¼ of a degree wee shall make theire true declinations agree very well with y^t of Hipparchus & o^r present annuall recesse collected from his observation compared with Tychoes

This to some may seeme but a strained fetch to helpe the ancients out & to excuse theire faultes. I confesse I am not at all fond of the conceit but since so faire a Correction offered it selfe I could not with justice to them or o^r selves omit it, those that cannot allow it must therefore grant mee that the Ancient observations were not made neare then to one quarter or third p[ar]te of a degree as I asserted last lecture. & confirmed both by theire differences of ye longitudes of severall bright fixed stars compared with later observations & also from the numbers of theire tables which never give the place of any fixed star not its latitude nearer y^n to ye sixth part of a degree. & tis probable theire Numbers were exacter & smaller y^n the divisions of theire Instruments

Seeing therefore these observations thus considered will agree very well with that of Hipparchus falleing betwixt them wee may reasonably conclude that hee used all necessary precaution & consideration in his observation & that it was as accurate as his instrum^t would allow. wee will therefore admit it since wee find no other better & ye annuall motion of the fixed stars or recesse of ye aequinoctiall points 51″.20‴ the Noble Tycho subtle Kepler & learned Bullialdus all make use of this observations & with one consent make ye Annuall recesse praecise 51″. which is something lesse then it really gives it but not materially for the omission of ye odde 20‴ make not a difference of above 10 minutes in the places of ye fixed stars in 1800 yeares which is a small error & lesse I am apt to beleive then the divisions on ye Ancient Instruments. Which how they were contrived & of what bignesse as yet wee know not but may ere.long there being some bookes lately brought out of ye East[2] of which one an Arabick Manuscript[3] describes the Instruments used both by the ancient Greekes and Arabians purposely to shew that the latter had much the Preheminence. had wee this worke it might serve to cleare or present doubt. I hope in some time to be able to give an account of it in the Interim wee must acquiesce where wee are in ye bare observation of Hipparchus.

Haveing done with this observation of Hipparchus I will make one triall more with ye place of Regulus observed by him accordeing to Tycho 128 yeares before the Incarnation in ♋. 29°-50′. the declination of this star is not set downe nor if it had would it have beene convenient to employ it as I have done that of Spica ♍s this being but one signe removed from ye solstitiall Colure & there altering its declination but halfe so fast annually as Spica ♍s then did yt is 12′ in one degree change of longitude whereas Spica alters neare 24′.

Tychoes place of this star is ♌ 24°-17′. in consequence of ye Hipparchean 24°-27′ or 88020″. which divided by ye Intervall of yeares 1728 gives the annuall progresse 50″. $56'''$-15^{IV}. some little lesse yn it was found by Considereing ye declination observed by Hipparchus with the place given by Tycho, but insensibly.

The reason of this agreemt proceedes from hence that the place of Spica derived from its declination with the Tychonick latitude agrees much better with ye Tychonick intervall of longitude betwixt Spica & ye Lions heart then that which Hipparchus himselfe makes it. for his difference of Longitude is 54°-30′. but the place hee gives us of the lions heart & of spica correct by its observed declination is but 54°-07′. Tychoes 53.59 onely 8 minutes lesse.

Should I apply Ptolemys place of ye Lions heart to Tychoes & thence enquire the Annuall recesse of the Earths aequinoctiall pointes I should find much more. for first place in Spica deduced from his declination observed & uncorrect differs considerably from Hipparchus. Next hee makes ye difference of longitude betwixt ye lions heart & spica ♍s bigger then Hipparchus does by 20 minutes which is yet considerably more yn it ought to be. so that Wee cannot at all confide in Ptolemys determination from this star wee will pitch on another & that shall be the Bulls south eye which being a very large & remarkeable star wee may reasonably conjecture was as exactly observed as any by him.

Its place in his Catalogue following the Greeke Copy is ♉ 12°-20′[4] The Tychonick ♊ 4°-12′½[5] theire difference is 21-52 or 78720″ which divided by 1464 the Intervall of yeares gives the annuall motion neare 54″.

But if wee take ye place of ye ★ in ♉ 12-40 as the latin editions have it the said recesse will be found onely 53″ p[er]Annum.

Riccioli[6] takes the declination of this star as observed by Ptolemy

11°-00′ & with it & its true latitude & obliquity of the ecliptick finds its true longitude ♉ 14°-38′ which is neare 2 degrees forwarder y^{n} Ptolemy places it & hence hee deduces the Annuall motion of the fixed stars onely 49″. but wee cannot suppose him to have erred lesse in y^{e} declination of this star then I have shewne hee did in Spica which is $\frac{1}{4}$ of a degree alow this the place of the star will be found still more faulty & the motion of Riccioli lesse & unwarranted.

Betwixt Hipparchus & Ptolemy there lived Agrippa in Domitians time. Menelaus in Trajans who in Bythnia and at Rome observed y^{e} Moones appulses to the brightest of y^{e} Pleiades. Spica ♍^{s} & the southerne in ye Front of ye Scorpion[7] from which severall observations each of them pretended to correct ye place of the star to which they saw ye Moone applying but theire Corrections are of little worth they ought rather by the places of these stars to have corrected theire Theoryes of the Moone for which purpose these notes may be of good use but to goe about to correct the places of the stars by them was preposterous her true theory being still wanteing & every dayes experience shewing us that shee needs further Correction.

Waveing these therefore I come to those who followed Ptolemy of whom the first is Albategnius. who in his Booke de scientia stellarum[8] tells us that in ye yeare of Nabonassar 1627 of Christ 879. hee observed the lions heart in ♌ 14°. whether hee collected this from his declination or not Riccioli[9] from whom I copied his place informs mee not but tis probable that hee made use of the Ptolemaick place & latitude in his determination. for hee adds that by Ptolemys tables the place of ye star is ♌ 2. 10 therefore y^{t} to make theire places for his owne time 11°-50′ must be allwayes added to ye places in Ptolemys catalogue. Hence it followes that ye Annuall Motion of ye fixed stars is above 57″. by ye yeare. & that they alter theire places 1 degree in 66 yeares & one quarter[10]

But if wee compare the place however observed by Albategnius with Tychoes wee shall come nearer the truth. the difference of his place from Tychos observed 721 yeares after him is 10°.17′ or 37020″. which divided by ye Intervall of yeares leaves ye Annuall motion 51″-21‴ which agrees exactly with ye correct Hipparchean observation compared with ye Tychonick & the Motion deduced from it.

Next after Albatani wee find that Ebennen Sophim an Arabian[11] in ye yeare of ye Hegira 325 of Christ 936 almost compleat rectified the places of the fixed stars in this Catalogue ye place of Regulus or ye Lions heart is ♌ 15°-12′ short of Tychoes 9°-05′ or 32700 which divided by 664 ye Intervall of yeares gives ye Annuall Motion of ye stars 49″.15‴. above 2 secondes less y^{n} ye Hipparchean & Tychonick places compared togeather make it 180 yeares after him In the 509th yeare of ye Hegira or 1115 of Christ the Persian Mathematicians restored the places of the fixed stars from theire owne observations. I suppose of theire declinations rather then otherwise except where the error was very evident for they seeme to retaine every where the Ptolemaick latitudes. the distances of y[e] Bulls eye y[e] Lions heart Spica ♍[s]. Antares y[e] bright star of y[e] Vulture Arcturus & Markab in Pegasus being exactly the same from the Ecliptick that they are in Ptolemys Cataloug. the differences of longitude sometimes altered but so as onely to keepe to y[e] next 10th minute for they give not the place of any nearer then to y[e] sixth part of a degree or 10 minutes of these I shall first examine Spica ♍[s] whose latitude south differs not above 1 minute from Tychoes & therefore if its place was deduced from its observed declination and assumed latitude with y[e] obliquity of the Ecliptick which they had not much amisse in those dayes its place thence got by calculation will in all likelihood be the truest of any wee could pitch upon.

In these tables[12] the place of Spica ♍[s] is ♎ 11°-40′ which Tycho 485 yeares after them has in ♎ 18°-16′ the intervall of longitude is 6°.36′ or 23760″ which divided by the Intervall of yeares leaves theire Annuall Motion 49″ exactly. that is above 2″ p[er] annum lesse y[n] the Hipparchean observation rendered it but agreable to y[e] Motion deduced from the places of Regulus given by Albattani. so y[t] if wee had but such Circumstances of these 2 observers Notes as might confirme the deduced places wee might here conclude but wee shall first trie one or two more of y[e] Persian places to see if y[e] give the same Annuall Motion & first wee will trie y[e] south eye of y[e] Bull, then y[e] Lions heart

The Place of the Bulls eye in these tables is ♉ 27°-00′. which is short of Tychos 7°-12′ or 25920″. which divided by 485 the Intervall of yeares betwixt them leaves y[e] annuall Motion 53″.40‴. bigger y[n] y[e] former by 4″ at least

The place of y[e] Lions heart in them is ♌ 17°.30′ in antecedence of y[e] Tychonick place 6°.47′ or 24420″. which divided by y[e] aforementioned Intervall Gives y[e] annuall Motion onely 50″.24‴ still bigger y[n] y[e] first

The Reason of the discoherence of these observations is that the difference of longitude betwixt Spica & Regulus is made lesse y[n] it ought to be by 16 minutes but betwixt Regulus & the Bulls eye greater by 24. whereas in old Ptolemaick Catalogues the first distance is on the contrary too big by 24 minutes the latter too little by 13.

Tho o[r] printed Authors & theire transcriber Riccioli make no mention of any observers betwixt Al Battani & Ebenen Sophim yet theire lived some very accurate ones betwixt them Thebet Ebn Corah. Abuhamet. Abul Rihan, & Abu Mahomet Alcogandi, of whose observations about y[e] obliquity of the Ecliptick I gave y[u] an account in my former Readeings on that subject;[13] it is not to be thought but that these Ingenuous p[er]sons employed some part of theire paines upon the fixed stars tho nothing of theires on that subject has come to our hands I onely mention them because if any of theire workes be happily met with by any diligent person it will be worth his while to enquire & search if they have any thinge relateing to this subject. Nor is Arzachel to be omitted nor Almeon the first wrote severall treatises which wee want and much desire[14] & tis to be hoped hee may informe us better then his Contemporarys for it appeares by his method of finding ye excentricity of ye solar systeme that hee was no despicable Geometrician.

These were succeeded by Alphonsus ye 10th of y^t name king of Leon & Castile. who by the helpe of some Jewes rectified ye place of the fixed stars to ye yeare 1256[15]. at first hee enterteined theire opinions who maintained the trepidation of ye 8th sphere & that the fixed stars moved unequally sometimes faster at others slower but comeing to see the booke of Albatani on this subject[16] hee changed his opinion & became of the Contrary that theire motion was regular & that they proceeded one degree in 66 yeares. if to Ptolemys places of ye fixed stars y^u add 17.08 y^u will have the Alphonsine alwayes. so that tis evident hee made no particular correction of each single star but onely trieing how much was requisite in any one hee added the like Arch in the rest & considering his processe I am persuaded hee or his observers for him made use of Spica ♍^{s}. for hee instances in y^t star & no other.

After these followed severall small attempts of An Anonymous Author in ye yeare 1364 an account of whose worke is given by the learned Gassendi in one of his Epistles to Wendelin[17] then of Ulug Beig in ye Yeare 1436.[18] of Nicolas Cardinal Cusa one of ye Patrons of Copernicus of the yeare 1440[19] of Bernard Walter one of the best p[er]haps of all that went before him in ye yeare 1504[20] but all theire determinations are found so full of repugnancys when they come to be compared togeather & with the late more accurate Method of the Noble Tycho that little credit can be given to them before ye Noble Lantgrave of Hassia set upon this rectification[21] which hee in good measure effected through the helpe of stipendiarys Rothman & Instrument Maker Justus Birgius with that Good successe that the Tychonick places are found not above 5 minutes different from his & there are some very excellent p[er]sons in the World esteeme them better. betwixt whom I shall not pretend to determine the Controversy[22] that small quantity be as little as in my opinion it was possible for them to attaine to with all such helpes as that Age Afforded. I examine no tables later then the Persian nor compare them with oures by reason I esteeme all since rather too neare or too rude for this examen. To conclude Tho I have formerly made ye Annuall recesse of the Earths aequinoctiall pointes praecise 50 seconds which I thought a bold attempt considering that a very

Judicious p[er]son of our nation & acquaintance had made it 2″ seconds lesse.[23] yet now for the reasons I have given yu in this readeing I thinke it may very well be encreased to 51″ which is the quantity stated by Noble Tycho the Judi[ci]ous Kepler & the learned Bullialdus & wherein with them I acquiesce till time the hand mayd of truth shall bring such observations to light as may give me cause to amend this limitation. the Method whereby Tycho Rectified his places shall be shewed yu with the reason of this change of theire Longitudes God Willing next opportunity.

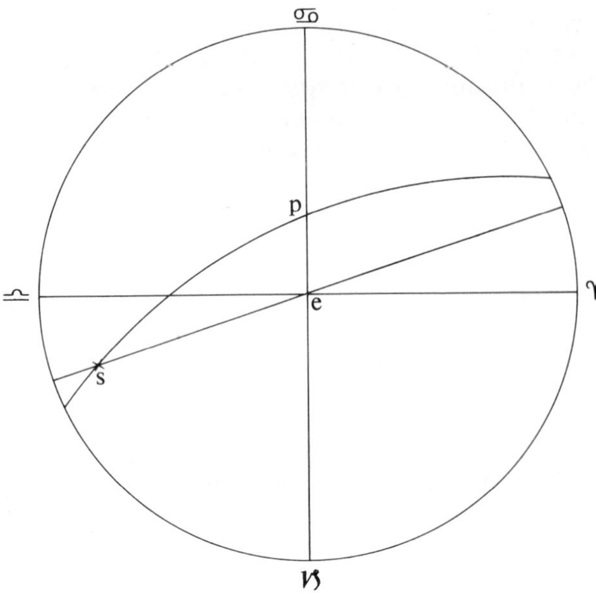

FIG. 31.1

NOTES

31.1. Flamsteed's information for this lecture was extracted primarily from Riccioli 1, pp. 440–8 and Riccioli 3, pp. 254–9.

31.2. Presumably including those edited by John Greaves and Thomas Hyde specified in note 10.6.

31.3. This is probably 'the Manuscript Paper' referred to in Lecture 11, f. 115v, p. 213. Cf. note 11.5.

31.4. This figure is tabulated among those of the stars extracted from Ptolemy's catalogue in Riccioli 3, p. 211. The entry bears the inscription 'Graecus Codex'.

31.5. This figure is the entry for 'Aldebara ♉' in the extract from Tycho's catalogue for 1660 in Riccioli 3, p. 239.

31.6. Riccioli 3, pp. 257, 259.

31.7. Riccioli 1, p. 441.

31.8. Alias Al-Battani, or Muhammed ibn Jābir. His treatise 'De Motu Stellarum' is published in Al-Farghāni or Ahmed ibn Muhammad ibn Kathīr, *Continentur in hoc libro Rudimenta astronomica Alfragani* (Norimbergae, 1537). Cf. cap. 51.

31.9. Riccioli 1, p. 257.

31.10. Marginal note, in Flamsteed's hand: 'Ptolemys Catalogue has y[e] place of Regulus ♌ 2.30 take this the Annuall motion is onely 55″.48‴ something nearer truth.'

31.11. Riccioli 3, p. 257: 'Azophi Astronomus *Persarum*, seu Ebennen Sophi . . .'.

31.12. In this and in the foregoing paragraph, Flamsteed is referring to the tables of the Persian mathematicians in 1115 A.D. Cf. op. cit. (note 10.13).

31.13. Lecture 10.

31.14. A manuscript copy of Arzachel's Toledan Tables, dating from the fourteenth century, is contained in the Crawford Room of the Royal Observatory, Edinburgh. It closely resembles the tabular part of that described in Joseph delambre, *Histoire de l'astronomie du moyen âge* (Paris, 1821).

31.15. These star places were incorporated into the Alfonsine Tables. Cf. note 30.8.

31.16. Op. cit. (note 31.8).

31.17. Cf. note 30.10.

31.18. Op. cit. (note 28.13).

31.19. Op. cit. (note 30.11).

31.20. Walther's observations (op. cit. [note 12.1]) are commented upon by Kepler, op. cit. (note 1.25), pp. 150–4.

31.21. Op. cit. (note 12.1).

31.22. Viz. that between Tycho and William IV, Landgrave of Hesse-Cassel, concerning the accuracy of their respective observations (cf. note 29.13) referred to by Flamsteed in Lecture 29, f. 269[v], or p. 378.

31.23. Probably a reference to Thomas Streete who, in op. cit. (note 2.46), adopts a value of 48″ for the precessional constant.

Lecture 32

21 *November* 1683

In my last lecture I shewed y[u] how the true correct place of Spica ♍[s] might be got from its declination & compareing the ancient places so correct of Timocharis Hipparchus & Ptolemy with y[e] later of Tycho that the annuall progresse of y[e] fixed stars was not more y[n] 51″-20‴. but chuseing the meane amongst those observations which were most accurate that probably it was no more y[n] 51″. upon which quantity I resolved as the best any observations wee have yet met with could afford us this being the same y[t] Tycho Kepler & Bullialdus had chose tho were it not for y[e] old observation of that great lover of truth & Industry Hipparchus I had resolved on a lesse with some of my cotemporarys.

I promised in the conclusion of y[t] lecture a larger explication or description of the Noble Tychoes method whereby hee rectified y[e] Catalogue but haveing mentioned then an opinion enterteined not many yeares agonie [i.e. ago] by some learned & curious p[er]sons & perhaps not well eradicated out of the Minds of those that have seene or read theire workes, but have lesse judgment to determine what is sound & what not, I shall first (to level our way & make all as plaine as I can) shew y[u] what this doctrine was whence it proceeded & its falsity whereby the truth of theire regular & equable motion will be made evident then I shall p[er]forme what I promised y[u] & lastly if I have time I shall adde the true reason of this progresse in my opinion for I would not have either the truth of the ingenuous conception of any skilfull p[er]son prejudiced by any praedetermination

Hipparchus was the first who tooke notice that the fixed stars altered theire longitudes which he found by Compareing his owne observations of theire longitudes & declinations with those of Aristillus & Tymocharis & that they were moved forward but that there latitudes were still the same yet durst hee not determine how much this motion was but onely grossly about one degree in an hundred yeares for it appeares by Ptolemy that hee had a very meane opinion of Tymocharis *Fixarum obbservation*[e]*s* (so y[e] interpreter of Ptolemy translates) *quas Aristillus & Tymochares conscripserant*

Notes on pages 406–7.

parum explanatae nec non & ambiguae sunt.[1] Those observations of the fixed stars which Aristillus & Tymochares have delivered downe to us are ill Circumstanced & ambiguous. & in another place *Ambigebat tamen ut asserit quoniam nec observationes Tymocharidis tempore factas certas putabat, eo quod minis ruditer captae fuerint nec tempus quod interea fluxit ad perfectam rei hujus Intelligentiam sufficiebat.*[2] hee doubted or fluctuated as hee affirmes because hee esteemed not the observations taken in Tymocharis his age exact enough nor the time lapsed betwixt him & them sufficient for the p[er]fect understanding of this matter. so that Ptolemy himselfe will excuse us for not adhereing to ye judgment of Hipparchus or his owne for it is evident by what I told y^u last lecture that his observations were worse then those of Hipparchus & his Catalogue no better it being at best but a Copy of what that Famous person had done before him & in some places rather vitiated then any wayes amended.

Al Battani was the first that encreased this quantity of ye annuall motion I have given y^u his observation before from which compared with Ptolemys hee determined that the progresse of ye fixed stars was one degree in 66 yeares this was a good limitation at that time hee made use of Regulus or ye lions heart but had he considered others with it tis probable hee had not resolved on this quantity especially if hee had employed Spica ♍^{s} whose place was more exactly given by Hipparchus then any other but Ptolemys fame was then as high amongst the Arabs for Astronomy as Aristotles for philosophy and no wonder since what wee have of any more ancient has beene conveyed to us through his eesterne & I could wish it had come cleare without corruption.

Regiomontanus tells us that there were some before Albatani who enterteined the opinion of the trepidation of the eight sphere. or that the fixed stars were carried sometimes in antecedence at others in consequence of theire meane places that they moved 8 degrees forward in 840 yeares & then returned as many back or westwards in the same space of time Arzachel enlarged this quantity to 10 degrees in 750 yeares or 1 in 75. So ye period betwixt ye limits or from east to West is compleated in 1500 yeares. Erasmus Reinholds tells us that Thebit Ebn Corah enlarged this space to 22 degrees but affirmed it could be no more. & Riccius that hee was persuaded to it because hee had found that Hermes who was 1985 yeares before Ptolemy had placed the bright star of ye Harpe in 24 deg of ♐ which Ptolemy has but in 17 of ye same signe. & Hydraes heart in 7 deg of ♌ which Ptolemy makes but in 29 deg of ♋. to which Riccius addes that Arzachel esteemed the Rams head to be in the same longitude with ye vernall aequinox in ye yeare of Christ 560. that thence it moved eastwards for 750 yeares so that in ye yeare 1310 it ought to be retrograde & returneing westward, But if y[t] were so in y[e] yeare 1474 in which Abraham Zaguth (a skillfull p[er]son & Riccius his Master) wrote, it should be onely 7°-11′ in consequence of the equinoctiall point whereas hee then found it in 13°-52′ of ♈. & Spica ♍[s] in ♎ 17°-10′ thus Riccius & Regiomontanus refuce Thebut attributeing the difference betwixt Ptolemy & Hermes to the ambiguity of the stars names whose longitudes hee has given. Alphonsus also findeing in his time that the stars were proceeded beyond those limits that Thebit had assigned rejected this once by them receaved opinion.

This opinion being exploded another was started & the ground of it taken from Ptolemys assertion that y[e] Annuall Motion of the Fixed stars in his age was so slow that they moved onely one degree in an 100 yeares and of Albattani that they had moved one degree in every 66 yeares betwixt Ptolemy, & him, those who beleived these true contrived to represent them by admitting that they were by theire proper motion carried allwayes Eastward but that there was a libration of it so that sometimes it was swifter at others slower, Rabbi Issack the Singer of the Synagouge of Toledo whom King Alfonso Cheifly employed in Makeing his Astronomicall tables states the period of theire revolution in 1000 Jubelies or 49000 yeares & the period of this libration in 1000 sabatticall or 7000 yeares. which those Authors who have delivered it downe to us laugh at as onely a Rabbinnicall Conceite but as King Alfonso himselfe when hee had seene a booke of Abu Hosein wherein the doctrine of Albatani was defended then rejected this opinion & mainteined the contrary that theire apparent progresse was alwayes aequable, yet this opinion so well fixed that after this both Turrian & Fracastor mainteined the unequall motion of the heavens they confest that it was highly probable the fixed stars moved alwayes forward upon the poles of the Eclipstick it not being found that since Metons time any of them had ran retrograde but yet they asserted that this motion was irregular the Meane being one degree in 100 yeares That From y[e] Incarnation to Albatani it was swift increaseing and the swiftest then being one degree in 66 yeares. but from that time to y[e] yeare 1800 is the swift decreasing at which time it arrives againe at the meane but from thence forward for 900 yeares decreased till it come to a station & then it increases againe till it comes to the meane in the yeare 3600. this is fracastors peculiar invention which I shall say more to by & by.

So far had this conceipt prevailed in those times that the Incomparable Copernicus tho hee allowed No motion at all to the fixed stars but defended it that they alwayes kept the same places & stations in y[e] vast expansion of the Fluid aether, and that all the motion they seemed to make in longitude onely was caused by the recesse of y[e] earths equinoctiall pointes in the plane of y[e] ecliptick, yet allowed of this irregular Motion & to salve it contrived to have it sometimes swifter at others slower. the Meane equinox & the true hee makes to coincede 63 yeares before our Saviours time its period to be 1717 yeares. 5 dayes. the period of y^{e} revolution of y^{e} aequinoctiall pointes 25798 yeares & therefore the Annuall recesse 50″.12‴$\frac{1}{4}$ this inaequality hee so order that sometimes the fixed stars may appeare to move one degree in 54 yeares which is there swiftest motion. at others but one degree in 102 yeares which is there slowest. since Tycho None that I know of have admitted this Inaequality save Lansberg who with Copernicus makes the period 1717 yeares, & Longomontanus who proposes it 3600 yeares but Landsberges Astronomy has beene already sufficiently confuted by Phocillides[3] & o^{r} Countrimen[4] his partiality taken notice of & this amongst his other errors layd open I shall not regard him therefore or the better deserveing Longomontanus but onely from those observations of the fixed stars I produced last lecture I shall shew you that theire annuall motion or rather ye recesse of the Earths aequinoctiall pointes is equable & that in reason it ought not to be otherwise admitted.

If wee should begin with Hipparchus or Ptolemy and compare theire observed places with those of Tymocharis by ye observations of Spica ♍^{s} the annuall recesse would be found slower then it is in o^{r} dayes. but if other fixed stars much swifter I shall therefore wave them because of this incerteinty Hipparchus haveing judiciously concluded them too course & over neare his time for this comparison

Wee will therefore begin & lay our foundation with Hipparchus on ye observation of Spica ♍^{s} whose declination hee has determined 0°.36′ whence I determined its true place then in ♍ 23°.38′ which Tycho in ye yeare 1600 makes to be in ♎ 18°-16′ whence I concluded the annuall recesse 51″-20‴ nearely. but considering that the Ancient observations were very course & p[er]haps no nearer then within the third p[ar]te of a degree in longitude if so neare I resolved upon 51″. as a round number upon which by Tycho Kepler & Bullialdus had pitcht before me

If now I make it evident that the observations of Hipparchus & AlBatani as also of Albattani & his followers also compared with Tychos shew neare the same Annuall Recesse tho in some stars it be lesse in others considerably greater I doubt not but all reasonable p[er]sons will Conclude it to be Regualar aequable & free from all such inaequalitys as the Arabians before Al Battani Turrian Fracastor Copernicus Landsberge or Longomontanus have introduced since

Compareing of Ptolemys place of Regulus or the Lions Heart with that of Al Battani wee found the Annuall recesse 57″. but Albattani tooke the place of this star in Ptolemy to be 2°-10′ of ♌ which o^{r}s and the Greeke Copys have 2°.30′[5] of ye same signe y^{t} is 20′

forwarder therefore the Motion from him to Ptolemy is 20′ lesse & the Annuall recesse onely 55″.48‴ nearer o[r] first found by the Hipparchean observation of Spica ♍[s] compared with the Tychonick.

But the place of Regulus observed by Albattani in 14° of Leo Conferd with Tychoes in ♌ 24°-17′, 721 yeares after him gives the Annuall recesse onely 51″.21‴ altogeather the same with y[e] first collation whence wee may gather that from Hipparchus to Tycho, & betwixt Albattani & Ptolemy as also from Albatani to Tycho the recesse was equable & regular.[6]

After the observation of Albattani I considered the places of y[e] fixed stars left us in a small Treatise of Ebennen Sophim[7] dated onely 57 yeares after him & by compareing his place of Regulus with Tychoes collected the Annuall recesse 49″.15‴

The Persian Mathematicians corrected the Ptolemaick Catalogue 180 yeares after Ebennen Sophim by compareing the place of Spica ♍[s] with Tychos I found the Annuall recesses 49″ exactly.

But conferreing theire place of Regulus with his it Was 50″.24‴.

Againe by y[e] place of the Bulls south eye 53″.40‴. So chuseing a meane betwixt them it will be 50″½ not much lesse then the Hipparchean observation of Spica collated with Ptolemys produced it

I compared not y[e] Alphonsine Copernicus Ulug Beig Walthers Werners or any later defined Places with Tychoes because they promise no greater accuracy then the ancient & besides the spaces of time both his and theires are, as Hipparchus well observed of his owne & Tymocharis his observations, too narrow for this purpose.

But from these it appeares that the Annuall recesse has not beene unequall as was supposed by the Arabs even before Al Battani & many eminent persons since I could have made more collations if I had found it needfull & pickeing those which would have served my turne (as has beene usuall wth some persons of remarke) have confirmed by them the quantity of the Annuall recesse of the Earths aequinoctiall pointes to have beene exact 51″. which I have fixed upon. had I not found many errors & such discoherence in the places of the old tables & such disagreement in the differences of longitude from what they have beene found & certeinely are by good instrum[ts] & late observation. I have therefore onely chosen such as seemed the least suspicious from which I conclude chuseing the Meane amongst them that the Motion of this recesse is equable & as above stated

I may adde one more to close all which I had before omitted that beside the place of Spica deduced from its declination observed by Hipparchus wee have also the longitude of Regulus given by Ptolemy from his [Hipparchus's] Workes to his [Ptolemy's] owne age by Compareing it with Tychoes the Annuall Motion of recession is given 50″-56‴-15[IV]. So little different from what is was determined by Spica that it confirmes us in y[e] stated quantity of 51″.

Resolveing on this it followes that the fixed stars change theire places and are promoted in longitude 51′ in 60 yeares. in 100 yeares 1°.25′ in 1000 14°-10′ in 5000, 2^{s}-10°-50′ & that from Hipparchus his time to oures in 1800 yeares they are promoted 25°-30′. From Ptolemy in 1546 yeares 21-54. from AlBattani in 800 yeares 11°-20′ whence y^e Tychonick Catalogues[8] May easily be reduced & the places of the fixed stars knowne in theire severall dayes

Or if wee would reckon y^e yeares in which they change theire places in even degrees. according to this reckning they are alterd 1 degree in 70 yeares & $\frac{3}{5}$ of a yeare. y^t is to avoyd Fractions 5 degrees in 353 yeares and a whole Signe in 2118. a quadrant in 6354. & a whole Circle in 35416. So y^t accounteing with Helvicus[9] the Age of the World to be 5632 yeares the fixed stars have changed theire places neare 80 degrees or 2^{s}. 20° since its creation.

It remaines now onely that I shew the reason of this alteration which I conceave I cannot doe better then I have alreadv done In a small treatise I wrote & published some yeares[10] almost in these Words. The distances & positions of the fixed stars have beene found the same in all ages whence tis more y^n probable they are subject to no motion but what appeares in them by ye recesse of ye Earths aequinoctiall pointes as Copernicus long before us conceaved the manner & reason whereof will appeare from ye figure before y^u [Fig. 32.1]. in which let ♈ ♋ ♎ ♑ represent the greate orbe in which the Earth is carried round the sun once in a yeare & those severall pointes its places at ye times when ye sun appeares entring ye 4 Cardinall pointes ☉ the suns place in one of the foci of this large Ellipsis ♎ the earths place at ye Vernall Equinox pe ye line connecteing its Northerne pole with the pole of ye Ecliptick projected on it or its solstitiall Colure. the Earths motion round her Axis or her Rotation is from West to East or from d by i to t, in ye Illuminate p[ar]te of the globe or that next ye sun. But ye motion of the aetheriall Matter is also from Weast to East or from α to β directly contrary to this in ye same illuminate part of ye Globe & that too something stronger then in ye remoter or obscure part by reason the p[ar]tes neare ye sun are something swiftlier moved then ye p[ar]tes more remote. whence it proceedes that ye line of direction t.p.d.e. connecteing ye poles is reflected & borne back contrary to ye Earths & Vortex motion. 51″ Annually so y^t whereas about the begining of Nabonassars era it cut the aecliptick & consequently made ye solstice neare ye cloudy place in ♋ it is now carried back to ye heele of Castor & about 250 yeares hence it will make ye solstice when ye earth passes under Propus

That point of the Orbite where when ye Earth Arrives the line of direction or solsti[ti]all colure Coincedes with ye horizon ye dyke And ye North pole passeing out of the obscure be to enter ye illuminate haemisphere of ye globe is the place where ye Vernall aequinox is made & ye ☉ enters ye first point of ♈. this point at ye Commencemt of Nabonassars era was at N but since by the Continuall impression of ye aethereall Matter & its renitency to ye earths motion is reflected aequally with ye solstitiall Colure which is altered from nea to ted recedeing continually. hence ye recesse of the Æquinoctiall pointes which if shee endure some ages longer the aequinoxes may happen in time under those stars neare which the solstices were celebrated at ye Creation. Such a Change requireing onely some few yeares more yn 6300 to compleate it[11] of which 5600 & odd are already past by or account[12] by ye septuagint more if not more yn ye necessary time.[13]

And since this impression of the aethereall Matter is allwayes equable & the same in every revolution its effect or the recesse of the aequinoctiall pointes caused by it Must also be granted equable so that I see no reason why the excellent Copernicus should conceave it otherwise besides what I mentioned before too great a respect for antiquity proceedeing perhaps from a better opinion of the ancients performances then they deserved. Which yet I speake not as a reflection on that great p[er]son to whom wee are indebted not onely for or reformed Astronomy but p[er]haps for a great p[ar]te of or Philosophy too. it rather shewes the excellency of his temper & ingenuity that hee thought better of others then himselfe & could find out the readiest meanes of answering what others pretended to have observed. Thus I have now and formerly shewne yu the reasons of those apparent Motions taken notice of in ye fixed stars which necessarily follow from theire unchanged positions & the Motions of ye earth in its orbite round ye sunne Annually & round its owne Axis dayly. the not necessary or accidentall appearances of them caused by ye sphere of Vapors encompasseing or Earth. & theire parallaxes will be the subjects of or discourses when Heaven affords us another opportunity.

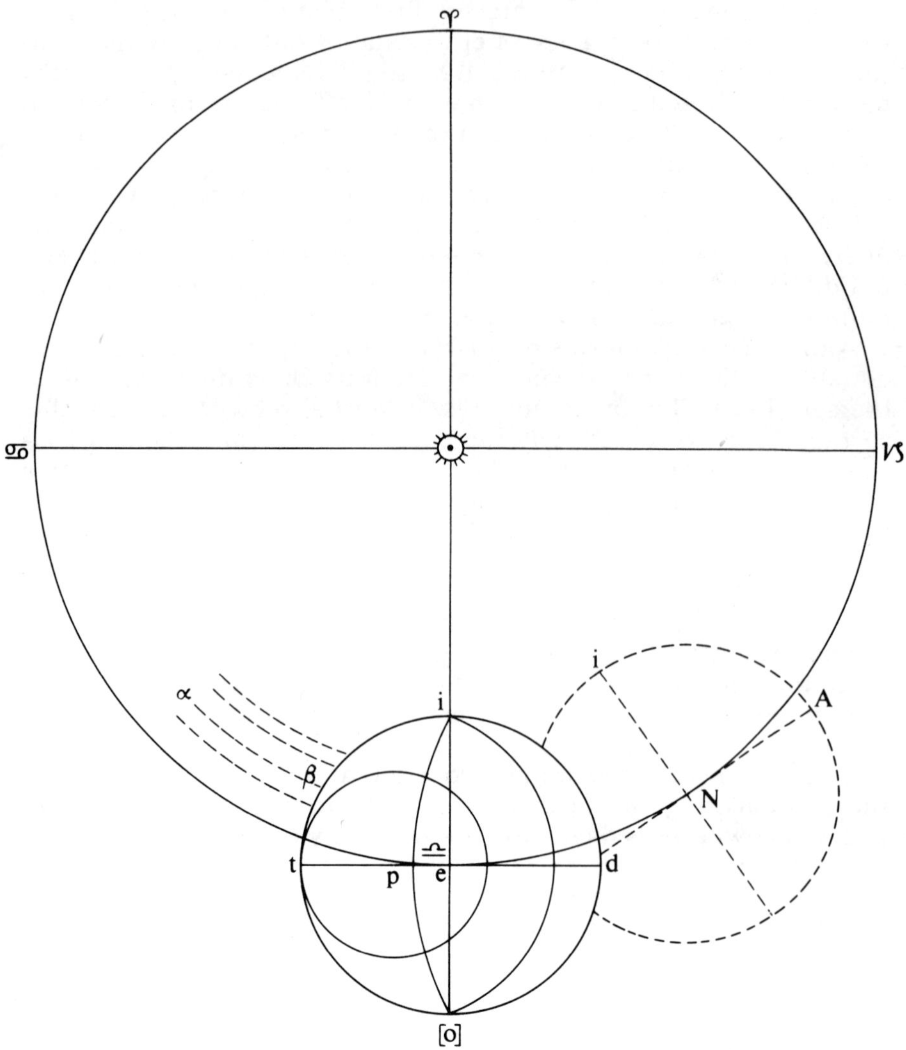

FIG. 32.1

NOTES

32.1. Riccioli 3, p. 201.
32.2. Ibid.
32.3. Op. cit. (note 24.38).
32.4. E.g. Jeremiah Horrox, in op. cit. (note 1.26) (cf., in particular pp. 130–55).
32.5. The appropriate entry is tabulated as no. 8 under 'Leo' in Riccioli's excerpt from Ptolemy's star catalogue. Cf. Riccioli 1, p. 212.
32.6. Four lines have been crossed out by Flamsteed at this point.
32.7. Ebbennesophim's tract on the positions of the fixed stars is referred to by Bullialdus, op. cit. (note 30.7), pp. 224–5.

32.8. Op. cit. (note 2.36).
32.9. Op. cit. (note 1.7).
32.10. Op. cit. (note 15.1).

32.11. Since $90°/51'' = \frac{90 \times 60 \times 60}{51} = 6{,}353$.

32.12. We know from Lecture 34, f. 312v, or p. 421, that Flamsteed took this figure from Christopher Helwich's chronology (op. cit. [note 1.7]).

32.13. The Septuagint, or Alexandrian version of the Old Testament, was first printed in the Complutensian Polyglot (1514–17); but two years prior to its publication in 1521, Aldus published another edition. The Textus Receptus issued by Pope Sixtus V (Rome, 1587) was based mainly on manuscripts in the Vatican Library and at Venice.

Lecture 33

2 *April* 1684

In my former Readings I have discoursed of the nature of ye fixed stars & shewed y^{u} how it comes to passe that some of them appeare & disappeare in stated periods of time that others have appeared in places where none have beene observed of the ancients suddenly but have gradually lost theire light & totally disappeared againe. I gave an account of the notable Nebulosae one in ye Girdle of Andromeda another in ye sword of Orion & suggested the reason of them on the same principles whereby I salved the appearances of those stars that are seene & disappeare by turnes From which I also shewed that probably after some time o^{r} sunne might have its light extinguished & the planets of o^{r} vortex deserteing theire Ellipticall paths might run of[f] in ye vast expansion of ye aether. here I told you my opinion of ye Nature of Comets which being proved to be bodys no whit lesse y^{n} o^{r} Earth but commonly much bigger I could not conceave to be other then planets lapsed from some ruined vortex & thus carried through the vast expansum of ye ether lastly I set to examine the quantity of y^{t} motion whereby wee find ye fixed stars promoted almost a signe in longitude from those places they have assigned them in Hipparchus his Catalogues & haveing well examined the best observations & determinations of ye ancients I proved this progresse to have beene alwayes equable contrary to ye opinion of some ingenious p[er]sons who make it anomalous & have given us severall contrivances for answering the Inaequality I proved ye annuall progesse to be 51″. nearely. & enquireing into ye cause of it I shewed y^{t} ye fixed stars being placed at a vast distance from us & keepeing theire latitudes & mutuall distances allwayes invariable it could not probably be assigned to any change made amongst them or any reall motion of theires, but that it was really to be attributed to ye recesse of the earths equinoctiall pointes or a change of the direction of her Axis caused by ye motion of the matter of ye vortex swifter & stronger neare ye sun weaker where more remote. Nor doe I doubt but this is the very reall cause whereby ye great builder of the world Who has created all things in number weight & measure produceing all the visible effects &

Notes on pages 415–16.

motions in ye universe from proper causes, has answered this in his vast mind. Since wee see the Node in ye Moones orbe. & in ye satellit[e]s of ♃ too are both carried the same way contrary to the succession of ye signes Not so slow as ye earths aequinoctiall pointes recede. because ye smallnesse of ye earths body & the violence of ye Earths diurnall turbination yeilds not so much to the impression & renitency in ye contrary motion of ye aethereall matter of ye vortex as ye wider & more fluid orbes of ye Moone & ♃ satellit[e]s, in ye latter of which the Motion of the Node tho very sensibly retrograde is so much slower y^n ye Node of ye Moones orbe, that it much confirmes o^r opinion & reason produced for ♃ being 5 times more remote from ye sun y^n ye earth is, the Motion of ye aethereall matter must be more then simply proportionably slower, but since the orbes of all the satellit[e]s seemed to have the same common node it seemes also probable that there should be some fixed proportion betwixt the revolutions of ye Nodes of ye Moone & Jupiter concerneing which it is not my business to determine at present I onely make use of these two instances to confirme the assertion of ye recesse of ye aequinoctiall pointes of o^r earth. for as I have often intimated all o^r knowledge is but comparative & of analogies so y^t where wee see like or ye same effects produced in similar matter wee may safely conclude they come from ye same or like causes In the conclusion of my last lecture I told y^u that haveing thus explained the motion attributed to ye fixed stars & shewed y^u also the reason of theire scintillation caused by theire raies passeing through a sphere a vapors encompasseing o^r earth & continually in agitation in my next I should shew y^u ye nature & effects of caelestiall refractions when by the true angles of theire distances are diminished more or lesse accordeing to theire positions which I now proceede to p[er]forme.

Which that I may ye better doe I shall shew y^u what Authors have wrote on this subject. whereby it will appeare how little ye ancients knew of it & how from small beginings wee have arrived to ye knowledge of Refractions and theire effects. I shall afterwards shew y^u in what proportions they vary & endeavor to salve ye appearances caused by them they rule.

It is not probable that the ancient Greeke Mathematicians knew any thinge of them since they are not so much as mentioned that I can learne either by Hipparchus or Ptolemy the former of which was most likely to have beene informed of them from his caelestiall observations his silence is therefore an argument to me that his Instruments were small & but Coursely divided else hee must needs have p[er]ceived them in the distances taken betwixt any two stars lieing neare ye same verticall circle observed at the riseing or setting of one of them and againe neare ye Meridian, for the distance when one was neare ye horizon would appeare much lesse then when they were both neare ye meridian. Archimed[es] says not a Word of them nor Appollonius Pergeus. So that wee may safely conclude the ancient Greekes wholly ignorant of & unacquainted with this subject which had they understood tis so easy & pleasant some Memoriall of it at least some transient one had beene transmitted to us either in a Worke on ye subject it selfe, or some account of it. Euclid I must confesse Recites immediately before his Catoptricks[1] that knowne experiment that if an object be placed in the bottom of a concave vessell, & the eye recede till ye sides of ye vessell just cover ye object from ye sight y^t if afterwards the same vessell be filled with water ye object againe becomes visible, but I find not that hee makes any use of it nor that wee have any thinge of Dioptricks or refracted vision. before Alhaysen a learned Moore. whose age is unknowne but Frederick Reisner who published him[2] conjectures that hee lived about y^e yeare 1100 & that hee was contemporary to Avicenne Averroes & Zoara all learned men of that nation amongst whom Ingenuous knowledge Flourished at that time. In y^e 51.52.53^d & 54$^{[th]}$ propositions of his 7th booke[3] hee asserts that Refractions change y^e true places of y^e luminarys & stars & in his 53 & 54 propositions[4] p[ar]ticularly enough that theire distances are contracted by them, more, near y^e horizon lesse as they approach y^e vertex. which wee shall enquire hereafter. but for the measures of Caelestiall refractions or indeed any other I find not any thinge of them in him tho in ye very first Chapter of this 7th booke hee describes an instrument purposely contrived for measureing ye refractions of transparent substances.[5]

The same Instrument is described againes by Vitellio another writer of Opticks about 170 yeares after Alhazen for it is evident by severall passages in his worke[6] that hee lived about ye yeare 1270, in ye 49th prop. of his 10th booke hee affirmes. *stellas caeli, et lunam secundum refractionem a visibus comprehendi Instrumentaliter*[7] in ye monkish latin stile of those dayes whereby hee intended to have said. that ye refraction of ye luminarys & stars might be perceaved by observation, for sayes hee "Si instrumento Armillarum nocte observetur aliqua stellarum fixarum Magnarum que cum pervenit ad meridianum sit transiens per Centrum capitis experimentantis aut prope. et consideretur illam orta suo dum elevatur super superficiem horizontis et tunc revolvatur armilla revolubilis in Circuita poli mundi qui est polus aequinoctialis donec fiat equidistans circulo magno caeli transeunti per polos aequinoctialis et per centrum corporis illius stellae, et certificetur locus stellae ex armillae ita ut habeatur distantia ejus a polo mundi. deinde observetur stella donec veniat ad Circulum meridie moveaturq[ue] Armilla mobilis donec fiat aequidistans circulo stellae ut prius et sit in superficie circuli meridiani, et tunc iterum habebitur distantiae stellae a polo mundi cum stella fuerit in Zenith Capitis aut prope invenieturque distantia stellae a polo mundi in tempore ortus et elevationis stellae minor ipsius distantia ab eodem Polo tempore quo est in Zenith capitis vel prope."[8] y^t is if in ye Night time some one of the bright fixed stars passeing ye Merid[ian] near ye earth be observed on its riseing or setting upon ye horizon wth ye Armilla, & its distance then taken from ye next pole, and afterwards againe with ye same instruments when it passes the Meridian neare ye Zenith. its distance from ye pole taken in its riseing or setting will be found lesse y^n when it passes ye zenith, whence it is evident sayes hee that the light of the caelestiall bodyes come not to o^r eyes directly in streight lines but refracted this is that which Bernard Walter in his observations of the Sun & ♀ 1489 March ye 5th6th & 7th tells us hee found expressly asserted by Alhayzen & Vitellio in theire opticks & that hee found the same by like observations before hee became acquainted with them. that is hee observed the distance of ♀ from the Sun contracted & her distance from ye pole allso lessened in her setting whence her visible place was considerably altered from what it was otherwayes found by the Suns true place & her distance from him observed neare ye Meridian. I shall not therefore transcribe what hee wrote on this subject or others on this occasion it will be sufficient that I have quoted the place it selfe in his workes, or if a comment be wanteing to explaine them the best is what I have met with in Keplers Paralipomena ad Vitellionem pag 151. 2. 3. & 4.[9] where those who desire to be More fully satisfied may find abundant satisfaction

But tho Alhayzen proved that the distances of any two stars were contracted by refraction, Vitello mentions an experiment whereby it may be proved & Bernard Walter gives an observation wherein the effects of refractions are very evident yet have none of them given us the quantity of ye horizontall refraction or the proportion whereby it diminishes as the appearances ascend this wee are indepted to the noble Tycho for who lived an hundred yeares after Walter & at first seemes no better informed then hee, but from these notes it seemes evident that Alhayzen was the first who suggested the effects of refractions in caelestiall appearances. that his doctrine was onely confirmd by Vitellio who gives us tables of the refracted angle to every 10 deg of ye Inclination of rayes passeing from aire to water or glasse, or from Water to glasse,[10] that Bernard Walter was the more positive on his assertion because hee found it corroborated by Vitellio's testimony but that nothing materiall as to Caelestiall refractions was yet discovered but at most determined by any of these three. so that for what wee have at present wee are almost wholly indebted to ye Noble Tycho & his successor or Disciple the subtle & Indefatigable Kepler.

The observations of Tycho related in his Historia Caelestis[11] begin with ye yeare 1582. but by reason either of Cloudy weather at ye Winter solstice or rather for that his instruments were not yet compleate as hee himselfe seemes to intimate wee find no determination of the latitude of his Uraniburge this yeare. but in the next yeare 1583. in ye solar observations of Dec. pag. 48[12] hee gives ye suns greatest & least heights correct by his parallax but without any consideration or mention of refraction whence hee collects his latitude lesse by 4 minutes then hee determines it afterwards in his observations of ye fixed stars pag 65:[13] to which hee confesses in his Progymnasmata[14] hee durst not trust till haveing made other more Curious instrum^{ts} & built his Armillas hee found the latitude sought by ye Merid. heights of ye fixed stars the same hee had first collected by them. The first time I find any observations made with his Armillaes is October ye 31 1584[15] when hee began to observe ye declinations of some fixed stars with them but he applyed them not to the sun till ye 9th of June following 1585[16] & till this time wee beleive him to have strove with ye effects of refraction for from some observations of narrow distances of the Pole Star from others hee concludes boldly that the distances of the fixed stars are not contracted when neare ye horizon more y^{n} when they are higher as Alhayzen & Vitellio taught especially if they are above 20 degrees high. Whereby it seemes intimated that hee was stil dubious concerneing Refractions tho hee Continued not long so, for at ye following Æquinox on the tenth of September hee subjoynes to his observations of that day Has itaque observationes [in ☉] idoneo instrumento feci[,]ut constaret ratio Refractionis Radiorum ipsius.[18] that is He made those solar observations with a Convenient instrument purposely for finding, ye ratio of ye Refractions of the suns raies. His method seemes to be thus

From ye 9th of June till ye 18th I find y^{t} hee onely noted the suns heights & that scarce nearer then to the 6th p[ar]te of a degree[19] at such time as hee observed his visible declinations but from this time forward finding no doubt the Want of it when hee came to examine his observations hee gives also the true times of each (either got by his accurate movements well rectified, or by the helpe of his Armillaes) togeather with the suns height then observed to ye halfe, quarter, or sixth p[ar]tes of a degree[19] & his visible declinations, not omitting what hee had found it the noone precedeing or following: by the time interlapsed betwixt noone & the Moments of his observation hee knew readily how much the suns true declination was more or lesse then hee had observed it in ye Meridian. hence haveing the hour from Noone with ye latitude of ye place & the suns true declination given, the angle which the Azimuth or verticall Circle formed with ye hourcircle passeing through the sun was found with no great difficulty the difference of ye true & observed declinations gave him ye refraction in declination. whence it was not difficult to find ye refraction in y^{e} verticall Circle or y^{e} refraction in Altitude to that declination a figure will make all plaine looke on that Before y^{u} [Fig. 33.1][20]

In which let y^{e} semicircle HÆZP☉ reprezent one halfe y^{e} Meridian. Z y^{e} Zenith P the pole PA the Axis of y^{e} world ÆA y^{e} Equator,

1$\odot$, a parallel of Northerne $\lambda\odot$ of Southerne declination. let Z$\odot$. be an Azumuth or verticall circle falleing from y^{e} Zenith on the Sun in which conceave him elevated by refraction from $\odot$ to C & let y^{e} hourcircles Pb$\odot$. P$\beta\odot$ be drawne as also Pc and PX in either of y^{e} triangles PZ$\odot$ accordeing to y^{e} Noble Tychoes method are given y^{e} Constant side ZP y^{e} Complem^{t} of y^{e} latitude P$\odot$ y^{e} suns true distance from y^{e} Visible Pole. with y^{e} Angle ZP$\odot$ y^{e} true hour from. Noone that is two sides of an oblique angled sp[h]aericall Triangle with y^{e} Angle included to find the other Angle P$\odot$Z which is commonly called the Angle of position but may as properly be named the parallactick or refraction angle because it determines the difference of aspect caused by y^{e} refraction raiseing y^{e} suns visible species from $\odot$ to c or x.

Which being obteined in y^{e} triangle cp$\odot$. are knowne p$\odot$ y^{e} Complem^{t} of y^{e} $\odot^{s}$ true declination. Pc or Px the Complement of y^{e} observed declination with y^{e} Angles p$\odot$c or P$\odot$X. before found whence y^{e} side c$\odot$ or x$\odot$ y^{e} refraction in altitude may be easily had. & from the observed altitude of the sun subtracteing y^{e} refraction thus got y^{u} have the true altitude to which it belongs.

But to avoyd the resolution of an oblique angled sphaericall triangle. on y^{e} hourcircles P$\odot$ from the pointes c and x let fall the perpendiculars cb. xβ[21] then may $\odot$b or $\odot\beta$ be taken for the refraction or declination, & then in y^{e} small right angled triangles c$\odot$b. x$\odot\beta$ are knowne. the said sides b$\odot$. $\beta\odot$ with y^{e} Parallactick or refractive Angles c$\odot$b x$\odot\beta$ whence the hypothenusa c$\odot$ or x$\odot$ will easily be found. & considering that the sides of the said Triangles are all of them small Arches of great Circles insensibly differing from streight lines, they may be taken for such and resolved as streight lined triangles without any materiall or indeed sensible error.

This was the method of the Noble Tycho who from his observations of y^{e} luminaryes & fixed stars concludes the horizontall refraction of y^{e} $\odot$ 34 min of y^{e} moone 33. of y^{e} fixed stars but 30. not that all his observations as he managed them gave him these precise quantityes but that hee chose them as a meane betwixt the extremes found by his observations which if severall of them be examined togeather will seldome be found to consent in ye sam[e] horizontall quantity nearer then within ye compasse of 6 or 8 or perhaps 10 minutes. for such differences I have often found in ye riseing & setting sun & I can not conceave the Aire more unequall here then in Denmarke

The reason why Tycho make ye horizontall refractions of the sun to exceed that of the stars is because hee supposed the suns horizontall parallax 3 min. Now hee found that without any consideration of this the solar refractions were equall to if not bigger then those of the fixed stars but the sun being depressed in his opinion 3 min in his riseing or setting the refractions must of necessity be 3 minitues greater in him then in them to compensate this defect but why hee first make the Moones lesse then the suns I see no reason in Nature nor will it be found by any experience to be so. but I am apt to beleive The Noble observer gave us the result of his Calculation fairely in which it is to be Wondered that hee differed or erred no more since hee made use of a faulty Theory of the Sun as ill of the Moone & Tables of the fixed stars Which tho they were the best his Age or Instruments could afford Yet by the help of better expedients[22] (discovered since hee put of[f] Mortality) are found to erre considerably but one error often compensates another. & by such a ballance his determinations are found not much wide of the truth. Next lecture I shall God Willing shew y^{u} a better method & some deductions from it & then p[er]haps wee may have a further opportunity of mentioning o^{r} Noble restorer of Astronomer, & examineing his Table of refractions which with a just Correction will be found to agree much better then could be expected with later & better Trialls.

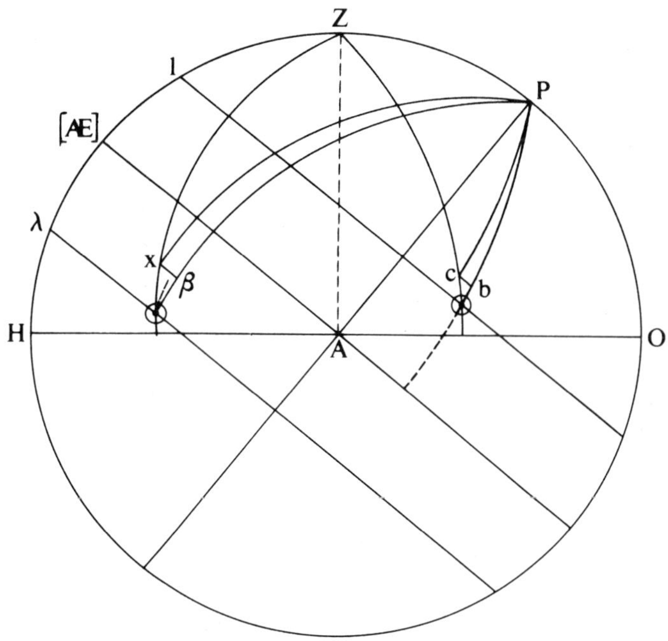

FIG. 33.1

NOTES

33.1. *Euclidis Optica & Catoptrica* (Parisiis, 1557), p. 46.

33.2. Cf. Federico Reisner's edition of Alhazen's *Opticae Thesaurus* (Basileae, 1572).

33.3. Ibid., pp. 278–80.
33.4. Ibid., pp. 279–80.
33.5. Alhazen's 'instrumentum rotundum' is actually described in ibid., pp. 231–3.
33.6. Reisner's edition of *Vitellonis Thuringopoloni Opticae libri decem* was published together with Alhazen's work (op. cit. [note 33.2]) but separately paginated.
33.7. Ibid., p. 444.
33.8. Ibid. Flamsteed himself supplied the first three words of this quotation.
33.9. This, and the following information concerning Bernhard Walther is cited in op. cit. 125, Cf. note 31.20.
33.10. Op. cit. (note 33.6), p. 412. This table is reproduced in Riccioli 2, p. 664.
33.11. Op. cit. (note 2.36).
33.12. Ibid., p. 48, where the latitude is found to be 55° 50′ 10″.
33.13. Ibid., p. 65, where the latitude is found to be 55° 54′ 30″.
33.14. Op. cit. (note 2.5) (1.)
33.15. Op. cit. (note 2.36), p. 102.
33.16. Ibid., p. 107.
33.17. Ibid., p. 98. Cf. Brahe's comment under the observations of 6 January 1584.
33.18. Ibid., p. 111. The word 'ipsius' was introduced by Flamsteed.
33.19. Flamsteed should have written 'minute'—not 'degree'.
33.20. There is a rough diagram on f. 304^{v}, which may be a first draft of that which has been reproduced here.
33.21. Flamsteed wrote 'xb' instead of 'xβ'.
33.22. Flamsteed certainly had in mind the telescope, pendulum clock, and micrometer when making this remark.

Lecture 34

16 *April* 1684

In my last lecture here I proved that the Annuall motion of ye fixed stars or more proper ye yearely recesse of the earths equinoctiall pointes was neare 51″. this I did by compareing the Ancient observation of theire places made by Hipparchus Ptolemy & Mahomet Albatani with ye later of Tycho Brahe. I suggested that this motion was equable. and alwayes the same tho the Contrary has beene asserted by some eminent persons amongst the ancients. I shall now prove that it really has beene so but it will be first requisite to acquainte y^{u} Who they were that held this inequality of theire motion of ye fixed stars of the Trepidation of the 8th sphere for so they called it, what reasons they had to suppose it theire invalidity, & then which will confirme the opinions of all o^{r} late Astronomers. that if the observations be duely managed & considered as they ought to be, it cannot reasonably be inferred from them that theire is any reall inequality or trepidation in the Motion of ye 8th sphere or in the recesse of ye Earths aequinoctiall points.

I shall shew y^{u} the reason of this recesse drawne from ye nature of the susteme. & afterwards if I have leasure the faults of Tychos methods in determineing the true places of them & the aequinoctiall pointes & the Method I employ to determine them more accurately. if this may not be done at present I shall take an opportunity to fall upon this subject ere long & discourse it more fully.

There were then six severall opinions amongst the Ancients concerneing the Motion of the fixed stars[1] the first was that of Hipparchus & Ptolemy who conceaved theire progresse to be one degree in one hundred yeares, this they found by compareing theire owne observations of ye place of Spica ♍^{s} & other stars that lay neare or not far removed from ye aequinoctiall points with the more ancient determinations of Tymocharis, & this motion they conceaved equable but as Hipparchus well observes of Tymocharis his notes they were too course for a precise determination & the intervall of yeares betwixt them too narrow for this purpose & therefore tho this opinion was enterteined by some of theire followers yet it held not

Notes on pages 423–4.

long but gave occasion however to those who came after & attributed too much to theire authority when they saw it was too slow for the following ages to Conceave the motion was unequall if not irregular

For Albatani. who lived 783 yeares after Menelaus Compareing his observed place of the bright star in ye Scorpions front with his owne collects thence its Motion in the yeares interlapsed betwixt them 11°-50′ or 11°-55′. whence hee concludes theire progresse to be one degree in 66 yeares, but Riccioli[2] compareing other fixed stars in Albatanis catalogue with theire places in the Ptolemaick collects theire motion thence one degree in something more y^{n} 64 yeares. & the Annuall motion 54″.32‴

With Albatani King Alphonso agreed in his latter dayes & the last edition of his tables tho at first hee had enterteined a Cabbalisticall Conceit of Isaack Hazan the Praecenter of ye Sinagogue at Toledo whom the Jewes of Senil affirme to have beene the Author of the Alphonsine Tables. Who because the 49th yeare was accordeing to theire law the yeare of Jabil or rest conceaved that the fixed stars made one revolution in 49000. yeares togeather with the Auges of the planets which Ptolemy had fixed under ye same fixed stars. p[er]petually hence it followes that the Annuall Motion of the fixed stars was onely 26″.26‴.54^{IV}. which is also the meane motion of the sunne in 10′.44″ ye space of time ye Alphonsine solar year wants of 365 dayes 6 houres compleate. but this opinion being founded onely on an idle conceipt was soone discarded to make roome for others as false tho not so apparently Idle as it selfe.

For theire lived some p[er]sons before Albatani as hemselfe reports & Regiomontanus from him who conceaved that the fixed stars moved forward accordeing to the succession of ye signes one degree in 80. or 84 yeares till by this motion they had runne over 8 degrees & returned toward ye west or in antecedence as many. this was a very silly conceipt & easily convinced of error. therefore Arzachell a learned Moore, rather thought that ye fixed stars moved Eastwards or accordeing to ye succession of the signes 10 degrees in 750 yeares. absolveing one degree in every 75 yeares, & in as many yeares returned the same Number of degrees Westwards or back againe hence theire motion from the Easter[n]most point to ye Wester[n]moste of theire is p[er]formed in 1500 yeares but theire whole period in 3000 Thebit Ebn Corah affirmed theire greatest remove on each side to be Not more y^{n} 11 degrees. & Augustin Riccio out of a book of Isaack Israelita concerneing the Creation of ye World[3] reports that Arzachel & Thebit were removed to be of this opinion, by reason that Hermes who was 1985 yeares elder then Ptolemy had observed the fixed stars in more longitude from ye Vernall Equinox then Ptolemy afterwards had found them. As for example the bright ★ of ye Vulture in 24 deg of ♐ which Ptolemy found onely in 17 of ye same signe. & Hydraes heart in 7 deg of Leo which Ptolemy places in 30 gr of ♋. Riccio addes that from the same Rabbi Isaack, Arzachel concluded that in ye yeare of ye Incarnation 560. the first star of ♈ ought to be on the Vernall Equinox. & that it ought to move forward 10 degr. in 750. yeares, so that in the yeare 1310 in which Isaack Israelita wrote it ought to begin to returne towards the West. granteing this in ye yeare 1474. the Head of ♈ ought to be onely 7°-11′ distant from ye Vernall equinox which yet Abraham Zaguth Riccius his Master a good astronomer found then in 13°.52′ p[er]haps it should be 23.52 & Spica ♍^s^ in 17.10′ of ♎ wherefore hee deservedly confuted & rejected this opinion of Thebit & his followers assigneing the difference betwixt the places of the stars observed by Hermes & Ptolemy to the incerteinty of the ancient names of the constellations.

Tho ye Compiler of ye Alphonsine Tables[4] made the meane so slow that the whole revolution should be compleated onely in 49000 yeares that is in 1000 Justinian Cycles yet to help it out & that hee might salve the appearances hee superadded a motion of trepidation whose period was 1000 sabbaticall or 7000 yeares. this Turrian[5] & Fracaster in his Homocentricks[6] rejects. affirmeing that the motion of the fixed stars is allwayes direct but anomalous and sometimes swifter at others slower then one degree in 100 yeares. which they esteeme the meane, that from ye Incarnation to Albatani it was swifter encreasing till at last they made a degree motion in 66 yeares. which is swiftest wherto they attain in ye yeare 900. that from thence it was swift still but decreasing to ye yeare 1800 in which it ought to arrive at the Meane againe & so the period of velocity is complete in 3600 yeares

Copernicus who allowes no motion in the fixed stars yet allowes of the like inaequality in the recesse of the Earths aequinoctiall pointes whereby this appeareince is made. Hee supposes this recesse therefore to be anomalous. its period to be compleat in 1717 Egiptian yeares.[7] halfe ye time of ye period of ye Eclipticks obliquity. the greatest aequation of theire motion 1°.10′ that the Meane & true aequinox coinceded in ye 63^d^ yeare before o^r^ Savior. the Meane recesse hee makes 50″.12‴.15^IV^. hence the period of their Meane revolution compleat in 25816 Egiptian or 25798 Julian yeares. the apparent motion of the fixed stars being when swiftest one degree in 54 yeares when slowest a degree in 102 yeares. wherein hee is followed by Lansberge wholly & Erasmus Reinholds. & by Longomntanus in part who makes ye Anomaly of ye Obliquity equall to y^t^ of ye aequinoxes & both to be compleate in 3600 Julian yeares. The reason of all which contrivances seemes to be nothing but the extraordinary & unreasonable veneration they had for ye assertions of the ancients. Hipparchus & Ptolemy haveing made the fixed stars to move a degree in 100 yeares Albattani in 66. & they being found in o^{r} dayes to proceed a degree in neare 72 yeares. these periods of inequalitys were contrived by them to Answer these assertions. which if they had duely considered the observations themselves they would have found wholly needlesse. Hipparchus & Ptolemy as appeares by the place I have quoted from them both esteemeing the observations of Tymocharis & Aristillus too neare them and too course for an accurate determination which objection I shewed y^{u} & sufficiently proved held good against theire owne for it is not probable that theire instruments were divided or theire observations made nearer then to ye 4th or 6th p[ar]te of a degree & such an error in the declination of a fixed star would give its place halfe a degree faulty. adde to this the error of theire obliquity of the Ecliptick assumed 20 minutes at least bigger then theire owne observations will warrant. the suns theory ill correct & the Moones (by whose transits over them they oft pretended to correct theire places) much worse, her p[ar]allax not well or allwayes duely applyed wee may justly conclude the places of the fixed stars delivered downe to us from Hipparchus by Ptolemy to be very dubious & tis well if they be lesse then a degree faulty.

A part of these objections will hold also against Albattani for tho hee had got the Obliquity of the Ecliptick much truer then his ancients yet had hee made it bigger then hee ought & then the observations of the next ages would warrant by a tenth p[ar]te of a degree, which shewed some fault in his instruments as I have proved there really was in my lectures concerneing the obliquity of the Ecliptick. so that his places of the fixed stars are suspicious 1st upon this account, 2 the Method of his processe makes them further dubious tho hee tooke ye best hee could have pitcht upon. for hee compared his observed declination of Regulus or the Lions head with Ptolemys. & by the difference. Supposeing the latitudes of the fixed stars the same with Ptolemys allwayes invariable hee collected how much change of places the difference of declinations required. Now if either Ptolemy or hee erred but so much as wee may justly conclude they might in ye observed declination of this star the fault would be much more then it would have beene in Spica ♍^{s}. for Spica being then not far from ye Autumnall equinox altered its declination as fast againe as Regulus who was but about a signe distant from ye Tropick. it were to be wisht therefore hee had rather made use of or left us the declination of Spica ♍^{s}. this being still one of the most convenient stars for this purpose both because of its remarkeable splendor, position in ye dayes of Hipparchus & for that its declination is not yet so much encreased but that it may still be very well applyed for this purpose.

Haveing duely weighed the observations of the ancients & Compared severall of them with o^{r} owne I find no reason to admit of these inequalitys or that motion of trepidation with its periods according to any of the Authors I have mentiond. tis evident to me that even from Metons time to o^r^s the fixed stars have kept still proceedeing and as far as humane conjecture or observation can assure us, aequably. theire Annuall motion or Rather recesse of ye Earths aequinoctiall points being constantly 51″ or neare it. for it will be impossibly to determine the true quantity of this recesse exactly without better observations of the Ancients then wee have yet met with. or y^e^ experience of future ages.

Admitting therefore the Annuall recesse to be praecise 51″ it followes. that the aequinoctiall pointes recede or the fixed stars appeare promoted in longitude one degree in 70 yeares. & neare $\frac{3}{5}$ of a yeare. but avoydeing fractions more nearly 5 degrees in 353 yeares. a whole signe in 2118. So that y^e^ head of ♈ which about y^e^ Commencement of Alexander the greats Æra or y^e^ Æra Dulkarnain was neare Our vernall aequinox is now removed a whole signe from it. Helvicus[8] the Chronologer accounts ye present age of ye world above 5600 yeares. in which space of time the Recesse is 80 So y^t^ if ye World continue & the sun be not extinguished within the space of 900 yeares future the aequinoxes will be celebrated neare those stars in ye heavens where the solstices were made at the Creation. the whole period of the Revolution accordeing to this account will be 25412 yeares,[9] to which it ye world be sustained by ye Power y^t^ created it, the aequinoxes & solstices Returne to ye same pointes on which they were made at ye Creation.

Haveing thus limited the motion of the recesse of ye earths aequinoctiall points whereby the progresse of the fixed stars is caused I come now to shew the true reason of it.

The Ancients supposed a vast orbe beyond the orbe of ♄ in which all the fixed stars were placed at that distance from o^r^ earth that theire p[ar]allax might be insensible. this they supposed to be moved slowly the same way with the orbes of the other planets or from west to East so that it might compleate its period in the stated time of theire Revolution which I have shewed you to be 25412 yeares. without which the[y] placed the primum Mobile whereby all were carried the Contrary way or from East to West round the Earth in 24 houres, the Necessary but improbable violence of this motion gave occasion to Copernicus first to suspect after wards totally to reject it. & substitute the recesse of ye Earths aequinoctiall pointes whereby the same appearances would be solved without any such strained & improbable Conception I have explained the Pythagorean susteme revived by him in a small tract lately printed[10] wherein I have given the Naturall Cause of this recesse as I conceaved it which worke being yet but in few hands I shall give y^u^ the reason I there assigne for it nearely in the printed wordes. premiseing first.

That I conceave the earth togeather with all the other planets to be carried round the sun in a large vortex each in theire stated Periods. of time & proportionall to theire distances from him.

That besides that Motion whereby the Earth is carried round ye sun sub fixis once in a yeare I suppose it to have another round its owne center once in 24 houres. on an Axis inclined 23½ from ye perpendicular to ye plane of ye Annuall orbe which keepes nearely[11] parallell to it selfe in every revolution

That all the space betwixt the sun & ♄ or beyond him far is filled with a very subtle aetheriall matter which serves to helpe to carry each planet round him in its orbe & to transmit ye light of the sun both to them & back from them every way.

That this matter is more dense or constipate neare the sun rarer & thinner at a distance which I conceave ye reason why light requires some time to passe. as the observed eclipses of ♃^s satellit[e]s prove it does & yet not more y^n one houre betwixt ye orbe of ♃ & the sun[12] as probably it must doe if there were no such interjacent matter whose motion might helpe to convey it for to me at least tis wholly inconceveable how light should be transmitted from that vast distance at which ♃ is placed from o^r earth. or any other motions made in ye narrow space of an hours time without ye helpe of some matter placed betwixt us by which I may be assisted in its transmission or in a manner handed to us. but I may discusse this perhaps more fully hereafter this hint may serve for my present purpose which is to shew the reason of the recesse of the Earths aequinoctiall pointes. to which the apparent motion of ye fixed stars is aequivalent & answerable by it.

In ye figure before y^u [Fig. 32.1] let ABCD. represent the Annuall orbe of ye earth in which shee is carried yearely round ye sun. ☉ his place neare the center of this orbite. ♈ ♋ ♎ ♑. the Earths place on ye 4 Cardinall pointes of the Ecliptick. that line discribed round the earth by the plane of ye Ecliptick encompasseing it or todi Call the Ecliptick itselfe e. its pole, P. the pole of the earth. & the line Pe. connecteing these two poles *the line of direction or solsti[ti]all Colure.* the sun illuminates but the one halfe of the earth. the line seperateing the illuminate from ye obscure part ted. I call *the horizon of ye diske.* the line of direction ye Pe. must be conceaved carried alwayes parallel to it selfe. & whose space not onely within the orbite to ye sun but outwards also filled with ye aethereall matter.

Veiw ye earth at ♎ whence ye sun is seene with stars of ♈. & the Vernall aequinox is celebrated. the Earths motion round her Axis is from West to East or from d̲ by i̲ to t̲ in ye illuminate p[ar]te of ye Globe. next the sun. but the Motion of the Ethereall Matter is also from West or from α to β directly contrary to this in the same illuminate p[ar]te of ye Globe & that too something stronger then in the remote or obscure p[ar]te by reason that the matter near ye sun is swiftlier moved then y^t farther distant as is evident by the revolutions of ♀ & ☿ round him p[er]formed in much lesse time then o^r earths, Hence it proceedes that the motion of rotation being Checkt the line of direction or plane Connecteing the poles tped. is reflected and borne back contrary to the Motion of ye Earth & Vortex 51″. yearely. so that whereas about ye begining of Nabonassars Æra it cut the Ecliptick & consequently made the solstice neare ye cloudy place in ♋ it is now carried back to ye heele of Castor & about 250 yeares hence the solstice will be made When the earth passes under Propus for the Solstices then happen when the Horizon of the diske becomes perpendicular to the line of Direction, the aequinoxes are then made when they coincede with each other.

Were it not for this reflection the Earths axis might be said to keepe all wayes its parallelisme to it selfe as well as the same constant inclination to ye plane of the Orbis Annuus. but this variation being so insensible as to make no more then 51″ p[er] Annum it is not regarded when onely the annuall & diurnall motions are discourst of.

However from this small Annuall recesse it is that the places of the aequinoxes being put back amongst the fixed stars they seeme to move forward slowly but equally from ye Colures or Cardinall points. about 5 degrees in 353 yeares as I before told y^u^. so that Spica ♍^s^ which in the time of Hipparchus was neare ye Autumnall aequinox is now found neare 20 deg. in Consequence of it Haveing thus shewed y^u^ in severall Lectures the Nature of the fixed stars theire light & substance why some are misseing in the heavens others haveing beene seene to shine for a time & then againe be extinguished haveing also intimated to y^u^ how o^r^ sunne in processe of time may come to lose his light o^r^ systeme to be ruined by it & by the way suggested something of the Nature & reason of Magnetisme. Haveing also examined the Ancient observations of most eminent Astronomers corrected them & by compareing them with later limited the Annuall motion apparent in them but really caused by the recesse of ye aequinoctiall points on o^r^ Earth & assigned a proba[b]le if not ye reall true Cause of it. I shall now leave this subject. & in my next lecture shew y^u^ to what accidents theire light transmitted through the Ether to us is subjected. the Nature & measure of refractions in o^r^ aire. & perhaps I may find occasion to divert by ye way to salve some appearances which however the Common people can give no account of yet they may probably be salved with facility upon due consideration of ye effects of Reflection & refraction. but of these God Willing Next lecture.

NOTES

34.1. These are summarized in Riccioli 1, pp. 440–2.

34.2. Ibid., p. 444.

34.3. Augustinus Riccius, *De motu octavae sphaerae* (Tridini, 1513). Isaac Israelita, *alias* Ishāk ibn Sulaimān (ninth/tenth centuries) wrote a work entitled 'De Fundamento Mundi', contained among his *Omnia Opera* ([Lyons,] 1515).

34.4. Isaac Hazan (cf. bottom of f. 308) or p. 418 alias Isaac ibn Sid (cf. note 11.12).

34.5. Probably a reference to Andreas Thurinus, the editor of Isaac Israeliti's posthumous works (cf. note 34.3).

34.6. Girolamo Fracastoro, *Homocentrica* (Venetiis, 1538).

34.7. An Egyptian year consisted of exactly 365 solar days. Its relation to the Julian year, stated below, is as 25816:25798.

34.8. Cf. note 1.7.

34.9. The modern accepted value, based upon a precessional constant of 50″·2 per annum, is 25,800 years.

34.10. Op. cit. (note 15.1).

34.11. The precessional effect is responsible for the departure from parallelism, as explained in Lecture 32 and stated explicitly on p. 423 (cf. Fig. 32.1).

34.12. Olaus Römer (1644–1710) was the first to appreciate that delays in the predicted times of eclipses of Jupiter's satellites could be accounted for on the assumption that light travels with a finite speed. In his 'Demonstration touchant le mouvement de la lumière', *Journal des Scavans, de l'An 1676*, iv (Amsterdam, 1677), 276–9, he claims that it required only 10 minutes to traverse the distance between Jupiter and the Sun. A report of this article appeared in *Phil. Trans.* xii, no. 136 (25 June 1677), 893–4.

Lecture 35

4 *June* 1684

In my last lecture of Easter terme[1] I began the History of Celestiall refractions whose nature & effects tho some have thought a discovery of the last Age I proved to y^{u} were well knowne & demonstrated by Alhayzen & Vitello, the two first & ancientest writers of Opticks but more especially to ye latter who affirmes that it might be found by ye Armillaes an instrument much used in those dayes, his Notions & knowledge seemes clearer then Bernard Walters who is the next wee find to mention them after him or betwixt him & Tycho who neverthelesse with Copernicus and all the Astronomers of that age inquired the latitude of theire habitations from theire observed least & greatest meridionall heights of the sun without any allowance for refractions or Consideration of it which rendred both theire obliquitys of the Ecliptick & latitudes of places sensibly faulty, nay that Tycho himselfe was as little as others acquainted with it when hee began his observations, for hee determined both the obliquity of the Ecliptick & his latitude after the same manner but liveing in a place more northerly then any of the eminent p[er]sons before him & haveing his Winter sun much more refracted then they, when hee came to compare his latitude thus gained with what his observations of the pole star gave it hee found a very sensible difference, this about the yeare 1583 hee attributes to refractions finding all his different & curious Instruments consented to give ye same heights & declinations of y^{e} ☉ & pole star both. His instruments were not all of them yet contrived & perhaps ye necessity of a better way of determineing ye quantitys of caelestiall refractions in all altitudes occasiond him to hasten the perfection of some of them for hee found that the Azimuths of the Luminarys & stars were difficultly taken by his quadrants & not very exactly. besides the Method of finding the true height by the observed azimuth & the refraction by compareing the observed height with it was no good one by reason his Azimuths were not exactly given by his Instruments. however hee grew cleare in his notions of refractions but I find not that hee attained the true measures of them till ye Summer Solstice of ye yeare 1585 or p[er]haps before ye Autumnall aequinox when hee

Notes on pages 432–3.

gives us in his historia Caelestis his observations made as hee says idoneo instrumento[. . .]ut constaret ratio refractionis radiorum.[2] with a proper instrum^t for finding the proportionall decrease of ye refraction of the raies of light as ye Luminary was removed from ye horizon. and from this time forward wee find him often observeing the suns visible declination as hee ascended from ye horizon in ye his riseing or drew neare it in his setting togeather with his visible heights & the true times of his observations. whence hee collects the refractions in every degree of height his Method I shewd y^u last lecture & with it I concluded. I shall now continue the history of his progresse & by the way I shall occasion given of discusseing the Nature of Refractions which will lead mee on to better methods for determineing theire quantitys then were knowne in o^r Noble observers dayes & more accurate determinations then it was possible for the knowledge of those times to attaine to.

In an Epistle of his to ye famous Lantgrave of Hesse dated ye 1st of March 1585/6 Wherein hee gives him an account of his endeavors hee asserts Refractiones insuper Radiorum solarium evitandae sunt.[3] i.e. the refractions of the suns raies which make his apparent heights greater then they ought to be and really are (by reason y^t ye Diaphanum of the aether is some little more transparent then the Diaphanum of the aire) must also be avoyded & allowed for especially if the observed heights be lesse then 30 or p[er]haps 40 degrees for by a peculiar contrivance hee had found that to these heights the sensible refractions reached but in lesse the Refractions were greater so the sun in his riseing & setting appeard at least halfe a degree higher then hee really was & that hee had found the same in ye moone planets & fixed stars. here it is remarkeable that hee affirmes the refractions of the sun sensible till he be elevated 30 or 40 degrees above ye horizon but scarce beyond that ye horizontall refraction is halfe a degree & the Cause of refractions is the difference of the Mediums of ye aether & the aire or of theire diaphaneity as hee calls it.

The Lantgraves answer is dated the 14 of April following 1586 wherein hee gives him an extraordinary & remarkeable observation of the effects of Refractions. which omitting his latin for brevity I shall give y^u in English[4] Once sayes hee I viewd ♀ setting & saw her stationary above ye horizon for almost a quarter of an houre in all which time shee appeared nothing carried forward by ye Motion of the Primum Mobile, shee appeared then more y^n 2 degrees above the horizon when shee was really as many beneath it, afterwards shee disappeared in a moment, which I had lookt upon as a miracle had not Vitellio in his opticks acquainted us with ye reason of.[5] I shall not examine this observation [of] Tycho out of respecte to ye Prince passes is as tis given but proceed to ye Answer of Christopher Rothmanne the Lantgraves stipendiary Mathematician. who by his

M[aste]rs order wrote a p[ar]ticular answer to Tychos letters[6] wherein hee tells him that hee could never find any sensible refraction in y^e^ sun when he was above 30 deg. high. nor that the horizontall refraction was $\frac{1}{2}$ a degree but commonly 15 or 20 minutes at most that y^e^ bright side of Perseus in 9° 42′ height was refracted $2\frac{1}{3}$ min. & y^e^ Swans tayle in 5° 18′ height $5'\frac{2}{3}$ these heights were the lowest of those stars in the North Meridian or under the pole. at Cassells where they were observed & from these it is evident that the lantgraves Mathematicians had no other instruments but theire quadrants for observeing refractions but in Rothmans next epistle[7] wee find them erroneous for hee confesses that the height of the pole was 1 min lesse then they had formerly esteemed it as hee found by examineing theire instruments over againe. & this allowed it will follow the refractions were two minutes greater then hee had esteemed them or if as later observations warrant wee allow a refraction in ye Pole star it selfe p[er]haps 4 minutes. which will make them agree better with Tychoes determinations.

But Tycho haveing attributed these refractions to ye different diaphenity of the aether & aire. Rothman[8] thinkes this can not be the onely cause of them. hee will have them made onely in ye grosse sphere of vapors encompasseing o^r^ earth. & argues that upon the other supposition Refractions must Not cease in ye altitude of 30 or 40 degrees, as Tycho makes them, but he continued up even to ye vertex. in which hee concluded aright but observed not that the argument held as well against himselfe. that the aire encompasseing the earth was of no great height, & might therefore be the Cause of these refractions hee argues from the Crepuscula begining long before sun rise & ending alike long after his setting, This argument Tycho could not evade hee acknowledges therefore that refractions may be continued up higher but that then they are so small that when ye appearances were elevated above 40 degrees or p[er]haps sooner they became insensible at least in his Instruments.[9] but haveing duely weighed the discourse on both sides I find the difference to consist rather in words & termes then otherwise. Tycho Calls the Matter that fills thc vast expansum of o^r^ vortex *Æther* that part of it filld with the exhalations of o^r^ earth & next it *Aire* Rothman will have the Matter y^t^ fills ye Wide vortex called *Aire* that p[ar]te of it which next encompasses o^r^ earth a sphere of vapors. that the refractions were made by the raies of light passeing from a more rare to a denser Medium both seeme to grant, both were alike ignorant of any rule or proportion whereby theire graduall decrease as the luminary arose from the horizon or decrease as it descended towards it might be determined except what was derived from simple experience. onely Rothman seemes to affirme that the Refractions were lesse at Cassells then Uraniburge, for no other reason that I can find but that hee had hastily determined so in ye begining of theire Correspondence some mens minds are of that weake constitution that they can not beare ye imputation of haveing changed theire once receaved opinions tho for better. Such seemes Rothmans temper to have bin, who tho hee might have satisfied himselfe by his later observations that the refractions were not lesse at Cassells then at Uraniburge, yet haveing once asserted that they were so hee shuns the application of such experiments as might convince him, & at last onely pretend that they must be lesse because Cassells standeing on ye continent, Could not but have a purer aire then Uraniburg y^{t} was compassed round with ye sea, whereas hee might have considered that probably the land exhalations and vapors are grosser then the sea, & then it will follow that his refractions at Cassells were greater y^{n} Tychoes at Uraniburg I have sometimes found them much greater in the riseing sun then his tables makes them, & I cannot impute it to any other cause.

Tychoes Empeiricall Table of Refractions so I call it because onely grounded & derived from observation & trialls was first published in his progymnasmata,[10] in ye begining of this century without the least mention of any Rule or law to which they were subject. or whereby the refraction in any one or more knowne heights being given the refraction in any other required height might be found out. hee thought it sufficient that hee had shewed what they were, posterity must take ye paines to find it & salve the appearance

Propitious heaven seemes to have provided & raised the subtle thoughtfull Kepler for this purpose who considdering that the Great Creator had order'd all appearances in the universe from proper & constant causes conceaved that these refractions must be subject to some law & sets himselfe to seeke it. His paralipomena ad Vitellionem seems purposely wrote to cleare ye businesse of refractions wherein haveing beate the bush & tried severall wayes to answer them hee punishes his reader for the paines hee has taken by leadeing him the same thorny way through his many & not allwayes needfull propositions. to his desired end. the Cheife of these I shall repeate. his 17 Prop. is *Plus difficultatis fuit in principijs constituendis.*[11] the Greatest difficulty is to find & fix on Principles, a great truth, in all sciences but not very materiall for they being found this might have beene omitted, the next & more necessary following is that The greater angles of incidence have the greater refractions. & that the raies of light passeing from a rare medium to a dense or thicker are alwayes refracted towards the Perpendicular[12] as if [in Fig. 35.1] $\underline{bk}$ were the superficies of Medium denser then the free aire. eb a ray of light passeing from y^{e} thinner medium of aire into this denser at b, $\underline{ba}$ y^{e} perpendicular which let be continued through it to $\underline{i}$ then is y^{e} angle abe y^{e} angle of incidence if y^{e} dense medium were not interposed the ray eb would proceede in a streight line towards $\bar{g}$. but by y^{e} occurse of y^{e} different medium. it is refracted or borne inwards towards y^e perpendicular bi & the angle which y^e refracted ray hb[13] formes with y^e way of y^e unrefracted ray bg is y^e angle of refraction gbh Now by how much greater y^e angle of incidence abe[14] is so much greater also is ye Angle of refraction gbh. this is the sense of his proposition whence hee deduces this Corollary that if the density of the medium onely were Considered ye angles of refraction would be proportionall to ye angles of Incidence.

And yet hee denyes that the Corpulence or density of ye Medium makes the refractions more or lesse for then the deepest medium should have the greatest refraction whereas it is found by experience that in ye same medium whether deeper or shallower the same angles of Incidence have the same refracted angles.

Wherefore in ye 2^d proposition of his 4th Chapter[15] hee assignes the cause of refractions to ye resistance the raye meetes with in its transit from one transparent Medium to another. & affirmes that by how much the Angles of Incidence are the More oblique so much the greater is the resistance of the Medium to ye ray of light falleing into it, the resistance hee proposes in the proportion of the secants of ye Angles of Inclination. for sayes hee [Fig. 35.2] let A be the light bc ye superficies of the denser Medium ab . km two raies of light comeing parallel to each other from ye same point of ye sun, whose distance from each other is lm seeing now y^t ye Angle blm is right & lbm an oblique acute angle, ye angle lbm shall be lesse y^n ye angle blm, & lm ye side opposite to ye lesser angle b lesse y^n ye side bm opposite to ye greater angle l but lm measures ye breadth of ye light bm of the Medium resisteing its incidence. Wherefore there is more density in bm then in lm & more resistance on this account

Hence his third Proposition followes, *that the Angles of refraction encrease proportionally faster then ye angles of incidence.*[16] for by the aforerecited corollary if the density were considered alone or separately by it selfe the Angles of refraction would be proportional to ye angles of Incidence, but by the last recited proposition the proportion of lm to mb mixes itselfe with them, Therefore ye angles of refraction are Composed of something that is proportionall to ye incidences & of something proportionall to ye lines bm, but ye lines bm encrease slowly in ye begining or neare ye perpendicular, in lower incidences faster, as sayes hee the table of secants shew. where as the angles encrease the secants still grow greater & greater Wherefore Addes hee part of the Angle of refraction is proportionable to ye incidences, part encreases swifter & therefore the whole faster as was asserted

Haveing thus resolved of the proportion where Refractions were increased accordeing to the Inclination of ye ray, the next thinge to be inquired was how the intire refraction was to be parte & how it might be knowne what p[ar]te of it was measured by ye inclination & what p[ar]te proportionall to ye secant of the Inclination. his method of Inquiry was all tentative by Analogies or Comparisons *Amo Analogias* sayes hee *fidelissimas meas Magistras.*[17] I love Compareing my faithfull mistresses & guides the reavelers of truth &c. hitherto they had helped him & Now hee makes a further step with them & not unprosperous.

Vitellio in the 10th booke of his opticks gives us a table of the refracted angle & refraction to every 10th degree of ye inclination of rayes of light passeing from aire to water,[18] the refraction of ye raies of light passeing from the aether to or aire hee supposes & that Certeinly to hold proportion or Analogy with these, & therefore proposes to him selfe this inquiry whether hee can find any Geometricall figure that may shew him these refracted angles to every degree of Inclination, those hee pitches upon are the Hyperbola & Ellipsis & hee tries, if haveing described such an one of each as that the angle of Inclination being formed on ye Axis by a ray proceedeing from one of its foci to some point in the Curve & the refracted Angle at ye other focus by a line drawne to the same pointe. whether afterwards another Angle of Inclination being formed at ye proper focus the line drawne from the Contrary focus to the pointe where ye line of Inclination cuts the Curve will give ye observed refracted Angle hee finds the experiment unsucces[s]ful but it will be worth the while to illustrate it with his, example, because it gave ye hint to de Chartes demonstration of ye Ellipticall & Hyperbolar glasses Collecteing ye rayes of light parallel to ye axis all into one point[19] & also that no more may be expected from these Curves then they really p[er]forme & Answer for.

In Vitellioes table of ye Refractions of Water the Inclination being 80 d. the refraction is 30 the Angle refracted 50 degrees. in ye 3d figure [Fig. 35.3][20] let ab be the two foci of an hyperbola. at one of them b draw ye line bc formeing ye Angle abc. 80 degrees and at ye other a the line ac formeing ye angle bac 50d. these meeteing in c. the Angle Acb shall be also 50 deg. & the sides ab bc equall wherefore if ab be 10000 bc shall be also 10000 & ac 12856. & δa the difference of the 2 sides bc and ac. shall be equall to da[21] the distance of ye vertices of ye two opposite sections or Hyperbola's which taken from ab 10000 leaves db[22] 7144. halfe of which be = ad 3572 shall be the distance of ye vertex of each Curve from its focus by ye 51 of ye 3 of Apollonius,[23] which being given the hyperbola it selfe is easily described. admitting it done hee proceedes to trie if any other of Vitellios observations will be represented by the same Curve as hee thinkes they ought to be.

In ye Inclination of 70d. Vitellios Angle of Refraction is 24.30 therefore the refracted angle 45°.30′ Makeing therefore the angle Abf 70d. & supposeing the side ab 10000 as before hee seekes what the angle afb[24] which ye line af intersecting the former where it cuts the Curve shall be by ye rule of false position,[25] hee knew no better method being ignorant or rather haveing forgot one affection of this Curve I shall shew a true Geometricall method & then tell y^u his conclusion.

Produce the line fb to g makeing gb = da[26] y^e distance of y^e vertices of y^e opposite sections, & draw y^e line ag then because δa = bg[27] by y^e knowne property of the Curve the line fg[28] shall be = to fa,[29] & y^e ∠fga = ∠fag.[30] and in y^e △abg[31] are knowne the side ab[32] y^e distance of the foci 10000 bg = δa = da[33] 2856 with y^e angle comprehended abg.[34] 110.00 whence y^e agf = gaf[35] will be found 56°.15′ & y^e angle gab[36] 13°.44′ which taken from it leaves y^e angle fab 42°. 31′ but this if the Curve had answered the refractions as was expected should have beene 45.30 whence it is evident that the hyperbola is not the Curve that will answer the measures of refractions but that whatever it is it lies without this & is more convex, a like triall he makes with the Ellipsis but with the same small successe. so that hee concludes that the measures of Refractions are not to be found either at ye foci of the hyperbola or Ellipsis hereafter I shall shew the locus proper for them but it will be convenient first in o^r progresse to shew you his method of Calculateing y^e table of refractions & y^e reason of it which with Gods assistance shall be the worke of my next lecture.

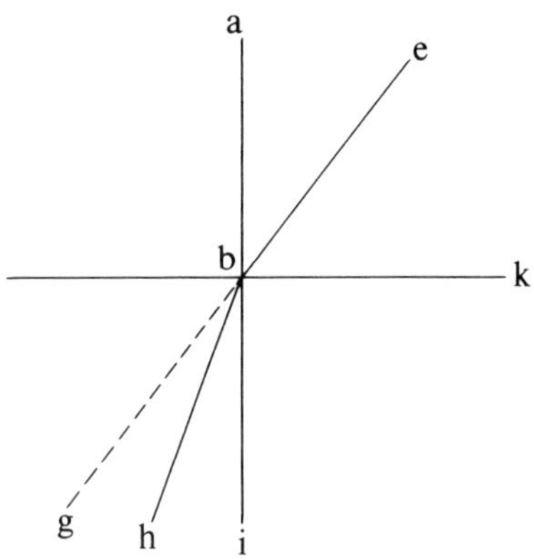

FIG. 35.1

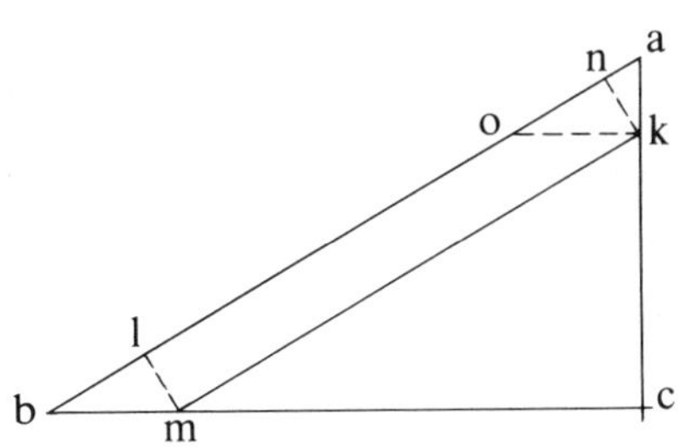

FIG. 35.2

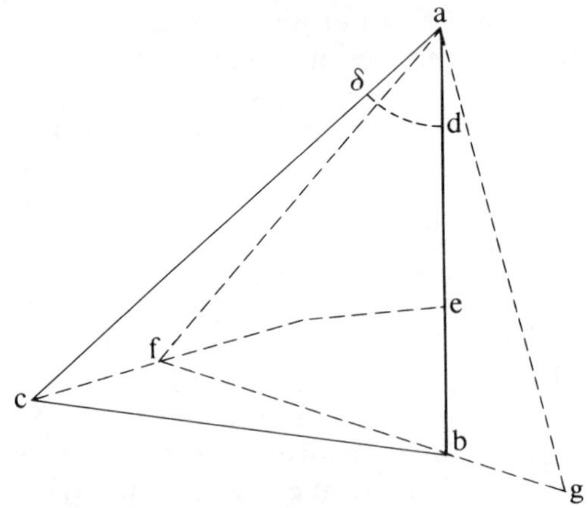

FIG. 35.3

NOTES

35.1. It is clear from Flamsteed's remarks that he is referring to Lecture 33 (2 April 1684) and *not* to Lecture 34 dated two weeks later. It would therefore appear that, in order of their delivery, Lecture 34 preceded Lecture 33.

35.2. Op. cit. (note 2.36), p. 111.

35.3. Op. cit. (note 29.13), pp. 5–12. For the sake of brevity, Flamsteed is omitting part of the Latin text (which he quotes from ibid., p. 10) between the words 'solarium' and 'evitandae', but he gives the English translation of the full sentence.

35.4. Ibid., pp. 21–23. The sentence which follows is published in both Latin and German in ibid., p. 22.

35.5. The significance of this remark is that the Landgrave appreciated the fact that the phenomenon in question was an astronomical application of the principle of refraction discussed at length in Vitello's work (op. cit. [note 33.6], pp. 403–74).

35.6. Op. cit. (note 29.13), pp. 27–31. Rothmann's reply to Tycho's letter of 1 March actually bears the date '18 Calend. Maij, Anno Christi 1586'.

35.7. Ibid., pp. 33–34. This letter is dated '7 Calend, Septembris, Anno 86'; that is, 26 August 1586.

35.8. Flamsteed now cites the gist of the argument developed in two later letters from Rothmann to Tycho, dated 'II Calend. Octob.' (or 30 September) and '14 Calend. Novemb.' (or 19 October) 1587. Cf. pp. 81–91 and 91–93 respectively.

35.9. The whole problem is discussed by Tycho in his letter to Rothmann of 17 August 1588, ibid., pp. 105–19.

35.10. Op. cit. (note 2.5) (1), p. 79.

35.11. Op. cit. (note 1.25), p. 68.

35.12. Cf. ibid., p. 110 (prop. 1) and p. 113 (prop. 7).

35.13. Flamsteed wrote gb instead of hb.

35.14. Flamsteed wrote acb instead of abe.

35.15. Op. cit. (note 1.25), p. 110. 'Cum lux obliquius incidit, maior fit resistentia ab eodem medio, quam in rectiori incidentia, respectu ipsius etiam medii.'

35.16. Ibid., p. 111. 'Refractionum anguli crescunt maioribus rationum incrementis, quam obliquitas incidentiae.'

35.17. Ibid., p. 95. 'amo analogias, fidelissimos meos magistros.'

35.18. Cf. note 33.6.

35.19. Descartes, op. cit. (note 4.4), pp. 89–121 ('Discours Huictiesme').

35.20. The notation, and values for the angles, are as given in op. cit. (note 1.25), p. 97; the only difference being that Flamsteed rounds off these figures to four decimal places.

35.21. Flamsteed wrote de instead of da.

35.22. Flamsteed wrote δb instead of db.

35.23. Apollonius's Conics (op. cit. [note 11.8]). The reference to ibid., lib. 3, prop. 51 was given by Kepler in op. cit. (note 1.25), p. 95.

35.24. Flamsteed wrote abf instead of afb.

35.25. Op. cit. (note 1.25), pp. 100–1.

35.26. Flamsteed wrote gf = de instead of gb = da.

35.27. Flamsteed wrote fg instead of bg.

35.28. Flamsteed wrote bg instead of fg.

35.29. Flamsteed wrote ba instead of fa.

35.30. Flamsteed wrote ∠bga = ∠bag instead of ∠fga = ∠fag.

35.31. Flamsteed wrote △afg instead of △abg.

35.32. Flamsteed wrote af instead of ab.

35.33. Flamsteed wrote fg = δa = dc instead of bg = δa = da.

35.34. Flamsteed wrote afg instead of abg.

35.35. Flamsteed wrote gad instead of gaf.

35.36. Flamsteed wrote gaf instead of gab.

Lecture 36

18 *June* 1684

In my last lecture I shewed y^u that the Angles of Incidence & refraction were not to be found at the foci of an ellipsis thence I proceeded to Keplers method of finding y^e refractions in any given height where in y^e prime requisite was y^e height of the atmosphere which from his data[1] I collected 2 English miles & $\frac{1}{3}$: but this not agreeing either with y^e height found by the Crepuscula according to y^e method of the ancients nor the later exper[i]ment of Torricelli,[2] and being the maine postulatum in y^e calculation of caelestiall refractions I thought worthy a more serious disquisition especially considering that the Ancients Alhayzen & Vitellio who shew ye method of finding the height of ye aire by ye observed duration of the Crepuscula or twilight omit all consideration of refractions tho they naturally intrude themselves & would have necessary followed from theire owne notions closely prosecuted. here therefore I shall shew y^u that a much lesse height of the aire then theire probleme resolved gives will serve to answer ye Crepuscula afterwards I shall proceede to Consider the height given by the Torricellian Experiment, & then to the later determinations of o^r Contemporarys & the better grounded rulles of refractions derived from the old & later observations.

The Ancients from theire observations of the duration of ye twilight after sun set concluded the height of ye Atmosphere, or the greatest distance of ye grosse aire encompasseing our earth from the surface to be about 50 miles as I proved to y^u in my last lecture from theire crepuscular probleme which I explained and solved as they did without takeing notice of any fault committed by them in the application of it. but now I shall examine & shew y^u what was wanteing in theire considerations, which will prove my first assertion that a far lesse then 50 miles will answer very well to all the appearances of ye Crepuscula or Twilight

The fault then that they committed was this, they allowed the earth to have a sphere of vapors. & they asserted that when the raies of light passed from the rarer medium of ye aether into this denser of

Notes on pages 440–1.

or Atmosphere that they ought to be & necessaryly were refracted towards the perpendicular falleing from ye earths center on ye points of theire Incidence this both Alhayzen & Vitellio assert. & this latter shewes a way for Measureing the Angles of refractions & had a clearer knowledge of them then Bernard Walter who lived at least 200 yeares after him. yet Neither of them consider what must be the effect of Refractions or how they had theire effect in formeing ye Crepuscula though theire figures the same with my 4[3] doe naturally suggest it. for if ☉ g a ray of light fall into ye atmosphere at g, it ought not to nor can proceede from thence in a streight line to e from whence it should be reflected to ye eye of ye observer in b. but on ye contrary it ought & really is accordeing to theire owne rules refracted towards ye perpendicular at its very entrance in ḡ & bending below the line fg. strikes upon ye earth suppose at H.[4] & never can reach the point e to which they suppose it to be directed.

But if ye ray falls into ye Atmosphere at some point above ḡ at k[5] γ the said raye may be so refracted that it may touch the superficies of ye atmosphere nearer ye observer then the point e & may consequently be thence reflected back or refracted so as to fall in his view. but here againe wee distinguish not betwixt the parts of ye Atmosphere which wee are not to suppose every where of an equall density but rather grosser & more constipated neare or earth rarer & thininer as it is more removed from its surface for which strong & irrefragable arguments may be drawne from or sensible vapors fogges & mists ye small height of ye cloudes, ye variablenesse of the observed horizontall refractions and the Aurora borealis or great lights seene in ye Night time when ye sun has beene too far depressed beneath or horizon to illuminate ye Aire. from each of these I shall fetch arguments to prove the unaequall density of or Atmosphere not but that I beleive every ingenuous & thoughtful p[er]son well satisfied of it but that I shall thereby prevent some mistakes yn might easily run upon if I did not touch upon each, & thereby cleare the way to ye proofe of my assertion that ye raies of light came to or eyes through the Atmosphere in curved lines. it may seeme strange to those that are not well acquainted with ye heavens, that I should thus engage my selfe in philosophicall considerations where they expect notions abstract from Matter & pure Mathematicks, but the Astronomers businesse is cleare[ly] different he cannot conclude without a deepe insight in Nature & Philosophy. Nor can any one be a good Philosopher (I affirme it positively) except he bee also well versed in Mathematicks & a good Geometrician. The learned & subtle Kepler layes the foundations of his excellent Theorys & accurate Tables[6] in philosophicall considerations. & the great wits of Galileo & De Chart had never improved Philosophy so much if they had not beene excellent Geometricians, The weight of ye Aire is one of ye most important discoverys of this age & that wee owe to Torricelli[7] an excellent Mathematician but I am not to give y^{u} the history of such improvem^{t} as have beene made by Mathematicians in Naturall knowledge I beg y^{r} pardon for this digression & returne to my enquiry concerneing the state of o^{r} aire or ye Composition of o^{r} Atmosphere from which I had deviated.

The aire that envests o^{r} earth seemes to be a Compound of severall sort of particles Watry nitrous & sulphureous saline p[er]haps too some terrestriall & to be formed by continuall emanations from it & as it spends it selfe to nourish us & the plants & animals upon it so to reaceave fresh recruits by continuall fresh p[er]spirations Now these particles if the[y] be grosse cannot be carried up high into the aire but are moved to & fro neare its surface; while the finer & more spirituous rise higher. but have still a propensity to fall back upon y^{r} Earth hence it is that the grossest partes are nearest the Earth & the crassitude continually decreases as ye ascent is higher which also the appearances of the Clouds confirme of which When y^{u} see two sorts at once in ye aire the grossest darkest & thickest lie alwayes lowest & under the more refined & thinner & clearer whose colour shew they have lesse grosse matter in them & therefore swim higher from ye earth in ye Liquid of ye aire. Now this different density of o^{r} aire as it ascends from the Earth is not Nor can be the same in all places for in marish grounds or places that emit plentifull vapers the aire next the Earth will be more constipated then in drie places that emit but fire & againe in such places the aire will have different constitutions according to ye season of ye yeare or time of ye day. the Morneing aire is generally more thick & misty then the eveining the vernall & autumnall. then ye solstitiall. & therefore the refractions of ye severall seasons of the yeare & times of the day must be something different according to ye different density of ye aire:[8] that they are I shall make good by experiment by & by at present I proceed a little further.

The reason why heavy bodyes cast up into the Aire descend & the weightiest fastest[9] seemes to be & most probably is because the earth attracts all bodyes consisteing of ye same p[ar]ticles that compose it, how this is p[er]formed I have hinted formerly[10] at present it will follow from this consideration that the height of ye atmosphere should be always the same in all places of the earth where this vertue of attraction is the same, & equall; & why it should be more in one place then another I apprehend Not the greatest depths of valleyes on o^{r} earth the heights of the hills above the surface of ye seas or the depeest places in them & ye commonly esteemed height of ye aire beareing but a very small proportion to ye semidiameter of ye earth. but Now the attractive power of o^{r} earth seemes to extend itselfe to or beyond ye lunar orbe[11] & wee have no need to suppose o^{r} aire or sphere of vapors so high; the heights of ye cloudes are uncerteine & no doubt but its limits are beyond them wee must therefore seeke for a limit elsewhere wee have gained nothing hence but my latter assertion that y^{e} density of y^{e} aire is unequall greatest neare y^{e} Earth lesse upwards. as y^{e} attractive vertue is strongest neare y^{e} Earth feebler upwards or at a distance from it.

The Weight of y^{e} aire suspends the quicksilver in y^{e} Toricellian Experiment to y^{e} height of neare $29\frac{1}{2}$ inches, and it has beene found by frequent experiment that if y^{e} tube be carried higher into y^{e} aire from y^{e} earth the quicksilver descendes proportionably to y^{e} ascent.[12] whence some ingenuous persons have conceaved the height of o^{r} aire might be collected.[13] I shall give you one of the most remarkeable & accurate experiments that I have met with of. In y^{e} yeare 1650 on y^{e} 26th of January Gassendus[14] tried the Torricellian experiment on y^{e} top of some hills neare Dignea in France. the height of the hill from y^{e} Level was 490 yards English Measure y^{e} height of y^{e} Quicksilver at y^{e} bottom or level $29\frac{44}{100}$ at y^{e} top it descended to $27\frac{82}{100}$ so that in 490 yards ascent the quicksilver descended $1\frac{62}{100}$. Now if with those who make use of this experiment to find y^{e} height of y^{e} aire wee suppose y^{e} pillar of quicksilver in y^{e} tube suspended by a pillar of aire whose density is in all partes the same, wee may hence collect its length or height very easily. for as 1./62 inche y^{e} descent of ye quicksilver, is to 490 yardes ye ascent of ye hill from ye level into ye aire so is 29/44 inches the height of y^{e} ☿ in ye tube on ye level to 8904. yardes the ascent in ye aire at which the quicksilver would totally descent out of ye tube into ye vessel. which is equall to 5 English miles & 104 yards. to which height & no further ye Atmosphere or aire ponderateing against ye pillar of quicksilver ought to be extended from o^{r} earth. but since as I tould y^{u} before the aire neare o^{r} earth is crowded with grosse vapors & exhalations. and as in the Magnet its vertue is strongest nearest it feebler at a distance. So it cannot be doubted but that the grossest parts are by its attraction retained at a small distance from o^{r} globe whiles the more rare & spirituous if I may use that terme are by a necessary repercussion of parts from its surface & the vis centrifuga[15] forced to a great distance. So that ye Torricelian experiment onely shewes us how high a pillar of aire of equall diameter & weight with that in the tube & of equall density in all its parts would be not what it really is because ye pillar of the Atmosp[h]ere is unequally dense more neare the earth lesse at a distance from it, were there any where a large & convenient ascent of 2 or 3 miles & could humane nature live in it which I much question I doubt not but that in the first 5 hundred yards the quicksilver in this experiment would be found to descend more y^{n} in ye next or 2^{d}, 500 & in the 3 & 4th still lesse but since this experiment cannot be made wee must sit downe content with ye suggestions of o^{r} owne reason & such conclusions as are deducible from that part of it which has beene tried whence wee can onely collect that the least possible height of ye Atmosphere is above

5 miles but that probably it is very much more tho scarce possible to determine how much:

For by ye crepuscular probleme wee found the height of the Atmosphere 49 miles but now I shall shew yu an error in that processe of Alhazen & Vitellio & that a grosse one too & contrary to theire owne receaved principles. whereby it will be evident that a much lesse height of ye aire will serve to answer ye appearance of twilight but that still how much lesse can not be determined.

I returne now to ye 4th figure of my last lecture [Fig. 37.4]. in which bcf represented the sup[er]ficies of ye earth deḡ of ye Atmosphere ā ye common center of both ⊙ ḡ a ray of light proceedeing from ye sun when 18 deg. beneath the horizon this they supposed to proceed through the Atmosphere in a streight [line] till touching its superficies again in e it was thence reflected to ye eye the Angle aeb = aef being ye complemt of $\frac{1}{2}$ ye depression to a Quadrant which being given with ye side ba = ac[16] ye earths semidiameter. the angle eba being right. the side ae was easily found & consequently ce 49 miles as I then determined but here they[17] had forgot theire owne dioptricall principle *that the rayes of light comeing from a rare medium into a dense would be refracted towards* ye *perpendicular* & consequently ye ray ⊙ ḡ should not proceed in ye streight line ⊙ g continued close by ye superficies of ye earth in f to e but upon its first entrance into ye Atmosphere at ḡ should be refracted toward ye perpendicular āḡ, and passeing under ye line ḡē should strike upon ye earth at h so yt it could never possibly reach the eye of ye observer at b whence it will follow that theire probleme was of no use nor determined any thing more then that the height of ye aire they deduced from it was p[er]haps the greatest that could reasonably be pitcht upon. but that a lesse may serve will not be difficult to prove if wee consider what I have often intimated that the extreme parts of ye atmosphere are purer & refract lesse yn the interior, wherefore ye raies of light entring into it with large inclinations are little refracted at theire first entrance. more as they are deeper [i]mmerged & proceed not in streight but curve lines through or aire contrary to what the Ancients imagined & some unthinkeing people of our owne age will allow.

And that they are really so carried will be made evident when wee come to give an account of that strang[e] appearance of light in ye North part of ye heavens which Gassendus & from him or Moderne Latins call ye Aurora borealis.[18] which is rare in England but frequent in Scotland Sweden Denmarke & ye Northerne Countryes nearer ye pole. but never seene in ye south which shewes it proceed not from any lucid substance slideing to and fro in the aire, but from ye very raies of ye sun carried through ye grosse Atmosphere of ye Polar part of ye World into theire horizon where meeteing with pellaicid vapors such as constitude raine & mists (for in other weather it appeares not) they come to ye eye some reflected but most by refraction. the account of these I may p[er]haps give y^u^ if heaven afford me further opportunitys, when I shall also confirme my assertion by an account of ye inaequality of ye horizontall refractions from which I shall proceede to a Theory of Refraction from which I shall give a New table of them & so conclude this subject, for which May heaven p^r^serve us all

Amen

NOTES

36.1. Cf. notes 37.2 and 37.9.

36.2. An interesting article by W. E. Knowles Middleton on the first known experiments on air pressure is contained in *Isis*, liv (1963), 11–28. It is suggested there that Torricelli's discovery had little to do with his skill as a mathematician.

36.3. Figs. 37.1–37.4.

36.4. Flamsteed wrote i instead of H.

36.5. Flamsteed wrote γ instead of k.

36.6. Flamsteed is probably thinking of Kepler's planetary theories and tables as well as his optical theories and tables of refraction.

36.7. Cf. note 36.2; also W. E. Knowles Middleton, *The History of the Barometer* (Baltimore, 1964).

36.8. This was Flamsteed's interpretation of observations of refractions made in June 1678, the results of which he later communicated to Newton in a letter dated 31 December 1694. Cf. J. F. Scott (ed.), *The Correspondence of Isaac Newton*, iv (Cambridge, 1967), p. 64.

36.9. It should be noted that this statement, based upon practical experience, is not necessarily in conflict with Galileo's demonstration that all terrestrial bodies fall with the same acceleration *in vacuo*. Flamsteed's commitment to his viewpoint stemmed from his belief in the magnetic origin of attraction.

36.10. This is probably a reference to his use of the postulate of a magnetic force of attraction to account for the phenomena of precession (Lecture 34) and cometary motion (Lecture 3).

36.11. John Wallis, in his theory of the tides published in *Phil. Trans.* i, no. 16 (6 August 1666), 280, had drawn attention to the fact that such a belief is consistent with an observed anomaly in the Moon's motion of which Horrox (op. cit. [note 1.26]) had already taken account when representing that motion geometrically.

36.12. This famous experiment was first carried out at Descartes' suggestion by Blaise Pascal's brother-in-law on the Puy de Dôme on 19 September 1648. The results were initially published in Paris that same year in a tract entitled 'Recit de la grande expérience de l'équilibre des liqueurs'. Flamsteed obtained Richard Towneley's co-operation in repeating the experiment in the localities of their respective homes near the end of 1673, and verified his belief that equal drops in the height of a mercury column need not correspond to equal differences in altitude. Cf. Flamsteed to Towneley; Derby, 25 November 1673, R.S. MS. F.1.2.

36.13. E.g. Hooke, op. cit. (note 4.16), p. 228 induces from the results of his barometric experiments and Boyle's Law that the atmosphere 'may extend it self . . . perhaps indefinitely, or beyond the Moon'.

36.14. Gassendi's Torricellian experiments are described in his letter from Digne to his assistant Bernier, dated 6 August 1652, in op. cit. (note 13.15) (vi), pp. 317–19; but it is difficult to say whether this was the actual source of Flamsteed's knowledge of them.

36.15. J. W. Herivel, 'Newton's discovery of the law of centrifugal force', *Isis*, lii (1961), 410–16. Christian Huygens, *Horologium Oscillatorium* (Parisiis, 1673), derived the same formula independently of Newton and applied it to pendulum clocks. Since Newton had not published his discovery at this time, Flamsteed may have been indebted to Huygens (whose book he possessed) for his knowledge of this important physical concept.

36.16. Flamsteed wrote bc instead of ac.

36.17. That is, Alhazen and Vitello. Cf. notes 33.2 and 33.6.

36.18. This phenomenon is described in Gassendi's 'Syntagma Philosophica' (pars 2, liber 2, caput 7). Cf. op. cit. (note 3.15) (ii), pp. 107–11.

Lecture 37

11 *June* 1684

In my last lecture[1] I informed y^u of Keplers notions concerneing refractions who was the first person that I can learne endeavored to bring them to rule His considerations are perplexed & tedious but haveing accordeing to my usuall method begun theire history I could not avoyd the takeing them in I must make the same plea for my excuse that hee did for his, y^u must be content to be lead the same rough & thorny way I have trod to attaine the same pleasant enjoyment the knowledge of the same measures & proportions of refraction

Keplers first attempt Was to trie whether or no such an hyperbola might being described that the Angle of incidence be formed at ye Axis by a line drawne from one of the foci to ye curve, another line drawne from the other focus to the same point in the hyperbola would forme the refracted angle at ye Axis: his method was tentative & unworthy of a Geometrician, I shewd y^u a direct Geometricall method, which to mention it by the Way also Answered this probleme. one side of a triangle with ye difference of the other two sides & one of ye angles adjacent to ye knowne side being given to find ye other angle, this I did both Geometrically & by Calculation whereby it was evident that the angle at the other focus was lesse then the refracted angle & therefore measured not the refractions but that the point lay without ye Curve & probably in a line more bent then the Hyperbola. Kepler had Concluded that ye Measures of refractions were not to be found betwixt the foci of this figure. & therefore tries next if they might not be placed at the Foci of an Ellipsis, for hee was resolved to trie all concludeing that if the Curve that would Answer them were once discovered, the method of Calculateing the Angles of refraction to any angle of Incidence could not long lie hid, nor the true Causes of them. Proceedeing in his method hee proposes it thus [Fig. 37.1].[2] let ab. be the distance of ye 2 foci of an Ellipsis the Angle pbf the incidence upon water 80 degrees make the angle at ye other focus baf 50°.00′ which is the refracted angle to that incidence in water. in the triangle abf both ye angles are given wth ye side ab which hee proposes

Notes on page 450.

10000. whence hee finds af 19796. & bf 15301. whose summe 35117 according to ye nature & property of ye Ellipsis is allwayes equall to its transverse diameter. this being got hee proceedes next to trie whether if the Angle pbc be proposed 70 deg. the angle bac formed by ye line ac drawne to ye point c in ye Curve will be equall to 45°- 30′ the refracted angle to that Inclination in Vitellio's table[3] & workes againe tentatively whereas there is a direct Geometricall method for resolveing the Case thus [Fig. 37.2]. there is one side of a triangle abc given with ye summe of the other two sides & an angle adjacent to ye knowne side, the other two sides required to effect it geometrically.

At y^e^ side given ab produced and on y^e^ point b draw y^e^ line bd formeing the Angle pbd[4] 70 degr. make bd equall to the Transverse diameter of the Elipsis & from the other focus a draw y^e^ line ad. from y^e^ same focus a draw allso the line ac formeing the angle d̲a̲c̲ = to adc. the point c̲ where it intersects the line bd is the point in the Curve of y^e^ ellipsis, & the Angle c̲a̲b̲ shall be the refracted angle if the Ellipsis be the Curve that measures refractions & now to worke the same Arithmetically in the triangle abd are given the side ab. 10000, bd 35117. with ye angle included abd. the complement of pbc 110 deg. whence the angle bad will be found 56°-18′. and adb[5] 13°-42′ whose difference 42°-36′ is y^e^ ∠c̲a̲b̲. for ye point c̲ is in ye center of a Circle that passes through d̲ and a̲ but this angle is lesse then 45°.30′ whence it is evident that the Measures of refractions are not to be found in ye ellipsis but in some wider or lesse bent curve.

When in the last lecture[6] I examined the Hyperbola I found the Curve that would represent refractions was more bent then it, here I find tis lesse bent then the Ellipsis, the locus therefore for representeing them must lie in a figure betwixt these which seemes probably & argueing accordeing to Keplers method by Analogies should be a Circle that it is such I shall prove hereafter at present I proceede to Keplers method of Calculateing his Table of Refractions

The cause of refractions is attributed by him and all since him to ye sphere of grosser vapors encompasseing o^r^ earth, whereby the raies of light comeing from the rarer medium of the aether & falling into this atmosphere, a denser, are bent in theire passage through it towards the perpendiculars, the ratio or proportionall increase as ye luminary descends or increase as it ascends from the Horizon Kepler partes & supposes distinguished proportionably to the Angle of Inclination & the secant of it for which y^u^ may remember I shewed y^u^ his reason in my last lecture.[6] tis no cogent one I confesse but since it helpes to answer the appearance it must be allowed at present tho' I hope to make it appeare hereafter there is little need for it when wee shall have found the true proportion whereby any one angle of Inclination & refraction answering it being given the refractions to any other inclination in the same medium may be determined.

And now the first thing he next requires is the height of this sphere of vapors above the superficies of o^{r} earth. which hee seekes after his usually method by Analogies or comparisons as follows[7] and indeed hee had no other.

First hee gives us 9 observations of the suns visible declinations made by Tycho at Uraniburge in ye yeare 1586. Jan 16. afternoone whence hee deduces his true refraction when 3°.50′ high 14′.22″ but when hee was onely 0^{d}.35′ above ye horizon 31′.10″.

now sayes hee in ye 3^{d}. figure [Fig. 37.3] let a represent ye center of ye earth bc its periphery de ye superficies of the atmosphere. whose height must be first inquired in order to determine the true angle of incidence of the raies of light upon it to every degree of the apparent height of the radiateing body for accordeing as this is proposed more or lesse so much greater or smaller will this inclination be & consequently the refractions will be alterd proportionably swifter or slower. that is the higher the atmosphere db or ce is proposed, the greater will the angle of inclination be & the lower ye lesse.

To ye refraction 31.′ 10″ in 0°.35′ of altitude let y^{e} angle bea be proposed 87°.30′ or geh 2°.30′. let fe be the unrefracted ray in y^{e} pure aether subtracting y^{e} refraction gef 31′.10″. there remaines feh 1°.58′.50″ or its complement leg 88°. 01′.10″ then hee argues. as y^{e} secant of y^{e} Angle bea = gel = 87°.30′: is to y^{e} Radius : : so is y^{e} refraction 31′.10″ or 1870″ to 82″$\frac{1}{2}$ that part of it due to y^{e} angle of Inclination which divided into 88 p[ar]tes gives the simple refraction to one degree of inclination 56″$\frac{1}{2}$. And because it is : s∠bea; ba : : s ∠eba : ea. therefore ea shall be 10009 such partes as ba is 10000

Againe since in 3°.50′ height the refraction was found 14′.22″ let the Angle eba be 3°.50′ more y^{n} 90 deg : ea: s∠eba : : ba : s ∠bea = gel which will therefore be 86°.59′ halfe a degree lesse y^{n} the former, to which addeing the angle of refraction gef 14′-22″ the ∠lef is made 87°.14′. three quarters of a degree lesse then y^{e} former which requires 42‴ decrease of the p[ar]te of the refraction proportioned to y^{e} Inclination this therefore will now be 1′.22″. then hee argues againe as the Radius to ye secant of ye Inclination 86°.59′ : : so is 1′.22″ (or 82″). to 25′.58″. the whole refraction. but this is above 11′ minutes more y^{n} it was observed. therefore this position of the ∠bea 87°. 30′ is false, the decrease of the secants being too small, it must therefore be proposed greater that the secants & consequently the refractions Measured by them may diminish swifter.

The angle eba being 35′ above 90 degrees let the angle bea = gel be proposed 89°. so y^{e} inclination of y^{e} unrefracted ray lef will be 89°.31′.10″ Now as y^{e} secant of 89 is to y^{e} Radius so 31′-10″ or 1870″ to 32$\frac{1}{20}$″ the simple refraction answerable to y^{e} Inclination, and as 89^{d}: 32″$\frac{1}{20}$: : 1°: 22‴ y^{e} part of the simple refraction answering 1 deg of Inclination, s ∠bea. 89°.00: ba 10000 : : s ∠eba 89°. 00′ : ea 10001 Againe if y^{e} ∠eba be 3°.50′ more y^{n} 90 deg. bea will be found 86°.05′ and addeing y^{e} refraction gef 14′.22″ to it the ∠^{e}fel[8] 86°.20′ differing from y^e former 2°.40′, & so many times 22‴ taken from 32″½ leaves 31″½. but as ye R: to ye secant of 86°.05′ : : so is 31″½ to 7′.40″ to ye refraction. which ought to be 14′.22″

Therefore this position of ye angle bea 89^d. is also false the secants differ too much, & wee are now stepd beyond ye truth as much as we were short of it before, let a triall be made betwixt them. & the said angle bea be proposed 88 degrees.

Then will the side ea be found 10006. the simple refraction or that part of the whole assigned by him to ye Inclination will be 1′.05″ divided by ye Inclination 88½ gives the refraction proportioned to 1 deg onely 44‴. Next supposeing y^e ∠eba 93°.50′ bea will be found 85°.43′ to which inclination the simple proportionall p[ar]te of ye refraction is therefore 63″-10‴. then R: sec. 85°.43′ : : 63″.10‴ : 14′.49″: agreeing very nearely with the observation. so that now hee has hit the marke neare enough. & concludes that hee may fix on the Angle bec which the horizontall ray makes with the p[er]pendicular from ye earths center to ye superficies of ye Atmosphere 88.00 but in his Table of Refractions simple & compound in ye very next page of his paralipomena[9] he makes it 88°.05′ not Materially different.

After this Manner Kepler teaches us to find the height of the aire & the true quantity of refractions in all altitudes togeather in his aforementioned Paralipomena ad Vitellionem. pag 123 & 214[10] his method is wholly tentative. & is againe repreated by him in his Epitome Astronomiae Copernicae pag 66.[11] something differently but more breifly & lesse complex the ration of his worke & the proportions hee seemes to have concealed whether studiously and on set purpose or not I doe not enquire however it bee I shall give y^u it Clearely & breifly

First hee supposes = ce = bd, 6 such partes as the semidiameter of y^e earth ac[12] or ab is 100000 whence it followes that the inclination of the horizontall ray bḡ or y^e angle bea = leg is 88°-01′ this being given to find y^e angle of inclination of any other ray whose height is proposed above y^e horizontall

let y^e angle kbg which y^e horizontall makes with another be 3 degrees y^e angle kba is 90 degrees more or 93.00 it will hold as y^e Radius or sine of y^e angle eba[13] 90.00 to ye sine of ye angle bea. 88°.01′ : : so is ye sine of ye angle kba 93°.00′: to ye sine of the angle bka. 86°. 24′¼

Now because the p[ar]tes of the refraction due to ye angle of Inclination are proportionable to it haveing the horizontall refraction further given hee argues

As ye secant of y^e horizontall Inclination bea multiplyed by y^e angle it selfe.

: is to y^e secant of any other Inclination multiplyed by its ownē Angle

: : So is y^e horizontall refraction

: to y^e refraction agreeable to y^e angle proposed

Thus the angle of the luminarys elevation kbe being proposed $3^{gr}. 0'$. the refraction in that height will be found $16'. 15''$. the horizontall being $30'$. had the horizontall beene p^r^oposed $34'$ as Tycho makes it the refraction in that height had beene found $18'\frac{1}{2}$. something bigger

Tho in reason I thinke it ought to be rather lesse the Cause I shall shew y^u^ hereafter & observations y^t^ will warrant it

This by trial what will answer the refractions observed by Tycho Kepler gropes out a rule for calculateing them wherein it is observable

that if y^e^ height of the Aire be altered the Angles of refraction & consequently theire proportionall decrease from ye horizon upwards is changed, for if the height of the Aire ce be proposed lesse then hee makes it the secants of ye Angles bea and bka differ faster then they ought being neare the extremity of ye Quadrant & consequently ye refractions but if ce be proposed larger then the said Angles are lesse, the secants & consequently refractions measured by them differ lesse & slower & are bigger in y^e^ encreaseing altitudes then they ought to be as Kepler allso by his calculation found them.

Accordeing to whose position if wee suppose the Angle bea $88°.01'$ and the semidiameter of the earth ab = ac. 3953 English miles,[14] the height of the Atmosphere ce = bd will be found 2 English miles & $\frac{37}{100}$ partes of a Mile or something better then 2 English miles & $\frac{1}{3}$. which height nearely is still retained by those who have calculated refraction on better grounds & demonstrative principles not knowne in Keplers dayes.[15]

But here will be a controversy raised concerneing the true height of the Aire. the Ancients Considering the Crepuscula or how long before sun rise the aire began to be illuminated & the day to dawne & againe how long it continued sensibly light after sun set concluded the height of it no lesse y^n^ 50 miles which granted it would follow that the refractions ought to decrease much slower accordeing to this Theory of Refractions then Kepler makes them.

Againe since the Invention of the Torricellian tube[16] whereby the weight of the Aire is measured tis argued from the descent of the quicksilver in the tube as tis carried up into somc elevated place from the earths superficies, that the height of y^e^ aire can not be lesse then 5 miles, which is yet too big for the Theory of refractions, both these must be answered the latter especially seeing that it seemes a very cogent Argument & infers certeinly a greater height of the Aire then $2\frac{1}{3}$ English miles

I shall begin with y^e^ Argument drawne from y^e^ length of y^e^ Crepuscula or twilight that is the time interlapsed betwixt the first appearance of y^e^ light upon y^e^ Easterne horizon & the suns riseing or betwixt his setting & the passing of the light of from the Westerne horizon Tycho Brahe makes the suns light then to touch the horizon when hee is 16 degrees beneath it. Rothman will have it 24 degrees, the first is the least the last the greatest that have beene proposed

Alhayzen & Vitellio 19 Clavius Kepler & Gassendi 18. but Riccioli will have the evening Crepuscula longer then ye Morneing.[17] amongst these different opinions tis difficult to determine which is nearest truth for tis evident to me that the Crepuscula ought not to be allwayes made at the same depression of ye sun but different accordeing to ye different constitution of ye Aire. which if Filled with Watery vapors receaves & retaines & transmits ye suns raies longer then when tis drie pure I have observed my selfe & therefore take the Arch of ye crepusculum to be nearest 18 degrees that is I suppose the sun depressed 18 degrees beneath ye horizon at ye begining or end of ye Crepuscula or twilight the raies of his light then onely towcheing the supreme parte of yt sphere of vapors encompasseing or earth As in ye 4d figure [Fig. 37.4] if wee suppose bcf to represent the earths superficies a its center deg ye superficies of ye Atmosphere b ye place or seate of ye observer beh ye horizontall ray intersecteing the Atmosphere in e. from which pointe let the line ēḡ be drawne a tangent to ye earths superficies in f and formeing the Angle heg with the horizontall line 18 degrees. draw also the line ae. & to ye touch pointe f, af. because ba = af & ye angles eba, efa both right and ēā common the angles bea and fea are both equall in those two triangles. Now ye angle heg being 18 degrees feb will be 162. & the halfe of it 81 deg. shall be y[e] ∠bea = fea. whose complement shall be ye angle bae = eaf 9 degrees which angle being knowne with ye side ba = af the Earths semidiameter 3953 miles the side ae the radius of the atmosphere will be found $4002\frac{3}{10}$ miles from which subtracteing ac the semidiameter of the Earth there remaines ce the height of the Atmosphere above ye earths superficies $49\frac{3}{10}$ miles exceeding Keplers determination & the postulata for salveing refractions at least 20 times. this was the method whereby Alhayzen & Vitellio inquired & determined the height of the aire not considering that the raie of light passeing from the rare medium of the aether into ye denser of ye Atmosphere ought to be refracted at its entrance in ḡ & therefore accordeing to theire owne doctrine a much lesse height of the aire might serve to answer the appeareance of ye Crepuscula. but this subject will require to be more amply discusst I shall therefore proceede with it next Lecture[18] with Gods p[er]mission

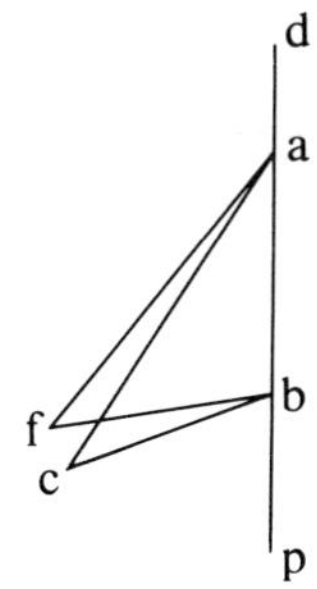

FIG. 37.1

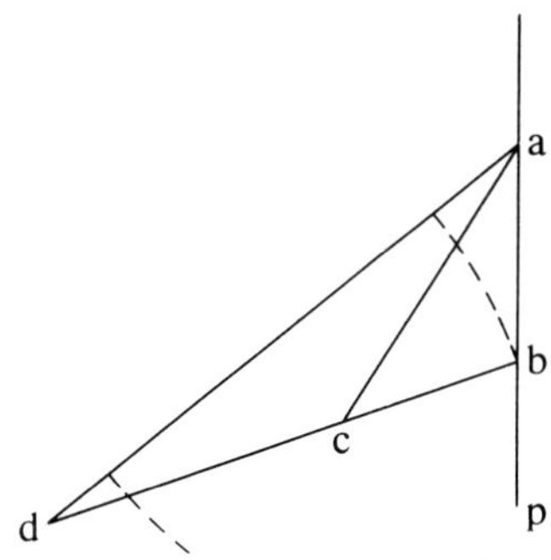

FIG. 37.2

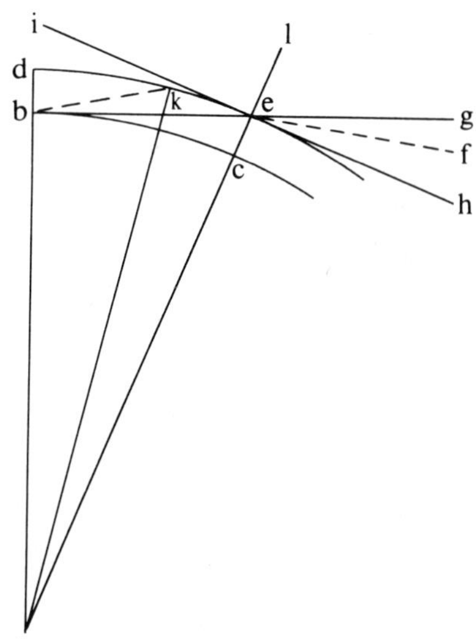

FIG. 37.3

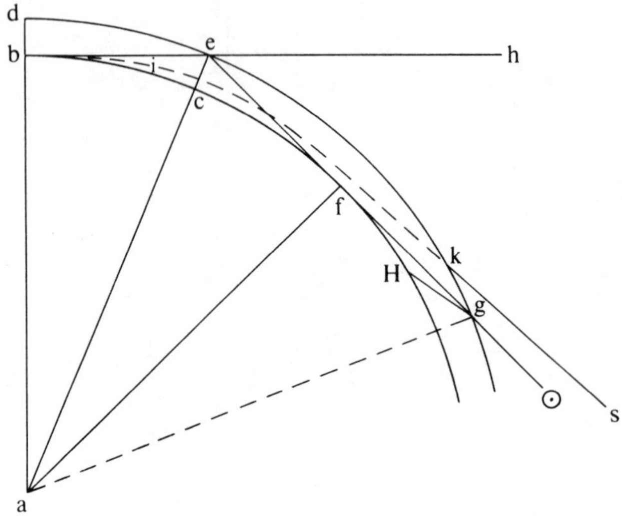

FIG. 37.4

NOTES

37.1. Lecture 35 (4 June 1684)—*not* Lecture 36, which was actually delivered one week after Lecture 37.
37.2. Op. cit. (note 1.25), pp. 98–99.
37.3. Cf. note 33.10.
37.4. Flamsteed mistakenly wrote bpd instead of pbd.
37.5. Flamsteed mistakenly wrote abd instead of adb.
37.6. Cf. note 37.1.
37.7. Flamsteed's description of Kepler's method, and the accompanying illustration (Fig. 37.3), are taken from op. cit. (note 1.25), pp. 123–4, as he later remarks on f. 338ᵛ, or p. 446.
37.8. Flamsteed wrote fea instead of fel.
37.9. Kepler, op. cit. (note 1.25), p. 125.
37.10. This last figure should be 124.
37.11. Op. cit. (note 2.15), p. 66.
37.12. Flamsteed wrote ae instead of ac.
37.13. Flamsteed wrote ebo instead of eba.
37.14. This figure for the Earth's radius is based upon Picard's result (op. cit. [note 15.4]) of 57,060 toises = 69·1 statute miles for the length of 1° of a meridian arc on the Earth's surface.
37.15. This is presumably an allusion to Descartes' deductive demonstration and mathematical application of the correct law of refraction in 'La Dioptrique' etc. (op. cit. [note 4.4]), and to those such as Gascoigne who had subsequently adopted that law as the basis of their own theories of geometrical optics.
37.16. The currently accepted date for the invention of the mercury barometer is 1641. Cf. note 36.2.
37.17. The various estimates cited here are conveniently tabulated in Riccioli 1, p. 39.
37.18. Flamsteed does not, in fact, develop this theme any further in Lecture 38. A summary of the main points in Lecture 37 is contained on f. 334ᵛ.

Lecture 38

19 *November* 1684

Dura mihi quidem et anceps contigerit provincia auditores cordatissimi, lectionem hic habui ante hebdoma latinam aderant plurimi quos amicos semper habere volui, quos honestis omnibus officijs studui demereri et certe habuissem si meritis conciliari potuissent qui verba mea et nonnullas sententias e reliquis violenter. decerptas malo in illos animo me contra adseverante, prolatas voluere. Effusior in fundatores laudes domesticorum Artes nonnulla levito ficta fabula p[er]strinxi, rapiant verba quaedam & in seipsos detorquent etiam me nolente. quid hisce faciam. Negligum? reum inde me dicent, excusem meipsum alienum a negotio praete[r]sunt Omittam ergo. ut infirmis animis praeditos donec facti sapientiores seipsos dididerunt non sauciare. vobisque quibus, majores spiritus dei bonitas dederit concilij mei rationem breviter exponam.

Dioptrica fuse in hoc loco et sermone vernaculo antehac tractavi,[1] tandem a refractionibus quae in duro vitri diapliano fiunt at eas quae in nostri aeris tenuiori p[er]spicuo p[er]aguntur descendi historiam Refractionum contexui, & ad earum mensuras indagandas paratus eram cum ex improviso latinas lectiones p[er] viginti et amplius Annos intermissas incipere coactus sum. ultima Termini elapsi lectione[2] in problemata duo Geometrica incideram quarum solutio etiam problematum in Trigonometria pulcherrimam solutionem mihi praebuit, alteram tetigi ultima, sermone Anglicano habita lectione,[3] aliam pomeridiana proxima prolaturus, ut pomeridiana lectio matutinae[4] respondeat, huius loci leges[5] reposcunt quibus ut quam maxime possim obtemporem breviarum earum quae de refractionibus differui vobis primum dabo deinde ad problemata Trigonometrica solvenda accedam itaque gressus moderare in posterium studebo ut vobis praesentibus semper idiomate proprie habita lectio pomeridiana Matutinam sermone doctiore habitam exponat.

De refractionibus radiorum lucis e sole luna planetis fixis in aerem nostrum transmissae, ne sommasse quidem veteres invenio. Notum quidem fuit Euclidi ut e propriis catop[t]ricis[6] suis praemissa constat, lucis radios ab aere nostro in aquam receptos Refractionem pati, sed

Notes on page 459.

nec inde ultra struxit. Hipparcho illarum effectus ignotos fuisse Miror siquidem ab observatis fixarum in orte aut occasu declinationibus itemque in Meridio certe ipsi innotuissent, nisi minora fuissent instrumenta quibus sidera observaverit, aut auctiora graduum spatia quam ut accuratas in hoc negotio necessarias divisiones admitterent. Arabes autem refractiones non lature. fuse satis de ijs differuit Alhayzen Murus qui ad annum 1100 vixerit juxta conjecturam Frederici Reisneri[7] Avicenna averrois et Zoarae coaevus. asserit ille luminarum & fixarum loca a Refractionibus mutari propositionibus 51.52.53 & 54. libri 7.[8] contractioresque illarum reddi distantias juxta horizontem, minus tamen quo vertici fiunt aut Meridiano propiores. Mensuras attamen illarum prorsus quantum assequi possum ignoravit. Instrumentum[9] refractionibus in vitro aut liquidis factis metiendo satis aptum describit. sed experimentorum illius ope factorum nullam nobis histor[i]am consignaverit. Annos 170 post Alhayzenum floruit Vitellio. Polono-Thuringus Nobilis Optices scriptor. Eadem ille asserit idem describit pro refractionibus dimetiendis Instrumentum, et prop. 49 libri 10. Insuper affirmat sermone illius aetatis semibarbara

"Stellas caeli et lunam secundum refractionem comprehendi instrumentaliter. et deinde quod ope Armillarum etiam refractiones caelestes sentiri possent. Addit enim si instrumento Armillarum nocte observetur aliqua fixarum Magnarum quae cum pervenit ad Meridianum sit transiens per Centrum capitis exper[i]mentantis prope et consideretur illa in ortu suo dum elevatur super superficiem horizontis, et tunc revolvatur Armilla revolubilis in Circuitu poli mundi qui est polus aequinoctialis donec fiat aequidistans circulo magno caeli tran[s]eunti per polos aequinoctialis et p[er] centrum corporis illius ★ et rectificetur locus stellae ex Armilla ita ut habeatur ejus distantia a polo mundi. deinde observetur stella donec veniat ad Circulum Meridiei moveaturque Armilla mobilis donec fiet aequidistans circulo stellae ut prius et sit in superficie Circuli meridiani et tunc iterum habebitur distantia stellae a polo Mundi cum stella fuerit in Zenith capitis aut prope. invenieturque (attendite) distantiam stellae a polo mundi in tempore ortus et elevationem stellae minor ipsius distantia ab eodem polo tempore quo est in zenith capitis aut prope," [10]

sic ille loco supra citato quem integrum transcripsi quia diserte adeo refractionum vires perceptibiles esse asserit. hujus etiam quippe loci meminisse videtur qui ducentos post annos Noribergae Claruit Walther Narrat enim ille quod anno 1489 Martij diebus 5.6.7. Armillis ♀[s] a sole distantias juxta horizontem a refractionibus contractiores redditas comprehendit omnino ut Alhayzen & Vitello in suis Opticis asservissent Waltheri observationes explicat Keplerus in suis edictis Vitellionem paralipomenis. pag 152[11] & deinceps ubi plura de his scire cupientibus abunde satisfiat.

A Walthero viscere Regiomontanus Copernicus Peurbachius alijque plurimi Astronomiae studiosi inter quos nulla fit refractionum mentio omnes locorum quos habitarunt latitudines a summis et imis Meridianis solis altitudinibus deduxere unde postea Copernicum in latitudine Frueburgi ubi ille habitaverat, errasse Tycho deprehendit integra vigesima parte gradus vel 3′ premissis et ipse Tycho secus in initio observationum Hertzuadij[12] sui latitudinem investigatum iverit, sed deprehenso postea Majoribus organis completis latitudinem per fixas acquisitam minime convenire cum ea quam a tropicis solis Merid. altitudinibus deduxerat, solem mox in brumali Tropico humiliorem a refractionibus attolli suspicatur. hoc primum evenit Anno 1582 & procul dubio ipsi occasionem administrarunt Armillarum structuram adproperandi hactenus enim Quadrantibus solis aut sextantibus usus fuerit, atque erat subtilis ingenij nunc praevidebat opus esse aliquo instrumento cujus ope ab horizonte ad verticem observari potuere refractiones.[13] at commodiora Armillis dari tunc temporis non potuere. quarum ope ab horizonte ad verticem in declinationibus observatae a refractionibus ortae mutationes ipsas in qualibet altitudine refractiones exhiberent. Ultima Octobris die Anno 1584 primum fixarum aliquot declinationes Armillis observatas in Historia caelesti[14] nobis tradit solis autem ijsdem animadversas declinationes apparentes habemus non ante diem 9 Junij Anni sequentis 1585 et hucusque refractionum ipsum rationes latitisse valde probabile etiamsi e brevioribus fixarum polo adjacentium distantijs cum supra tum intra eum diremtis concluserat[15] radiorum refractiones supra vigesimum altitudinis gradum utique esse imperceptibiles quippe suis haud promte tractandis instrumentis eas vix percipere potuit. A 9°. Junij ad 18. solis altitudines crassiuscule ad quadrantes solummodo graduum captas in quaque observatione dederit exinde tamen scrupulior horologio vel aliter acquisitum cujusque observationis connotat tempus, quia eo opus fore sagax & forsan. ab experientia admonitus praeviderit. Methodus autem qua utebatur ad altitudinis refractiones a refractionibus observatis declinationis depromendas ut ingenis & subtilis fuerit sic vobis auditores forsan scire acque jucudum erit aperiam ergo.[16] primum ex observata solis altitudine Meridiana quantum ipse ab aequinoctiali circulo disteterit facile cognoverit et ab intervallo temporis a qualibet observatione ad Meridiem quantum interea sol suam mutaverit declinationem facile investigaverit & proinde praecisam in qualibet observatione ab aequatore distantiam. his autem collectis haud difficile fuerit angulum circuli horarij & verticalis centrum solis transeuntis & deinde a refractione declinationis ipsam in ea altitudine refractionem investigare. Esto enim in exposito schemate [Fig. 38.1] HZq circulus Meridianus P polus PA. Axis mundi Z. Vertex loci alicujus Æ A. Æquator l ⊙ vel λO. paralleli duo alter boreae alius meridi[a]nae declinationis sint ⊙ et O duo loca solis p[er] quae scribantur circuli duo horarij P ⊙.PO simul et verticales duo Z ⊙.λo in triangulis ZP⊙.ZPO datis temporibus observationum dantur Arcus duo ZP latitudinis comple[me]ntum et P ⊙ vel PO solis a polo distantiae cum Angulo Incluso temporia Meridia aequale ZP ⊙ vel ZPO unde nota Trianguli resolutione investigari possit angulus utoque Z⊙P vel ZOP. quem parallacticum nuncupat refractivum propter proprietatem dicere mallem. esto nunc in circulo horario ⊙ b. vel Oß declinationis refractio. per b vel ß. aequatori parallelus scribatur circuli minoris Arcus cb vel xß.[17] in triangulis minoribus c⊙b vel xOß. innotescunt ⊙ b vel Oß cum angulis refractivis ad ⊙ et O anguli autem ad b et ß recti sunt unde facile triangulorum resolutione indagari possint latera c ⊙ vel xO altitudinis refractiones distantiae a vertice Z ⊙ vel ZO competentes quibus si addantur ipsas exhibent solis vel phamomeni a vertice distantias quibus illae sigillatim refractiones fuint. attamen quandoquidem duo latera horum triangulorum angulum refractivum complectentia Majorum circulorum Arcus sunt p[er]exigui, nec a rectis lineis sensibiliter discrepant, pro meris rectis assumantur & perinde ac si fuere resolvantur absque periculo erroris

Methodo Nobilissimus Braheus insistens horizontalem fixarum statuit refractionem 30 minutorum. Solis autem majorem quatuor Minutorum quoniam tantam fere putavit ejus parallaxin Non quia semper eam r[efractione]m in horizonte expertus est longe enim diversam habuit aliquando majorem quandoque et Minorem, sed medium inter extrema eligendo addita quam supposuerat parallaxi statuit ut dixi varias enim ego habui in horizonte aliquando triginta & sex fere minutorum nonnumquam vix plus sex supra viginti propter variam aeris constitutionem et tamen minorem veriorem existimo & in condenda tabula potius eligendam. cujus opinionis post hac orationem vobis reddam. In praesentiarum $\alpha\pi\varepsilon\chi\omega$.[18] trigonometrica quae destinaveram in aliud tempus differens ut tandem pomeridianae lectiones Antemeridianis respondeant, quod spero post aliam lectionem. praestem vos interea Auditores humanissimi aequo haec accipiatis animo.

Translation

It proved a difficult and hazardous task for me, dear sirs, to deliver a Latin lecture here a week ago. On that occasion there were several people present whom I have always wished to consider as friends, and whose friendship I have always striven by every honest means to deserve, and certainly would have won, if only they could have judged me on my merits, instead of being determined to construe some words and opinions, which had been taken entirely out of context, as being directed against them, despite my assertions to the contrary. Being somewhat more lavish in my praise of domestic [i.e. British] Arts, I casually mentioned some stories [that were] false,

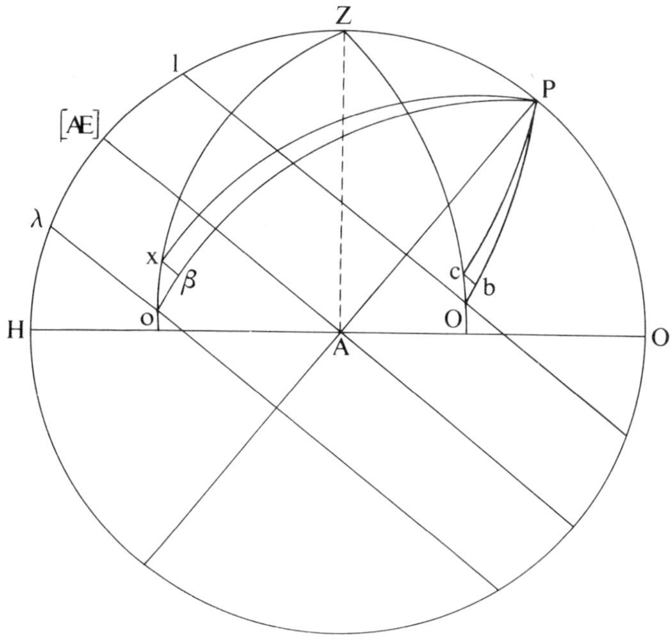

FIG. 38.1

whereupon they seized upon certain words and applied them to themselves, quite against my intention. What, then, am I to do, when men accuse me of negligence, and moreover claim that I excuse myself from my duty when they are not present? I shall therefore avoid such people as possess weak intellects, until they have grown in wisdom and pledged themselves not to wound, and to you, to whom God in his goodness has given greater understanding, I shall give a brief account of the subject matter of my lecture.

I have previously lectured on dioptrics in the vernacular at considerable length in this very place.[1] On that occasion, I proceeded from a discussion of the refractions which occur in hard clear glass to those which are produced in our rarer transparent atmosphere, and with this I linked a history of refractions. I was ready to investigate measurements of refractions when I was unexpectedly obliged to undertake Latin lectures after a gap of more than twenty years. In the last lecture of this past term[2] I dealt with two geometric problems whose solution also provided me with an admirable solution to problems in trigonometry. I touched on one of the problems in the last lecture delivered in English;[3] the other I shall mention in the next afternoon lecture,[4] so that the afternoon lecture may correspond to the morning one. The rules of this place[5] demand that I first give as brief an outline as possible of what I told you about refractions, and then progress to the solving of trigonometric problems. Thus, I shall try to modify the subject matter later, so that, if you are present, the afternoon lecture, which is always delivered in idiomatic terms, explains the morning lecture which is delivered in a more learned manner.

Concerning the refraction of rays of light entering our atmosphere from the Sun, the Moon, the planets, and the fixed stars, I find that the ancients have not been surpassed. Indeed, it was known to Euclid, as is clear from the premisses in his *Catoptrics*,[6] that rays of light undergo refraction when entering our aequeous atmosphere, but he did not draw any further inferences from this fact. I am amazed that these effects were unknown to Hipparchus, seeing that they became known from observing the positions of fixed stars and occasionally these might also be observed in the meridian itself, unless the instruments with which he observed the stars were smaller, or the spaces between the degrees too large, to record the divisions with the accuracy necessary in this matter. Moreover, the Arabs did not differ to any great extent from these authorities on the subject of refractions. Alhayzen, a Moor, who was alive in 1100 according to the conjecture of Frederick Reisner,[7] and contemporary with Avicenna, Averroes, and Zoara, maintains in propositions 51, 52, 53 & 54 of Book 7[8] that the positions of the fixed stars are altered by their refractions, and that they are closer together when near the horizon, but become less so the more vertical or closer to the meridian they are. However, as far as I can see, he was quite ignorant of their actual sizes. He describes quite accurately the instrument for measuring refractions made in glass or water,[9] but he has left us no record of any experiments conducted with the aid of this instrument. Witelo, a Pole from Thüringen [and] a notable writer on optics, flourished 170 years after Alhayzen. He made the same assertions, and also described an instrument for measuring refractions; and moreover in proposition 49 of Book 10 he states in the semi-barbarous language of his age: "By means of instruments I have gained an understanding of the stars and the Moon as regards their refraction, and also with the help of Armillae I have come to know what can be perceived of the celestial refractions. For, in addition, if the armillary instrument is used at night to observe whichever of the bright fixed stars reaches the meridian, provided that the star passes close to the observer's zenith and is observed at the place where it appears above the horizon, then the rotatable armilla may be turned round in the [great] circle containing the celestial pole which is the pole of the equinoctial [circle], until it becomes equidistant from the great circle of the sky passing through the equinoctial poles and through the centre of that stellar body, and the position of the star may be determined from the armilla.

The result is that we obtain its distance from the celestial pole; then, if the star is observed until it reaches the circle of the meridian, and the moveable armilla moved until it is at the same distance from the circle of the star as before, and is on the circumference of the meridian circle, we obtain the distance of the star from the celestial pole, since the star will be on or near the observer's zenith, and thence we can discover (mark this) the star's distance from the celestial pole at the time of its rising and the star's elevation less the distance of the same from the same pole at the time when it is at the observer's zenith or near it." [10]

Thus writes Witelo in the work cited above. I have transcribed it in its entirety, because he describes so eloquently the perceptible strength of refractions, & he also seems to have recorded in this place what Walther clarified 200 years later at Nuremberg. For he relates that in the year 1489, on 5th, 6th & 7th March it was shown by an armilla that the distances of Venus from the Sun near the horizon were more contracted by refractions, just as Alhayzen & Witelo asserted in their works on optics. Kepler sets out Walther's observations on p. 152 of his *Paralipomena* on Witelo,[11] and there he gives ample satisfaction to those desirous of more knowledge on these matters.

After Walther came Copernicus, Regiomontanus, Peuerbach and several others who studied astronomy. None of them makes any mention of refractions. All calculated the latitudes of the places where they lived from the highest and lowest altitudes of the meridian Sun, and afterwards Tycho discovered that Copernicus was mistaken in the latitude of Frauenburg where he had lived, by a whole twentieth part of a degree or 3′.

Tycho himself, on the other hand, set out at the start of his observations to discover the latitude of his own Hertzuadius [i.e. Herritzvad Castle][12] but on finding after the completion of his major instruments that the latitude obtained from the fixed stars corresponded hardly at all to what he had deduced from the meridian altitudes of the tropical Sun, he suspected that the Sun, then in a lower [position] in the winter tropic, had been subject to refractions. This first happened in 1582 and without a doubt these events were instrumental in hastening his construction of the armilla, for so far only quadrants or sextants had been used, and it is a mark of his keen intellect that he now foresaw that there was a need for some instrument with whose help the refractions from the horizon to the zenith could be observed.[13] However, at that time he was unable to incorporate those improvements into the Armilla which would enable changes in declinations observed from the refractions between horizon and zenith to give information about the refractions themselves in any given altitude. On the last day of October 1584 it is recorded[14] that for the first time in the history of astronomy several declinations of fixed stars, and also of the Sun, were observed with the help of the armilla, but we do not have any actual note of the declinations before 9th June in the following year 1585, and it is highly probable that he concealed his calculations of the refractions up to this point; although, taking the shorter distances of the fixed stars adjacent to the pole both above and below it, he had concluded[15] that the refractions of rays above the 20th degree of altitude were definitely imperceptible, since he could handle the instruments well, and yet was scarcely able to perceive them. From the 9th to the 18th June he took very crude readings of the altitudes of the sun, calculating the altitudes only to quarters of degrees in each observation; thereafter, however, he was more careful in noting the time of any given observation, obtained from the clock or some other means, because he realised that he would have to be mentally alert, and perhaps he had learned from experience. The method which he used to work out the refractions of the altitude from the observed refractions of the declination is ingenious and complicated, and, as you my listeners will perhaps be interested to know it, I shall therefore expound it.[16] First of all, through observing the meridian altitude of the Sun, he easily discovered how far it was from the equinoctial circle; and from the interval of time which elapsed between any given observation and 12 noon, he could readily calculate how much the Sun had changed its declination in the meantime, and accordingly its precise distance from the equator in any given observation. When this data has been assembled, it is not difficult to investigate the angle of the hour-circle and of the vertical [circle] passing through the centre of the Sun, and then from the refraction of the declination to investigate the refraction itself in that [body's] altitude. Expressing this in diagrammatic form [Fig. 38.1], let HZQ be the meridian circle, P the pole, PA the polar Axis, Z the zenith of any given place, Æ A the equator; l ⊙ or λ O two parallel lines, the one of northern, the other of southern declination; ⊙ and O two positions of the Sun through which are drawn two hour-circles, P ⊙, PO. At the same time the two verticals Z ⊙, λ O in the triangles ZP ⊙, ZPO are known, the times of the observations being given. If the two arcs ZP correspond to the co-latitude, and P ⊙ or PO corresponds to the distance of the Sun from the pole, then the contained angle is equal to the time from the meridian. When ZP ⊙ or ZPO are known through the resolution of the triangle, either angle Z⊙P or ZOP can be worked out, which he calls parallactic, but which I prefer to call refractive as being more appropriate. Now, in the hour-circle, let ⊙ b or Oß be the refraction of the declination. Let the smaller circle arc cb or xß[17] be drawn parallel to the equator through b or ß to form the lesser triangles c ⊙ b or xOß. b or Oß are known, with the refractive angles at ⊙ and O, and moreover the angles at b and ß are right angles. Once the triangles have been solved, it is easy to find out the sides c ⊙ or xO, the refractions of the altitude corresponding to the zenith distance Z ⊙ or ZO; and if they are added, they give the actual zenith distances of the Sun or the phenomenon at which these individual refractions occur. However, since the two sides of these triangles containing the angle of refraction are small arcs of great circles, and are not noticeably different from straight lines, they [i.e. the triangles] may, without any risk of errors, be resolved by assuming them [i.e. the small arcs] to be straight lines.

The distinguished Brahe, applying this method, established the horizontal refraction of fixed stars as 30 minutes, but reckoned that the Sun's refraction was 4 minutes more, since he thought that its parallax would be so great. This was not because he had always looked for that refraction on the horizon, for he had long recognized that it changed and was sometimes bigger and sometimes smaller; but by choosing a mean value, he determined what he imagined the parallax would be, as I have said. For I myself had variations on the horizon, sometimes of almost 36 minutes, sometimes of scarcely more than 26 minutes, because of the differing constitution of the atmosphere, and yet I think that the smaller, truer reading should be selected in drawing up a table. I shall give you a lecture on this opinion at a later stage. For the present I shall defer to another occasion the trigonometric proof which I had intended, in order that the afternoon lectures may correspond to the morning ones, and I hope that I shall prove my case to you after another lecture; but in the meantime, my very gracious audience, you must just take what I say on trust.

NOTES

38.1. Lecture 33 (2 April 1684).

38.2. Probably Lecture 39, which bears no date; or an English version of it.

38.3. The lecture in English (if ever written) would not appear to have been preserved. Flamsteed may here be referring to the theorem stated at the top of f. 350^r, or p. 465.

38.4. This probably refers to Flamsteed's proof of the trigonometrical analogue described in Lecture 39. No English version of it has been found.

38.5. Cf. Introduction, pp. 3–4.

38.6. Cf. note 33.1.

38.7. Cf. note 33.2. Riccioli 1, p. xxii says of Alhazen: 'floruitque Anno Christi 1072, vel circa 1100, ut conjectura est Federici Risneri'.

38.8. Cf. note 33.3.

38.9. Cf. note 33.5.

38.10. Cf. note 33.7.

38.11. Op. cit. (note 1.25). The correct page reference is p. 151.

38.12. Herritzvad Castle. Cf. note 29.7.

38.13. Viz. the armillary sphere. Cf. Lecture 33, f. 302^v, or p. 411–12.

38.14. Op. cit. (note 2.36), p. 102.

38.15. Ibid., p. 98.

38.16. The method is contained in op. cit. (note 2.5) (i), pp. 91–96.

38.17. Flamsteed wrote 'cß vel xb' instead of 'cb vel xß'.

38.18. Cf. note 38.4.

Lecture 39

Dixi in hoc discrimen adduxerit ut vel meipsum Malevelorum immeritis dicterijs differri, aut Meritissimi fundatoris ultimum violari testamentum[1] et quodammodo gravissime illudi manes p[er]mitterem cui quando hujus loci possessionem Lectores debent. ejus potius voluntati obtemperandum quam quieti propriae consulendum duxi, fidem frangant aut excusent aut datam evadant alij, mihi etiamsi alterius solummodo operam praesto, nihil antiquius quam loci legibus et mandatis vestris obtemporare, neque unquam aliquid egregium majus aut laude dignum meipsum p[er]fecisse putabo quam si officium juxta legatoris testamentum me praestasse agnoveritis, sed praefatum satis est. Nunc ad institutum porro procedo.

Muneris mei est Astronomica vobis exponere quem ad modum motus fiunt caelestes, & quibus potissimum rationibus ad eorum cognitionem pervenere possumus, quod fuse satis lectionibus sermone vernaculo habitis p[er] triennium anteactum ostendi, Trigonometrica ubique opus erat scientia, etenim juxta illud Tychonis Nobilissimi.[2]

Cuncta Trigonus habet satagat que docta Mathesis ille apperit clausum quicquid Olimpus habet Triangulorum autem doctrinam fuse tractarunt authores plurimi, inter priscos Hipparchus Agrippa et Ptolemeus. Sed hi solummodo Chordis aut semichordis loco sinnuum utebantur.[3] Annos post Mille et trecentos. Johannes Mullerus[4] (Regiomontani melius notus Nomine) sinuum tabulam confecit, qua Radius erat unitas cum adjunctis cyphris sex, ad vitandas divisiones aut Multiplicationes sexagenarias in veteri forma inevitabiles & laboriosas ubi Radius vel sinus totus locum habuit in solutione trianguli Addidit nonnulla Theoremata Copernicus, Rheticus, Ursinus, Clavius, Pitiscus Vieta, quisque suum symbolum contulit;[5] sed omnium Maxime futura saecula devinxerit Baro Marchistonius[6] Logarithmo reperto, ejusque excultores optimi Briggius[7] olim in hoc loco honoratissimus Geometriae & Gillibrandus[8] Astronomiae, praelectores Nec penitus obliviscendus est qui illorum opera pulchre impressa orbi edidit Adrian Vlacque Goudanus,[9] his successor Oughtredus[10] Nostras, at Quantus homos Cavalerius[11] alijque plures

Notes on pages 467–8.

quibus denumerandis $\frac{1}{2}$ hora fere impenderetur. at quid ego post hos? ut dies docet sic scientarum inexhaustum Thesaurum semper aliquid inde promendum posteris praebere. At Non ego post hos ipsa Trigonometria fundamenta exponere suscipiam, inutile profecto esset opus praestare vobis Trigonometriae peritissimis, faciliora hujus scientiae principia sunt quam ut opus sit ea vobis recitare recondita vestra Magis est scientia, Missis ergo vulgaribus Trigonometriae Novas vobis demonstrationes problematum difficilorum solummodo exhibebo quas vel Marte proprio indagavi aut ab alijis non infimae classis aut vulgaribus hausi geometris. utilissimum est in Astronimia problema. Datis Trianguli rectilinei cujusvis duobus lateribus cum Angulo ab Ijsdem comprehenso Angulos reliquos investigare. Duas hujus problematis solutiones dedere authores antenominati familiarem alteram & frequentissime demonstratam. Aiunt enim ut Summa laterum datum

Angulum includentium; ad eorum differentiam: : Ita cotangens dimidij dati Anguli ad tangentem semidifferentiae Angulorum reliquorum.

Addita quae dimidio complementi Anguli dati ad semicirculum, majorem e quaesitis angulis exhibet, ab eadem sublata Minorem petitum relinquet. cujus quilibet fere author propriam dat demonstrationem. meam ego ni fallor mox etiam dabo

Attamen ex quo logarithmorum inventio orbi erudito inmotuit Astronomicarum Tabularum conditores loco distantis ipsius planetae a sole, commodius duxere ejus in tabulis dare Logarithm[or]um et loco analogiae antedictae aliam excogitarum quae abs summa aut differentia laterum angulos quaesitos exhibet aiunt enim

ut Latus e datis minus. ad latus majus se habet: : Ita est Radius. ad tangentem arcus quarti.

e quo ablato angulo semirecto vel gradibus desumtis 45°. iteram aju . .

ut Radius ad tangentem arcus relicti: : Ita cotangens dimidij dati anguli: ad tangentem semidifferentiae angulorum quaesitorum.

et hujus etiam proportionis nobis olim demonstrationes dederunt sed p[er]plexissentur Andersonus[12] nostrus, & ab eo alij. quamobrem e nostratibus nonnulli faciliorema me non raro nec modeste valde petierunt, illorum diu protervitatem neglectam habui quoniam utcunque p[er]petuo vera tamen ea demonstratio apparuit. Sed tandem postea cum ipsa solutio liquidiorem promittere videbatur, retentata liquidissima utriusque proportionis demonstrationem inveni, quibus uno solummodo schemate paucis lineis nec pluribus literis opus esset, quas vobis ut possim breviter exponam.[13] praemisso solummodo hoc facile Theoremata

Duorum quorumbilet arcuum vel binarum linearum [Fig. 39.1 et 39.2] ea est semidifferentia quae addita semi summa earum, majorem confic[i]t; ablata vero minorem relinquit

Sint lineae duae, vel bini arcus AB. BC. in linum juncta, quorum semisumma sit Ae = eC. auferatur a parte A longitudo vel arcus Ad = minori Bc tunc est DB differentia et. de = eb eorum semi differentia, addita quae semi summae Ae majorem conficit lineam vel arcum AB: quod si a semi summa ec = ae. auferetur semi differentia eb = de manet linea vel arcus minor cB: quod non minus in angulis (ut a schemate ipso patet) verum est. ut etiam in superficiebus, et corporibus: sed procedendum.

Esto Nunc triangulum obliquangulum abc. cujus duo dantur latera AB.AC. angulum datum CAB Includentia [Fig. 39.3]. Centro A. Radio AB scribatur $\frac{1}{2}$ Circulus xrz. producto tunc cA donec peripheriae occurrat in z. erit cz summa, et cx differentia datorum laterum.

A punctis z et x. ad b ductantur lineae zb. xb. Angulus externus ZAB equalis est duobus internis et oppositis ad B et C: sed Angulus AXB (quoniam ad circuli peripheria[m] est dimidium anguli zAB. et aequalis $\angle^{\circ}$ ABX, et Angulus xbz rectus est, ergo zB est tangens $\angle^{i}$ zxB sive $\frac{1}{2}$ summae angulorum quaesitorum Radio XB.[14]

Ducatur etiam abx linea xd aequidistans lateri CB erit Angulus zxd = angulo minori quaesito ACB & angulus dxB semidifferentia angulorum ignotorum cujus Bd tangens erit ad eundem XB. sed similes sunt $\triangle^{i}$. zxd et zcb, ergo ut summa laterum datorum zc. ad eorum differentiam xc tangens semi summae Angulorum quaesitorum zB, ad tangentem semi differentiae illorum Bd.[15]q.e.d.

Ad posteriorem & supputatoribus utilissimam proportionem demonstrandam A puncto A. p[er]pendicularis ad A ducatur linea Ar. & a puncto r quo peripheriam secat lineae rz et rx. ab x etiam ducatur linea xt ad zr aequidistans. erit AB = Ar. ergo ut Ab latus minus ad Radium ita se habet Ac latus scilicet majus. ad tangentem *anguli* ArC. e quo sublato semirecto Arx vel gradu manet angulus residuus Xrt. cujus xt tangens est Radio rx = rz. ergo propter similitudinem triangulorum czr. cxt:erit ut cz:cx: : zr(= xr):xt: sed antea ostendimus esse ut cz:cx: :zB:Bd. Ergo abjectis 2^{s} prioribus utriusque proportionis aequalis terminis erit ut. Radius rx. ad tangentem anguli residui xt. sic tangens semisummae angulorum quaesitorum zb. ad tangentem semidifferentiae eorundem Bd[16] q.e.d.

Esse autem $\angle^{m}$ dxb (cujus Bd tangens est) semidifferentiam angulorum B et C inde patet: quod si semisumma incognitorum $\angle^{\circ}$ Axb[17] addatur ei aequali xbc fit major angulus ABC. dempto autem dxB ab angulo semisummae Axb, manet angulus Axd $\angle^{\circ}$ ACB e quaesitis minore, juxta prius recitatum theorema. Euclidis propositiones in his demonstratibus adhibendas non recitavi quoniam e vulgaribus et trit fuere talesque vos Geometris esse credo ut inutilis omnino iste cum vobis tum mihi esset valete

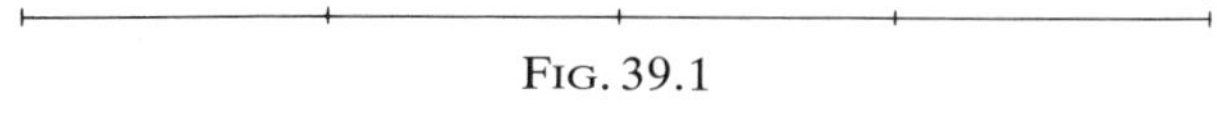

FIG. 39.1

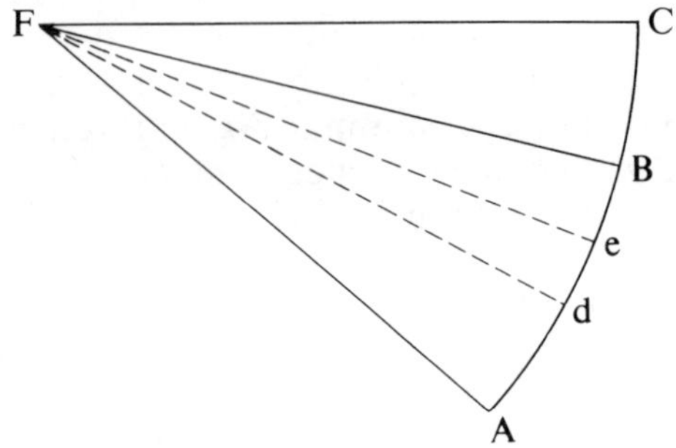

FIG. 39.2

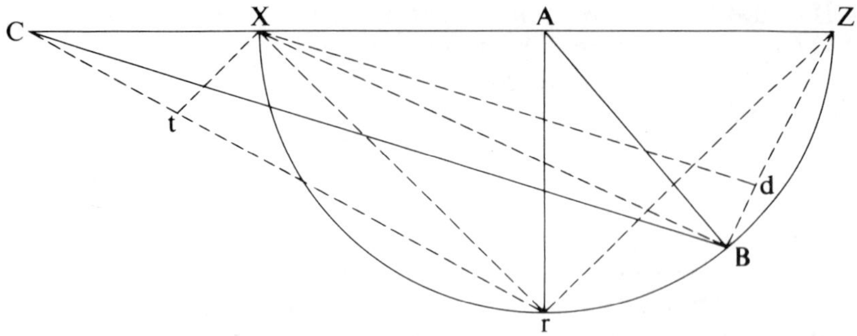

FIG. 39.3

Translation

I have said [already] that in this dispute I am obliged either to dissociate myself from the undeserved taunts of ill-wishers, or to violate the last wish of our worthy founder,[1] and allow a shameful mockery to be made of his memory, when lecturers owe their position on this rostrum to him. I chose to abide by his wishes rather than to have regard for my own comfort. Let the others break faith, or excuse themselves, or evade their duty; as for me, even if I only fulfil no more noble task than obeying the laws of the place and your injunctions, I shall consider myself never to have done anything more distinguished or praiseworthy than to obtain your acknowledgement of my having acted in accordance with the wishes of the testator. But enough in the way of introduction. I shall proceed with my lecture without more ado.

It is my duty to teach you astronomy—how the celestial bodies move, and in particular how we can arrive at a knowledge of them. All this I expounded at considerable length in my lectures delivered in the vernacular three years ago and, wherever knowledge of trigonometry was needed:—

"Trigonus has everything with which to satisfy learned mathematicians; he reveals whatever Olympus has hidden." to quote the distinguished Tycho.[2]

Moreover, several authors have covered in detail the doctrine of triangles; among the ancients, Hipparchus, Agrippa and Ptolemy, but they only used chords or semichords instead of sines.[3] 1300 years later Johannes Müller[4] (better known by the name of Regiomontanus) drew up a table of sines to an accuracy of six [decimal] places, in which the Radius was unity in order to avoid the inevitable and laborious sexagesimal divisions or multiplications in the old formula, where the radius or whole sine was involved in the solution of the triangle. Copernicus added some theorems, also Rheticus, Ursinus, Clavius, Pitiscus, Vieta, each of whom made his own contribution;[5] but it was the Baron of Merchiston[6] [i.e. John Napier] who placed future generations most in his debt by the discovery of logarithms, along with the excellent teachers Briggs[7]—formerly held in esteem in this place for his skill in geometry—and Gellibrand[8] the astronomer; nor should we entirely forget Adrian Vlacq of Gouda, who published a beautifully printed edition of his works for the world.[9] And after them our Oughtred:[10] moreover, what a splendid man Cavalieri[11] was, and many others, whom it would take nearly $\frac{1}{2}$ hour to name. What need, then, is there of my efforts, after such as these? Simply to show that there is an inexhaustable treasure-house of knowledge which will always yield some gem for posterity. But I shall not undertake to teach the elements of trigonometry after these men; it would surely be a useless exercise to teach you, who are skilled in trigonometry, the more elementary principles of this science. It will be more profitable to run over the more obscure points. Therefore, by passing what is commonly known about trigonometry, I shall provide you only with new demonstrations of more difficult problems, which I have investigated myself or derived either from others of equal ability or from common geometricians. The following is a very useful problem in astronomy. Given two sides of any plane triangle, with the angle contained by them, the problem is to discover the remaining angles. The authorities mentioned above provide two solutions to this problem; one of them is well-known and frequently demonstrated. Its champions maintain that:

As the sum of the sides containing the given angle is to their difference: : so also is the cotangent of half the given angle to the tangent of the half-difference of the remaining angles. If this is added to half of the given complementary angle to the semicircle, it will provide us with the greater of the unknown angles, and when it is subtracted from the same, the remainder will be the lesser of the unknown angles. Almost any authority you care to name has his own particular demonstration of this. Therefore, it is highly probable that I too shall presently give you my version!

However, ever since the discovery of logarithms became known to the learned world, people who draw up astronomical tables have found it more convenient, instead of recording the actual distance of a planet from the sun, to give it in logarithmic tables, and instead of working from the above-mentioned analogue, to devise another which gives the required angles in terms of the sum or difference of the sides; for they claim that: As the lesser of the given sides is to the greater: : so also the radius is to the tangent of a quarter circle from which half a right angle has been taken away, or 45 degrees removed. Again, as the radius is to the tangent of the remaining arc: : so also is the cotangent of half the given angle to the tangent of the half-difference of the unknown angles. Demonstrations of this analogue were once given by us, but our Anderson[12] and others were puzzled by it. Wherefore several of our fellow countrymen quickly launched a relentless attack upon me. I have long disregarded their impertinence, however, since that demonstration has at length been proved true. But since it has subsequently occurred to me that the same solution might be more clearly presented, I retained the most evident parts of both proportions, but discovered a demonstration where only one diagram, a few lines, and no more letters were necessary, which I shall set out for you as briefly as possible.[13] I presume only this simple theorem.

Half the difference of any two given arcs or two lines, when half their sum is added, gives the greater of the two, and when subtracted, leaves the lesser.

Suppose [Figs. 39.1 and 39.2] we have two lines, or two arcs AB,BC joined together, half of whose total is Ae = eC. Suppose the length or arc Ad(= the smaller BC) is subtracted from the section A[B], then dB is the difference, and half of their difference dc = eB, and when this is added to half the sum Ae, the result is the greater line or arc AB. But if from half the total eC = Ae one subtracts half the difference eB = de, the smaller line or arc CB remains. This is equally true in the case of the angles (as is clear from the actual diagram) and also in the case of surfaces and bodies [i.e. areas and volumes] But to proceed:—

Let us suppose now an oblique angled triangle ABC whose two sides AB & AC are given, enclosing the given angle CAB [Fig. 39.3]. If with centre A, radius AB, is drawn a semicircle XrZ, then if CA is produced until it meets the circumference at Z, CZ will be the sum, and CX the difference of the given sides.

From the points Z and X, let lines ZB, XB be drawn to B. [In $\triangle$ABC] the exterior angle ZAB is then equal to the two interior opposite angles, $\hat{B}$ and $\hat{C}$, but the angle $\hat{A}$XB (since it is at the circumference of the circle) is half of the angle ZAB and equal to

∠ABX, and the angle XBZ is a right angle; ZB to the line XB is therefore the tangent of ∠ZXB, or of the half-sum of the unknown angles.[14]

Moreover, if a line Xd, parallel to the side CB, is drawn from X, the angle ZXd = the smaller unknown angle ACB, and the angle dXB = half the difference of the unknown angles, whose tangent will be [the ratio] Bd to the same [line] XB. But the △ˢ ZXd and ZCb are similar, therefore:

As the sum of the given sides ZC to their difference, so also the tangent of half the sum of the unknown angles ZB [÷XB] to the tangent of half the difference of those [angles] Bd [÷XB][15] q.e.d.

In order to demonstrate the last most useful proportion from calculations, let a perpendicular Ar be drawn from point A; and from point r where it cuts the circumference, let the lines rZ and rX be drawn. From X also let the line Xt be drawn parallel to Zr; then AB = Ar, so that:

As the lesser side AB is to the radius, so also the greater side AC is to the tangent of the angle ArC, and when the half right-angle ArX or [45] degrees is subtracted from it [viz. A$\hat{r}$C], the angle Xrt is left, whose tangent is Xt to the radius rX = rZ. Therefore, on account of the similarity of the triangles CZr. CXt, it follows that CZ:CX: :Zr(= Xr)Xt:

But previously we proved that CZ:CX: :ZB:Bd. Thus, having eliminated the two equal terms of both proportions, it is clear that just as the radius rX is to the tangent of the residual angle Xt[÷rX] so also is the tangent of half the sum of the unknown angles ZB[÷XB] to the tangent of half their difference Bd[÷XB] q.e.d.[16]

Moreover, ∠dXB (whose tangent is Bd[÷XB] is half the difference of the angles B and C; thus it is clear that if the half-sum of the unknown ∠AXB[17] is added to the equal ∠XBC, the result is the larger angle ABC, and if d$\hat{X}$B is taken from the angle of the half-sum AXB, the angle AXd remains, which is the lesser of the unknown angles ACB, as stated in the above theorem. I have not repeated Euclid's propositions in these demonstrations, since they were drawn from simple everyday things and, I imagine, are such as would be useless to you geometricians . . . And with this I take my leave of you.

NOTES

39.1. Thomas Gresham's will is published in John Ward, *The Lives of the Professors of Gresham College* (London, 1740), pp. 19–25.

39.2. Cf. note 27.20.

39.3. An account of these early Greek contributions to the foundations of trigonometry may be found in, for example, A. von Braunmühl, *Vorlesungen über Geschichte der Trigonometrie* (Leipzig, 1900) and in other similar works.

39.4. Cf. note 29.4.

39.5. Cf. note 39.3.

39.6. John Napier (1550–1617). Cf. *op. cit.* (note 29.5).

39.7. Henry Briggs (1556–1630), author of *Arithmetica Logarithmica* (Londini, 1624). Cf. note 29.5.

39.8. Henry Gellibrand (1597–1637), editor of Briggs's *Trigonometria Britannica* (Goudae, 1633) and discoverer of the secular change in magnetic variation.

39.9. Adrian Vlacq, *Trigonometria artificialis; sive magnus canon triangolorum logarithmicus* . . . (Goudae, 1633).

39.10. William Oughtred (1574–1660), author of works on dialling, navigation, and trigonometry. Flamsteed possessed Oughtred's *Clavis* (op. cit. [note 15.12]) and R. Stokes's English translation of his *Trigonometrie* (London, 1657).

39.11. Bonaventura Cavalieri (1598–1647), author of *Geometria indivisibilibus* (Bononiae, 1635) and *Trigonometria Plana, et Sphaerica*, etc. (Bononiae, 1643).

39.12. Robert Anderson's own demonstration of the same two analogies as are proved here were first published in the third edition of Thomas Streete, *Astronomia Carolina* (London, 1716), pp. 58–59.

39.13. In order to make the demonstrations easier to follow, we have rationalized Flamsteed's notation by introducing—in the English translation only—capital and small letters where they ought to be used.

39.14. That is:

$$\frac{\mathrm{ZB}}{\mathrm{XB}} = \mathrm{Z\hat{X}B} = \tfrac{1}{2}\mathrm{Z\hat{A}B} = 90^\circ - \tfrac{1}{2}\mathrm{C\hat{A}B}$$
$$= 90^\circ - \tfrac{1}{2}(180^\circ - \overline{\hat{\hat{\mathrm{B}}} + \hat{\mathrm{C}}}) = \tfrac{1}{2}(\hat{\mathrm{B}} + \hat{\mathrm{C}}).$$

39.15. That is:

$$\frac{\mathrm{dB}}{\mathrm{XB}} = \mathrm{d\hat{X}B} = \mathrm{Z\hat{X}B} - \mathrm{Z\hat{X}d} = \mathrm{A\hat{X}B} - \hat{\mathrm{C}} = \mathrm{A\hat{B}X} - \hat{\mathrm{C}} = (\mathrm{A\hat{B}C} - \mathrm{X\hat{B}C}) - \hat{\mathrm{C}}$$
$$= (\hat{\mathrm{B}} - \hat{\mathrm{C}}) - \mathrm{d\hat{X}B} = \tfrac{1}{2}(\hat{\mathrm{B}} - \hat{\mathrm{C}}).$$

39.16. That is:

$$\frac{\mathrm{ZC}}{\mathrm{XC}} = \frac{\tan \mathrm{Z\hat{X}B}}{\tan \mathrm{d\hat{X}B}} = \frac{\mathrm{ZB}}{\mathrm{dB}} \Rightarrow 1 - \frac{\mathrm{XC}}{\mathrm{ZC}} = 1 - \frac{\mathrm{dB}}{\mathrm{ZB}}$$
$$\Rightarrow \frac{\mathrm{ZX}}{\mathrm{ZC}} = \frac{\mathrm{Zd}}{\mathrm{ZB}} \Rightarrow \frac{\mathrm{ZX}}{\mathrm{Zd}} = \frac{\mathrm{ZC}}{\mathrm{ZB}}.$$

39.17. Flamsteed wrote $\mathrm{A\hat{B}C}$ instead of $\mathrm{A\hat{X}B}$.

Index